Lecture Notes in Chemistry

Volume 95

The series Lecture Notes in Chemistry (LNC), reports new developments in chemistry and molecular science - quickly and informally, but with a high quality and the explicit aim to summarize and communicate current knowledge for teaching and training purposes. Books published in this series are conceived as bridging material between advanced graduate textbooks and the forefront of research. They will serve the following purposes:

- provide an accessible introduction to the field to postgraduate students and nonspecialist researchers from related areas,
- provide a source of advanced teaching material for specialized seminars, courses and schools, and
- be readily accessible in print and online.

The series covers all established fields of chemistry such as analytical chemistry, organic chemistry, inorganic chemistry, physical chemistry including electrochemistry, theoretical and computational chemistry, industrial chemistry, and catalysis. It is also a particularly suitable forum for volumes addressing the interfaces of chemistry with other disciplines, such as biology, medicine, physics, engineering, materials science including polymer and nanoscience, or earth and environmental science.

Both authored and edited volumes will be considered for publication. Edited volumes should however consist of a very limited number of contributions only. Proceedings will not be considered for LNC.

The year 2010 marked the relaunch of LNC.

Yi Chen

Surface Plasmon Resonance Imaging

Basic Theory and Practical Methodology

Springer

Yi Chen
Institute of Chemistry
Chinese Academy of Sciences
Beijing, China

ISSN 0342-4901 ISSN 2192-6603 (electronic)
Lecture Notes in Chemistry
ISBN 978-981-99-3117-0 ISBN 978-981-99-3118-7 (eBook)
https://doi.org/10.1007/978-981-99-3118-7

This Springer imprint is published by the registered company Springer Nature Singapore Pte Ltd.
The registered company address is: 152 Beach Road, #21-01/04 Gateway East, Singapore 189721, Singapore

Preface

This book was originally (in 2000) written to transfer my own understanding and research experience of SPRi to my graduate students. Later, Dr. June Tang, the Springer Beijing Representative Office Editor, contacted me, hoping to publish it after supplement of related content. I thus scheduled to complete the writing in 2017 or 2018. Unfortunately, I could not finish it until 2023 due to my heavy burden of teaching and researching works. Also, I have been hesitant to write a book on SPRi because there are already many excellent reviews and books on this topic. A consequence of my delay is that some written content may be outdated. Nevertheless, I hope that the core principle, methodology and some fantastic ideas presented in the book will keep their value for readers to read or refer.

It is due to the original writing purpose and history that the weight of each chapter in this book and the style of narration are not really consistent or may vary greatly. Despite all this, I have persisted my own writing norms: The narration of theory and principle goes always from the easy to the difficult, the introduction to devices or instruments goes for dissection to enable fabrication, the presentation of methodology goes with protocols (exactly, approaches) to facilitate replication and the prospect of future research goes for inspired ideas to increase the reference value. I hope my attempt can help some novices and they can understand and master SPRi from theory to methodology and their applications, so easily as to avoid the further reading of other books or documents and further to become interesting in SPRi. It is also hope that graduate students and even specialists would like to skip through this book to have a reference or a new perspective on the history, current situation and future development of SPRi, or even to copy or apply the approaches and techniques introduced in their research works.

Chapter 1 was designed for a neophyte who either want to have an overview of the past, present and future of SPRi or even hope to learn whole the story of SPRi from soup to nuts. The designed content thus consists of the history to discover and understand the SPRi-related events, the concepts and theory to explain the discoveries and the methodology and devices aiming at applications. The challenges concerning with the future developments are also mentioned for readers who would not want to read throughout the full book. For readers who know very well the SPRi history,

Contents

Abbreviations

a	Acceleration
A	Optical absorbance; constant symbol; area
A431	Human epidermal carcinoma cells
Aac	*Acidovorax avenae* subsp. *citrulli*
Ab157	Anti-*E. coli* O157:H7 IgG polyclonal antibody
Ab145	Anti-*E. coli* O145 IgG polyclonal antibody
AD	Atopic dermatitis
AFP	Alpha-fetoprotein
Ag_{10}	Silver decahedra nanoparticles
AgNPs	Silver nanoparticles
AI	Artificial intelligence
AKI	Acute kidney injury
Alloc	Allyloxycarbonyl
APS	Ammonium persulfate
APTMS	(3-aminopropyl)trimethoxysilane
ARF1	ADP-ribosylation factor 1
ATIR	Attenuated total internal reflection
ATR	Identical to ATIR
ATRP	Atom transfer radical polymerization
AuNPs	Gold nanoparticles
B	Constant symbol
B	Magnetic field or magnetic flux density (Wb/m^2)
BiB	Bromoisobutyryl
BiBB	Bromoisobutyryl bromide
Boc	t-butyloxy carbonyl
BPy	2,2′-bipyridine
BSA	Bovine serum albumin
B16-F1/10	A mouse melanoma cell line
Bz	Benzoyl; Benzyl
Bzoc	Benzyloxycarbonyl (= Cbz)

c	Light speed in free space (= 2.99792 × 10^9 m/s); concentration (mol/L)
cisPt	Cisplatin or exactly *cis*-diamminedichloroplatinum(II)
C	Number density or counts per unit surface or per unit volume
CAE	Capillary array electrophoresis
Cbz	Benzyloxycarbonyl (= Bzoc)
CC	Cyanuric chloride
CCD	Charge-coupled device
CCM	Capture-culture-measure
CCT2	Chaperonin containing t complex polypeptide 1 subunit 2
CDC42	Cell division control protein 42
CDI	N,N′-carbonyldiimidazole
CDs	Clusters of differentiation antigen molecules, uniformly name rule for the leukocyte and other leukocyte differentiation antigens series recommended, in 1986, by the Nomenclature Committee of the World Health Organization
CE	Capillary or chip-based electrophoresis, counter electrode
CEA	Carcinoembryonic antigen
CFSE	5(6)-carboxyfluorescein diacetate succinimidyl ester
CFU	Colony forming units
CIEF	Chip-based or capillary isoelectric focusing
CMOS	Complementary metal oxide semiconductor
Con A	Concanavalin agglutinin
COVID-19	Corona virus disease 2019
[Cp·RuCl]	Pentamethylcyclopentadiene ruthenium chloride
CVD	Chemical vapor deposition
CXCL12	Also known as stromal cell-derived factor-1 (SDF-1), being a pre-molecular cytokine, belonging to the chemokine protein family
CYTOP	A trade name of perfluoro(1-butenyl vinyl ether) polymer
d	Thickness; Diameter
d_a	Thickness of an adlayer
d_g	Grating constant
d_m	Membrane thickness
d_s	Spacing distance
ddNTP	2,3-dideoxy-ribonucleoside triphosphate, where N is a variable of A, G, T and C
dis-O_2	Dissolved oxygen
dNTP	Deoxy-ribonucleoside triphosphate, where N is a variable of A, G, T and C
dsDNA	Double strand DNA
D	Dimension(al)
$\boldsymbol{D}$	Electric displacement field or electric flux density (C/m^2)
D_{dif}	Diffusion coefficient
DART	Direct analysis in real time

DCC	Dicyclohexylcarbodiimide
DEA	Diethylamine
DEPC	Diethyl pyrocarbonate
DESI	Desorption electrospray ionization
DIC	Diisopropyl carbodiimide
DIEA	N-methylmorphorphine or diisopropyl ethylamine
DIPEA	N,N-diisopropylethylamine
DMAP	4-N,N-dimethylpyridine
Dmb	2,4-dimethoxybenzyl
DMF	Dimethylformamide
DMSO	Dimethylsulfoxide
DN	Diabetic nephropathy
DTT	Dithiothreitol
e	Charge of an electron
$\boldsymbol{E}$	Electric field, V/m
E_v	Energy of a frequency
$\boldsymbol{E}_{sc}$	Electric field of scattered wave
EA	Ethanolamine
EC	Electrochemistry
ECDF	Empirical cumulative distribution function
EDC	N-ethyl-N′-(3-dimethyl-aminopropyl)carbodiimide
EDTA	Ethylenediaminetetraacetic acid
EGF	Epidermal growth factor
EGFR	EGF receptor
ELISA	Enzyme linked immunosorbent assay
EMW	Electromagnetic wave
EOA	Ethanolamine
Eoc	Ethoxycarbonyl
EpCAM	Epithelial cell adhesion molecule
ESI	Electrospray ionization
EtOH	Ethanol
EWs	Evanescent waves
$\boldsymbol{F}$	Force
F_{am}	Amplification factor
FcεRI	The Fc segment receptor of immunoglobulin, being the high-affinity receptor of IgE, belonging to the antigen receptor superfamily
FFT	Fast Fourier transform
FK506	Tacrolimus or [3S-[3R [E (1S,3S, 4S)],4S,5R,8S,9E,12R,14R,15S,16R,18S,19S,26aR]]—5,6,8,11,12,13,14,15,16,17,18,19,24,25,26,26a-hexadecyl-5,19-dihydroxy-3-[2-(4-hydroxy-3-methoxycyclohexyl)-1-methylvinyl]—14,16-dimethoxy-4,10,12,18-tetramethyl-8—(2-propenyl)-15,19-epoxy-3H-pyridino [2,1-c] [1,4] oxaza-cyclopentadecane-1,7,20,21 (4H, 23H)—tetraketone monohydrate

FKBP12	FK506-binding protein 12 or protein kinase C inhibitor 2
Fmoc	Fluorene methoxycarbonyl
FT	Fourier transformation
FTIR	Fourier transform infrared
g	Green's function
G	Wave admittance ($= \sqrt{\varepsilon\varepsilon_0/\mu\mu_0}$)
GAPDH	Glyceraldehyde-3-phosphate dehydrogenase
GC	Gas chromatography
GM1	A type of gangliosides
Gun	Guanidine
GPCR	G protein-coupled receptor
h	Planck constant ($= 6.62607 \times 10^{-34}$ J s)
$\hbar$	Reduced Planck or Dirac constant ($= h/2\pi = 1.05457 \times 10^{-34}$ J s/rad)
hAdVs	Human adenoviruses
H	Hamiltonian
$\boldsymbol{H}$	Magnetizing field (A/m)
$H_0^{(2)}$	The zeroth order Hankel function of second kind
H1	3′-HS-tagged hairpin DNA chain complementary to H2
H2	Hairpin DNA chain complementary to H1, with NH_2 terminal at the 3′-end
HAECs	Human aortic endothelial cells
HCR	Hybridization chain reaction
HEMA	Hydroxyethyl-2-methacrylate
HEPES	4-(2-hydroxyethyl)-1-piperazineethanesulfonic acid
HER2	Human epidermal growth factor receptor-2
HINI	N-hydroxy-5-norbornene-2,4-dicarboximide
HL-60	A human promyelocytic cell line
HMG	High mobility group. A protein contains HMG domain is also shorted as HMG that can be further divided into three families of HMGA, HMGB and HMGN. The HMGB family has three members of HMGB1, HMGB2 and HMGB3, which have 80% consistency in amino acid sequences, where HMGB1 is the most abundant, with about 106 molecules in typical mammalian cells, and an average of 10–15 nucleosomes can contain one HMGB1 molecule
HMG1	High mobility group 1 protein
HOAt	1-hydroxy-7-azabenzotriazole
HOBt	1-hydroxybenzotriazole
HOSu	Equal to NHS
HPLC	High-performance liquid chromatography
HS578T	A breast cancer cell line
HSPs	Heat shock proteins
i	Imaginary sign
iSPR or ISPR	Imaging SPR

$\boldsymbol{I}$	Current strength, signal intensity
$\bar{I}_0$	Averaged imaging intensity over a negative probe spot or a probe-free location
I_{in}	Incident intensity
$\bar{I}_p$	Averaged imaging intensity over a probe spot
I_{re}	Reflected intensity
I_{tr}	Transmitted intensity
j	General numbering symbol
$\boldsymbol{J}$	Current density ($= \boldsymbol{J}_f + \boldsymbol{J}_p$) (A/m^2)
J_0	Zeroth order Bessel functions
$\boldsymbol{J}_D$	Displacement current density associated with a single particle
$\boldsymbol{J}_f$	Conduction current density induced by external field only (A/m^2)
J774	A murine macrophage cell line
k	Wave number
$\boldsymbol{k}$	Wave vector
$\hat{\boldsymbol{k}}$	Unit wave vector
k_B	Boltzmann's constant
k_{ec}	Elastic constant
$\boldsymbol{k}_g$	Shifting value of a wave vector
k_{off}	Desorption or dissociation kinetic constant
k_{on}	Adsorption or binding kinetic constant
k_{spp}	Wave number of propagating SPP
K	Equilibrium constant
K_a	Adsorption or binding equilibrium constant
K_d	Desorption or dissociation equilibrium constant
L	Length
L_{SPP}	Propagation length of SPP
LC	Liquid chromatography
LDG	Lactate dehydrogenase
LED	Light emitting diode
LOD	Limit of detection
LOQ	Limit of quantification
LRSPRi	Long-range SPR imaging
LSPPs	Localized SPPs
LSPR	Localized surface plasmon resonance
LS102.9	Mouse B-type lymphocytes
LWs	Longitudinal waves
m	Numbering symbol
miRNA	MicroRNA
M	Mass of a vibrator, a molecule or a particle
M_e	Electron mass
M_+	Mass of a positive ion
MALDI	Matrix-assisted laser desorption and ionization
MCF-7	A human breast adenocarcinoma cell line
MCH	6-mercapto-1-hexanol

MDS	Minimum detectable size
MeOH	Methanol
MES	2-(N-morpholino) ethanesulfonic acid
MFGE8	Milk fat globule EGF factor 8
MHC-II	Major histocompatibility complex-II
MHCC97H/L	A human hepatoma cell line
Moc	Methoxycarbonyl
MRi	Magnetic resonance imaging
MS	Mass spectrometry
MSi	Mass spectrometry imaging
MSPPs	Magnetic surface plasmon polarizations
MUA	Mercaptoundecanoic acid
$MUNH_2$	11-mercaptoundecylamine
MUOH	Mercaptoundecanol
n	Relative refractive index, number density of charges
$\hat{\boldsymbol{n}}$	Unit vector normal to an interface
n_e	Number density of electrons or charges
n_p	Photon number
n_{pe}	Number of photoelectrons
n_{pix}	Pixel number in an observed spot area
N	The upper limit of a general number code
N_a	Avogadro constant
N_{eri}	Equivalent refractive index
N_P	The total specific sites within the penetrating depth of SPPs
NA	Numerical aperture
$NaBH_3CN$	Sodium cyanoborohydride
$NaBH(OAc)_3$	Sodium triacetoxyborohydride
NDBA	N,N-dicarbonamide
NHPI	N-hydroxyphthalimide
NHS	N-hydroxysuccinimide
NMR	Nuclear magnetic resonance
NPs	Nanoparticles
Ns	o- or p-nitrobenzene sulfonyl
NSCLC	Non-small-cell lung cancer
OG	Octylglucoside
OVA	Ovalbumin
$\boldsymbol{p_e}$	Dipole moment of electron-ion pair
$\boldsymbol{p}_p$	The effective dipole moment of a single particle
$\boldsymbol{P}$	Polarization intensity
P_r	Actual polarizability along axis r
$\boldsymbol{P_v}$	Momentum
PAT	Process analytical technology
PB	Phosphate buffer
PBS	Phosphate buffered saline
PC4	Human nuclear protein positive cofactor 4

PC12	A cell line from Rattus norvegicus adrenal pheochromocytoma
PCR	Polymerase chain reaction
PDA	polydopamine
PDMS	Polydimethylsiloxane
PEG	Polyethylene glycol
PEI	Polyethyleneimine
PFPOH	Pentafluorophenol
PFU	Plaque forming unit
PGK1	Phosphoglycerate kinase 1
PIPES	1,4-piperazinediethanesulfonic acid
PKC	Protein kinase C
PKD	Polycystic kidney disease
PKM2	M2-type pyruvate kinase
PLL	Poly-L-lysine
PLP1	Sphingomyelin lipid protein 1
PMA	Phorbol-12-myristate-13-acetate that is a PKC activator
Pmb	p-methoxybenzyl
PMMA	Polymethyl methacrylate
PNA	Peanut agglutinin
4-PPY	4-pyrrolidinopyridine
PS	Polystyrene
PSF	Point source function
Pth	Phthaloyl
PVA	Polyvinyl alcohol
PW	Plasma wave
q_e	Charge amount of a particle
qRT-PCR	Quantitative reverse-transcription real-time polymerase chain reaction
Q	Flux
r	Radius; Spatial coordinate; Reflection coefficient
r_B	Bohr's radius (0.529 Å)
r_e	Equivalent radius of a free valance electron
r_p	Reflection coefficient of p-light
r_s	Reflection coefficient of s-light
rhEpCAM	Recombinant human EpCAM
R	Reflectance; resistance; linear correlated coefficient; the universal gas constant (8.314 J mol^{-1} K^{-1})
$\boldsymbol{R}_0$	Amplitude vector of exponential wave function
RBL	Rat basophilic leukemia
RBL-2H3	An RBL cell line
Re	The real part of a complex function
RE	Reference electrode
RI	Refractive index
RIU	Refractive index unit
RPS3	Ribosomal protein S3

RSD	Relative standard deviation
RT-PCR	Reverse transcription-PCR
sCMOS	Scientific CMOS
ssDNA	Single-strand DNA
S	Slope or sensitivity
S	Poynting vector
SAM	Self-assembly membrane
SCAOS	Stepwise cyclic amplification of signals
SDCR	Strand displacement cyclic reaction
SDF-1	Stromal cell-derived factor-1
SEC	Size exclusion chromatography
SELDI	Surface-enhanced laser desorption/ionization
SELEX	Systematic evolution of ligands by exponential enrichment, which is developed for screening of aptamers
SEM	Scanning electron microscope
SERS	Surface-enhanced Raman scattering
SKBR3	Human breast adenocarcinoma cells
S/N	Signal to noise
SP	Surface plasmon; Surface plasma
SPPs	Surface plasmon polarizations
SPR	Surface plasmon resonance
SPRS	Surface plasmon resonance sensing
SPRi	Surface plasmon resonance imaging
SPRI	Identical to SPRi
SPRiD	SPRi detection or detector
SPRi-MS	SPRi-mass spectrometry
SPRM	SPR microscopy
SPW	Surface plasma wave
Sulfo-NHS	N-hydroxysulfosuccinimide
Sulfo-SMCC	Sulfosuccinimidyl-4-(N-maleimidomethyl)cyclohexane-1-carboxylate
t	Time; transmission coefficient
transPtTz	*Trans*-[$PtCl_2$·NH_3·thiazole]
T	Transmittance
T_e	Temperature
T_ν	The time to complete a round vibration termed period
TAMRA	Tetramethyl rhodamine
TCTA	2,4,6-trichloro-1,3,5-triazine or CC
TE	Transverse electric field (mode)
TEA	Triethanolamine
TEM	Transmission electron microscope
Teoc	Trimethylsilylethoxycarbonyl
Tfa	Trifluoroacetyl
TFA	Trifluoroacetic acid
THF	Tetrahydrofuran

TIR	Total internal reflection
TIRF	TIR fluorescence
TIRFM	TIRF microscopy
TLC	Thin layer chromatography
TM	Transverse magnetic mode
Tos	P-toluenesulfonyl
Tris	Tris(hydroxylmethyl)aminomethane
Trt	Trityl
Tv	Tervaleryl
TWs	Transverse waves
u	Flow rate
ULOD	Upper limit of detection
UV	Ultraviolet
V	Volume; Voltage
VED	Vacuum evaporation deposition
VMS	Vacuum magnetic sputtering
VPF	Volume plasma frequency
VPP	Volume plasma polarization
VU1D9	A monoclonal antibody (IgG1 subtype) specific to EpCAM
w	Weight quantity
WE	Working electrode
WHO	World Health Organization
Y	Equivalent admittance
z	Valence, coordinates and axis
α	Ratio or coefficient
α_{ext}	Optical extinction coefficient, ratio
β	Artificial enlarging factor (e.g., 100)
β_{a}	The ratio of adsorption sites occupied by a substance
γ	Ratio of $\boldsymbol{E}_{sc}$ over $\boldsymbol{E}_{spp}$ = the scattered fraction of $\boldsymbol{E}_{spp}$ at time t = timely proportion
γ_{a}	Adsorption ratio coefficient with its largest limit at 1
δ	Effective phase thickness
δ'	Effective optical thickness
δ_D	Debye screening length or Debye length
δ_P	Penetrating depth of VPP
$\delta(\boldsymbol{r})$	The delta function
$\delta_{\mathrm{SPP},\, j}$	Penetrating depth of SPP, j = any substance
ΔG	GIBBS free energy change
ε	Relative permittivity
ε_0	Permittivity of free space
η	Modified admittance
θ	Angle
θ_{c}	Critical angle of TIR
θ_{B}	BREWSTER'S angle
θ_{in}	Incident angle

θ_r	Resonant angle of excitation light with SP
θ_{SP}	SP resonant angle
$\boldsymbol{\kappa}$	Decaying constant of SPP
λ	Wavelength
λ_{in}	Incident wavelength
λ_p	Wavelength of plasma oscillation
λ_r	Resonant wavelength of excitation light with SP
λ_{sp}	SP resonant wavelength
Λ	Plasma parameter
μ	Relative permeability
μ_0	Permeability of free space
ν	Frequency or the number of vibrations per second
ν_e	Electron oscillation frequency
ν_p	Plasma oscillation frequency
ρ	Volume density of an adsorbate in unit of g/cm^3
ρ_f	Volume density of free charge (C/m^3)
ρ_m	Mass density of metal (kg/m^3)
$\boldsymbol{\rho}_s$	Surface charge density
σ	Electric conductivity; Standard deviation
τ	Averaged time; free travel time or relaxation time between two collisions
υ	Velocity
φ	(Initial) phase
$\Phi(r)$	Wave function
χ	Polarizability
$\boldsymbol{\Psi}$	Universal symbol of vector function
ω	Angular frequency
ω_e	Angular frequency for collective oscillation of valance electrons
ω_p	Angular frequency of plasma oscillation
ω_+	Angular frequency of collective oscillation of positive ions
ω_{sp}	Angular frequency of surface plasma oscillation
∇	Nabla symbol or three-dimensional gradient operator
∇^2	A second-order partial differentiation operator

Chapter 1
Introduction

1.1 Overview of Plasma

You should know about ice that may melt under sunlight and changes to water. The water can further change to vapor by heating. They form three common states of mater around the world, that is, the solid, liquid and gas. Nevertheless, do you know what may happen when a gas is further heated to an extremely high temperature such as that in the sun? Science tells us that the gas there will not be stable any more but immediately collapse. The electrons escape from the atomic orbits to become free, leaving behind the bare atoms as positively charged particles. The sun is thus full of negatively charged electrons and positively charged ions but they exist as a spherical whole. That is called plasma, the fourth state of matter, in addition to the solid, liquid and gaseous states.

Different from the sun, natural plasmas do not often appear everywhere on our earth. You have to find them in some special cases such as aurora in the arctic sky. Away the two polar regions, you have also chance to observe plasmas in the lightening areas. In universe, it is believed that over 99% of the visible substances (not including dark matter), such as stars and interstellar dusts, are in the state of plasmas. In modern world, we people are lucky to have various chances for watching artificial plasmas and to use plasma-based techniques in daily life, such as electric arc, glow discharge, plasma boll, fluorescent light bulbs, neon light tubes and plasma display TV and computer screens.

1.1.1 Definition of Plasma

A plasma can be defined more exactly as a collection of unbound positive and negative particles, and neutral particles as well. The same as other states of matter, a pure plasma can return to its lower energy state as its inner energy is dissipated to a proper level.

Y. Chen, *Surface Plasmon Resonance Imaging*, Lecture Notes in Chemistry 95,
https://doi.org/10.1007/978-981-99-3118-7_1

The definition stresses that a plasma is a collection or it behaves collectively. By this concept, we can understand that the so-called unbound particles are not really "free" but somewhat restricted. In fact, the charged particles in a plasma must be close enough to each other so that each member in the collection can influence many its neighboring particles, rather than only the closest one as in a molecule. In the microscopic, the movement of a charged particle in the plasma will affect and be affected by many neighboring particles, causing all neighboring and then whole particles to move. It is these neighboring particles that form a Debye screening sphere governing the collective behavior of a plasma in various degree [1, 2]. Although having charged particles, the plasma is roughly neutral or quasi-neutral as a whole if it is generated from a neutral initial, but it is conductive. The charged particles will be pulled together or pushed apart by an external electric and/or magnetic field. A plasma can hence be held in isolation by a magnetic field. This makes a plasma distinguishable from a gas that must be held or stored in a container rather than a magnetic field.

The collective behavior is also called plasma approximation in many other papers. It is valid when the number of charge carriers within the Debye screening sphere of a particular particle is more than a unity. The number is averaged by the so-called *plasma parameter*, Λ [3]:

$$\begin{cases} \Lambda = 4\pi n_e \delta_D^3 \\ \delta_D = \sqrt{\frac{\varepsilon k_B T_e}{n_e q_e^2}} \end{cases} \tag{1.1}$$

where n_e is the number density of charges or electrons, δ_D the Debye screening length or Debye length, ε permittivity, k_B Boltzmann's constant, q_e charge amount of a particle, and T_e its temperature. The statistical character of the collective plasma behavior is observed on a length larger than the Debye length and a time longer than the plasma period.

1.1.2 Sort of Plasma

The classic plasmas are dissociated from gases at a high temperature and hence termed high temperature plasmas. A plasma can also be created at a low temperature that is called a low temperature plasma. In fact, there are nowadays generalized plasmas based on the concept of "charged collective" such as non-neutral plasma, electrolyte solutions, melt salts, semiconductors and metals. Nevertheless, they have intrinsic differences:

(i) A non-neutral plasma contains significantly unbalanced charges. In extreme cases, it can be composed of only a single species of charge where the electric field plays a dominant role. Examples are ion beams, a cooled electron cloud (in such as a Penning trap) and positron plasmas. These plasmas are commonly hard to prepare and difficult to maintain. They are not our focus and will not be further discussed in this book.

(ii) Electrolyte solutions and melt salts are easy to prepare but we have not yet obtained sufficient plasmonic knowledge on this type of plasmas and cannot give a clear picture about them. They will also be skipped in later chapters.

(iii) Conductive metals especially those noble metals contain free valance electrons and positively charged atoms that construct a crystal frame immersed in the electron cloud or "Fermi electron sea". They can thus be considered a special class of plasmas that are the theme of this book associated with the concepts of plasmon and surface plasmon (SP).

(iv) Gaseous plasmas may have lower conductivity than metal or electrolyte solutions.

(v) Ion beams, electron cloud, dissociated gases, melt salts and electrolyte solutions are all fluidic, while the semiconductors and metals are solid with exploitable surfaces. Metals are easier to process and have hence been explored in nearly all methods of SP resonance (SPR) and SPR imaging (SPRi).

1.1.3 Basic Property of Plasma

According to the composition, the basic property of a plasma can be deduced as follows:

(i) **Existence of long range Coulomb interaction**: The long range Coulomb force makes plasma behave collectively.

(ii) **Collective action**: Plasmas act collectively, with abundant collective effects and moving modes.

(iii) **Effect of external electromagnetic field**: The long range Coulomb force makes plasma extremely easy to synchronize its oscillations with an external electromagnetic filed. As a result, the shape and related properties of plasmas can easily be alternated by the external field.

(iv) **Influence of boundary conditions**: Plasmas and their wave type and oscillation modes are all sensitive to boundary conditions.

(v) **Obeying the classical mechanics and electrodynamics**: The movement of charged particles in a plasma can simply be treated by classical theories such as Maxwell's equations, without the consideration of quantum effects and the relativistic effects. The electrons in a plasma collide so frequently that they will soon de-cohere into tiny wavelet packets at a size within a de Broglie wavelength, much smaller than particle-to-particle distance. The quantum effects responsible for electronic diffraction and interference are too small to be considered. Another type of quantum effect in respect of electron–positron annihilation is even not needed to touch because it happens only at the extremely high level of energy. Electrons can hardly keep their energy above keV because they will lose their energy by colliding the atoms and emitting X-rays to penetrate and escape the plasma region, and the energy loss increases with system temperature. In reality, we are still unable to create a plasma on earth at a temperature beyond 100 keV.

Similar to classic vibration, the electrons in plasmas oscillate at a given frequency ν_e, usually regarding as the plasma oscillation ν_p:

$$\nu_p \cong \nu_e = \frac{1}{2\pi}\sqrt{\frac{e^2 n_e}{\varepsilon_0 M_e}} \doteq 8.979\sqrt{n_e} \tag{1.2}$$

where M_e electron mass, e its charge and ε_0 the permittivity in free space.

(vi) **Quasi-neutral**: Unless original imbalance of charges, most plasmas are quasi-neutral as a whole because its physical size is much larger than the Debye screening length, so that the bulk interactions are more important than the edges. Macroscopically, plasmas are homogeneously charged fluids except for their boundaries or sheaths; while in microscopic scale at a size ~R_D, a plasma looks like groups of particulates, where the electric field around a charge within the Debye screening sphere is much close to Coulomb field in a free space. It decays quickly and soon becomes equal to the total electric field.

(vii) **Thermal imbalance**: The temperatures of electrons and ions in a plasma may be very different because their energy needs to be exchanged through collisions between electrons and ions. Because electrons reduce their collision cross section at high energy, the high energy electrons easily maintain their energy level while the low energy electrons not, which in turn leads to sharp imbalance. As consequence, electrons can keep their temperature at about 10^4 K in different plasmas while the ions often at far below 10^4 K in a low temperature plasma. Undoubtedly, the collision will also vary the distribution of electrons, for example, slow collisions may retain some extra-speeded electrons, showing a long distribution tail.

(viii) **High chemical activity**: In plasmas, the numerous bare ions, electrons and collision-excited atoms and molecules are very corrosive, able to corrode nearly all substances. In a plasma, even a commonly inert element can blow out the atoms from a solid surface, or can weaken chemical bonds to catalyze various chemical reactions.

1.1.4 Brief History

Plasma as a phenomenon was first observed in the vacuumed cathode area and called "radiant matter" by Crookes [4] and Preston [5]. The "cathode ray" matter was identified to associate with electrons by Thomson [6]. However, the term "plasma" was coined by Langmuir [7] to stress moldable nature of the glowing discharge inside the Crookes tube. The English word of plasma was in fact adapted from an ancient

Greek word of $\pi\lambda\acute{\alpha}\sigma\mu\alpha$ for meaning "moldable substance" or "jelly" [8]. Plasma has since then been running into a fast track of development, finding wider and wider applications which cover from daily life to industry and have been studied in sciences from earth to universe.

1.2 Plasmon

Plasmon is used to denote the energy quantum of a plasma oscillation, just the same as the photon for electromagnetic vibration. This term was initially proposed in 1952 [9] and was gradually accepted after Ritchie who used plasmon concept in the study of energy loss of electrons after passing through a metal membrane [10]. The plasmon was soon shown to associate with the long-range electron–electron correlations that is the second term of Hamiltonian [11]:

$$H = \sum_j \frac{P_j^2}{2M} + \frac{1}{2}\sum_{j \neq m} \frac{e^2}{|r_j - r_m|} = H_{\text{free}} + H_{\text{long}} \tag{1.3}$$

The first term at the right side of the equation represents free electrons while the second represents the long range Coulomb interaction of the electrons corresponding to a plasmon. Interestingly, most of the plasmon properties or the energy quantization of a plasma oscillation can be derived directly from Maxwell's equations that are going to be discussed in Chap. 2.

It is now clear that plasmon correlates with free electron density oscillation around positive ion points. Under an external electric field, free electrons will move away the ions to compensate for electrostatic force; and they will move back once the external electric field is cancelled due to the ion attraction. The back electrons will not stop at their origin but go farther through the origin due to inertia. The electrostatic attraction reverses direction immediately after the electrons pass back through the ions. This direction alternation makes the electrons oscillate back and forth at about the plasma frequency (Eq. 1.2). The oscillation will gradually stop if there is resistance.

1.3 Surface Plasmon

1.3.1 Basic Concepts

The concept of plasma what we have talked has not yet been limited to any dimension. It is applicable commonly to volume plasma or 3D space. When we focus our view on a surface or shrink the volume plasma to a very thin layer, the on-surface plasma oscillations occur. An energy quantum of such an oscillation is called a surface plasmon first proposed in 1960 [12] and its travelling wave is called SP wave (SPW).

Different from the volume plasmons that are induced by longitudinal electron density oscillations, surface plasmons contain also transverse oscillations components that can hence interact with lights [13]. To visualize the difference between volume and surface plasmons, one can go to a still pool and drop a stone into the water. You will see a transverse surface wave spreading out on the water surface, but beneath the surface wave, longitudinal waves are also created that travel down along the stone sinking direction due to the press of the stone, *though*, you cannot watch the travel.

Similar to the water pool, SPs do not necessarily occur on any a surface but at some special interfaces that are formed by two different materials with opposite signs of relative permittivity. Examples are interfaces between a conductor (e.g., metals and doped semiconductors) and a dielectric material (e.g., air, solutions, glass, plexiglass and other dielectrics including free space). Permittivity is in general a complex varying with optical wavelength. The permittivity of dielectric materials and vacuum have a positive real part while that of metals can have a negative real part at some given optical frequencies. To excite SPs, the magnitude of the real part of the negative permittivity has to be larger than that of the positive permittivity [14].

SPs can exist in the range of visible lights, where metals like silver and gold are commonly used to form the interface with either air or solutions. SPs can also be observed in the X-ray emission spectra of metals as expected in theory [15]. Importantly, SPs occur at variable interfaces ranging from flat to sharply curl surfaces, including slides, cylinders, v-grooves and nanoparticles or other shaped structures. On-particle SPs are responsible for surface-enhanced Raman spectroscopy and gold-colored glass in various churches, and can be used to control colors of materials. SPs on strips or gratings explain the famous Wood's anomaly [16, 17]. SPR with visible lights opens a novel way to construct SPR sensors (SPRS that is used in this book instead of SPR to avoid the confusion of SP resonance mechanism with SPR sensing). Recently, doped graphene was shown able to accommodate SPs in the infrared range [18, 19, 20].

1.3.2 *Application and Interpretation*

The practical use of artificial SPs is not really new in our human society. It can be traced at least back to the date of the fourth century AD when the Lycurgus cup of the Byzantine Empire appeared. More recent applications of SPs to real life might be back to the date of building glorious churches. Travel in Europe, you can easily find that the windows in the historic Churches are always installed with vibrantly colored glasses. Some early pastors and artists knew how to color glass and artifact glass handcrafts with metallic "dyes" which are nowadays explained as localized SPR (LSPR) adsorption phenomenon.

The scientific studies of SP phenomenon were however initiated after the work of Wood in 1902. He observed a narrow dark band in the diffraction spectrum from a metallic grating that could not be interpreted clearly in theory at that time [16, 17].

Two years later, Maxwell–Garnett tried to interpret the colors of metal-doped glass wares by Drude model [21], a theory on metal conductors [22]. In 1907, Rayleigh tried to give a theoretical explanation by light scattering [23] and Mie also developed a theory of light scattering from spherical particles [24]. Fano gave a better interpretation in 1941 by use of the concept of electromagnetic surface wave [25], first proposed by Sommerfeld [26] and then by Zenneck [27]. The understanding at that time was that an optical diffraction wave vector could match and resonate with the surface wave vector at a certain angle, depending on the surface property (e.g., the grating constant).

In 1956, Pines gave a review on the energy losses of fast electrons after bombardment transmission through a metal foil, and attributed the energy losses to the collective oscillations of the free electrons in the metal [28]. In the same year, Fano discussed the "polarization" associated with the coupled oscillation of electrons [29]. A years later, Ritchie studied the energy losses of electrons in thin metals and showed for the first time that the "plasmon" can exist just near the metallic surfaces [10], which helps to coin the concept of "surface plasmons". Firm experimental proof of SP concept was done by Powell and Swan [30, 31]. At nearly the same time, Stern and Ferrell revealed the resonant conditions to utilize the "Surface plasmon" in respect of the oscillation energy quantum of the surficial free electrons [12]. They also found that oxides on the metal film surfaces could significantly shift the frequency of SPs and suggested the use of the oxides to control the characteristic energy losses. It took 66 years when Ritchie and his coworkers could successfully explain the Wood's observation by use of the concept of SPs and SPR [32]. This is the first milestone in studying the SPs and opened the door to use SPs for sensing substances, but it was not really aware of by the scientists at that time.

In 1970, Kreibig and Zacharias compared, for the first time, the electronic and optical responses of gold nanoparticles (AuNPs) and silver nanoparticles (AgNPs) in terms of SPs [33]. Cunningham and coworkers thus created the term of surface plasmon-polarization (SPP) to denote the electron oscillation in metal nanoparticles [34], and Fleschmann et al. [35] could enhance the Raman spectra of pyridine on a roughened silver surfaces. Due to not awareness of the SP effect, their works, though initiating the blooming of surface-enhanced Raman Scattering, did not actually light up the study of nanoparticles-based optics like plasmonics [36]. These are out of the scope of this book and will not be discussed further.

1.4 Surface Plasmon Resonance and Basic Features

It is now clear that SPs are the oscillators of surface electrons and can be polarized under a proper external electric field to cause SP polarizations (SPPs). The SPPs will then travel along the surface to produce SPWs. According the electromagnetic theory, the external electric field can be created by an electromagnetic wave or a light. This suggests a possibility to initiate or excite SPWs by a light. It is conceivable that the excitation needs a way to couple SPs with the light, that is to make an SP resonates with a light, i.e., what we often call SPR. Theoretical studies have revealed all the conditions for SPR but its realization needs to explore some unusual coupling configurations or devices to have insights into the features of SPR methodology.

One unique feature to mark SPR is the energy loss of the exciting light at a certain wavelength called resonant wavelength (λ_{sp}) or at a certain excitation angle called resonant angle (θ_{sp}). The devices able to interrogate either λ_{sp} or θ_{sp} have been invented that is the second milestone in the studies and development of SPs and SPR. Two core devices that have made the most significant contributions to the exploration and development of SPR methodology are the prism-based Otto configuration [37] and Kretschmann configuration [38, 39]. They all work based on the attenuated total internal reflection (ATIR or ATR) phenomenon realized in a prism. Their major difference lies in how to position the metal sensing film: Otto configuration sets a gap between ATR prism and metal sensing film for a sample to pass through, while Kretschmann configuration cancels the gap, allowing a sample to flow through the upper surface of the metal surface (detailed in Chap. 3). Kretschmann configuration is now dominantly used in the commercial SPRS and SPRi instruments, which may also be attributed to the commercial initiation of Biacore® [40].

The utilization of SPRS as a new sensing technology could be dated back to 1960 when the term of SP was first proposed and found to be changeable by oxides on the metal surface [12]. A more practical technique was founded by Cordon et al. who found that an organic film deposited on a gold film could also impact on SPs [41]. In 1980, Gordon and Ernst demonstrated that SPs could also serve as a probe of electrochemical interface [42]. A year later, the real applicability of SPRS to sense gases was demonstrated, and it is also applicable to monitoring the antigen-antibody interactions on a silver sensing film [43, 44]. SPRS then started to boost in 1990s and becomes now a typical or reliable tool to study and verify the interactions of biomolecules.

Inspired by the successful exploration and applications of SPRS, more and more SPs-based methodologies have been exploited, including plasmonics, SP-enhanced Raman scattering [35], LSPR [45, 46], SP-enhanced fluorescence spectroscopy [47], plasmon enhanced near-field scanning optical microscopy [48] and of course SPRi [49–51] which forms the topic of this book.

1.5 Surface Plasmon Resonance Imaging

1.5.1 Definition

SPR mechanism enables the exploration of novel optical imaging methodology that has been termed SPRi, SPR microscopy (SPRM), imaging SPR (iSPR or ISPR), two dimensional SPR (2D SPR), multichannel SPR or plasmoinic imaging. To simplify and unify terminology, only SPRi is used as the standard abbreviation in this book while plasmonic imaging is used occasionally. Clearly, it can be defined as an optical surface imaging methodology that works on the root principle of SPR-caused optical adsorption to reveal the subtle interfacial features by refractive index contrast. The excitation mechanism is the same as SPRS but the excitation beam is much wider than that of SPRS to cover all imaging areas. It should also be notated that the connotation of SPRi can be much broader than what we defined, involving at least the following aspects;

(i) Image formed by optic projection of molecules or particles on flat surfaces onto a recording medium through SPR mechanism;
(ii) Image formed by use of plasmonics and/or LSPR to "light up" or "color" an object, for example, use of AuNPs as a "dye" to image cells;
(iii) Image formed by use of SPR or SPP superlenses to break through the optical diffraction limit for better spatial resolution or to extract the feature details of an object commonly hindered in the evanescent waves (EWs) which decay within a very short distance of about a wavelength and cannot be obtained by a far field optical techniques like a common optical microscopy;
(iv) Use of SPs as a novel means to display image information or to design and fabricate next generation of displayers with high contrast and non-time delay.

This book discusses only the first type of imaging methodology, but not the other three, unless particularly necessary.

1.5.2 Basic Features

In order to implement practical SPRi assays with high fidelity, the excitation beam must be collimated to uniformly cover an imaging surface. This simple variation makes SPRi able to acquire 2D images, therefore, SPRi can tremendously improve the usage of a limited sensor surface to increase the analytical throughput compared with SPRS that uses only a point of the sensing surface. SPRi allows to have an insight into the variation of refractivity along the metal film, and because the film refractivity can be affected by either the thickness of the film or the amount of adjacent materials, SPRi is in turn applicable to the direct determination of film thickness and related concentration under required condition, without any chemical label.

In history, SPRi was first used to measure the thickness of a metal film or a thin film deposited on a metal (e.g., silver or gold film) sensing film. Later it is more and more utilized to sense the variation of permittivity or concentration of analytes on or adjacent to the sensor surface. The analysis can be performed in a static or dynamic state. The static measurement can be realized by changing the surface step by step while the dynamic by passing a sample across the sensor surface or by dotting/patterning samples on the surface. The measurement can be performed *in situ* and in real time. All concentration- or permittivity-dependent events, such as on-surface-localized chemical reactions, captures of molecules, deposition and peeling off matter, can simultaneously be observed under the conditions what you set or like. SPRi is hence applicable to the high throughput analysis of native bio-molecules under physiological conditions, which is hardly realized by many ordinary methods. Furthermore, SPRi can be applied to the direct determination of opaque, turbid and even dark samples, and can reveal the subtle interfacial features by refractive index contrast. The high sensitivity of SPRi has made it a good tool to read biomolecular binding events, in a label-free fashion. These features are unique and highly favored in conducting bioassays, which has gradually made SPRi stand on the top for choice over many other analytical methods. SPRi is at present coming to a phase of fast development and quick spreading its applications.

1.5.3 *Development History*

The first set of SPRi system was reported in 1987 and termed SPRM [49]. About one year later, a similar setup showed up in a letter to Nature [50]. Corn group has since 1995 (and our group since 1997) bent on the analytical applications of SPRi and published numerous research papers, including long-range SPRi [52]. The existence of long-rang SPs was first predicted in 1981 [53] and demonstrated in 1983 [54]. It is now shown that the long-range SPs can be created in several different multiple-layered structures ([55–58], Toyama et al. 1991), similar to the combination of Otto configuration with the Kretschmann. The long-range SPRi opens a way to image large particles on the sensor surface.

Prism-based SPRi limits the numerical aperture (NA), its magnification and lateral spatial resolution. It easily causes distortion and position shift of images as the incident angle or wavelength is scanned. In 2004, cylindrical prism was used to conduct angle-resolved SPRi of spots with size larger than 200 μm [59], thus the lateral resolution SPRi is not more critical. In 2005, an imaging SPR apparatus was patent in USA [60]. In 2007, a configuration of total internal reflection fluorescence microscopy [61] was adapted into SPRi by use of a high NA microscopic objective to excite SPR with high magnification [62]. The configuration could image objects on 50-nm gold surface without shape distortion, and had angle-resolved imaging capability to extract detailed information on surface properties, with nearly a diffraction-limited spatial resolution.

1.6 Hyphenation Aspects

In addition to the independent use, SPRi can hyphenate to or be hyphenated by other instrumental methods. In general, there are two basic coupling strategies: Combination with other similar analytical principles/methods to confirm the SPRS and SPRi interpretation, and use of SPRS/SPRi to detect the separated substances. The first strategy has made several useful coupling methods and techniques become true, for example, SPRi-mass spectrometry (SPRi-MS), SPRi-fluorescent and/or Raman microscopy, SPRi-optical microscopy. The second strategy is to adjust or modify either the SPRi or SPRS device or the separation tools to perform *in situ* or online detection of separated bands. We have tried to fabricate micro-channels on SPRi-sensing surface and perform fast electrophoresis. A more interesting example is to use SPRi as quasi-CCD (charge-coupled device) to detect low level molecules with the aid of only fairly common or cheap CCD. This method can largely reduce the cost in the establishment of high throughput separation system based on channels or columns arrays. In our laboratory, capillary array electrophoresis coupled with a SPRi detector (SPRiD) was also designed and fabricated, and it is applicable to the direct separation and detection of many non-UV (ultraviolet), non-fluorescent chemicals.

Considering that SPR relies on the evanescent optic wave, its inherent coupling should point to other optic methods. The most significant coupling is to use SPR principle to enhance optical detection sensitivity by fully use of near-field optical component and effective reduction of background. SPR-induced fluorescence is an excellent example which has special spatial distributions. This can be achieved by prim configuration and excites the fluorescence at both sides of the SP film. The coupling of SPs with Raman spectrometry which has been mentioned is also a famous example. (Refer to Sect. 9.5 for more information).

1.7 Prospect of Developments

SPRi as a newly established instrumental method is not yet matured but remains to develop in respects of its devices, methodological development and applications. It is presently not sensitive enough to perform bioassay of many trace biomolecules and their recognition kinetics. This is becoming a bottleneck in expanding the applicability of SPRi. The current efforts are laid on two main streams: exploitation of strategy and methodology to magnify the SPRi signals; and design and fabrication of better instruments by melting in better working principle, including exploration of novel resonance mechanism.

Signal amplification is the easiest strategy with abundant chemical reactions waiting for mining. There have already been studying on this strategy with various significant progresses, but the way is still long to our goals. Many challenges are lying ahead together, waiting us to conquer with various chances. (Refer to Chap. 9 for deeper insight into the challenges).

Instrumentation is a globe stream to push forward in instrument-based analytical chemistry. Although this costs high and takes time, near none manufacturer world like to give up this effort. As a result, new instruments, at least new models will show up from time to time. In a common laboratory, hardware modifications or renewal can be considered but not a good choice. A more effective approach is to level up the theoretical understanding and relative software for data mining.

Development of novel SPRi method and expanding its application fields are two key studying frontiers. Applications, the depth and width, determine the ultimate fate of SPRi or how far it can reach. The present studies on SPRi are mainly on biomolecular interactions and screenings, including reaction kinetics. Actually, SPRi is much more useful, various vast fields are in front of it.

References

1. Sturrock PA (1994) Plasma physics: an introduction to the theory of astrophysical. Geophysical and laboratory plasmas. Cambridge University Press, Cambridge
2. Hazeltine RD, Waelbroeck FL (2004) The framework of plasma physics. Westview Press, Boulder
3. Chen FF (2006) Introduction to plasma physics and controlled fusion. Springer, New York
4. Crookes W (1879) On radiant matter; a lecture delivered to the British Association for the Advancement of Science, at Sheffield, Friday, August 22, 1879. Am J Sci 318(106):241–262
5. Preston S (1881) On some points relating to the dynamics of "Radiant Matter." Nature 23:461–464. https://doi.org/10.1038/023461a0
6. Thomson JJXL (1897) Cathode Rays. London Edinburgh Dublin Philos Mag J Sci 44:293–316. https://doi.org/10.1080/14786449708621070
7. Langmuir I (1928) Oscillations in ionized gases. PNAS 14:627–637. https://doi.org/10.1073/pnas.14.8.627
8. Goldston RJ, Rutherford PH (1995) Introduction to plasma physics. Taylor & Francis, New York, pp 1–2
9. Pines D, Bohm D (1952) A collective description of electron interactions: II. Collective vs individual particle aspects of the interactions. Phys Rev 85:338–353
10. Ritchie RH (1957) Plasma losses by fast electrons in thin films. Phys Rev 106:874–881
11. Bohm D, Pines D (1953) A collective description of electron interactions: III. Coulomb interactions in a degenerate electron gas. Phys Rev 92:609–625. https://doi.org/10.1103/physrev.92.609
12. Stern EA, Ferrell RA (1960) Surface plasma oscillations of a degenerate electron gas. Phys Rev 120:130–136
13. Zeng S, Yu X, Law W-C, Zhang Y, Hu R, Dinh X-Q, Ho H-P, Yong K-T (2013) Size dependence of Au NP-enhanced surface plasmon resonance based on differential phase measurement. Sens Actuat B Chem 176:1128–1133. https://doi.org/10.1016/j.snb.2012.09.073
14. Raether H (1988) Surface plasmons on smooth and rough surfaces and on gratings. Springer, New York
15. Harsh OK, Agarwal BK (1988) Surface plasmon dispersion relation in the X-ray emission spectra of a semi-infinite rectangular metal bounded by a plane. Phys B+C 150:378–384. https://doi.org/10.1016/0378-4363(88)90078-2
16. Wood RW (1902) On a remarkable case of uneven distribution of light in a diffraction grating spectrum. Phil Mag 4:396–402
17. Wood RW (1902) A suspected case of the electrical resonance of minute metal particles for light-waves. A new type of absorption. Proc Phys Soc Lond 18:1478
18. Chen CY, Chang CC, Yu C, Lin CW (2012) Clinical application of surface plasmon resonance-based biosensors for fetal fibronectin detection. Sensors 12:3879–3890
19. Fei Z, Rodin AS, Andreev GO, Bao W, McLeod AS, Wagner M, Zhang LM, Zhao Z, Thiemens M, Dominguez G, Fogler MM, Castro Neto AH, Lau CN, Keilmann F, Basov DN (2012) Gate-tuning of graphene plasmons revealed by infrared nano-imaging. Nature 487:82–85. https://doi.org/10.1038/nature11253
20. Yan H, Low T, Zhu W, Wu Y, Freitag M, Li X, Guinea F, Avouris P, Xia F (2013) Damping pathways of mid-infrared plasmons in graphene nanostructures. Nat Photonics 7:394–399. https://doi.org/10.1038/nphoton.2013.57

21. Maxwell-Garnett JC (1904) Colors in metal glasses and in metallic film. Philos Trans R Soc London 203:385–420
22. Drude P (1900) Zur Elektronentheorie der metalle. Ann Phys 306(3):566–613. https://doi.org/10.1002/andp.19003060312
23. Rayleight L (1907) Note on the remarkable case of diffraction spectra described by Prof. Wood. Phil Mag 14:60–65
24. Mie G (1908) Beiträge zur Optik trüber Medien, speziell kolloidaler Metallösungen. Ann Phys 25:377–455
25. Fano U (1941) The theory of anomalous diffraction gratings and of quasi-stationary waves on metallic surfaces (Sommerfeld's waves). J Opt Soc Am 31:213–222
26. Sommerfeld A (1899) Über die Fortpflanzung elektrodynamischer Wellen an längs eines Drahtes. Ann der Physik und Chemie 302(2):233–290
27. Zenneck J (1907) Über die Fortpflanzung ebener elektromagnetischer Wellen längs einer ebenen Leiterfläche und ihre Beziehung zur drahtlosen Telegraphie. Ann der Physik 23:846–866
28. Pines D (1956) Collective energy losses in solid. Rev Mod Phys 28:184–198
29. Fano U (1956) Atomic theory of electromagnetic interactions in dense materials. Phys Rev 103:1202–1218
30. Powell CJ, Swan JB (1959) Origin of the characteristic electron energy losses in aluminum. Phys Rev 115:869–875
31. Powell CJ, Swan JB (1960) Effect of oxidation on the characteristic loss spectra of aluminum and magnesium. Phys Rev 118:640–643
32. Ritchie RH, Arakawa ET, Cowan JJ, Hamm RH (1968) Surface-plasmaon resonance effect in grating diffraction. Phys Rev Lett 21:1530–1532
33. Kreibig U, Zacharias P (1970) Surface plasma resonances in small spherical silver and gold particles. Z Phys 231:128–143
34. Cunningham SL, Maradudin AA, Wallis RF (1974) Effect of a charge layer on the surface-plasmon-polarization dispersion curve. Phys Rev B 10:3342–3355
35. Fleschmann M, Hendra PJ, McQuillan AJ (1974) Raman spectra of pyridine adsorbed at a silver electrode. Chem Phys Lett 26:163–166
36. Zia R, Schuller JA, Brongers ML (2006) Plasmonics: the next chip-scale technology. Mater Today 9:20–27
37. Otto A (1968) Excitation of nonradiative surface plasma waves in silver by the method of frustrated total reflection. Z Phys A Hadrons Nucl 216:398–410
38. Kretschmann E, Raether H (1968) Radiative decay of non-radiative surface plasmons excited by light. Z Naturf 23A:2135–2136
39. Kretschmann E (1971) Die bestimmung optischer konstanten von metallen durch anregung von oberfliichenplasmaschwingungen. Z Phys 241:313–324
40. Owen V (1997) Real-time optical immunosensors-A commercial reality. Biosens Bioelect 12:i–ii
41. Gordon JG II, Swalen JD (1977) The effect of thin organic films on the surface plasma resonance on gold. Opt Commun 22(3):374–376
42. Gordon JG II, Ernst S (1980) Surface plasmons as a probe of the electrochemical interface. Surf Sci 101(1–3):499–506
43. Nylander C, Liedberg B, Lind T (1982–1983) Gas detection by means of surface plasmon resonance. *Sens Actuat* 3:79–88
44. Liedberg B, Nylander C, Lunström I (1983) Surface plasmon resonance for gas detection and biosensing. Sens Actuat 4:299–304
45. Rycenga M, Cobley CM, Zeng J, Li W, Moran CH, Zhang Q, Qin D, Xia Y (2011) Controlling the synthesis and assembly of silver nanostructures for plasmonic applications. Chem Rev 111:3669–3712
46. Liu X, Swihart MT (2014) Heavily-doped colloidal semiconductor and metal oxide nanocrystals: an emerging new class of plasmonic nanomaterials. Chem Soc Rev 43:3908–3920

47. Toma K, Vala M, Adam P, Homola J, Knoll W, Dostálek J (2013) Compact surface plasmon-enhanced fluorescence biochip. Opt Express 21:10121–10132
48. Bao W, Staffaroni M, Bokor J, Salmeron MB, Yablonovitch E, Cabrini S, Bargioni AW, Schuck PJ (2013) Plasmonic near-field probes: a comparison of the campanile geometry with other sharp tips. Opt Exp 21:8166–8176
49. Yeatman E, Ash EA (1987) Surface plasmon microscopy. Elect Lett 23:1091–1092
50. Rothenhäuslar B, Knoll W (1988) Surface plasmon microscopy. Nature 332:615–617
51. Roberta DA, Spoto G (2013) Surface plasmon resonance imaging for nucleic acid detection. Anal Bioanal Chem 405:573–584
52. Wark AW, Lee HJ, Corn RM (2005) Long-range surface plasmon resonance imaging for bioaffinity sensors. Anal Chem 77:3904–3907
53. Sarid D (1981) Long-range surface-plasma waves on very thin metal films. Phys Rev Lett 47:1927–1930
54. Quail JC, Rako JG, Simon HJ (1983) Long-range surface-plasmon modes in silver and aluminum films. Opt Lett 8:377–379
55. Matsubara K, Kawata S, Minami S (1990) Multilayer system for a high-precision surface plasmon resonance sensor. Opt Lett 15:75–77. https://doi.org/10.1364/OL.15.000075
56. Yang F, Bradberry GW, Sambles JR (1991) Long-range surface mode supported by very thin silver films. Phys Rev Lett 66:2030–2032
57. Kessler MA, Hall EAH (1996) Multilayered structures exhibiting long-range surface exciton resonance. Thin Solid Films 272:161–169
58. Lyndin NM, Salakhutdinov IF, Sychugov VA, Usievich BA, Pudonin FA, Parriaux O (1999) Long-range surface plasmons in asymmetric layered metal-dielectric structures. Sens Actuat B 54:37–42
59. Shumaker-Parry JS, Campbell CT (2004) Quantitative methods for spatially resolved adsorption/ desorption measurements in real time by surface plasmon resonance microscopy. Anal Chem 76:907–917
60. Johansen K (2005) Imaging SPR apparatus. U.S. patent 6862094, March 1, 2005
61. Axelrod D (2001) Total internal reflection fluorescence microscopy in cell biology. Traffic 2:764–774
62. Huang B, Yu F, Zare RN (2007) Surface plasmon resonance imaging using a high numerical aperture microscope objective. Anal Chem 79:2979–2983

Chapter 2
Basic Theory

It needs various theories and principles to master SPRi such as wave theory, molecular interactions, analytical chemistry and so forth. Considering that the wave theory is the essential of SPRi and may be faded gradually from one's memory, it will be introduced from the most basic part by use of language what we think is easy to understand, while the other theories are considered to be common knowledge to readers and will not be touched unless necessary. Readers equipped with sufficient knowledge on photoexcitation of SPR can undoubtedly skip this chapter.

2.1 Vibration, Wave and Wave Function

Around the world, you can see and hear many types of waves. You may have experienced a pitch-changing sound from a by-passing train or car. That is a type of sound wave. You can also see waves spreading on a pond while a wind blows over the water surface or you happen to drop a piece of tiny stone into the water. A wave can also be created when you click a tighten elastic or press and release a spring. These two actions create two different types of waves, transverse waves (TWs) and longitudinal waves (LWs). TWs vibrate perpendicular to its propagating direction while LWs parallel to or along the spreading direction.

There is in this world an amazing type of wave called electromagnetic wave (EMW) that you use every day to view the world. It is the sun light. A visible light can also be emitted by an electric lamp, a torch or burning things. Some more abstract EMWs are γ-radiation, x-ray, ultraviolet, infrared, microwaves, radio waves and so forth. According to the general relativity theory of Einstein, gravity also acts through gravitational waves. Quantum theory tells us that a tiny particle is also a type of wave, and reversely, a wave such as lights may also be considered as a type of particles, that is, photons. These have been termed "wave-particle duality". The modern string theory suggests that our world is nothing but some space–time expanded tiny vibrating strings. The plasma what we will discuss in this book can also

Y. Chen, *Surface Plasmon Resonance Imaging*, Lecture Notes in Chemistry 95,
https://doi.org/10.1007/978-981-99-3118-7_2

run into a state of vibration, called plasma oscillation. While a plasma oscillation propagates, it creates a plasma wave (PW); correspondingly, a plasma vibrator is called plasmon. They are all attributed to a particular type of charge motions.

All the waves can be summarized into two categories: classical and non-classical waves. The classical waves include mechanical waves (e.g., sounds) and EMWs, while the typical non-classical waves are the modern quantum theory-treatable waves. PWs belong to the family of classical waves. All classical waves can be decomposed into various harmonic waves that are induced by harmonic vibrations. The so-called harmonic vibration is in fact a motion of a vibrator with its displacement proportional to a restoring force always pointing toward the balancing position. The vibration position (z) of a vibrator in a system at a time t is hence a sine or cosine function:

$$z = A_z \sin(\omega t + \varphi) = B_z \cos(\omega t + \varphi), \tag{2.1}$$

where A_z or B_z is the maximum displacing position or amplitude, indicating the vibration strength, φ is initial phase, and ω is incremental angle per unit time or angular frequency defined by:

$$\omega = \frac{2\pi}{T_\nu} = 2\pi\nu, \tag{2.2}$$

where T_ν is the time to complete a round vibration termed period, and ν is the number of vibrations per second, commonly shortened as frequency. It should be noted that ω is the natural frequency of a vibrating system, dependent only on the mass (M) of a vibrator and its elastic constant (k_{ec}), not on the external initial conditions:

$$\omega = \sqrt{\frac{k_{\text{ec}}}{M}}, \tag{2.3}$$

where T_ν and ν are hence intrinsically determined by the system. In general, the amplitude A_z or B_z, initial phase φ and frequency ν or angular frequency ω are three essential factors to describe a harmonic motion which is the base of all vibrations. A harmonic vibration spreading in space without energy loss creates also a harmonic wave. Commonly, a wave needs a medium to travel except for EMWs that can also propagate in free space as the sun light does. Wave propagation media in the world can roughly be sorted into three categories of dielectrics, semiconductors and conductors. More detailed classification can be made according to the conductivity σ (S/m):

$$\frac{\sigma}{\omega\varepsilon_0}\begin{cases} = 0, & \text{lossless media including free space} \\ < 10^{-2}, & \text{dielectric,low loss} \\ \approx 10^{-2} - 10^{2}, & \text{semiconductor,high loss} \\ > 10^{2}, & \text{conductor} \\ = \infty, & \text{lossless conductor,superconductor} \end{cases}, \tag{2.4}$$

where ε_0 is the permittivity in free space.

Once a wave travels from the vibration origin, at a velocity of υ, to a spatial point at $r = r(x, y, z)$ after a certain time of $t = r/\upsilon$, it will initiate the medium particle(s) at r to vibrate the same way as the vibration origin, with a phase drag of $\omega r/\upsilon$. Set a vibrator in a Cartesian coordinate and make it vibrate along the z-axis at the origin; starting from time 0, we can transform Eq. 2.1 into the following form:

$$z = A_z \sin\left[\omega(t-0) + \varphi - \omega\frac{r}{\upsilon}\right] = A_z \sin\left[\omega\left(t - \frac{r}{\upsilon}\right) + \varphi\right], \tag{2.5}$$

where z denotes the vibration position along z-axis at the point of (r, t). For a uniform plane wave (with phase unchanged along a plane) spreading along x-axis, that is, $r = r(x, 0, 0) = x$, combination of Eqs. 2.5 with 2.2 can generate several equivalent sine wave equations as follows:

$$\begin{cases} z = A_z \sin\left[\omega\left(t - \frac{x}{\upsilon}\right) + \varphi\right] = A_z \sin\left[2\pi\left(\frac{t}{T_v} - \frac{x}{\lambda}\right) + \varphi\right] \\ \quad = A_z \sin\left[2\pi\left(\nu t - \frac{x}{\lambda}\right) + \varphi\right] = A_z \sin(\omega t - kx + \varphi) \\ k = \frac{2\pi}{\lambda} = \frac{\omega}{\upsilon} = \frac{\omega T_v}{\lambda} \end{cases}, \tag{2.6}$$

where λ is wave length, k is wave number, $\omega t - kx + \varphi$ is phase angle where ωt determines time phase while kx describes spatial phase and φ is known to be the initial phase. At a given time t, if $\omega t - kx + \varphi$ is equal to constant, then kx and x are also constant (i.e., a plane with equal phase). Note, the wave can travel either positively along x-axis with time or negatively after its reflection. Keep this in mind, we will do partial differential on the wave functions to obtain a neat, general wave equation:

$$\frac{\partial^2 z}{\partial x^2} - \frac{1}{\upsilon^2}\frac{\partial^2 z}{\partial t^2} = 0. \tag{2.7}$$

As known, a sine function can also be transformed into a complex exponential function, and the time function of $z(t) = e^{i\omega t}$ can thus be separated from the spatial function of $z(x) = e^{-kx}$. By taking this advantage and $\partial^2 z(t)/\partial t^2 = -\omega^2$, Eq. 2.7 can be changed into a complex partial differential equation free of time variable:

$$\frac{\partial^2 z(x)}{\partial x^2} + k^2 z(x) = 0. \tag{2.8}$$

This is a second-order partial differential equation for harmonic waves, without energy dissipation. The so-called position function $z(x)$ can be generalized as $\Phi(r)$ in 3D space, and Eq. 2.8 becomes now:

$$\nabla^2\Phi(\boldsymbol{r}) + k^2\Phi(\boldsymbol{r}) = 0. \tag{2.9}$$

This is the homogeneous Helmholtz' equation, where ∇^2 is a second-order partial differentiation operator in 3D space. It must be noted that $\Phi(r)$ can either be a scalar or a vector function because the vibration is not only just a change in position but more changes of energy, concentration, density or intensity (e.g., electric and/or magnetic fields). In case of vector, it will be highlighted with an italic bolded letters, for example, use of $\boldsymbol{E}$ rather than $\vec{E}$ for convenience and conciseness of writing. The vector wave function will have the same form as Eq. 2.9 and a vector solution similar to Eq. 2.6 (by putting back time variable):

$$\begin{cases} \nabla^2\Phi(r) + k^2\Phi(r) = 0 \\ \Phi(r,t) = A_\Phi \sin(\omega t - k \cdot r) = B_\Phi \cos(\omega t - k \cdot r) = \mathrm{Re}[R_0 e^{i(\omega t - k \cdot r)}] \end{cases}, \quad (2.10)$$

where $\boldsymbol{k}$ is now called a wave vector whose direction is always parallel to wave propagation. The unit of the amplitude vector $\boldsymbol{A}_{\Phi}$, $\boldsymbol{B}_{\Phi}$ or $\boldsymbol{R}_0$ depends on the real definition of $\boldsymbol{\Phi}(\boldsymbol{r},t)$. The symbol of "Re" means to take only the real part of the complex exponential function. For simplicity, "Re" will be removed from the equation in later statement but kept always in mind. Equations 2.9 and 2.10 are extremely useful in the studies of vibrations and waves.

In short, a wave is the travel of a vibration initiated by a vibrator, a matter capable of reciprocating motion. The vibration is usually a periodic motion that can neatly be treated by mathematics. Specifically, a harmonic vibration and its wave can simply be described by a sine or cosine function, and more conveniently by a complex natural exponential function, $e^{i(\omega t - \boldsymbol{k} \cdot \boldsymbol{r})}$.

2.2 Electromagnetic Waves

EMWs are the traveling vibrations of electric and magnetic fields and lights are EMWs in the visible range. They are the accompanied phenomena of charge oscillations. Recall to the fact that PWs are also caused by collective oscillations of charges, we can imagine that light waves must be able to interact with PWs under certain conditions. For delivery convenience, we will first survey EMWs and then PWs.

2.2.1 *Basic Equations*

EMWs can be described by either quantum or Maxwell theory. In quantum theory, an EMW can be considered as both a wave and a particle with momentum $\boldsymbol{P}_\upsilon$ and energy E_ν:

$$\begin{cases} E_\nu = h\nu = \hbar\omega \\ \boldsymbol{P}_\upsilon = \hbar\boldsymbol{k} \end{cases}, \quad (2.11)$$

where h (=6.62607 × 10^{-34} J·s = 4.13567 × 10^{-15} eV·s) is Planck constant, and $\hbar$ (=$h/2\pi$ = 1.05457 × 10^{-34} J·s/rad = 6.58212 × 10^{-16} eV·s/rad) is the reduced Planck constant or Dirac constant.

As EMWs are able to travel in free space and in media, they are more often treated by the well-known Maxwell's macroscopic equations or Maxwell's equations in matter as follows:

$$\begin{cases} \nabla \times \boldsymbol{H} = \boldsymbol{J}_f + \frac{\partial \boldsymbol{D}}{\partial t}; & \nabla \cdot \boldsymbol{B} = 0 \\ \nabla \times \boldsymbol{E} = -\frac{\partial \boldsymbol{B}}{\partial t}; & \nabla \cdot \boldsymbol{D} = \rho_f \end{cases}. \tag{2.12}$$

The nabla symbol of ∇ is a three-dimensional gradient operator; thus, its dot product, $\nabla\cdot$, becomes a divergence operator, and its cross product, $\nabla\times$, is a curl operator; $\boldsymbol{E}$ (V/m) is electric field and $\boldsymbol{H}$ (A/m) magnetizing field; $\boldsymbol{D}$ (C/m^2) is electric displacement field or electric flux density and $\boldsymbol{B}$ (Wb/m^2) the magnetic field or magnetic flux density; ρ_f (C/m^3) is free charge density; $\boldsymbol{J}_f$ (A/m^2) is conduction current density induced by external electric field according to Joule law, $\boldsymbol{J} = \sigma \boldsymbol{E}$.

To solve the Maxwell's equations, we need to further clarify the relationships of $\boldsymbol{D}$ with $\boldsymbol{E}$, and $\boldsymbol{H}$ with $\boldsymbol{B}$. They are dependent on the features of media. In this book, we will focus more on the ideal dielectrics and ideal conductors that are normally linear, isotropic and homogeneous or inhomogeneous, with following relationships:

$$\begin{cases} \boldsymbol{D} = \varepsilon_0 \boldsymbol{E} + \boldsymbol{P} = \varepsilon_0[1 + \boldsymbol{P}/(\varepsilon_0 \boldsymbol{E})]\boldsymbol{E} = \varepsilon_0 \varepsilon \boldsymbol{E} \\ \boldsymbol{B} = \mu_0 \mu \boldsymbol{H} \\ \boldsymbol{J}_f = \sigma \boldsymbol{E} \end{cases}. \tag{2.13}$$

These are the so-called constitutive equations where ε_0 (=(1/36π) × 10^{-9} F/m) and μ_0 (=4π × 10^{-7} H/m) are the permittivity and permeability of free space, respectively, μ and ε are the relative permeability and permittivity of a medium, respectively, and $\boldsymbol{P}$ polarization intensity.

In case of neutralized media without the application of an external electric field, there will be no extra free charges and conduction current, that is, $\rho_f = 0$, and $\boldsymbol{J}_f = 0$. By taking the advantage of exponential form of $e^{i\omega t}$ in combination with the vector operation of

$$\begin{cases} \boldsymbol{a} \times (\boldsymbol{b} \times \boldsymbol{c}) = (\boldsymbol{a} \cdot \boldsymbol{c}) \cdot \boldsymbol{b} - (\boldsymbol{a} \cdot \boldsymbol{b}) \cdot \boldsymbol{c} \quad \text{or} \\ \nabla \times (\nabla \times \boldsymbol{\Psi}) = \nabla \cdot (\nabla \cdot \boldsymbol{\Psi}) - \nabla^2 \boldsymbol{\Psi} \end{cases}, \tag{2.14}$$

where $\boldsymbol{\Psi}$ denote a universal symbol of vector function, we can change Maxwell's equations into:

$$\begin{cases} \nabla^2 \boldsymbol{\Psi} + k^2 \boldsymbol{\Psi} = \nabla(\nabla \cdot \boldsymbol{\Psi}), & (\boldsymbol{\Psi} = \boldsymbol{E}, \boldsymbol{H}) \\ \nabla \cdot \boldsymbol{\Psi}_{\varepsilon\mu} = 0, & (\boldsymbol{\Psi}_{\varepsilon\mu} = \boldsymbol{D}, \boldsymbol{B}) \end{cases} \quad \begin{cases} k^2 = \frac{\omega^2}{\upsilon^2} = k_0^2 \mu \varepsilon \\ k_0^2 = \frac{\omega^2}{c^2} = \omega^2 \mu_0 \varepsilon_0 \end{cases}, \tag{2.15}$$

where c ($=1/\sqrt{\mu_0\varepsilon_0} = 2.99792 \times 10^9$ m/s) is the light speed in free space. By taking the mathematical advantages of $e^{i(\omega t-\boldsymbol{k}\cdot\boldsymbol{r})}$, following equivalent operators are derived:

$$\begin{cases} \partial/\partial t = i\omega \\ \nabla\cdot = -i\boldsymbol{k}\cdot \\ \nabla\times = -i\boldsymbol{k}\times \end{cases} \tag{2.16}$$

Apply them to Eqs. 2.12 or 2.15, we can obtain even neat and meaningful Maxwell's equations:

$$\begin{cases} (a)\ \boldsymbol{k}\times\boldsymbol{H} = -\omega\varepsilon_0\varepsilon\boldsymbol{E} \\ (b)\ \boldsymbol{k}\times\boldsymbol{E} = \omega\mu_0\mu\boldsymbol{H} \\ (c)\ \boldsymbol{k}\cdot\boldsymbol{B} = 0 \\ (d)\ \boldsymbol{k}\cdot\boldsymbol{D} = 0 \end{cases} \tag{2.17}$$

or

$$\begin{cases} (a)\ \boldsymbol{k}\cdot(\boldsymbol{k}\cdot\boldsymbol{\Psi}) = k^2\boldsymbol{\Psi} - \mu\varepsilon k_0^2\boldsymbol{\Psi}, & (\boldsymbol{\Psi} = \boldsymbol{E}, \boldsymbol{H}) \\ (b)\ \boldsymbol{k}\cdot\boldsymbol{\Psi}_{\varepsilon\mu} = 0, & (\boldsymbol{\Psi}_{\varepsilon\mu} = \boldsymbol{D}, \boldsymbol{B}) \end{cases}. \tag{2.18}$$

2.2.2 *Basic Features*

At the initial stage without any knowledge of EMWs, we can first assume that EMWs are LWs, TWs or both, and then test which assumption is correct. To this end, let us first suppose that an EMW is an LW to see if it can withstand the electromagnetic test. To test, let the vector function of $\boldsymbol{\Psi}$ in *a*) of Eq. 2.18 be replaced with $\boldsymbol{E}$. Considering $\boldsymbol{k}\|\boldsymbol{E}$ and hence $\boldsymbol{k}\cdot(\boldsymbol{k}\cdot\boldsymbol{E}) = (\boldsymbol{k}\cdot\boldsymbol{k})\cdot\boldsymbol{E}$ for an LW, we obtained $\mu\varepsilon {k_0}^2\boldsymbol{E} = 0$ from Eq. 2.18. Because $\boldsymbol{E} \neq 0, \boldsymbol{k} \neq 0, k_0 \neq 0$ and $\mu \neq 0$ for a propagating EMW, the only possibility is $\varepsilon \equiv 0$. Replace it back into either Eqs. 2.17 or 2.18, we get

$$\left.\begin{array}{r} \because \varepsilon \equiv 0 \\ \therefore \boldsymbol{H} \equiv 0 \end{array}\right\} \text{for LWs.} \tag{2.19}$$

This is against the preset that a propagating EMW must in general contain both of magnetic and electric fields rather than just a pure electric filed. Therefore, the assumption that EMWs are both of LWs and TWs is also not true. The only possibility is that EMWs are TWs and $\varepsilon \neq 0$, so that, $\boldsymbol{k}\cdot\boldsymbol{\Psi} = 0$. Eq. 2.17 can then change into another form:

$$\begin{cases} \boldsymbol{E}G = \boldsymbol{H}\times\hat{\boldsymbol{k}} \\ \boldsymbol{H} = \hat{\boldsymbol{k}}\times\boldsymbol{E}G \\ G = \sqrt{\varepsilon_0\varepsilon/\mu_0\mu} \end{cases}, \tag{2.20}$$

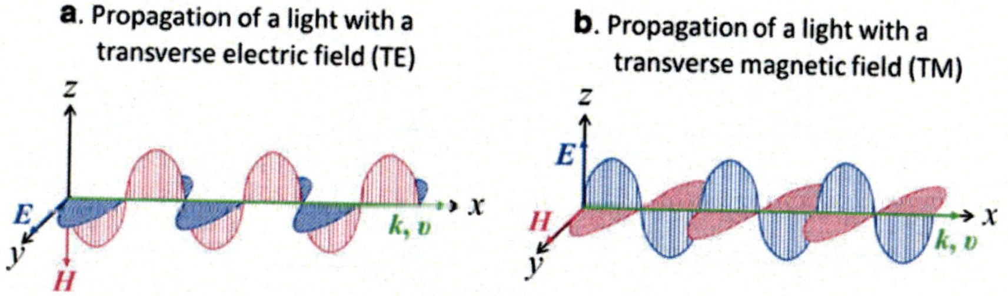

Fig. 2.1 Propagation features of polarized light. **a** Transverse electric field (TE) light and **b** a transverse magnetic field (TM) light propagating along x-axis always keep their electric field $\boldsymbol{E}$, wave vector $\boldsymbol{k}$ (or velocity $\boldsymbol{\upsilon}$) and magnetic field $\boldsymbol{H}$ perpendicular to each other, following the right-hand rule

where G is called wave admittance and $\hat{\boldsymbol{k}}$ is a unit wave vector equal to $\boldsymbol{k}\ \upsilon/\omega$. Through dot multiplication of Eq. 2.20 by the relevant nonzero vector, we get

$$\because \begin{cases} \boldsymbol{k}\cdot\boldsymbol{E} = \boldsymbol{k}\cdot(\boldsymbol{H}\times\hat{\boldsymbol{k}}) = \boldsymbol{H}\cdot(\hat{\boldsymbol{k}}\times\boldsymbol{k}) = 0 \\ \boldsymbol{k}\cdot\boldsymbol{H} = \boldsymbol{k}\cdot(\hat{\boldsymbol{k}}\times\boldsymbol{E}G) = \boldsymbol{E}\cdot(\boldsymbol{k}\times\hat{\boldsymbol{k}})G = 0 \\ \boldsymbol{E}\cdot\boldsymbol{H} = \boldsymbol{E}\cdot(\hat{\boldsymbol{k}}\times\boldsymbol{E}G) = G\hat{\boldsymbol{k}}\cdot(\boldsymbol{E}\times\boldsymbol{E}) = 0 \end{cases} .$$
$$\therefore \quad \boldsymbol{k}\perp\boldsymbol{H}\perp\boldsymbol{E} \tag{2.21}$$

Thus, $\boldsymbol{k}$(or $\boldsymbol{\upsilon}$), $\boldsymbol{H}$ and $\boldsymbol{E}$ are perpendicular to each other. It is normal that the following the right-hand rule is arranged for their directions ($\boldsymbol{E} \times \boldsymbol{H} // \boldsymbol{k}$, $\boldsymbol{H} \times \boldsymbol{k} // \boldsymbol{E}$, or $\boldsymbol{k} \times \boldsymbol{E} // \boldsymbol{H}$) as shown in Fig. 2.1.

2.2.3 Polarization

As known, EMWs can propagates in either free space or media. An EMW such as a visible light travels linearly (without the relativistic effect) and reversibly in space. It can be refracted and reflected while passes through an interface formed by two different media. Light will also diffract while passing a corner and/or interfere in case of optical path difference. Unless in free space, such propagation may result in energy loss, phase change and/or variation of vibration direction, depending on media. These lead to polarization and dispersion of light.

The polarization of a light is a phenomenon that the light propagates along a predominant vibration direction. If you can imaginarily take a photon as a vehicle to follow the light, you will observe that, in free space, photons change their $\boldsymbol{E}$ or $\boldsymbol{H}$ direction randomly around you while in a medium not: Some specific directions may be resisted. As a consequence, the summed $\boldsymbol{E}$ or $\boldsymbol{H}$ vector will then become predominant along other direction(s), where it is polarized. If you turn your photon vehicle to follow a planar light (with its frontal phase shaping a flat plane) that is traveling in a medium, you may see three dominant directions of $\boldsymbol{E}$ (or $\boldsymbol{H}$): It vibrates along a straight line while moving forward, which is called a linear polarization, or by turning its direction clockwise or anticlockwise, which draws an ellipse or a circle and are in turn termed an elliptical or circular polarization, respectively.

2.2.3.1 Elliptical Polarization

As a uniform planar EMW is traveling along x-direction, its vibration, which is normally represented by its $\boldsymbol{E}$, has in common a random direction but can be determined by its components along z- and y-axes in a Cartesian coordinate system:

$$\boldsymbol{E}(x,t) = \boldsymbol{E}_z + \boldsymbol{E}_y = \boldsymbol{E}_p + \boldsymbol{E}_s, \tag{2.22}$$

where $\boldsymbol{E}_y$ or $\boldsymbol{E}_s$ is a transverse electric field (TE, refer to Fig. 2.1a) and normally called s-polarized light or s-light, while $\boldsymbol{E}_z$ or $\boldsymbol{E}_p$ is vertical electric field and normally called p-polarized light or p-light. A p-light is also called TM mode of light to emphasize its transverse magnetic field (Fig. 2.1b). Take these into consideration, we have:

$$\boldsymbol{E}(x,t) = \boldsymbol{E}_{z\,\max}\cos(\omega t - k_x x + \varphi_z) + \boldsymbol{E}_{y\,\max}\cos(\omega t - k_x x + \varphi_y), \tag{2.23}$$

where φ_z and φ_y are the initial phase of $\boldsymbol{E}_z$ and $\boldsymbol{E}_y$, with amplitude of $\boldsymbol{E}_{z\max}$ and $\boldsymbol{E}_{y\max}$, respectively. Let

$$\begin{cases} \beta = \omega t - kx + \varphi_y \\ \varphi = \varphi_z - \varphi_y \end{cases} \tag{2.24}$$

we can have the component scalar equations from their vectors:

$$\begin{cases} E_z = E_p = E_{z\,\max}\cos(\beta + \varphi) \\ E_y = E_s = E_{y\,\max}\cos(\beta) \end{cases}. \tag{2.25}$$

Replace the second equation of $\cos(\beta) = E_y/E_{y\max}$ into the first formula in Eq. 2.25 and operate it based on $\cos(\beta + \varphi) = \cos(\beta)\cos(\varphi) - \sin(\beta)\sin(\varphi)$ and $\sin^2(\beta) = 1 - \cos^2(\beta)$, we obtain

$$\frac{E_z}{E_{z\,\max}} - \frac{E_y}{E_{y\,\max}}\cos(\varphi) = -\sqrt{1 - \frac{E_y^2}{E_{y\,\max}^2}}\sin(\varphi). \tag{2.26}$$

Square both side of it, the equation changes into:

$$\frac{E_z^2}{E_{z\,\max}^2} + \frac{E_y^2}{E_{y\,\max}^2} - 2\frac{E_z E_y}{E_{z\,\max}E_{y\,\max}}\cos(\varphi) = \sin^2(\varphi). \tag{2.27}$$

This is an elliptical equation on y, z-plane, with its long axis away the z-axis at an angle of

$$\theta_z = \frac{1}{2}\tan^{-1} 2\frac{E_{z\max}E_{y\max}}{E_{z\max}^2 - E_{y\max}^2}\cos(\varphi). \tag{2.28}$$

Thus, the composed vector $\boldsymbol{E}(x, t)$ turns its vibration arrow, around an elliptical track, anticlockwise at $0 < \varphi < \pi$ or clockwise at $-\pi < \varphi < 0$.

2.2.3.2 Circular Polarization

Equation 2.27 will change into a circular function

$$E_z^2 + E_y^2 = 2E_zE_y = E_r^2 \tag{2.29}$$

$$\text{at} \quad \begin{cases} E_{z\max} = E_{y\max} = E_r/\sqrt{2} \\ \varphi = \varphi_z - \varphi_y = \pm\frac{\pi}{2} \end{cases}. \tag{2.30}$$

The vibration now draws a circle with its vector turning clockwise " − " or anticlockwise " + " as time runs:

$$\theta_z = \tan^{-1}\frac{E_z}{E_y} = \pm\tan^{-1}[\tan(\omega t + \varphi_y)] = \pm(\omega t + \varphi_y) \tag{2.31}$$

2.2.3.3 Linear Polarization

In particular, at $\varphi = 0$ or $\pm\pi$, Eq. 2.27 changes into a linear function:

$$\left(\frac{E_z}{E_{z\max}} \mp \frac{E_y}{E_{y\max}}\right)^2 = 0 \quad \text{or} \quad E_y = \pm\frac{E_{y\max}}{E_{z\max}}E_z \tag{2.32}$$

It draws straight lines of the vibration at an angle:

$$\tan\theta_z = \begin{cases} \tan\frac{E_z}{E_y} = \tan\frac{E_{z\max}}{E_{y\max}} = \text{constant} > 0 & \text{for } \varphi = 0 \\ \tan\frac{E_z}{E_y} = \tan\left(-\frac{E_{z\max}}{E_{y\max}}\right) = \text{constant} < 0 & \text{for } \varphi = \pm\pi \end{cases}. \tag{2.33}$$

The composed vector $\boldsymbol{E}$ vibrates along a line in 1,3-quadrant for $\varphi = 0$, or in 2,4-quadrant for $\varphi = \pm\pi$.

2.2.4 Propagation Across an Interface

If a light beam travels across a flat interface between two half-infinitive isotropic media, it may transmit from one medium into the other by refraction, and/or be bounced back called reflection. More generally, both of refraction and reflection happen simultaneously no matter if the light is *p*- or *s*-polarized as shown in Fig. 2.2.

As indicated in Eq. 2.32, a linearly polarized plane EMW can be decomposed into two orthogonal components of TE mode (Fig. 2.3a) and TM mode (Fig. 2.3b). As mentioned, TE mode or *s*-light has a transverse $\boldsymbol{E}$ or vertical $\boldsymbol{H}$, while the TM mode or *p*-light has a vertical $\boldsymbol{E}$ and transverse $\boldsymbol{H}$. They are hence linearly polarized, respectively. Similar to $\boldsymbol{H}$, $\boldsymbol{E}$ and $\boldsymbol{k}$, the p- and s-components and $\boldsymbol{k}$ are also perpendicular to each other (Fig. 2.2).

2.2.4.1 Laws of Reflection and Refraction

The propagation of an EMW through an interface needs to satisfy following boundary conditions at $z = 0$:

$$\begin{Bmatrix} \hat{\boldsymbol{n}} \times (\boldsymbol{E}_2 - \boldsymbol{E}_1) = 0 \\ \hat{\boldsymbol{n}} \times (\boldsymbol{H}_2 - \boldsymbol{H}_1) = J_s \\ \hat{\boldsymbol{n}} \cdot (\boldsymbol{D}_2 - \boldsymbol{D}_1) = \rho_s \\ \hat{\boldsymbol{n}} \cdot (\boldsymbol{B}_2 - \boldsymbol{B}_1) = 0 \end{Bmatrix} \xrightarrow[\rho_s = 0]{J_s = 0} \begin{cases} E_{1xy} = E_{2xy} \\ H_{1xy} = H_{2xy} \\ \varepsilon_1 E_{1z} = \varepsilon_2 E_{2z} \\ \mu_1 H_{1z} = \mu_2 H_{2z} \end{cases}, \tag{2.34}$$

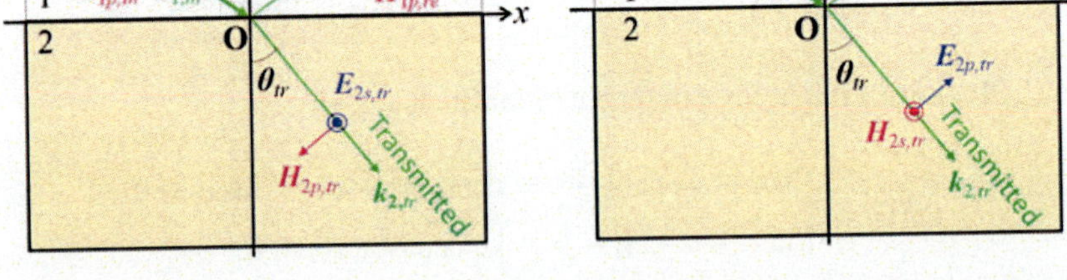

Fig. 2.2 Refraction and reflection of a light at an interface. **a** Refraction and reflection of TE mode or *s*-light; **b** refraction and reflection of TM mode or *p*-light. ⊙ vector pointing toward readers

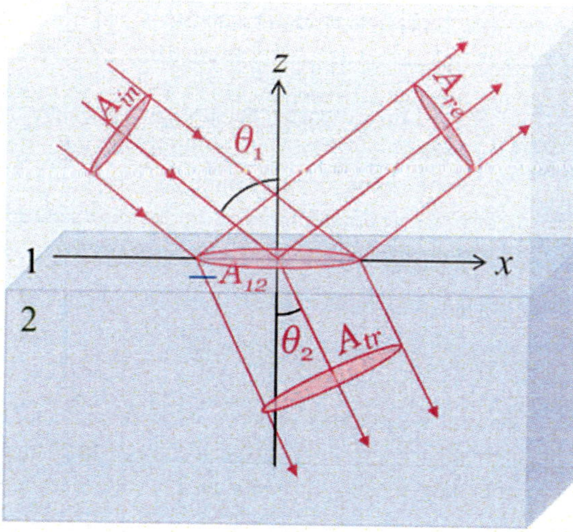

Fig. 2.3 Dependence of the cross sectional areas of incident (A_{in}), reflected (A_{re}) and refracted (A_{tr}) wave fluxes on their related angles and projected interface area A_{12}

where $\hat{\boldsymbol{n}}$ is a unit vector normal to the interface. The equations mean that the normal components of electric displacement $\boldsymbol{D}$ and magnetic induction strength $\boldsymbol{B}$ are continuous across the interface, and along the interface same continuous are the tangential components of electric and magnetic fields. If a plane EMW hits the interface (Fig. 2.2), the following space waves exist:

$$\begin{cases} \text{Medium 1}: \boldsymbol{E}_1(\boldsymbol{r}) = \boldsymbol{E}_{1,\text{in}}(\boldsymbol{r}) + \boldsymbol{E}_{1,\text{re}}(\boldsymbol{r}) = \boldsymbol{E}_{1,\text{in max}} e^{-i\boldsymbol{k}_{1,\text{in}}\cdot\boldsymbol{r}} + \boldsymbol{E}_{1,\text{re max}} e^{i\boldsymbol{k}_{1,\text{re}}\cdot\boldsymbol{r}} \\ \text{Medium 2}: \boldsymbol{E}_2(\boldsymbol{r}) = \boldsymbol{E}_{2,\text{tr}}(\boldsymbol{r}) = \boldsymbol{E}_{2,\text{tr max}} e^{-i\boldsymbol{k}_{2,\text{tr}}\cdot\boldsymbol{r}} \end{cases}, \tag{2.35}$$

where the subscripts of *in, re and tr* denote the incident, reflected, transmitted light, respectively, and max again denotes the amplitude. The corresponding tangential components along y-axis at $z = 0$ are:

$$\begin{cases} \boldsymbol{E}_{1,y}(\boldsymbol{r})|_{z=0} = \boldsymbol{E}_{1s}(\boldsymbol{r}) = \boldsymbol{E}_{1,in\,\text{max}} e^{-ik_1 x \sin\theta_{\text{in}}} + \boldsymbol{E}_{1,re\,\text{max}} e^{-ik_1 x \sin\theta_{\text{re}}} \\ \boldsymbol{E}_{2,y}(\boldsymbol{r})|_{z=0} = \boldsymbol{E}_{2s}(\boldsymbol{r}) = \boldsymbol{E}_{2,tr\,\text{max}} e^{-ik_2 x \sin\theta_{\text{tr}}} \end{cases}. \tag{2.36}$$

According to Eq. 2.34, there should be $\boldsymbol{E}_{1,y} = \boldsymbol{E}_{2,y}$,that is, $k_1 \cdot x \cdot \sin\theta_{\text{in}} = k_1 \cdot x \cdot \sin\theta_{\text{re}} = k_2 \cdot x \cdot \sin\theta_{\text{tr}}$ and $\theta_{\text{in}} = \theta_{\text{re}}$. Let $\theta_{\text{in}} = \theta_{\text{re}} = \theta_1$ and $\theta_{\text{tr}} = \theta_2$, we get:

$$k_1 \sin\theta_1 = k_2 \sin\theta_2. \tag{2.37}$$

Equation 2.37 describes the famous "Law of Reflection" where k_m ($m = 1, 2$) is in general a complex function of relative permittivity. For lights propagating in isotropic and homogeneous non-magnetic media, its $\mu \approx 1$, Eq. 2.37 can be changed to:

$$\frac{\sin\theta_2}{\sin\theta_1} = \frac{k_1}{k_2} = \sqrt{\frac{\varepsilon_1}{\varepsilon_2}} \doteq \frac{n_1}{n_2}. \tag{2.38}$$

This is the so-called Snell's law, or fairly, Ibn Sahl's law. Abu Said al-Ala Ibn Sahl from Baghdad discovered it in 984, exactly 637 years ahead of the discovery of Snell who never formally published his manuscript and claimed his discovery (refer to http://materiaislamica.com/index.php/ History_of_Islamic_Physics_(Snell%27s_Law)).

2.2.4.2 Fresnel's Equations

Fresnel's equations concern with the reflectance and transmittance of an incident light that is the basic in performing SPRS and SPRi. From Fig. 2.2, according to boundary conditions and Eqs. 2.37 or 2.38, there will be:

$$\begin{cases} TE\begin{cases} H_{1p,in}\cos\theta_1 - H_{1p,re}\cos\theta_1 = H_{2p,tr}\cos\theta_2 \\ E_{1s,in} + E_{1s,re} = E_{2s,tr} \end{cases} \\ TM\begin{cases} E_{1p,in}\cos\theta_1 - E_{1p,re}\cos\theta_1 = E_{2p,tr}\cos\theta_2 \\ H_{1s,in} + H_{1s,re} = H_{2s,tr} \end{cases} \end{cases}. \tag{2.39}$$

In combination with the scalar form of Eq. 2.20, $H = GE$, the reflection ® and transmission (t) coefficients of amplitude or amplitude reflectivity and transmittivity are derived:

$$\begin{cases} r_j = \frac{E_{j,re}}{E_{j,in}} = \frac{\eta_{1j}-\eta_{2j}}{\eta_{1j}+\eta_{2j}} \\ t_j = \frac{E_{j,tr}}{E_{j,in}} = \frac{2\eta_{1j}}{\eta_{1j}+\eta_{2j}}A_j \end{cases} \quad (j = s, p); \quad \begin{cases} \eta_{ms} = G\cos\theta_m \\ \eta_{mp} = G/\cos\theta_m \\ A_s = 1 \\ A_p = \frac{\cos\theta_1}{\cos\theta_2} \end{cases} \quad (m = 1, 2), \tag{2.40}$$

where η is termed modified admittance. It should be noted that theses amplitude-based coefficients have a relationship of $t_s = r_s + 1$ but $t_p \neq r_p + 1$because the s-light propagates in a same medium while the p-polarized through two different media. These two polarized light components also propagate independently; otherwise, a p-polarized incident light would generate both of p- and s-components in the reflected and refracted lights, which is against the boundary conditions since the corresponding magnetic field of the s-vector has opposite tangential components, $H_{1,\mathrm{re}}\cos\theta_1 = -H_2\cos\theta_2$. Clearly s- or p-component needs to keep unchanged during propagation or in a same plane of the incident. Specifically at $\theta_1 = \theta_2 = 0$, Eq. 2.40 changes to:

$$\begin{cases} r_s = \frac{n_1-n_2}{n_1+n_2} = -r_p \\ t_s = \frac{2n_1}{n_1+n_2} = t_p \end{cases}. \tag{2.41}$$

Thus, for an incident light from air ($n = 1$) into a glass ($n = 1.5$), we can have $r_s = -r_p = -0.20$ and $t_s = t_p = 0.80$; or reversely, from glass into air, $r_p = -r_s = -0.20$ and $t_s = t_p = 1.20$. The complex amplitudes are either negative or > 1. They are against the conservation of light intensity but can be corrected by replacing the coefficients with reflectance R and transmittance T, which corresponds to the real energy intensity $\boldsymbol{I}$ and in turn to the averaged Poynting vector $\boldsymbol{S}$:

$$\begin{aligned}\langle \boldsymbol{S}\rangle &\equiv \frac{1}{T}\int_0^T \mathbf{S}dt = \frac{1}{2}\mathrm{Re}[\boldsymbol{E}\times\boldsymbol{H}] = \frac{1}{2}\mathrm{Re}[\mathbf{E}\times(G\hat{\boldsymbol{k}}\times\boldsymbol{E})]\\ &= \frac{G}{2}\mathrm{Re}[(\boldsymbol{E}\cdot\boldsymbol{E})\hat{\boldsymbol{k}}-(\boldsymbol{E}\cdot\hat{\boldsymbol{k}})\boldsymbol{E}] = \frac{G}{2}\mathrm{Re}[\boldsymbol{E}^2\hat{\boldsymbol{k}}].\end{aligned} \tag{2.42}$$

Its components are as follows:

$$\begin{cases}\langle \boldsymbol{S}_{\mathrm{in}}\rangle = \frac{G_1\boldsymbol{E}_{1,in}^2}{2}\hat{z}\\ \langle \boldsymbol{S}_{\mathrm{re}}\rangle = -\frac{G_1\boldsymbol{E}_{1,re}^2}{2}\hat{z} = -\frac{G_1\boldsymbol{E}_{1,in}^2}{2}r^2\hat{z}\\ \langle \boldsymbol{S}_{\mathrm{tr}}\rangle = \frac{G_2\boldsymbol{E}_{1,tr}^2}{2}\hat{z} = \frac{G_2\boldsymbol{E}_{1,in}^2}{2}t^2\hat{z}\end{cases}, \tag{2.43}$$

where $\hat{z}$ is a unit vector of z-axis. The Poyting vector is referred to the density of energy flow of an EMW, and the flow strength is thus a product of the flow density and cross sectional area A_i (Fig. 2.3). We can thus define and calculate the reflectance (R) and transmittance (T) from the incident (I_{in}), reflected (I_{re}) and transmitted (I_{tr}) energy intensities as follows:

$$\begin{cases}R = \frac{I_{\mathrm{re}}}{I_{\mathrm{in}}} = \frac{\langle \boldsymbol{S}_{\mathrm{re}}\rangle\cdot\boldsymbol{n}A_{\mathrm{re}}}{\langle \boldsymbol{S}_{\mathrm{in}}\rangle\cdot\boldsymbol{n}A_{\mathrm{in}}} = \frac{2G_1A_{12}\cos\theta_1|\boldsymbol{E}_{1,re}|^2}{2G_1A_{12}\cos\theta_1|\boldsymbol{E}_{1,in}|^2} = |r|^2 = \left|\frac{\eta_1-\eta_2}{\eta_1+\eta_2}\right|^2\\ T = \frac{I_{\mathrm{tr}}}{I_{\mathrm{in}}} = \frac{2G_2A_{12}\cos\theta_2|\boldsymbol{E}_{2,tr}|^2}{2G_1A_{12}\cos\theta_1|\boldsymbol{E}_{1,in}|^2} = \frac{G_2\cos\theta_2}{G_1\cos\theta_1}|t|^2 = \frac{4\eta_1\eta_2}{(\eta_1+\eta_2)^2}\end{cases}. \tag{2.44}$$

Equation 2.44 complies now with the energy conservation law: $R_s + T_s = 1$ and $R_p + T_p = 1$. In other words, the interface-caused reflection and refraction have no energy loss unless absorption A happens:

$$R + T + A = 1. \tag{2.45}$$

It also suggests that the reflectivity of energy flow is the same as that of wave intensity, but the reflectance and transmittance of energy flow differ in general unless at $\theta_1 = \theta_2 = 0$.

2.2.4.3 Brewster's Law

For non-magnetic media with $\mu = 1$ and $G = \sqrt{\varepsilon} = n$, r_s and r_p can be reorganized into the following forms by use of the Snell's law,

$$\begin{cases} r_p = \frac{\tan(\theta_1 - \theta_2)}{\tan(\theta_1 + \theta_2)} \\ r_s = \frac{\sin(\theta_2 - \theta_1)}{\sin(\theta_2 + \theta_1)} \end{cases}. \tag{2.46}$$

Clearly, $r_p \to 0$ at $\theta_1 + \theta_2 = \pi/2$ since $\tan(\pi/2) \to \infty$. Thus, the reflected wave contains only *s*-polarized component. This is the so-called Brewster's law, and the corresponding incident angle is called Brewster's angle, θ_B, which can be calculated more easily by using the Snell's law with $\theta_2 = (\pi/2) - \theta_1$ and $\sin((\pi/2) - \theta_1) = \cos\theta_1$, that is, $n_1 \sin\theta_1 = n_2 \cos\theta_2$ or $\tan\theta_1 = n_2/n_1$, thus

$$\theta_B = \theta_1 = \tan^{-1}\frac{n_2}{n_1}. \tag{2.47}$$

Brewster's law is applicable to both cases of external and internal reflections. The external reflection is meant the reflection of a light in a thinner (low n) medium against a denser (high n), such as the sunlight strikes on the water surface, and the internal reflection is the reflection inside the denser medium against the thinner, like a light inside a glass projected onto its surface against air. The internal reflection concerns with the excitation of SPR in a more special case of total internal reflection or TIR.

2.2.4.4 Total Internal Reflection

In internal reflection, the transmission angle θ_2 is larger than the incident angle θ_1 due to the increase of υ_2 or decrease of n_2 according to Eqs. 2.38 and 2.15:

$$\frac{\sin\theta_2}{\sin\theta_1} = \frac{\upsilon_2}{\upsilon_1} = \frac{n_1}{n_2} > 1. \tag{2.48}$$

The transmission angle increases also faster than the incident one, and it will finally reach a critical value at $\theta_2 = \pi/2$ where the "transmission light" does not really pass into the thin medium but just travels along the interface. The corresponding incident angle is now called the critical angle θ_c:

$$\sin\theta_c = \sin\theta_1 = \frac{n_2}{n_1} \quad \text{or} \quad \theta_c = \sin^{-1}\left(\frac{n_2}{n_1}\right). \tag{2.49}$$

In this case, both $r_s = r_p = 1$. The incident light will totally be reflected back into the thick medium at $\theta_1 > \theta_c$. An excellent example is to project a light from inside water ($n_1 = 1.333$) into air ($n_2 = 1.000$). At $\theta_1 < \theta_c = 48.6°$, the light ray is

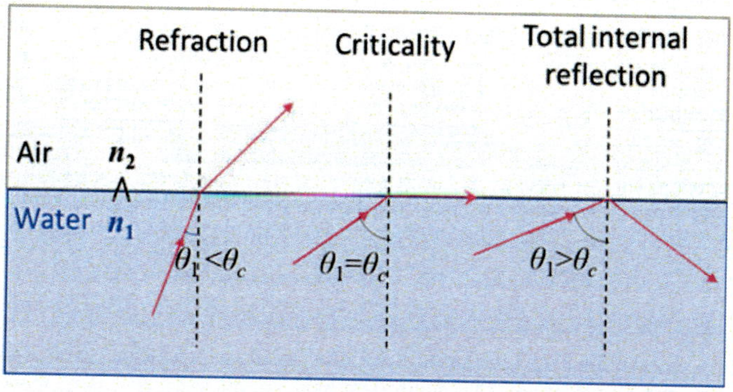

Fig. 2.4 Three typical propagating forms of a planar light wave from water into air

refracted into air, but at $\theta_1 = \theta_c = 48.6°$, it lights up only the water surface. Further increase of the incident angle at above θ_c, the light is totally reflected back into water. However, unusual data may be calculated at $\theta_1 > \theta_c$, for example, at $\theta_1 = 55°$, $\sin\theta_2 = 1.333 \cdot \sin 55° = 1.092$ that is against the trigonometric function rule! To correct, a complex transmittance has to be considered by use of the following formulae:

$$\cos\theta_2 = \sqrt{1 - \sin^2\theta_2} = -i\sqrt{\sin^2\theta_2 - 1} = -i\sqrt{\frac{\varepsilon_1}{\varepsilon_2}\sin^2\theta_1 - 1}. \tag{2.50}$$

A planar light wave can thus have three typical ways to propagate between two thin and thick media, refraction into the thin medium, traveling along the interface and having TIR (Fig. 2.4) that may initiate SPR to be discussed later.

2.2.5 *Propagation Across Multi-interface*

In the real world, a light easily travels through multiple membranes and interfaces. The light will subject to more than one times of reflection and refraction. Interference may also happen between different reflected or refracted beams depending on the path difference. These propagations can still be quantitatively analyzed by Fresnel's equations. We will start from the simplest one membrane system and expand to multiple membrane system later.

2.2.5.1 One Membrane System

The most useful optical propagation system in exciting SPR is basically composed of one membrane inserting between two media (e.g., a metal film in between glass and air or water). This one membrane system contains two interfaces that differs from the above-discussed one-interfacial system formed by two media of infinitive halves.

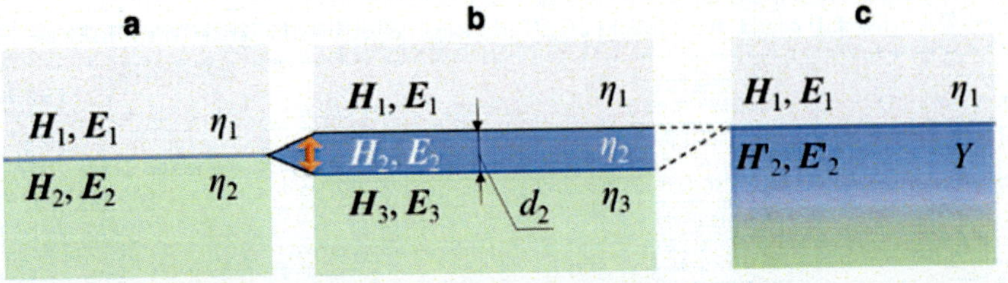

Fig. 2.5 Imagination model to expend one interface to two and to shrink back to one equivalent interface. **a** One interface formed between two semi-infinitive media; **b** two interfaces formed by inserting a membrane, at a thickness of d_2, into (**a**); **c** one equivalent interface shrunken from **b** to ease mathematic treatment. $\boldsymbol{H}'_2$. Equivalent magnetizing field; $\boldsymbol{E}'_2$. Equivalent electric fields; Y. Equivalent admittance

The reflectance and transmittance of an incident optical beam cannot be discussed directly with Eq. 2.44. Modification is required.

To modify, let us conceive of the interlayer of this system as a thickened interface of the two infinitive halves (from Fig. 2.5a, b), or reversely, think of shrinking the middle membrane into one equivalent interface (from Fig. 2.5b, c), Fresnel's equations similar to Eq. 2.44 can thus be derived by simply substitution of the modified admittance η_2 with an equivalent admittance Y_2:

$$\begin{cases} R = \left|\frac{\eta_1 - Y_2}{\eta_1 + Y_2}\right|^2 \\ T = \frac{4\eta_1 Y_2}{(\eta_1 + Y_2)^2} \end{cases}. \tag{2.51}$$

Now, the question is how to get the equivalent admittance Y_2. By definition of medium 2 and 3 as *one* combinatory or equivalent medium with an equivalent electric field $\boldsymbol{E}'_2$ and an equivalent magnetic field $\boldsymbol{H}'_2$, we can derive their tangent components according to their continuity across the equivalent interface (boundary conditions):

$$H_{1\,\tan} = H'_{2\,\tan}; \quad E_{1\,\tan} = E'_{2\,\tan}. \tag{2.52}$$

Replace G in Eq. 2.20 by Y, it gives the scalar form as follows:

$$Y = \frac{H'_{2\,\tan}}{E'_{2\,\tan}} = \frac{H_{1\,\tan}}{E_{1\,\tan}}. \tag{2.53}$$

The calculation is now directed to the ratio of tangent magnetizing field over the electric field in the first incident interface, which needs to go back the real system as illustrated in Fig. 2.6

In the real system, there will be numerous refractions and reflections as a light beam passes through the two interfaces. For easiness, we symbolize the light rays with " + " for up propagation while " – " for down propagation (Fig. 2.6). If numbering

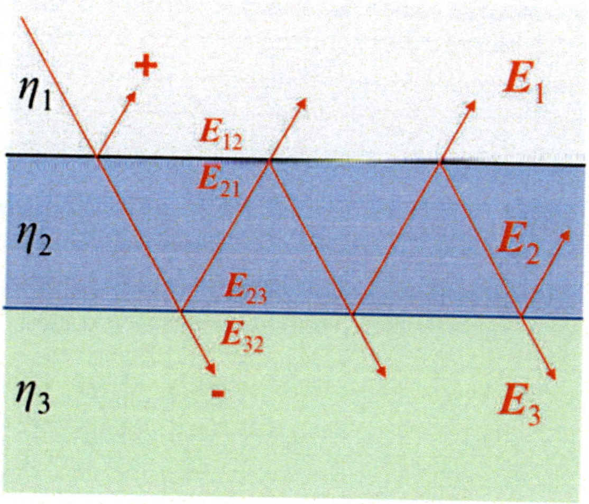

Fig. 2.6 Transmission and reflection of a light ray through a doubly interfaced system. " + ", Reflected ray; " – ", transmitted ray

the upper interface at the side of the medium 2 with 21 and the lower with 23, we can write the local electric fields of $\boldsymbol{E}_{21}^{+}$ for up direction and $\boldsymbol{E}_{21}^{-}$ for down, and similarly, we have $\boldsymbol{H}_{21}^{+}$, $\boldsymbol{H}_{21}^{-}$, $\boldsymbol{H}_{23}^{+}$, $\boldsymbol{H}_{23}^{-}$ and $\boldsymbol{E}_{23}^{+}$ and $\boldsymbol{E}_{23}^{-}$. Thus, following boundary conditions are available at the upper interface:

$$\begin{cases} \hat{\boldsymbol{k}} \times \boldsymbol{E}_1 = \hat{\boldsymbol{k}} \times (\boldsymbol{E}_{21}^{+} + \mathbf{E}_{21}^{-}) \\ \boldsymbol{H}_1 = \boldsymbol{H}_{21} = \eta_2(\hat{\boldsymbol{k}} \times \boldsymbol{E}_{21}^{+}) + \eta_2(-\hat{\boldsymbol{k}} \times \boldsymbol{E}_{21}^{-}) \end{cases}. \tag{2.54}$$

As the wave passes through the medium 2, reaching the upper surface of the second interface, its electromagnetic field changes from $\boldsymbol{E}_{21}$ to $\boldsymbol{E}_{23}$ due to potential decay or phase variation by a factor of $e^{i\delta_2}$:

$$\begin{cases} \boldsymbol{E}_{23}^{+} = \boldsymbol{E}_{21}^{+} e^{-i\delta_2} \text{ or } \boldsymbol{E}_{23}^{-} = \boldsymbol{E}_{21}^{-} e^{i\delta_2} \\ \delta_2 = \frac{2\pi}{\lambda}\delta_2' = \frac{2\pi}{\lambda} G_2 d_2 \cos\theta_2 \end{cases}, \tag{2.55}$$

where δ_2 is the effective phase thickness of medium 2 and δ'_2 its effective optical thickness, and the refraction angle θ_2 is determined by Snell's law. Combining the two sets of equations, we obtain:

$$\begin{cases} \hat{\boldsymbol{k}} \times \boldsymbol{E}_1 = (\hat{\boldsymbol{k}} \times \boldsymbol{E}_{23}^{+}) e^{i\delta_2} + (\hat{\boldsymbol{k}} \times \boldsymbol{E}_{23}^{-}) e^{-i\delta_2} \\ \boldsymbol{H}_1 = (\hat{\boldsymbol{k}} \times \boldsymbol{E}_{23}^{+}) \eta_2 e^{i\delta_2} - (\hat{\boldsymbol{k}} \times \boldsymbol{E}_{23}^{-}) \eta_2 e^{-i\delta_2} \end{cases}. \tag{2.56}$$

These two equations can be neatly expressed by a matrix form:

$$\begin{bmatrix} \hat{\boldsymbol{k}} \times \boldsymbol{E}_1 \\ \boldsymbol{H}_1 \end{bmatrix} = \begin{bmatrix} e^{i\delta_2} & e^{-i\delta_2} \\ \eta_2 e^{i\delta_2} & -\eta_2 e^{-i\delta_2} \end{bmatrix} \begin{bmatrix} \hat{\boldsymbol{k}} \times \boldsymbol{E}_{23}^{+} \\ \hat{\boldsymbol{k}} \times \boldsymbol{E}_{23}^{-} \end{bmatrix}. \tag{2.57}$$

The same relationship is obtained at the second interface supposing that the substrate does not cause any energy loss and phase variation,

$$\begin{cases} \boldsymbol{E}_{23}^{+} + \boldsymbol{E}_{23}^{-} = \boldsymbol{E}_{32}^{+} + \boldsymbol{E}_{32}^{-} = \boldsymbol{E}_{32} = \boldsymbol{E}_3 \\ \boldsymbol{H}_{23}^{+} + \boldsymbol{H}_{23}^{-} = \boldsymbol{H}_{32}^{+} + \boldsymbol{H}_{32}^{-} = \boldsymbol{H}_{32} = \boldsymbol{H}_3 \end{cases} \tag{2.58}$$

or

$$\begin{cases} \hat{\boldsymbol{k}} \times (\boldsymbol{E}_{23}^{+} + \boldsymbol{E}_{23}^{-}) = \hat{\boldsymbol{k}} \times \boldsymbol{E}_3 \\ \eta_2(\hat{\boldsymbol{k}} \times \boldsymbol{E}_{23}^{+}) + \eta_2(-\hat{\boldsymbol{k}} \times \boldsymbol{E}_{23}^{-}) = \boldsymbol{H}_3 \end{cases}. \tag{2.59}$$

Their matrix form is as follows:

$$\begin{bmatrix} \hat{\boldsymbol{k}} \times \boldsymbol{E}_{23}^{+} \\ \hat{\boldsymbol{k}} \times \boldsymbol{E}_{23}^{-} \end{bmatrix} = \begin{bmatrix} 1/2 & 1/2\eta_2 \\ 1/2 & -1/2\eta_2 \end{bmatrix} \begin{bmatrix} \hat{\boldsymbol{k}} \times \boldsymbol{E}_3 \\ \boldsymbol{H}_3 \end{bmatrix}. \tag{2.60}$$

Combining Eqs. 2.57 and 2.60, a connection of electromagnetic field between medium 1 and 3 is derived

$$\begin{aligned} \begin{bmatrix} \hat{\boldsymbol{k}} \times \boldsymbol{E}_1 \\ \boldsymbol{H}_1 \end{bmatrix} &= \begin{bmatrix} e^{i\delta_2} & e^{-i\delta_2} \\ \eta_2 e^{i\delta_2} & -\eta_2 e^{-i\delta_2} \end{bmatrix} \begin{bmatrix} 1/2 & 1/2\eta_2 \\ 1/2 & -1/2\eta_2 \end{bmatrix} \begin{bmatrix} \hat{\boldsymbol{k}} \times \boldsymbol{E}_3 \\ \boldsymbol{H}_3 \end{bmatrix} \\ &= \begin{bmatrix} \cos\delta_2 & i\eta_2^{-1}\sin\delta_2 \\ i\eta_2 \sin\delta_2 & \cos\delta_2 \end{bmatrix} \begin{bmatrix} \hat{\boldsymbol{k}} \times \boldsymbol{E}_3 \\ \boldsymbol{H}_3 \end{bmatrix}. \end{aligned} \tag{2.61}$$

By use of $\boldsymbol{H}_3 = \eta_3(\hat{\boldsymbol{k}} \times \boldsymbol{E}_3)$ and $\boldsymbol{H}_2' = Y(\hat{\boldsymbol{k}} \times \boldsymbol{E}_2') = \boldsymbol{H}_1 = Y(\hat{\boldsymbol{k}} \times \boldsymbol{E}_1)$ (refer to Eq. 2.53), the above equations can be uniformed to $\boldsymbol{E}$ form:

$$\begin{bmatrix} \hat{\boldsymbol{k}} \times \boldsymbol{E}_1 \\ Y(\hat{\boldsymbol{k}} \times \boldsymbol{E}_1) \end{bmatrix} = \begin{bmatrix} \cos\delta_2 & i\eta_2^{-1}\sin\delta_2 \\ i\eta_2 \sin\delta_2 & \cos\delta_2 \end{bmatrix} \begin{bmatrix} \hat{\boldsymbol{k}} \times \boldsymbol{E}_3 \\ \eta_3(\hat{\boldsymbol{k}} \times \boldsymbol{E}_3) \end{bmatrix} \tag{2.62}$$

or

$$\begin{cases} (\hat{k} \times E_1)\begin{bmatrix} 1 \\ Y \end{bmatrix} = \begin{bmatrix} B \\ X \end{bmatrix}(\hat{k} \times E_3) \\ \begin{bmatrix} B \\ X \end{bmatrix} = \begin{bmatrix} \cos\delta_2 & i\eta_2^{-1}\sin\delta_2 \\ i\eta_2 \sin\delta_2 & \cos\delta_2 \end{bmatrix} \begin{bmatrix} 1 \\ \eta_3 \end{bmatrix} \end{cases}, \tag{2.63}$$

where $\begin{bmatrix} \cos\delta_2 & i\eta_2^{-1}\sin\delta_2 \\ i\eta_2 \sin\delta_2 & \cos\delta_2 \end{bmatrix}$ is called the eigen matrix of the intermembrane. Thus, the equivalent admittance can be calculated whenever the parameters of the membrane and the substrate are given:

$$Y = \frac{X}{B}. \tag{2.64}$$

The reflectivity R can now be calculated by its definition.

2.2.5.2 Multi-layer System

The above-used equivalent thought is extendable to a multi-layer system. The basic idea is to reduce an $m+1$ interface system, step by step, to equivalent mono-interface system (Fig. 2.7) that remains obeying the same set of boundary conditions. The equivalent admittance Y can thus be mathematically calculated through deduction of electromagnetic fields layer by layer in condition that the related parameters of each membrane are given.

To calculate Y, we take the first three layers as a unit, so that Eq. 2.61 can be adopted by replacing $\boldsymbol{E}_3$ with $\boldsymbol{E}_{23}$ and canceling $\hat{\boldsymbol{k}}$ to simplify the formulae:

$$\begin{bmatrix} \boldsymbol{E}_1 \\ \boldsymbol{H}_1 \end{bmatrix} = \begin{bmatrix} \cos\delta_2 & i\eta_2^{-1}\sin\delta_2 \\ i\eta_2\sin\delta_2 & \cos\delta_2 \end{bmatrix} \begin{bmatrix} \boldsymbol{E}_{23} \\ \boldsymbol{H}_{23} \end{bmatrix}. \tag{2.65}$$

Starting from this point, we move down one more layer, that is, taking the second, third and fourth layers as a new unit of tree tier system. The characteristic matrix equation is

$$\begin{bmatrix} \boldsymbol{E}_{23} \\ \boldsymbol{H}_{23} \end{bmatrix} = \begin{bmatrix} \cos\delta_3 & i\eta_3^{-1}\sin\delta_3 \\ i\eta_3\sin\delta_3 & \cos\delta_3 \end{bmatrix} \begin{bmatrix} \boldsymbol{E}_{34} \\ \boldsymbol{H}_{34} \end{bmatrix}. \tag{2.66}$$

This operation is continued until the last two interfaces

$$\begin{bmatrix} \boldsymbol{E}_{(m-1)m} \\ \boldsymbol{H}_{(m-1)m} \end{bmatrix} = \begin{bmatrix} \cos\delta_m & i\eta_m^{-1}\sin\delta_m \\ i\eta_m\sin\delta_m & \cos\delta_m \end{bmatrix} \begin{bmatrix} \boldsymbol{E}_{m(m+1)} \\ \boldsymbol{H}_{m(m+1)} \end{bmatrix}. \tag{2.67}$$

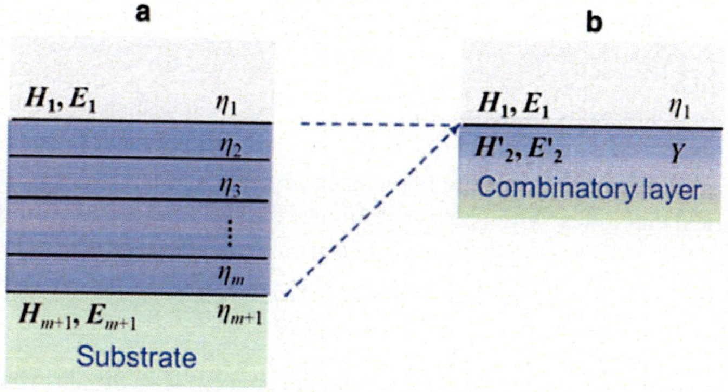

Fig. 2.7 Imaginative reduction of multi-interface into one equivalent interface. **A** Multi-interface form from m layers of membranes; **b** one equivalent interface between one pure (top) layer and one combinatory or equivalent layer. $\boldsymbol{H}'_2$. Equivalent magnetizing field; $\boldsymbol{E}'_2$. Equivalent electric field; Y. Equivalent admittance

Considering the substrate $m + 1$ does not reflect waves and dissipate energy, i.e., $\boldsymbol{E}_{m(m+1)} = \boldsymbol{E}_{m+1}$, the final equation from layer 1 to $m + 1$, after substitution, will be

$$\begin{bmatrix} \boldsymbol{E}_1 \\ \boldsymbol{H}_1 \end{bmatrix} = \left\{ \prod_1^m \begin{bmatrix} \cos\delta_m & i\eta_m^{-1}\sin\delta_m \\ i\eta_m\sin\delta_m & \cos\delta_m \end{bmatrix} \right\} \begin{bmatrix} \boldsymbol{E}_{m+1} \\ \boldsymbol{H}_{m+1} \end{bmatrix}. \tag{2.68}$$

Recall to $Y = H_1/E_1$ and $\eta_{m+1} = E_{m+1}/H_{m+1}$, we have

$$\boldsymbol{E}_1 \begin{bmatrix} 1 \\ Y \end{bmatrix} = \left\{ \prod_1^m \begin{bmatrix} \cos\delta_m & i\eta_m^{-1}\sin\delta_m \\ i\eta_m\sin\delta_m & \cos\delta_m \end{bmatrix} \right\} \begin{bmatrix} 1 \\ \eta_{m+1} \end{bmatrix} \boldsymbol{E}_{m+1}. \tag{2.69}$$

The eigen matrix for the multi-layer is as follows,

$$\begin{bmatrix} B \\ X \end{bmatrix} = \left\{ \prod_1^{m+1} \begin{bmatrix} \cos\delta_m & i\eta_m^{-1}\sin\delta_m \\ i\eta_m\sin\delta_m & \cos\delta_m \end{bmatrix} \right\} \begin{bmatrix} 1 \\ \eta_{m+1} \end{bmatrix}. \tag{2.70}$$

The equivalent admittance of the final combinatory layer is given by $Y = X/B$, the same as Eq. 2.64. In fact, Eq. 2.70 returns to Eq. 2.63 if let $m = 2$. Thus, Eq. 2.70 is the base of nearly all optical propagation-related calculations, and it is critical in exploration of SPRi methodology. It should be noted that only the symmetric membrane layers can be equivalent to one interface, while the asymmetric layers will be equivalent to two. The equivalence is not only physical but also strictly mathematic.

Replacing $Y = C/B$ into Eq. 2.51 and considering complex function, we have

$$\begin{cases} R = \left(\frac{\eta_1 C - B}{\eta_1 C + B}\right)\left(\frac{\eta_1 C - B}{\eta_1 C + B}\right)^* \\ T = \frac{4\eta_1\eta_{m+1}}{(\eta_1 C + B)(\eta_1 C + B)^*} \end{cases}, \tag{2.71}$$

where * indicates the conjugate complex number. Equation 2.71 shows that the transmittance is independent on the propagation direction no matter the membranes are adsorptive or not, but the reflectance depends on the optical adsorption property of the membranes, with phase variation of the reflected light given by:

$$\varphi = \arctan\frac{i\eta_1(CB^* - BC^*)}{\eta_1^2 BB^* - CC^*}. \tag{2.72}$$

It needs to note that, although we did not concern with the absorption of the media during the deduction, all the resulted equations are generally applicable, including adsorption metals.

2.3 Plasmon and Related Waves

2.3.1 Motion Equations for Electrons and Ions in Metals

Plasmon is a quasi-quantum of plasma vibration somewhat similar to a phonon. It is directly related to the collective oscillation(s) of charges. Similar to the case of dropping a stone into water to create waves, "dropping" some things (not yet clear what they are at this moment) into a "sea" of mobile charges may also create waves that spread in and on the "sea". Such a mobile charge "sea" can be found not only in a classic plasma but also in a metal full of mobile valence electrons. A macroscopic sold metal is microscopically a crystal of atomic ions immersed in the "sea" of valence electrons. This explains why a metal can immediately conduct electricity right after the application of a voltage. In fact, many noble metals like gold (Au) and silver (Ag) and some non-noble metals like copper (Cu) and aluminum (Al) are full of mobile valence electrons at an estimated concentration n_e of

$$n_e = N_a \frac{z\rho_m}{M} = 6.023 \times 10^{23} \frac{z\rho_m}{M}, \tag{2.73}$$

where N_a is Avogadro constant, z is here the valence of a metallic atom with a mass of M and ρ_m is the mass density of the metal. For pure Au, Ag, Al or Cu at $z = 1$, with $m = 169.97, 107.87, 26.98$ and 63.55, and $\rho_m = 19.328, 10.53, 2.70$ and 8.89 g/cm^3, their n_e is equal to $6.85 \times 10^{22}, 5.88 \times 10^{22}, 6.03 \times 10^{22}$ and 8.43×10^{22} cm^{-3}, respectively, about 1000 folds more concentrated than the particles in an ideal gas under the standard conditions. Having such a huge number of electrons gathering in a cube of only cm^3, the metals are really the matter of atomic frames submerged in an electronic sea. Instead of "sea", Drude used "electron gas" [1] to describe the metallic valence electrons (inspired by the kinetic theory for ideal gases). He took a metal as a matter of positive atomic crystal lattice inserted into the "electron gas", which is often called a Drude mode, in addition to the following approximations and assumptions:

(i) **Free motion of particles**: Each of the mobile valence electrons wanders independently and freely around the entire crystal lattice, and the electron–electron and electron-atom interactions are negligible (*though*, which is eccentric that the very crowded electrons could move freely in a positive ionic lattice). These electrons are hence called free electrons or just electrons.

(ii) **Rigid particles**: All the mobile electrons are regarded as rigid spheres, and their collisions are instantaneous, causing immediate change of velocity, much the same as ideal gaseous particles. Thus, the wandering border for such a rigid electron can be estimated by its equivalent radius:

$$\frac{4}{3}\pi r_e^3 = \frac{V}{N_a} = \frac{1}{n_e} \Rightarrow r_e = \sqrt[3]{\frac{3}{4\pi n_e}} = 0.6204 n_e^{-1/3} \approx 1\ \text{Å}, \tag{2.74}$$

where V denotes the volume of a particle. The radius r_e is about onefold larger than Bohr's radius ($r_B = 0.529$ Å for bound electrons).

(iii) **Collision property**: Regardless the position and velocity of electrons, there is a relaxation time, τ, between two collisions, which allows the disrupted velocity distribution of the collided electrons to return their equilibrium state. Its reciprocal, $1/\tau$, determines the collision probability or frequency. At an infinitesimal time, $\mathrm{d}t$, the collision probability is equal to $\mathrm{d}t/\tau$, while the collision-free probability is hence equal to $(1 - \mathrm{d}t/\tau)$.

(iv) **Existence of classic force**: The motion of electrons between two collisions is assumed to obey Newton's law, and the electrons are consistent with Boltzmann distribution according to their energy levels.

These approximations and assumptions in fact allow the long-range Coulomb force to act on the system. Thus, the so-called free electrons are not entirely free inside the metal. Based on the ratio of r_e/r_B, the electrons are in fact very close to each other. Hence, each single electron will influence and be influenced by many neighbors, rather than only the closest one (which is the case in a molecule). As a result, only collective motion of the whole electrons occurs rather than a single electron. This type of collective motion is static and characteristic.

Consider a tiny layer of electrons collectively shifts away the positive ions from the balanced position along z-axis, the Coulomb force of the positive ions will immediately pull the electrons back, as a restoring force $\boldsymbol{F}_p$. Owing to the inertia effect, the electronic layer will pass the original position and shifts to the other side, which induces a reversed restoring force. The electrons will thus oscillate around the positive position. Such an oscillation is far below the relativity speed so that it obeys Newton's motion law:

$$\boldsymbol{F} = M_e \boldsymbol{a} = \frac{\mathbf{d}\boldsymbol{p}_\upsilon}{\mathbf{d}t}, \tag{2.75}$$

where $\boldsymbol{F}$ is the total force acting on an oscillating electron with a mass of m_e, making the electron acquire an acceleration of $\boldsymbol{a}$ or a momentum of $\boldsymbol{p}_\upsilon$. The collective electron shift can also be considered as a type of medium polarization $\boldsymbol{P}$ (related to but different from the polarizations of EMWs discussed). $\boldsymbol{P}$ is a sum of each dipole moment $\boldsymbol{p_e}$ of electron–ion pairs:

$$\begin{cases} \boldsymbol{p}_e = -e z = -e \upsilon \mathrm{d}t \\ \boldsymbol{P} = \sum_n \boldsymbol{p}_e = n_e \boldsymbol{p}_e = -n_e e z \end{cases} \tag{2.76}$$

A planar restoring electric field $\boldsymbol{E}_p$ can also be conceived of a planar capacitor. Considering the restoring force comes from the positive ions with a charge density the same as electrons $\boldsymbol{\rho}_+ = \boldsymbol{\rho}_e$ for a neutral metal, we have

$$\begin{cases} \boldsymbol{E}_p = \frac{\rho_{s+}}{\varepsilon_0} = \frac{|-e|n_e A \boldsymbol{v} \mathrm{d}t}{\varepsilon_0 A} = \frac{en_e}{\varepsilon_0} \boldsymbol{v} \mathrm{d}t = \frac{en_e}{\varepsilon_0} \boldsymbol{z} \\ \boldsymbol{F}_p = -e\boldsymbol{E}_p = -\frac{e^2 n_e}{\varepsilon_0} \boldsymbol{v} \mathrm{d}t = -\frac{e^2 n_e}{\varepsilon_0} \boldsymbol{z} \end{cases}, \tag{2.77}$$

where $\boldsymbol{\rho}_{s+}$ denotes surface charge density. Considering an oscillation of electrons in a metal without any external force and attenuation, the Newton's motion equation

$$\boldsymbol{F} - M_e \boldsymbol{a} = 0 \Rightarrow -\frac{e^2 n_e}{\varepsilon_0} \boldsymbol{z} + M_e \frac{\mathrm{d}^2 \boldsymbol{z}}{\mathrm{d}t^2} = 0 \tag{2.78}$$

can be converted to scalar form along the z-axis

$$\begin{cases} \frac{\mathrm{d}^2 z}{\mathrm{d}t^2} + \omega_e^2 z = 0 \\ \omega_e = \sqrt{\frac{n_e e^2}{M_e \varepsilon_0}} \end{cases}. \tag{2.79}$$

This is again a homogeneous Helmholtz' equation, and ω_e is the angular frequency of the electronic oscillation. By a same procedure, a motion equation can be derived for the positive ions with mass of M_+ that have a charge density equal to the absolute value of electrons for a neutral plasma

$$\begin{cases} \frac{\mathrm{d}^2 z}{\mathrm{d}t^2} + \omega_+^2 z = 0 \\ \omega_+ = \sqrt{\frac{n_e e^2}{M_+ \varepsilon_0}} \end{cases}. \tag{2.80}$$

Because $M_+ >> M_e$ or $\omega_+ << \omega_e$, the oscillation of the positive ions in metals is normally in acoustic range that is negligible compared with ω_e. The electronic oscillation thus dominates the system, with its characteristic frequency and wavelength as follows:

$$\begin{cases} \omega_p = 2\pi \nu_p \equiv \omega_e + \omega_+ \cong \omega_e = \sqrt{\frac{n_e e^2}{M_e \varepsilon_0}} = 56.415\sqrt{n_e} \\ \lambda_p = \frac{c}{\nu_p} \cong \sqrt{\frac{4\pi^2 M_e}{\mu_0 n_e e^2}} = 3.339 \times 10^{16} \frac{1}{\sqrt{n_e}} \ (/\mathrm{nm}) \end{cases}, \tag{2.81}$$

where ω_p is the plasma frequency of a metal block. It describes the collective vibrations of whole electrons and ions in a metal and hence is a volume plasma frequency (VPF). Given a metal having an electron density at about 1.00×10^{22} cm^{-3}, its ω_p will be 5.64×10^{15} Hz or λ_p= 334 nm and quantum energy at $\hbar\omega_p$= 5.95 eV. This is thus in the range of UV lights. These data also show that the electrons inside a metal are nearly disturbance-free!

In a real metal, attenuation-free electron motion is not possible but dissipates energy against at least the collisions. As mentioned, during a relaxation time of τ between two collisions, the electrons, after traveling for a tiny time of dt, can have a chance of dt/τ to collide and $(1 - \mathrm{d}t/\tau)$ to move freely. Because the collisions are completely randomized, their averaged momentum is zero or negligible, the net increase of total momentum, $\boldsymbol{p}_v(t + \mathrm{d}t)$, after d$t$ time, is due to the non-collision motion of the electrons:

$$\boldsymbol{p}_v(t+\mathrm{d}t) = \left(1 - \frac{\mathrm{d}t}{\tau}\right)\left[\boldsymbol{p}_v(t) + \boldsymbol{F}_p(t)\mathrm{d}t\right] \tag{2.82}$$

After re-arrangement it becomes

$$\frac{\boldsymbol{p}_v(t+\mathrm{d}t) - \boldsymbol{p}_v(t)}{\mathrm{d}t} = \boldsymbol{F}_p(t) - \frac{1}{\tau}\boldsymbol{p}_v(t) \tag{2.83}$$

Let $\mathrm{d}t \to 0$, we obtain a general motion equation for the electronic oscillation:

$$\frac{\mathrm{d}\boldsymbol{p}_v(t)}{\mathrm{d}t} = \boldsymbol{F}_p(t) - \frac{1}{\tau}\boldsymbol{p}_v(t) \tag{2.84}$$

This is more or less a general electronic motion equation. It shows that the collisions act as a resistance or friction to block the motion of the electrons.

2.3.2 *Volume Plasma Polarization and Propagation*

It is a common knowledge that you cannot feel any metallic plasma oscillations when you touch a metal, can you? Therefore, we need to think of some unusual methods in odor to reveal their existence. The most reasonable way is to start with the electron and ion oscillations. Electromagnetics tells us that the electron and ion oscillations can be affected by an electromagnetic filed and/or charged particles like photons and/or electrons. This means that we can try to "drop" photons into a metal to induce a volume plasma oscillation. Unfortunately, we are still unable to observe any plasmonic emission from a metal just illuminated by lights. We need a theory to have insight into the reason(s).

Assume that an attenuation metal system is hitted with a type of photons having electric field at $\boldsymbol{E} = \boldsymbol{E}_0 \cdot \mathrm{e}^{i(\boldsymbol{k}\cdot\boldsymbol{r}-\omega t)}$, a modified term of $e\boldsymbol{E}$ must be further added into Eq. 2.84 to give:

$$M_e \frac{\mathrm{d}^2 z}{\mathrm{d}t^2} = -\frac{M_e}{\tau}\frac{\mathrm{d}z}{\mathrm{d}t} - e\boldsymbol{E}_0 e^{i(\boldsymbol{k}\cdot\boldsymbol{r}-\omega t)}. \tag{2.85}$$

It can have a solution of

$$z = \frac{e}{M_e}\frac{1}{\omega^2 - i\omega/\tau}\boldsymbol{E} = \frac{\varepsilon_0}{n_e e}\frac{\omega_p^2}{\omega^2 - i\omega/\tau}\mathbf{E}, \tag{2.86}$$

where z denotes now an electric field function. Combining Eq. 2.86 with Eq. 2.76, we have

$$P = -n_e e z = -\varepsilon_0 \frac{n_e e^2}{M_e \varepsilon_0} \frac{1}{\omega^2 - i\omega/\tau} E = -\varepsilon_0 \frac{\omega_p^2}{\omega^2 - i\omega/\tau} E. \tag{2.87}$$

The relative permittivity of a metal can further be derived according to Eq. 2.13:

$$\varepsilon = 1 + \frac{P}{\varepsilon_0 E} = 1 - \frac{\omega_p^2}{\omega^2 - i\omega/\tau}. \tag{2.88}$$

It is a complex function of frequency and relaxation time. It is clear now that, although a volume plasma may oscillate inside a whole metal, the intrinsic oscillation at ω_p can never be observed in a statistic state. However, if we project an appropriate type of photons into a target metal, a mixed mode of an oscillation called volume plasma polarization (VPP) can be expected to occur at an angular frequency of ω. To "feel" the plasma oscillation, we have to resolve out the ω of VPP. To this end, we need to combine Eq. 2.18 with Eq. 2.88 at $\boldsymbol{k}\cdot\boldsymbol{E} = 0$ (for transverse waves). For a non-magnetic metal ($\mu = 1$), we have

$$k^2 = k_0^2 \left(1 - \frac{\omega_p^2}{\omega^2 - i\omega/\tau}\right). \tag{2.89}$$

Again Eq. 2.89 is a complex function. In general, all the ε-dependent functions and parameters are complexes, for example, the admittance G, modified admittance η and equivalent refractive index, N_{eri}. Let

$$\begin{cases} k = k' - ik'' \\ \varepsilon = \varepsilon' - i\varepsilon'' \\ N_{\text{eri}} = n - i\alpha \end{cases}. \tag{2.90}$$

Here α is an optical extinction coefficient and n relative refractive index. Equation 2.90 gives following relationships:

$$\begin{cases} \varepsilon' = n^2 - \alpha^2 = [(k')^2 - (k'')^2]/k_0^2 \\ \varepsilon'' = 2n\alpha = 2k'k''/k_0^2 \end{cases}. \tag{2.91}$$

In an isotropic linear medium of non-conductor, the complex relative permittivity changes to a spatial function, $\varepsilon = \varepsilon(r) = n^2(r)$. If the medium is homogeneous (more useful in SPRi), the relative permittivity will be a real constant, $\varepsilon = n^2$.

Let us go back to Eq. 2.89 that shows the possibility to excite VPPs with variable frequency ω by "drop-in" photons; it describes in fact the dispersion of VPPs. The variation of ε with ω formulated in Eq. 2.88 exhibits an increasing curve as shown in Fig. 2.8A. The curve passes across an important point at ω_p and finally reaches a limit at $\varepsilon = 1$. The cross point at $\omega = \omega_p$ indicates the resonance of the excitation photon(s) with the intrinsic plasma oscillation(s). Thus, the excitation photon(s) will lose the most energy at the resonance point, which opens a way to "feel" (academically, to measure) the intrinsic frequency ω_p of a metal.

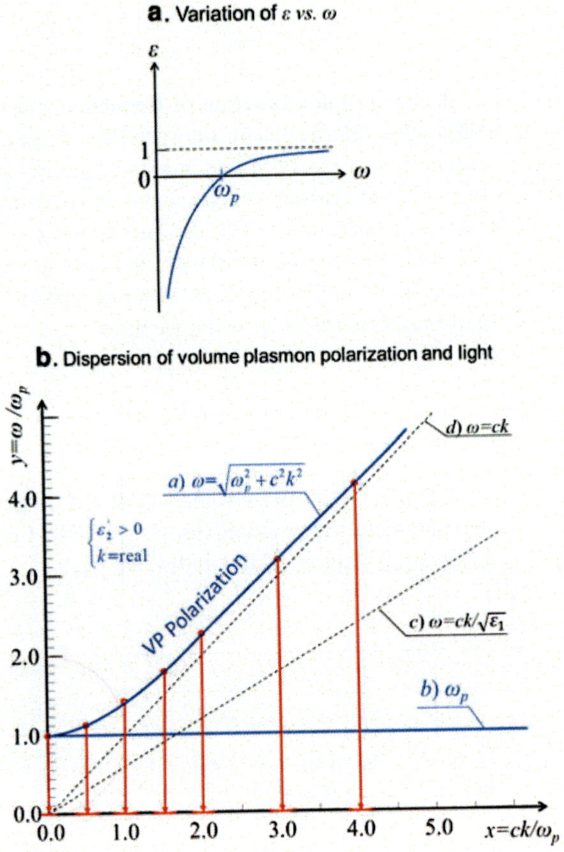

Fig. 2.8 Dispersion curves of different waves. **a** Plot between metal ε and photonic ω; **b** Plots between ω/ω_p and ck/ω_p. (*a*) Volume plasma polarization; (*b*) Volume plasma; (*c*) Light in medium; (*d*) Light in free space

A more complete set of dispersion curves to describe the propagation of excited VPPs can be plotted from Eq. 2.89. Figure 2.8b illustrates the positions and relationships among VPPs, volume plasma and photons in media and in free space. The wave vector of a VPP finds its minimum again at the limit of $\omega = \omega_p$, where the energy of the projected photons is maximally absorbed due to resonance.

It should, however, be noted that, at the resonance point of $\omega = \omega_p$, the relative permittivity becomes zero, $\varepsilon = 0$, which suggests that the volume plasmas oscillate longitudinally based on Eq. 2.19. More detailed discussions are achieved by taking off the time variable item in the optical wave equation and replacing the wave vector with following solutions:

$$\begin{cases} \varepsilon' = 1 - \frac{\omega_p^2 \tau^2}{1+\omega^2\tau^2} = n^2 - \alpha^2 \\ \varepsilon'' = \frac{\omega_p^2 \tau}{(1+\omega^2\tau^2)\omega} = 2n\alpha \end{cases}. \tag{2.92}$$

It gives

$$\begin{aligned} \boldsymbol{E} &= \boldsymbol{E}_0 e^{-i\left(\sqrt{\varepsilon' - i\varepsilon''}\right)\boldsymbol{k}_0 \cdot \boldsymbol{r}} = \boldsymbol{E}_0 e^{-i\left(\sqrt{1-\frac{\omega_p^2\tau^2}{\omega^2\tau^2+1} - i\frac{\omega_p^2\tau}{\omega(\omega^2\tau^2+1)}}\right)\boldsymbol{k}_0 \cdot \boldsymbol{r}} \\ &= \boldsymbol{E}_0 e^{-i\left(\sqrt{n^2-\alpha^2-i2n\alpha}\right)\boldsymbol{k}_0 \cdot \boldsymbol{r}} \end{aligned} \tag{2.93}$$

Roughly four typical cases can be inferred:

Case 1: High frequency at $\omega >> \omega_p$ and $\omega\tau >> 1$, there will be $\omega^3\tau >> \omega^2{}_p$ and $\varepsilon'' \approx 0$, $\alpha \approx 0$ or $n \approx \sqrt{\varepsilon'} > 0$, so that

$$\boldsymbol{E} = \boldsymbol{E}_0 e^{-i\left(\sqrt{\varepsilon'}\right)\boldsymbol{k}_0 \cdot \boldsymbol{r}} = \boldsymbol{E}_0 e^{-in\boldsymbol{k}_0 \cdot \boldsymbol{r}} = \boldsymbol{E}_0 e^{-i\sqrt{1-\frac{\omega_p^2}{\omega^2}}\boldsymbol{k}_0 \cdot \boldsymbol{r}} \xrightarrow{\omega=\infty} \boldsymbol{E}_0 e^{-i\boldsymbol{k}_0 \cdot \boldsymbol{r}}. \tag{2.94}$$

The electric field can freely propagate, without any decay, that is, the corresponding metal is now transparent to these waves at high frequency, such as in the range of x-rays.

Case 2: Medium frequency at $\omega << \omega_p$ but $\omega\tau >> 1$ where $\varepsilon < 0$ or $\sqrt{\varepsilon} = n - i\alpha$, which is possible only when $n \approx 0$ and $\alpha > 0$, so that

$$\boldsymbol{E} \doteq \boldsymbol{E}_0 e^{-i\sqrt{\varepsilon'}\boldsymbol{k}_0 \cdot \boldsymbol{r}} = \boldsymbol{E}_0 e^{-i\left(\sqrt{-\alpha^2}\right)\boldsymbol{k}_0 \cdot \boldsymbol{r}} = \boldsymbol{E}_0 e^{\alpha\boldsymbol{k}_0 \cdot \boldsymbol{r}}. \tag{2.95}$$

The VPP propagation decays exponentially and the metal is not more transparent to photons but highly reflective, with a penetrating depth (***E*** decaying to $1/e$) at

$$\delta_p = 1/\alpha k_0 = c/\alpha\omega. \tag{2.96}$$

As an ideal conductor of $1/\tau = 0$, the metal will become totally reflective.

Case 3: Low frequency at $\omega\tau << 1$ where $\varepsilon' << \varepsilon''$ or $\varepsilon \approx i\varepsilon''$, and $n \approx \alpha \approx \sqrt{\varepsilon''/2}$, so that

$$E = E_0 e^{-i(\sqrt{-i\varepsilon''})k_0 \cdot r} = E_0 e^{\sqrt{i2n\alpha}k_0 \cdot r} = E_0 e^{ink_0 \cdot r} e^{\alpha k_0 \cdot r}. \quad (2.97)$$

The propagation also exponentially decays by exp[$\alpha \boldsymbol{k_0} \cdot \boldsymbol{r}$] that will lead to reflection or even total reflection at a very low frequency, giving also $\delta_p = c/\alpha\omega$.

Case 4: Resonance frequency at $\omega = \omega_p$ which makes $\varepsilon = 1 - \omega^2{}_p/\omega^2 = 0$, so that longitudinal oscillations or longitudinal electron density fluctuation happens, giving a wave vector function in scalar form of

$$k^2 = k_0^2 \varepsilon = k_0^2(1 - \omega_p^2/\omega^2) \quad (2.98)$$

or $$\omega^2 = \omega_p^2 + c^2 k^2. \quad (2.99)$$

Equation 2.99 draws a circle with a variable radius of ω starting at $ck/\omega = 0$ and $\omega/\omega_p = 0$ (Fig. 2.8b). In a coordinate system of $y = \omega/\omega_p$ and $x = ck/\omega_p$, the circle (the red in Fig. 2.8b) increases its size from a unit at the center of $r = y = \omega/\omega_p = 1$, and circle center draws a dispersion curve for the corresponding VPP. This curve gradually approaches the dispersion curve of the excitation light. This clearly suggests that a photon-excited VPP is a mixed mode of the incident photon (ck) and volume plasmon (ω_p). A pure plasmonic mode is obtained at $\omega = \omega_p$ that is a horizontal straight line in Fig. 2.8b. This is also a sensitive way to determine the inherent value of ω_p as mentioned already. The ratio of ω_p in VPPs decreases reversely with ω, until the VPP infinitively approaches the dispersion line of the incident light.

It can thus be concluded that, although transverse electromagnetic waves can propagate in metals at $\omega \geq \omega_p$, they are prohibited at $\omega < \omega_p$; while at $\omega = \omega_p$, volume plasma resonance happens at the electric field of an incident wave parallel to the plasma vibration, whereas magnetic field is out of function so that the oscillations of volume plasma cannot emit electromagnetic waves and propagate away the metals. It should also be clarified that, although volume plasma oscillations are longitudinal, the volume plasma polarizations are a mixed mode containing both of electron density fluctuations and electromagnetic vibrations of the incident photons.

2.3.3 *Surface Plasma Polarization and Propagation*

2.3.3.1 Essential Concepts

Surface plasma polarization or SPP is a key to study the surface plasma, the same as the VPP to the volume plasma. SPR what we have defined is something related to SPP.

Let us recall to the scene to drop a stone into a calm pond, the water molecules in the pond will be pressed down and then bounce back after the sedimentation of the stone, which causes a longitudinal wave along the stone sinking way; at the same time, the pressed molecules will squeeze their lateral molecules to show up transverse waves that can easily be observed on the water surface. We can imagine that a similar scenario can reasonably occur when photons hit metal: The hitting can bring up not only VPPs but also SPPs. Similar to the easy observation of the transverse waves on water surface, we can also expect that a SPW is more easily measured through SPP than a volume plasma wave. In theory, at/in the very outer surfaces, some valence electrons are unbalanced, being more sensible to a light of lower energy than the volume electrons. That is why we can observe the color of metal nanoparticles (NPs), a type of SPPs, that is, localized SPPs or LSPPs.

2.3.3.2 Non-excitation Under *S*-Light

Theoretically, SPP is a mixed mode of a surface plasmon and an excitation photon, the same as a VPP. Clearly, the mobile valance electrons on a metal surface must oscillate also collectively at a characteristic frequency of ω_{sp} corresponding to an oscillating quantum SP. From an electrically neural point of view, the collective oscillation or fluctuation of valance electrons at the surface must have a structure as illustrated in Fig. 2.9a at a time. It can be guessed that the excited SPPs may propagate along the interface and into the space at both sides of the interface (Fig. 2.9b).

The essence is thus how to excite an SPP with a photon so that we can have an insight into the nature of the SPP and SPW and their relationship with the surface plasma or surface plasmons. By taking the advantage of EMWs, we can try TE and TM lights. Let us start with the TE or s-light.

Based on the orientation and orthogonality of TE mode, it will co-vibrate with a magnetic field along *z*- and *x*-axes, in addition to its electric field along only *y*-axis in the Descartes coordinate system as shown in Fig. 2.2a. In an infinitive interface, the propagation equations of an EMW are as follows:

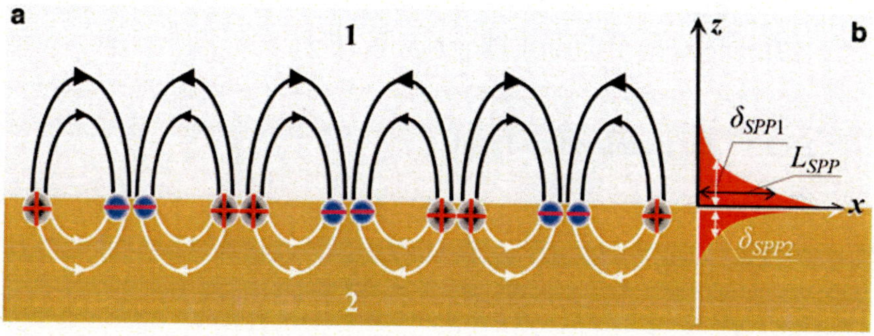

Fig. 2.9 Electron fluctuation and wave propagation between a dielectric-metal surface. **a** State of electron fluctuation; **b** Exponential decaying of the wave propagation along the interface (x), into the dielectric (1) and the metal (2); $\delta_{spp,m}$ ($m = 1, 2$), Penetration depth of SPP; L_{SPP}, Propagation length of SPP along the interface

$$\begin{cases} \boldsymbol{H}_m = (\boldsymbol{H}_{mx_0},\ \ 0,\ \ \boldsymbol{H}_{mz_0})e^{i(k_{mz}z-k_{mx}x+\omega t)} \\ \boldsymbol{E}_m = (0,\ \ \boldsymbol{E}_{my_0},\ \ 0)e^{i(k_{mz}z-k_{mx}x+\omega t)} \end{cases},\quad (m=1,2) \tag{2.100}$$

with boundary conditions of

$$\begin{cases} \left.\begin{array}{l} k_{1x}=k_{2x}=k_x \\ E_{1y}=E_{2y};\ \ E_{1x}=E_{2x} \\ H_{1y}=H_{2y} \end{array}\right\}\text{along interface} \\ \left.\begin{array}{l} \varepsilon_1 E_{1z}=\varepsilon_2 E_{2z} \\ \mu_1 H_{1z}=\mu_2 H_{2z} \end{array}\right\}\text{perpendicular to interface} \end{cases}. \tag{2.101}$$

From Eq. 2.17, there will be

$$\boldsymbol{k}\times\boldsymbol{E}_m = i\mu_0\mu\omega\begin{pmatrix}\boldsymbol{H}_{mx_0}\\ 0\\ \boldsymbol{H}_{mz_0}\end{pmatrix}e^{i(k_{mz}z-k_{mx}x+\omega t)} \tag{2.102}$$

and
$$\boldsymbol{k}\times\boldsymbol{E} = \begin{vmatrix}\hat{x} & \hat{y} & \hat{z}\\ k_{mx} & k_{my} & k_{mz}\\ 0 & E_{my_0} & 0\end{vmatrix} = i\begin{pmatrix}-k_{mz}\boldsymbol{E}_{my_0}\\ 0\\ -k_{mx}\boldsymbol{E}_{my_0}\end{pmatrix}e^{i(k_{mz}z-k_{mx}x+\omega t)} \tag{2.103}$$

thus
$$\begin{cases} \pm k_{mz}\boldsymbol{E}_{my} = \mu_0\mu_m\omega\boldsymbol{H}_{mx}=0 \\ -k_{mx}\boldsymbol{E}_{my} = \mu_0\mu_m\omega\boldsymbol{H}_{mz}=0 \end{cases} \tag{2.104}$$

Since we have assumed the existence of an SPP, that is, $k_{mx}= k_x\neq 0$ and $\mu_m\omega\neq 0$, we need to let

$$\begin{cases} \boldsymbol{H}_{mx} = \boldsymbol{H}_{mz} = 0 \\ \boldsymbol{E}_{my} = 0 \end{cases} \tag{2.105}$$

Therefore, the trial failed. Photons with TE mode or s-light excite nothing SPP.

2.3.3.3 Excitation with *p*-Light

A TM or p-light (Fig. 2.2b) has its field components of $\boldsymbol{H}_{my}$, $\boldsymbol{E}_{mz}$ and $\boldsymbol{E}_{mx}$ as follows:

$$\mathbf{E}_m = \begin{pmatrix}\mathbf{E}_{mx_0}\\ 0\\ \mathbf{E}_{mz_0}\end{pmatrix}e^{i(\mp k_{mz}z-k_x x+\omega t)} \text{ and } \mathbf{H}_m = \begin{pmatrix}0\\ \mathbf{H}_{my_0}\\ 0\end{pmatrix}e^{i(\mp k_{mz}z-k_x x+\omega t)} \text{ for } m=1,2. \tag{2.106}$$

$$\text{Thus}\quad \begin{cases} \boldsymbol{k}\times\boldsymbol{H} = -\varepsilon_0\varepsilon_m\omega\begin{pmatrix}\boldsymbol{E}_{mx_0}\\ 0\\ \boldsymbol{E}_{mz_0}\end{pmatrix}e^{i(\mp k_{mz}z-k_xx+\omega t)} \\ \boldsymbol{k}\times\boldsymbol{H} = \begin{vmatrix}\hat{x} & \hat{y} & \hat{z}\\ k_x & k_{my} & k_{mz}\\ 0 & H_{my} & 0\end{vmatrix} = \begin{pmatrix}\pm k_{mz}\boldsymbol{H}_{my_0}\\ 0\\ -k_x\boldsymbol{H}_{my_0}\end{pmatrix}e^{i(\mp k_{mz}z-k_xx+\omega t)}\end{cases} \tag{2.107}$$

$$\text{that is,}\quad \begin{cases}-\varepsilon_0\varepsilon_m\omega\boldsymbol{E}_{mz} = -k_x\boldsymbol{H}_{my}\\ -\varepsilon_0\varepsilon_m\omega\boldsymbol{E}_{mx} = \pm k_z\boldsymbol{H}_{my}\end{cases}. \tag{2.108}$$

Considering $\boldsymbol{H}_{1y}=\boldsymbol{H}_{2y}$ and $\boldsymbol{E}_{1x}=\boldsymbol{E}_{2x}$, we have

$$\frac{k_{1z}}{k_{2z}} = -\frac{\varepsilon_1}{\varepsilon_2}. \tag{2.109}$$

To confine the waves within the interface, k_{mz} should be a pure complex for the waves fast decaying away the interface, that is,

$$k_{mz}^2 = k_0^2\varepsilon_m - k_x^2 < 0 \tag{2.110}$$

or $k_{1z}/k_{2z} > 0$, so that

$$\varepsilon_1\varepsilon_2 < 0. \tag{2.111}$$

This means that the relative permittivity changes its sign across the interface to have SPPs. It is usual that a dielectric has only positive permittivity, which asks that the metal should have a negative permittivity. This is possible in use of some conductive metals under visible lights, such as the discussed Au, Ag, Cu and Al.

Although it looks complicated, we are now possible to "feel" the surface plasma with the TM mode of lights.

2.3.3.4 Relationship of Surface Plasmons with Their Polarizations

Similar to a volume plasma, we can expect to find the relationship of the intrinsic vibration frequency, ω_{sp}, of a surface plasmon with that of an optically excited SPP. From Eqs. 2.109 and 2.110, the scalar equations for the decomposed wave vectors can be derived along the directions of x- and z-axes:

$$\begin{cases}(a)\ \frac{k_{1z}^2}{k_{2z}^2} = \frac{\varepsilon_1k_0^2-k_x^2}{\varepsilon_2k_0^2-k_x^2} = \left(-\frac{\varepsilon_1}{\varepsilon_2}\right)^2\\ (b)\ k_x^2 = k_0^2\frac{\varepsilon_1\varepsilon_2}{\varepsilon_1+\varepsilon_2} = k_0^2\varepsilon_1\frac{\varepsilon_2'(\varepsilon_2'+\varepsilon_1)+(\varepsilon_2'')^2-i\varepsilon_1\varepsilon_2''}{(\varepsilon_1+\varepsilon_2')^2+(\varepsilon_2'')^2}\\ (c)\ k_{1z}^2 = k_0^2\varepsilon_1 - k_x^2 = k_0^2\frac{\varepsilon_1^2}{\varepsilon_1+\varepsilon_2} < 0\\ (d)\ k_{2z}^2 = k_0^2\varepsilon_2 - k_x^2 = k_0^2\frac{\varepsilon_2^2}{\varepsilon_1+\varepsilon_2} < 0\end{cases}. \tag{2.112}$$

According to the first equation in Eq. 2.90, we have

$$k_x^2 = (k_x' - ik_x'')^2 = [(k_x')^2 - (k_x'')^2] - i[2k_x'k_x''] \tag{2.113}$$

Comparing it with the equation (*b*) in Eq. 2.112 and omitting the high order term of ε''_{2r} by assumption of $\varepsilon''_{2r} < |\varepsilon'_{1r}|$, we have the special wave vector equations for SPPs propagating along the interface:

$$\begin{cases} (a)\ k_x = k_x' + k_x'' \\ (b)\ k_x' = k_0\sqrt{\varepsilon_{12}} \\ (c)\ k_x'' = k_0\sqrt{(\varepsilon_{12})^3}\frac{\varepsilon_2''}{2(\varepsilon_2')^2} \\ (d)\ \varepsilon_{12} = \frac{\varepsilon_1\varepsilon_2'}{\varepsilon_1+\varepsilon_2'} \end{cases}. \tag{2.114}$$

To detect ω_{sp} without attenuation, we only need to consider the real part of the propagation parameter. Combining Eq. 2.88 with the equation (*b*) in Eq. 2.114 and neglecting the attenuation item of the imaginary part, we have

$$k_{spp} = k_x' = k_0\sqrt{\varepsilon_{12}} = \frac{\omega}{c}\sqrt{\frac{\varepsilon_1(1-\omega_p^2/\omega^2)}{\varepsilon_1+1-\omega_p^2/\omega^2}}. \tag{2.115}$$

Its extreme value (pointing to resonance) is at

$$\begin{cases} (a)\ \varepsilon_1+\varepsilon_2' = 0 \ \text{ or} \\ (b)\ \varepsilon_1 + 1 - \omega_p^2/\omega^2 = 0 \end{cases}. \tag{2.116}$$

Let $\omega = \omega_{sp}$, the equation (*b*) in Eq. 2.116 gives

$$\omega_{sp} = \frac{\omega_p}{\sqrt{1+\varepsilon_1}}. \tag{2.117}$$

As expected, ω_{sp} is smaller than ω_p for at least 30% in the case with air as medium 1.

Another expectation is that surface plasma should oscillate longitudinally like the volume plasma. This is just proved by the equation (*a*) in Eq. 2.116 according to Eq. 2.19, that is,

$$\varepsilon_{sp} = \varepsilon_1 + \varepsilon_2' = 0. \tag{2.118}$$

The magnetic field long the interface accordingly loses its function. Nevertheless, it is still worth of notice that the longitudinal surface plasmons also couple with the TM mode of photons in SPPs. A brief explanation is as follows:

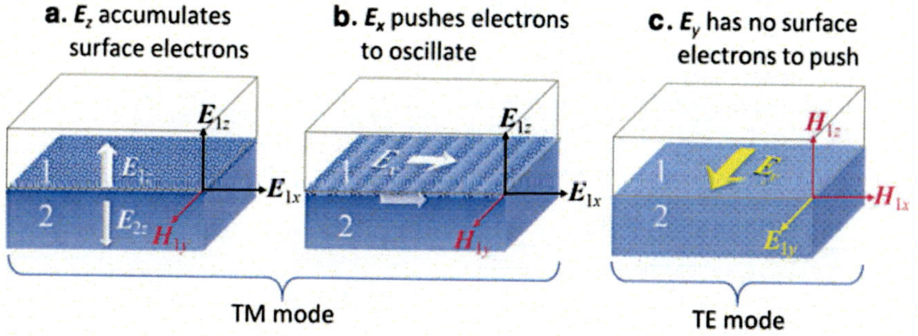

Fig. 2.10 Mechanism to induce surface plasma polarizations by different light modes.

TM mode has a discontinuous $\boldsymbol{E}_z$ across the interface but a continuous $\boldsymbol{E}_x$. The discontinuous $\boldsymbol{E}_z$ accumulates electrons at the interface (Fig. 2.10a), acting transversely, while $\boldsymbol{E}_x$ pushes the accumulated electrons away the balanced position to induce longitudinal oscillations along the interface (Fig. 2.10b). Differently, TE mode has only $\boldsymbol{E}_y$ component which has got no way to enrich electrons at the interface and then push them forward (Fig. 2.10c).

2.3.3.5 Dispersion of Surface Plasma Polarizations

By drawing Eqs. 2.88 and 2.117 together in Fig. 2.11A to illustrate the relationships of ε_2 with ε_1 and ω_{sp} together with ω_p, we can easily find that a metal always contains volume and surface plasmons if its relative permittivity is below 1 until negative infinity. Figure 2.11A also reveals the zonal relationships of the relative permittivity with frequency but not vividly. More detailed dispersion information is obtained by analyzing the equation (*b*) in Eq. 2.112. Three characteristic cases can be inferred:

$$\begin{cases} (1)\ \ k_x(0 > \varepsilon_2 > -\varepsilon_1) = k_0\sqrt{\frac{\varepsilon_1\varepsilon_2}{\varepsilon_1+\varepsilon_2}} = ik_0\sqrt{\left|\frac{\varepsilon_1\varepsilon_2}{\varepsilon_1+\varepsilon_2}\right|} \\ (2)\ \ k_x(\varepsilon_2 = -\infty) = \lim\limits_{\varepsilon_2\to-\infty}\left(k_0\sqrt{\frac{\varepsilon_1\varepsilon_2}{\varepsilon_1+\varepsilon_2}}\right) = k_0 \\ (3)\ \ k_x(\varepsilon_2 = -\varepsilon_1) = \lim\limits_{\varepsilon_2\to-\varepsilon_1}\left(k_0\sqrt{\frac{\varepsilon_1\varepsilon_2}{\varepsilon_1+\varepsilon_2}}\right) = \infty \end{cases} \quad . \tag{2.119}$$

As the first equation in Eq. 2.119 indicates, in the region just below 0 but above $-\varepsilon_1$, the wave propagation, if exist, will fast decay into zero since k_x is a complex which gives $e^{-ik_{sp}\cdot r} = e^{|k_{sp}\cdot r|}$. The second equation suggests that, in the region far below $-\varepsilon_1$ or at ε_2 approaching $-\infty$, the wave vector approaches the limit of k_0 (light in free space), where the SPPs weigh nearly only photons. The third equation tells that, at $\varepsilon_2 = -\varepsilon_1$, the SPPs obtain infinitively large momentum. This is a case of resonance of the excitation photons with SPs.

More detailed information on the dispersion of SPPs can be obtained by analyzing Eq. 2.115 after re-arrangement of it into:

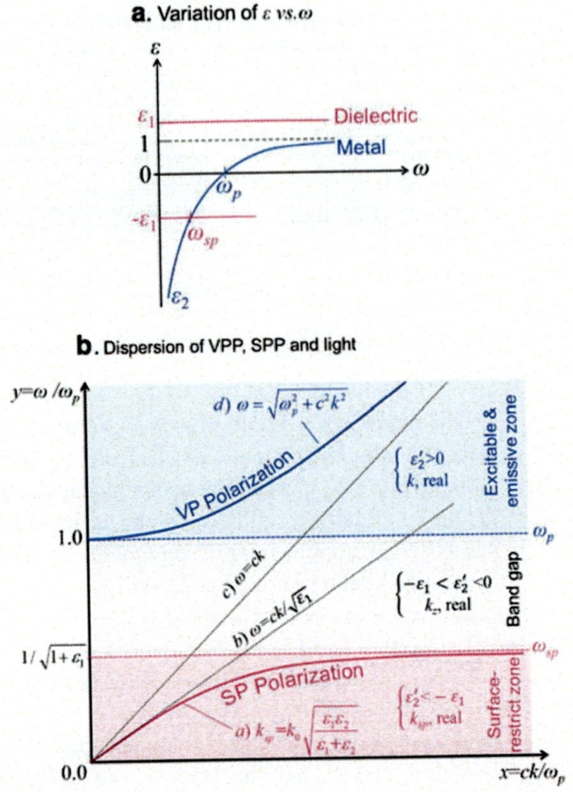

Fig. 2.11 Dispersion curves drawn **a** between ε and ω through different zones and **b** between ω/ω_p and ck/ω_p for (*a*) surface and (*d*) volume plasma polarizations, and (*b*) light in medium and (*c*) light in free space

$$\omega^2=\frac{\omega_p^2}{2}+\frac{\varepsilon_1+1}{2\varepsilon_1}k_{\text{spp}}^2c^2-\sqrt{\frac{\omega_p^4}{4}+\left(\frac{\varepsilon_1+1}{2\varepsilon_1}\right)^2k_{\text{spp}}^4c^4+\frac{\varepsilon_1-1}{2\varepsilon_1}\omega_p^2k_{\text{spp}}^2c^2}. \quad (2.120)$$

The corresponding dispersion curve is added to the bottom of Fig. 2.8b to give Fig. 2.11b where the dispersion space is divided into three parts:

(i) Excitable and emissive zone at $\omega \geq \omega_p$ or $\varepsilon'_2 > 0$, which concerns with VPPs;
(ii) Surface-trapped zone at $\omega \leq \omega_{\text{sp}}$ or $\varepsilon'_2 \leq -\varepsilon_1$, which is sure for SPPs; and
(iii) A band gap between $0 > \varepsilon'_2 > -\varepsilon_1$ where polarization by an external field is impossible and the excitation photons will be reflected by the surface.

As a mixed mode, SPPs change their weights between the excitation photons and surface plasmons as illustrated by the dispersion curve (the bottom line in Fig. 2.11b). At a frequency of ω far below ω_{sp}, the dispersion curve becomes closer and closer to the straight line of the excitation photons in dielectric, that is, SPPs increase their weight of electromagnetic field reversely with the excitation frequency (opposite to VPPs) until they meet with each other at $\omega = 0$ where the oscillators are simply the photons. Reversely, SPPs decrease their weight of photons while increase the ratio of SPs as the frequency increases, until reaching a limit of ω_{sp} (also opposite to

VPPs). Different from VPPs whose dispersion line locates always at the left side of the optical dispersion in free space, SPP dispersion curve is overall away the right side of optical dispersion. To the right side, the curve goes faster and faster away the optical line by bending down and gradually approaches a horizontal limit of ω_{sp}, leading to larger and larger momentum difference between SPPs and lights. This makes them unable to fuss together to exchange their energy by direct interaction.

2.3.3.6 Propagation Velocity and Wavelength

The propagation of SPPs in an interface is measured by their group velocity:

$$\boldsymbol{v}_{\text{spp}} = \frac{\omega}{\boldsymbol{k}_{\text{spp}}} = \boldsymbol{c}\sqrt{\frac{\varepsilon_1 + \varepsilon_2'}{\varepsilon_1 \varepsilon_2'}} < \boldsymbol{c} \xrightarrow{\text{SP}} \boldsymbol{v}_{\text{sp}} = \boldsymbol{c}\sqrt{\frac{\varepsilon_1 + \varepsilon_2'}{\varepsilon_1 \varepsilon_2'}}\Bigg|_{\varepsilon_2' = -\varepsilon_1} = 0 < \boldsymbol{c} \quad (2.121)$$

Because metals what we need have negative permittivity, the SPPs propagates at a velocity slower than its corresponding light velocity in free space. As mentioned, for an ideal metallic conductor at its infinitive surface, there will be $\varepsilon_2' = -\varepsilon_1$ or $\boldsymbol{v}_{\text{sp}} = 0$ as a limit. This means that the SPs do not really propagate but keeps silent. It is a quasi-static surface mode with divergence of surface charges. It can only be detected under the condition that the excitation photons can resonate with it.

The corresponding wavelength of a SPP can be calculated by normal definition:

$$\lambda_{\text{spp}} = \frac{2\pi}{k_{\text{spp}}} = \lambda_0\sqrt{\frac{\varepsilon_1 + \varepsilon_2'}{\varepsilon_1 \varepsilon_2'}} \xrightarrow[\varepsilon_1 = 1]{\text{in free space}} \lambda_{\text{spp}} = \lambda_0\sqrt{\frac{\varepsilon_2' + 1}{\varepsilon_2'}} < \lambda_0 \quad (2.122)$$

Clearly, SPPs have always shorter wavelengths than the exciting lights, which is determined not only by the relative permittivity but also by λ_0. By calculations with the data collected in Table 2.1, it can be found that the SPPs increase their wavelength with $|\varepsilon_2'|$ and $|\varepsilon_1|$ (Table 2.2) at $\lambda_0 = 633$ nm (that is an often used optical wavelength in SPR and SPRi). It should be noted that it is hard to excite SPPs at the interface of Cr/dielectric. It happens only at the interface of Cr/vacuum, giving $\lambda_{SPP} = 215$ nm, where the ε'_{Cr} is approaching -1. This means that for a given interface, there exists a lowest light wavelength, $\lambda_{0,\min}$, to make $\varepsilon'_m \approx -\varepsilon_d$ at a given temperature (since n and ε are the functions of temperature). The shortest wavelength is more exactly called the shortest cutoff wavelength. A very useful reference is the cutoff wavelength for Au/H_2O at below 475 nm at room temperature.

According to Eq. 2.122, the wavelength of SPPs can be adjusted trough reduction of λ_0. Take the interface of Au/H_2O and Au/SiO_2 as examples, their SPP wavelengths can approach 152 and 8 nm at $\lambda_0 = 490$ nm where the relative permittivity ε'_2 changes to -2.1530, 2.1401 and 1.7833 for gold, SiO_2 and water, respectively. It is striking that λ_{spp} can be shortened to X-ray range, even down to only 8 nm for Au/SiO_2 system. This is in fact a useful way to largely increase the photolithographic resolution.

Table 2.1 Refractive index and calculated relative permittivity of some potentially usable metals and dielectrics at $\lambda_0 = 632.8$ nm and 25 °C

Matter	n	α	$\varepsilon' = n^2 - \alpha^2$	$\varepsilon'' = 2n\alpha$	Cited from*
Al	1.4482	7.5367	– 54.704	21.829	(a)
Ag	0.056235	4.2760	– 18.281	0.48108	(a)
Au	0.18377	3.4313	– 11.740	1.2611	(a)
Cu	0.27002	3.4081	– 11.542	1.8405	(a)
Ti	2.7039	3.7651	– 6.8655	20.361	(a)
Cr	3.1399	3.3150	– 1.1305	20.818	(a)
BK7	1.5151	1.2122×10^{-8}	2.2955	3.6732×10^{-8}	(a)
SiO_2	1.457	0	2.123	0	(a)
MgF_2	1.383	0	1.913	0	(b), (c)
H_2O	1.3317	1.4680×10^{-8}	1.7734	3.9098×10^{-8}	(a)

*(a) https://refractiveindex.info; (b) Dodge M. J. (1984) Refractive properties of magnesium fluoride, *Applied Optics*, Vol. 23, Issue 12 doi:10.1364/AO.23.001980; and (c) http://cn.filmetrics.com/refractive-index-database

Table 2.2 SPP wavelength (/nm) calculated at the surfaces between metals (first line) and different dielectrics (first column) excited by a light with $\lambda_0 = 632.8$ nm

	Al	Ag	Au	Cu	Ti	Cr
BK7	409	390	375	374	342	–
SiO_2	428	408	394	392	361	–
MgF_2	449	433	419	418	388	–
H_2O	467	451	438	437	409	–
Vacuum	627	615	605	604	585	215

2.3.3.7 Propagation Length and Depth

In theory, a wave can propagate infinitively in space and time if without any energy attenuation, while in practical systems, attenuation always happens and may be very serious, leading to a sharp decrease of waves after only a short length/depth. These can be figured by analyzing the related wave function together with wave vector(s). For example, the propagation of SPPs into both sides of an interface can be pictured by the equations (*c*) and (*d*) in Eq. 2.112. Since $k_{mz}{}^2 < 0$ or $k_{mz} = i| k_{mz} |$, the corresponding wave equation can thus be written as:

$$\boldsymbol{E} = \boldsymbol{E}_0 e^{-ik_x x} e^{-ik_{mz} z} = \left(\boldsymbol{E}_0 e^{-ik_x x}\right) e^{|k_{mz}| z}. \tag{2.123}$$

The electric field decays exponentially away the surface, making the waves concentrate at the interface, which is right the meaning for surface waves. Clearly, the greater is the value of k_{mz}, the faster will be the decay of the waves or the shorter will the waves travel. By comparison of equation (d) with (a), we know that $|k_{2z}| > |k_{1z}|$, so that the wave decay in metal is faster than in dielectric, which has been expected as shown in Fig. 2.9b.

At $\varepsilon_1 > 0$ and $\varepsilon_2 < 0$, there will be $|\varepsilon_1 + \varepsilon_2| < |\varepsilon_1 \varepsilon_2|$ or $k_x > k_0 > 0$; thus, SPPs have a greater momentum along the interface than the incident light as has been discussed. Similar to k_{mz}, k_x is also a complex with an attenuation item in the imaginary part. Consider the equation (c) in Eq. 2.114, we have the wave function along the x-direction:

$$\boldsymbol{E} = \left(\boldsymbol{E}_0 e^{|k_{mz}|z} e^{-ik_{sp}x}\right) e^{k_0 \frac{\varepsilon_2'' \sqrt{\varepsilon_{13}}}{2(\varepsilon_2')^2} x}. \tag{2.124}$$

The last term causes propagation decay. However, k_x" is much smaller than k_{mz}; SPPs will travel much a longer distance along the interface than into the side spaces from the interface (Fig. 2.9b).

In order to quantitatively discuss the propagation, a travel length is often defined by two ways. The first way defines a propagation length δ_I by energy losing to $1/e$, while the second way defines a propagation length δ_A by amplitude decreasing to $1/e$. The former definition is often used when a wave can propagate quite a long distance such as a SPP along an interface, while the latter is used to measure a wave how deep it can penetrate into a medium from its surface, just like the case to discuss how deep an SPP wave can decay away the interface.

By the definition on energy loss and consider only the imaginary part (responsible for loss), we have:

$$\frac{\boldsymbol{E}^2_{x=\delta_{I,x}}}{\boldsymbol{E}^2_{x=0}} = \frac{\boldsymbol{E}^2_{\text{other}}\left(e^{i(i|k_x''|)\delta_{I,x}}\right)^2}{\boldsymbol{E}^2_{\text{other}}\left(e^{i(i|k_x'|)\cdot(0)}\right)^2} = \frac{e^{-2|k_x''|\delta_{I,x}}}{e^0} = \frac{1}{e}, \tag{2.125}$$

where $\boldsymbol{E}_{\text{other}}$ includes the amplitude and other variables of not impact on the propagation. We have then the propagation length for SPPs along the interface L_{SPP}:

$$L_{\text{SPP}} = \delta_{I,x} = \frac{1}{2|k_x''|}. \tag{2.126}$$

The same, by the definition based on amplitude decay,

$$\frac{\boldsymbol{E}_{mz=\delta_{A,mz}}}{\boldsymbol{E}_{mz=0}} = \frac{\boldsymbol{E}_{\text{other}}\left(e^{i(i|k_{mz}''|)\delta_{A,mz}}\right)}{\boldsymbol{E}_{\text{other}}\left(e^{i(i|k_{mz}''|)\cdot 0)}\right)} = \frac{e^{-|k_{mz}''|\delta_{A,mz}}}{e^0} = \frac{1}{e} \tag{2.127}$$

we obtain the penetration depth $\delta_{\text{SPP}m}$ is

$$\delta_{\text{SPP}m} = \delta_{A,mz} = \frac{1}{|k_{mz}''|} = 2\delta_{I,mz}. \tag{2.128}$$

The penetration depth is also called skin effect for it is perpendicular to the surface. Note, the penetration depth defined by amplitude is twofold of the propagation depth calculated by energy loss. From the equation (*c*) in Eq. 2.114, the equations (*c*) and (*d*) in Eqs. 2.112, 2.126 and 2.128, there will be

$$\begin{cases} L_{\text{spp}} = \frac{1}{2k_x''} = \frac{\lambda_0}{2\pi}\frac{(\varepsilon_2')^2}{\varepsilon_2''}\sqrt{\frac{\varepsilon_1+\varepsilon_2'}{\varepsilon_1\varepsilon_2'}} \approx \frac{\lambda_0}{2\pi}\frac{(\varepsilon_2')^2}{\varepsilon_2''\sqrt{\varepsilon_1}} \text{ at } |\varepsilon_2'| >> \varepsilon_1 \\ \delta_{\text{SPP1}} = \frac{1}{|k_{1z}|} = \frac{\lambda_0}{2\pi}\sqrt{\frac{|\varepsilon_1+\varepsilon_2'|}{\varepsilon_1^2}} \\ \delta_{\text{SPP2}} = \frac{1}{|k_{2z}|} = \frac{\lambda_0}{2\pi}\sqrt{\frac{|\varepsilon_1+\varepsilon_2'|}{(\varepsilon_2')^2}} \end{cases}. \tag{2.129}$$

These equations determine the sensing thickness of dielectric and metal membrane and lateral resolution. By use of the data in Table 2.1, the three critical lengths are calculated at λ_0= 632.8 nm:

$$\begin{cases} \delta_{\text{SPP,Au}} = 27 \text{ nm} \\ \delta_{\text{SPP,H}_2\text{O}} = 179 \text{ nm} \\ L_{\text{spp,Au/H}_2\text{O}} = 5.7\ \mu\text{m} \end{cases}. \tag{2.130}$$

An SPP wave can propagate a length of micrometers along an interface, with much shorter penetration depth into the dielectric (ca. $\lambda_0/3.5$) and into metal (only ca, $\lambda_0/23$). The values of $\delta_{\text{SPP}m}$ indicate a clear condensation of SPPs in the interface, which confirms that SPPs are really surface waves. It is worth of mentioning that the propagation length of SPPs, L_{spp}, largely varies with metal, for example, silver can increase L_{spp} for about 10 folds but much less impacts on the penetration depths:

$$\begin{cases} \delta_{\text{SPP,Ag}} = 22 \text{ nm} \\ \delta_{\text{SPP,H}_2\text{O}} = 230 \text{ nm} \\ L_{\text{spp,Ag/H}_2\text{O}} = 49.9\ \mu\text{m} \end{cases}. \tag{2.131}$$

A comparative order of the propagation-related parameter is as follows:

$$L_{\text{SPP}} > \lambda_{\text{SPP}} > \delta_{\text{SPP,dielectric}} > \delta_{\text{SPP,metal}}. \tag{2.132}$$

Longer L_{spp} helps the construction of SPP-based devices for conversion between photons and electrons but also decreases the lateral spatial resolution. It has to be compromised in different researches. SPPs concentrated at the interface can increase the vertical spatial resolution but are not preferred for probing substances away the sensing surface.

2.3.4 Localized Surface Plasma Polarizations on Nanoparticles

If we increase the curvature of a metal by shrinking its volume together with bending the surface until it is enclosed on only a nanoparticle, the SPP will be localized on the enclosed surface and the way to polarize the charges becomes different, giving different dispersion relationship. Such localized SPPs can also exist in microdefects, wrinkles or nano-protrusions on surfaces and in turn couple with normal SPPs. LSPPs can be used independently to increase the sensitivity of SPRS, SPRi and other optical methods. Therefore, we have here a brief discussion on their basic theory.

A tiny particle shrinks its volume fast while expends its surface ratio reversely with diameter. On and in a particle with a diameter far below the wave length of an EMW, an excited LSPP will have no space to travel except for being trapped in/on some corners of the particle.

For more clarity, let us consider the case of a metal nanosphere irradiated with light: The optical electric field will act on and polarize the metal sphere, just the same as a spherical molecule to become a dipole. As known, there are various theories to treat such a dipole, of which Mie [2] gave the strictest solution from Maxwell's equations. The normal thoughts include the transfer of Maxwell's vector equations into equivalent scalar equations by an operation of:

$$E = -\nabla\Phi, \tag{2.133}$$

where Φ is scalar function normally called potential function that can be treated by Laplacian operator:

$$\nabla^2\Phi(r,t) = 0 \quad \text{for } |r| < d/2 \text{ and } |r| > d/2. \tag{2.134}$$

The symbol d denotes the diameter of the particle. Other related specifications are shown in Fig. 2.12 drawn according to Fig. 2.9 after bending a pair of vibrator. Under the boundary conditions of

$$\begin{cases} \Phi_2(\boldsymbol{r},t) = \Phi_1(\boldsymbol{r},t) \text{ at } |\boldsymbol{r}| = d/2 \\ r \overset{\lim}{\longrightarrow} \infty\Phi_1(\boldsymbol{r},t) = -\boldsymbol{E}_0 \cdot \boldsymbol{r} \end{cases}. \tag{2.135}$$

The equations can readily be solved in a spherical coordinate:

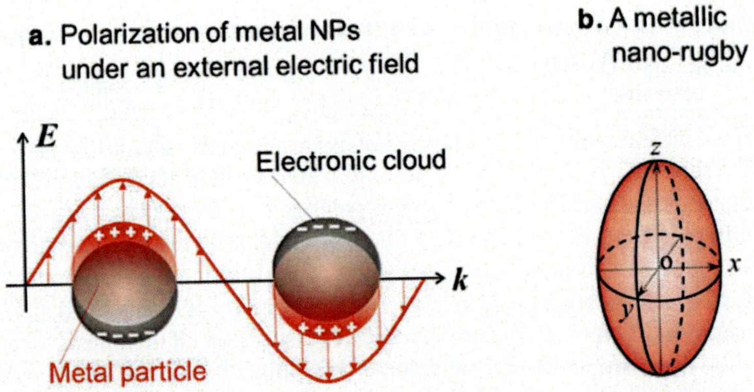

Fig. 2.12 Equivalent polarization of **a** aspherical metallic nanoparticle by an external electric field at an arbitrary direction and **b** an ellipsoidal particle at two independent directions, i.e., along its long and short axes

$$\begin{cases} \begin{cases} \Phi_1(r,\theta,\phi) = \sum\limits_{j=0}^{\infty}\sum\limits_{m=-j}^{\infty} a_{jm}\dfrac{Y_{jm}(\theta,\phi)}{r^{j+1}} \\ \Phi_2(r,\theta,\phi) = \sum\limits_{j=0}^{\infty}\sum\limits_{m=-j}^{j} b_{jm}r^{j}Y_{jm}(\theta,\phi) \end{cases} \\ \begin{cases} Y_{jm}(\theta,\phi) = (-1)^m\sqrt{\dfrac{(2j+1)(j+m)!}{4\pi(j-m)!}}f_j^m(\cos\theta)e^{im\phi} \\ \int\limits_0^{\pi}\int\limits_0^{2\pi} Y_{jm}(\theta,\phi)Y_{jm}^*(\theta,\phi)\sin\theta\mathrm{d}\theta\mathrm{d}\phi = 1 \end{cases} \\ \quad (j = 0, 1, 2, \cdots); (m = 0, \pm 1, \pm 2, \cdots) \end{cases} \cdot \tag{2.136}$$

At $r = d/2$, there exists an extremum:

$$\frac{\varepsilon_2}{\varepsilon_1} = \frac{j+1}{m}. \tag{2.137}$$

The quantum-like number j is the order of spherical harmonic function Y_{jm}, corresponding to polarization modes, for example, $j = 1$ is a dipole, while $j = 2$ and 3 are quadrupole and octupole, respectively. High order of polarization happens if a metal particle becomes larger and larger. However, if it is smaller than 30 nm, a dipole is sufficient to describe the electronic oscillations. In this case of $j = 1$, Eq. 2.137 changes to

$$\varepsilon_2 = 2\varepsilon_1, \tag{2.138}$$

and the solutions are given by

$$\left\{\begin{array}{l}\Phi_1 = -E_0 r\cos\theta + E_0\frac{\varepsilon_2-\varepsilon_1}{\varepsilon_2+2\varepsilon_1}\frac{d^3}{8}\frac{\cos\theta}{r^2}\\ \Phi_2 = -E_0\frac{3\varepsilon_1}{\varepsilon_2+2\varepsilon_1}r\cos\theta\end{array}\right\}\xrightarrow{E=-\nabla\Phi}\left\{\begin{array}{l}E_1 = E_0 - E_0\dfrac{\varepsilon_2-\varepsilon_1}{\varepsilon_2+2\varepsilon_1}\dfrac{d^3}{8}\dfrac{1}{r^3}\\ E_2 = E_0\dfrac{3\varepsilon_1}{\varepsilon_2+2\varepsilon_1}\end{array}\right. \quad (2.139)$$

Clearly, Eq. 2.138 is the extremum of Eq. 2.139. Replace $\varepsilon_2 = 1 - \omega_p{}^2/\omega^2$ into Eq. 2.137, we have

$$\omega = \omega_p\sqrt{\frac{j}{j+(1+j)\varepsilon_1}} = \left\{\begin{array}{ll}\frac{\omega_p}{\sqrt{1+2\varepsilon_1}} & (j=1)\\ \frac{\omega_p}{\sqrt{1+\varepsilon_1}} & (j=\infty)\end{array}\right. \quad (2.140)$$

This equation correlates LSPPs with SPPs: As a particle increases its diameter up to infinity to cause the polarization modes increase to also infinity, LSPPs go back to SPPs, or reversely, if a planar surface is bended and shrunken to a size of a dipole, SPPs shrink also to LSPPs. SPPs can also be considered as a film or surface tightly arrayed by infinitive nanoparticles (NPs at $p = \sum_{m=1}^{m=\infty}(x_m^0)$. In the case of LSPPs, a nanoparticle behaves really like an optically absorptive molecule, with the maximum absorption wavelength at

$$\lambda_{L_{SPP},\max} = \lambda_P\sqrt{1+\varepsilon_1(1+j)/j} \quad (1 \le j << \infty) \quad (2.141)$$

Thus the maximum absorption wavelength is dependent on the relative permittivity of medium where the particle is suspended. This implies that we can sense the medium property (e.g., concentration that is a function of permittivity) by measuring the absorbance of metal nanoparticles.

Another important feature is that a nanoparticle can be polarized by an unrestricted electric field acting on any a direction. The polarization remains a collective oscillation of whole the valence electrons in a particle but only surface can accumulate charges. The arbitrary excitable property not only makes metal nanoparticles themselves be easily manipulated in use but also endows SPPs with four traveling abilities: being reflected by and transmitting through nano-bulges, scattering and/ or emitting into far space as illustrated in Fig. 2.13. Except for the last case, the first two cases are near field propagations, and they may interfere with each other. LSPPs can also enhance Raman scattering and fluorescence emission due to the local concentration of polarized electric field.

The optical absorption of nanoparticles is size-dependent, normally appearing redshift with size. Thus, their shape will impact on the adsorption because differently shaped particles have different equivalent sizes. In fact, metal nanoparticles are more

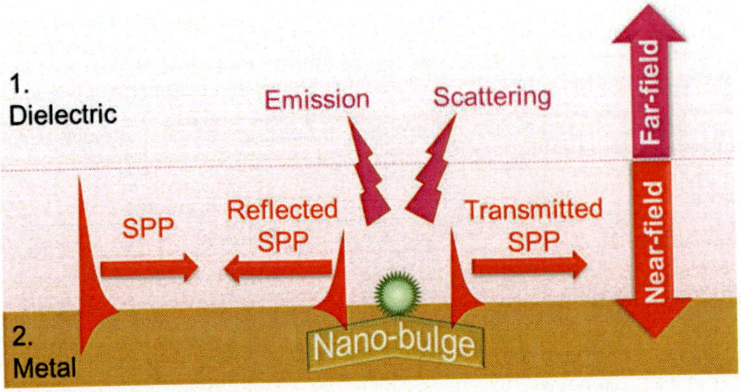

Fig. 2.13 Reflection, transmission, scattering and emission of a propagating SPP along a metal surface with a nano-bulge on its way

or less distorted in shape. A reasonable approximation to this deform and for nano-rods is to replace the nano-spheres with nano-ellipsoids (Fig. 2.12b). Their solutions can be obtained through modification of the spherical ones, which has been finished by Gans [3]:

$$\begin{cases} \alpha_{\text{ext}}(\lambda) = \frac{2\pi V\sqrt{\varepsilon_1^3}}{3\lambda}\sum_r \frac{\varepsilon_2''/P_r}{\left[\varepsilon_2'+\varepsilon_1(1-P_r)/P_r\right]^2+(\varepsilon_2'')^2} \quad (r = x, y, z) \\ P_{r=x} = P_{r=y} = \frac{1-P_{r=z}}{2} \\ P_{r=z} = \frac{\zeta^2-1}{\zeta^2}\left(1 - \frac{1}{2\zeta}\ln\frac{1+\zeta}{1-\zeta}\right) \\ \zeta^2 = 1 - \frac{b_x}{b_z} \end{cases}, \tag{2.142}$$

where α_{ext} denotes optical extinction coefficient, P_r is actually polarizability along an axis and b_x and b_z are the short and long radii of the ellipsoid (Fig. 2.12b). An ellipsoid has two characteristic polarizing direction along the short axis and long axis. Correspondingly, it has two maximal extinction wavelength while the longer one shift red as the long axis elongates.

2.4 Surface Plasmon Resonance

Waves, EMWs and optically excited plasma polarizations (e.g., VPPs, SPPs and LSPPs), are a series of vibration-generated propagation events we have introduced, of which SPPs are the focus for this book on the discussion of SPRi. The key is how to excite and resonate with the mixed mode of SPPs. In fact, the theoretical ways and conditions have been revealed already. Although there are different ways to excite SPPs, the visible photon-based excitation is more practical to establish SPRi methods.

2.4.1 Resonance with Light

As shown already, the excitation of SPPs needs a *p*-polarized or TM light, with a frequency and momentum able to match with SPPs. It is easy to select an optical wave with TM mode at a proper frequency because SPPs have lower energy to the light than VPPs (Eq. 2.117). However, as Fig. 2.11b illustrates, SPPs have always greater momentum than photons, $\boldsymbol{k}_{sp} > \boldsymbol{k}_0$. Unless we can either reduce $\boldsymbol{k}_{sp}$ or enlarge $\boldsymbol{k}_0$ to make them equal, the excitation and resonance are not probable.

Although there are means to shift $\boldsymbol{k}_{sp}$, for example, on a rough rather than smooth metal surface, it is better to adjust $\boldsymbol{k}_0$ in practice to have its *x*-component along a metal surface, $\boldsymbol{k}_{0,x}$, that can be changed by either rotation (Fig. 2.14a) or translation (Fig. 2.14b).

Rotation of a wave vector can be achieved through TIR mechanism or scattering of a wave from nanoparticles or nano-structures, while shifting of a wave vector is possible by making use of some regular nano-structure such as grating to cause different orders of optical scattering and diffraction. Once a wave vector changes its incident angle θ_{in} at above θ_c in a dense medium of ε_{in} or projection angle (i.e., θ_{in}) in a medium of ε against nanoparticles, its *x*-component will be enlarged by a factor of $\sqrt{\varepsilon_{in}}\sin\theta_{in}$. In general, the rotating and shifting regulations can be expressed in one equation according to Eq. 2.15 and $\omega = 2\pi\nu = 2\pi/T = 2\pi\upsilon/\lambda$:

$$\begin{cases} \boldsymbol{k}_{in,x} = \boldsymbol{k}_0\sqrt{\varepsilon_{in}} \pm j\boldsymbol{k}_g \ \text{ or} \\ k_{in,x} = k_0\sqrt{\varepsilon_{in}}\sin\theta_{in} \pm jk_g(x) = \frac{2\pi}{\lambda_0}\sqrt{\varepsilon_{in}}\sin\theta_{in} \pm jk_g(x)\,. \\ j = 0, 1, 2, \ldots \end{cases} \tag{2.143}$$

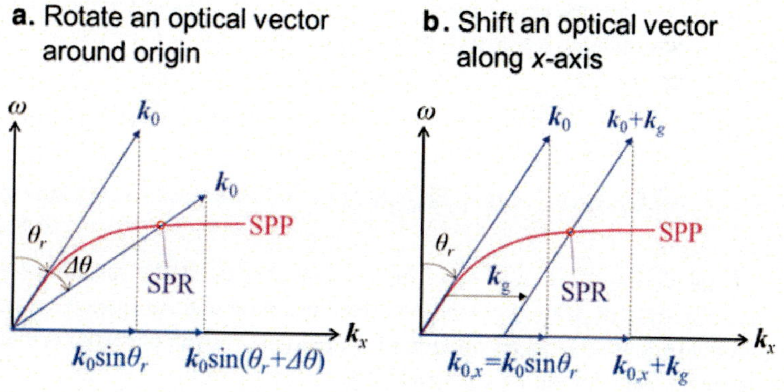

Fig. 2.14 Two theoretic ways to adjust an optical vector by **a** clockwise rotating the incident vector for an angle of $\Delta\theta$ or **b** parallel shifting the vector for a distance of k_g

2.4.1.1 Prism- and Lens-Based Resonance

A prism made of a dense transparent material such as glass can easily cause TIR of light, simply by adjusting the light incident angle at $> \theta_c$. It can be imagined that the evanescent wave of the incident light may have chance to interact with an SP just near the bottom of the prism. Once the SP is polarized and in turn the corresponding SPPs resonate with the excitation light, the energy may transfer between the light and SPPs, accompanied by attenuated TIR (ATIR or ATR) of the light. There are two basic theoretical ways to excite SPP through evanescent field component: excitation of SPPs through an optical tunneling gap of ε_1 (Fig. 2.15a) or direct excitation of SPPs on a metal membrane deposited on the prism bottom (Fig. 2.15b1). The former structure is feasible to directly study the features of a metal surface or a film on the metal surface, while the latter one is very feasible for building up various SPR sensors and imagers. This configuration can also be varied into another layout where a high refractive lens or a large numerical aperture objective is used instead of the prism. When an off-axis light beam is projected through the lens at an incident angle $>\theta_c$, its focused beam can also excite and resonate with SPPs just above the objective (Fig. 2.15b2). This type of resonance layout can reach an imaging resolution down to sub-micrometers and is hence called SPRM.

Different from the common TIR where the light energy or intensity keeps unchanged, once resonance happens, all the configurations will cause energy loss of the reflected light against the incident angle or wavelength, with the minimum at the perfect resonance position (Fig. 2.15c). Clearly, a resonance-absorption-caused ATR spectrum is similar to that of molecular absorption spectrum (after inversion of the curves) and will open a novel way to explore analytical ways. The key is how to find the resonance position, so that we can make use of the resonance adsorption.

To this end, we have to make $\boldsymbol{k}_{\text{in}} = \boldsymbol{k}_{\text{SPP}}$. From Eq. 2.143 where $j\boldsymbol{k}_g = 0$ for the ATR configurations, and from Eq. 2.122, we get the scalar equation by use of their parallel components (refer to Fig. 2.15b):

$$\begin{aligned} k_{\text{SPP}} &= \frac{2\pi}{\lambda_0}\sqrt{\frac{\varepsilon_1\varepsilon_2}{\varepsilon_1+\varepsilon_2}} = k_{\text{in},x} \\ &= |\boldsymbol{k}_{\text{in}}|\sin\theta_{\text{in}} = |\boldsymbol{k}_0|\sqrt{\varepsilon_{\text{in}}}\sin\theta_{\text{in}} = \frac{2\pi}{\lambda_{\text{in}}}\sin\theta_{\text{in}} = \frac{2\pi}{\lambda_0}\sqrt{\varepsilon_{\text{in}}}\sin\theta_{\text{in}}. \end{aligned} \quad (2.144)$$

Considering $\lambda_{\text{in}} = \lambda_{\text{SP}} = \lambda_r$ or $\theta_{\text{in}} = \theta_{\text{SP}} = \theta_r$ at the resonance point, where the subscripts of *SP* and *r* denote surface plasmon and resonance, respectively, two practical equations can be obtained for the interrogation of resonance angle θ_{SP} and resonance wavelength λ_{SP}, respectively:

$$\begin{cases} \theta_{\text{SP}} = \theta_r = \theta_{\text{in,nadir}} = \sin^{-1}\left(\frac{1}{\sqrt{\varepsilon_{\text{in}}}}\sqrt{\frac{\varepsilon_1\varepsilon_2}{\varepsilon_1+\varepsilon_2}}\right) \\ \lambda_{\text{SP}} = \lambda_r = \lambda_{\text{in,nadir}} = \lambda_{0,\text{in}}\sin\theta_{\text{in}}\sqrt{\frac{\varepsilon_1+\varepsilon_2}{\varepsilon_1\varepsilon_2}} \end{cases}. \quad (2.145)$$

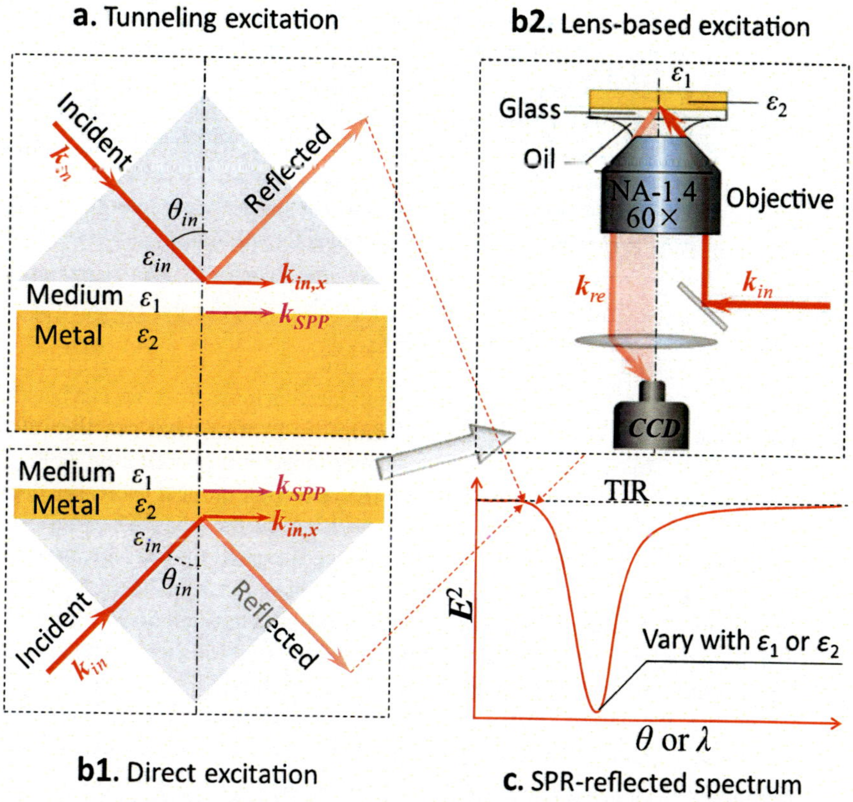

Fig. 2.15 Attenuated total reflection-based configurations to excite SPPs by light. **a** Otto configuration. **b1** Prism-based Kretschmann configuration. **b2** Lens-based configurations. **c** Attenuated total reflection (ATR) spectra with a negative peak variable with the permittivity of medium 1 and/or metal; $k_{\mathrm{in},x}$, x-Component of incident vector; k_{spp}, Vector of a photons-excited SPP

Equation 2.145 reveals the existence of the resonance angle and resonance wavelength, while Eq. 2.144 suggests the ways to interrogate the resonance nadir position from the reflectance curve plotted against the incident angle θ_{in} of monochromatic light for θ_{SP} or against the wavelength of white light (at a given incident angle above θ_c) for λ_{SP}. For a given configuration with fixed hardware, Eq. 2.145 indicates that the resonance nadir position is dependent not only on ε_1 but also on ε_2. For example, at a given ε_1, the tunneling excitation configuration can measure the change of ε_2 (i.e., metal); and at a given ε_2, the direct excitation configuration can sense the change of ε_1 that may be associated with movable or captured analytes (Fig. 2.15b).

It seems also possible to study the incident medium through measuring the change of $\varepsilon_{\mathrm{in}}$ or n_{in} by fixing ε_1 and ε_2 and relative geometric parameters. This is realizable on the condition that a liquid prism is available. Nevertheless, no report on this study has been published yet to the best of our knowledge. It may take time to explore it.

Practical analysis of analytes has been enabled with the direct excitation configuration by recording the reflectivity change against angle or wavelength [4, 5]. It is easy to find from Eq. 2.145 that the real part of the dispersion function is much more sensitive to the variation of the refractive index than the imaginary [6], and hence, real-time changes of $\mathrm{Re}\sqrt{\varepsilon}$ or refractive index n are focused in practice. Note, the change of propagation constant will also cause other variations of the reflected photons such as phase, amplitude, polarization or spectral distribution. They are also the intrinsic SPR mechanisms for methodological exploration.

2.4.1.2 Waveguide-Based Resonance

If we shrink the prism of Kretschmann configuration into a fiber, we can have another configuration for SPR that can be called optical fiber or more exactly waveguide-based SPR. This is the base of optical fiber biosensors working on SPR [7]. The resonance mechanism remains TIR but the reflection number increases depending mainly on the metal-coated length and position on the fiber. The shortest one is shown in Fig. 2.16a where only the tip surface is coated. In fact, the lateral surface can also be coated by metal (Fig. 2.16b) to increase the sensing surface and hence sensitivity. This lateral coupler can be used for pass through or online applications while the end-coupler suitable for point or point array sensing. It can also increase its sensitivity by lengthening the lateral part of coating.

Besides the optical fiber configuration, other types of waveguide coupler [4, 7] for multi-reflection can also be designed. Figure 2.16c illustrates one schematic configuration where the metal film contacts directly with the surface of waveguide material. It may feasible to establish some sensitive sensors or detectors.

2.4.1.3 Near Field Technology

Similar to Kretschmann configuration, Otto configuration can also be shrunken to a fiber size as shown in Fig. 2.16d where the prism becomes now a conical optical fiber. However, different from the shrinking of Kretschmann configuration, the shrunken Otto configuration must work on scattering and reflection mechanisms: The light inside the fiber will penetrate from the tip and spread out at all directions, and the momentum along the different direction varies. As the tip approaches the metal surface at a distance of near filed that is shorter than the optical wavelength, those photons with matching momentum will excite and resonate with SPPs, causing attenuated reflection and/or scattering of the light to yield a dip spectrum.

This near field SPR can be configured in two different ways: excite the SPPs by projection of the light from beneath or above the metal membrane, and collect the near field wave information by a vertical or inclined fiber. The fiber can now be either solid or hollow made of either glass or even metals.

Because near field coupling technology is normally used in scanning nano-optical microscopy. We will not more concern with this technique in later discussions.

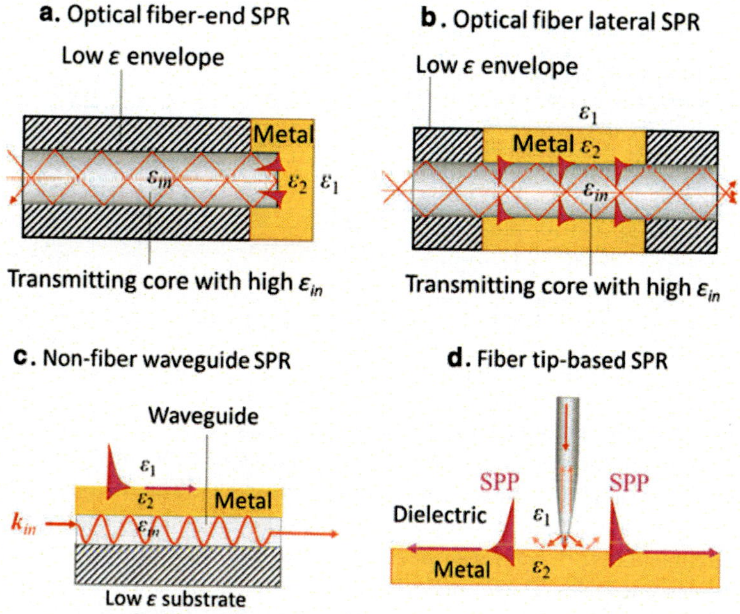

Fig. 2.16 Photon-excitation of SPPs through **a–c** waveguide-based configurations by use of **a**, **b** fiber and **c** non-fiber, or **d** near field technology

2.4.1.4 Localized Surface Plasmon-Bridged Technology

Recall to Fig. 2.13 and consider reciprocal inverse of light travel, we can have a novel way to excite and resonate with SPPs by making use of LSPPs: Projection of a light onto a rough surface of the metal or onto metal nanoparticles embedded on a vast metal surface, the light will first excite LSPPs that in turn excite SPPs and pass the optical energy to the excited SPPs. The excitation mechanism is somehow similar to the near field technology.

Photons projected on a rough or wrinkle surface of metals will be reflected and refracted into various wave vectors, including vector components able to excite and resonate with SPPs, which may enhance the LSPPs-excited SPPs as illustrated in Fig. 2.17. However, the portion of scattered photons are high in this configuration, leading to low excitation efficiency and low measuring sensitivity. Unless on purpose study of surface structure, this method is in theory not suitable for exploitation of chemical and biological sensors.

2.4.1.5 Grating-Based Resonance

Going further from the rough metal surface is the regularly structured metal surface. The typical one is grating that has led to the discovery of SPP phenomenon by Wood.

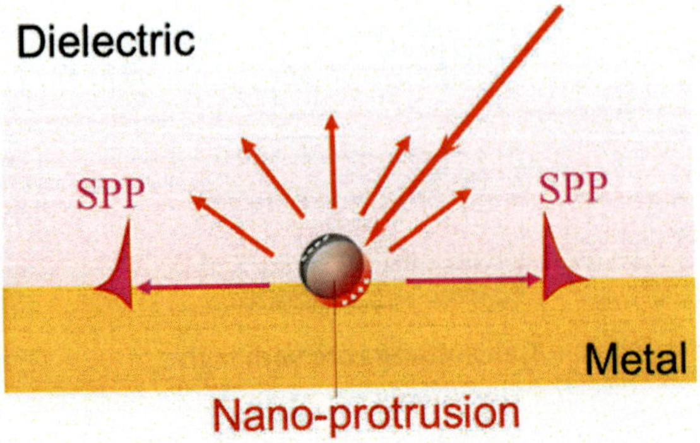

Fig. 2.17 Excitation of SPPs by LSPPs that are excited by a far field optical beam projected directly on a rough and/or nano-structured metal surface

In this world, gratings now include 1D, 2D and 3D formats. The most typical 3D gratings are various crystals which results in Bragg diffraction. Except for 3D and 2D gratings, 1D gratings are often used in spectrometry. The basic feature of a grating is its periodic structure along at least one dimension that diffracts light beam into multiple orders of optical stripes (Fig. 2.18a) with both of far field propagating and evanescent waves. The $\boldsymbol{jk_g}$ in Eq. 2.143 is now not necessarily be zero but is a function of grating constant d_g and diffraction order j (Fig. 2.18a). Note d_g is in general a position vector, but along x-direction, we can let $k_g = 2\pi/d_g$, the wave vector of the jth diffraction must adjust Eq. 2.143 to become:

$$\begin{cases} \boldsymbol{k}_{j,x} = \boldsymbol{k}_{\mathbf{in},x} \pm j\boldsymbol{k}_g = \boldsymbol{k}_{\mathbf{in}} \sin\theta_{\mathbf{in}} \pm j\frac{2\pi}{\boldsymbol{d}_g} \text{ or} \\ \boldsymbol{k}_{j,x} = |\boldsymbol{k}_{\mathbf{in}}| \sin\theta_{\mathbf{in}} \pm j\frac{2\pi}{d_g} = |\boldsymbol{k}_0|\sqrt{\varepsilon_{\mathbf{in}}} \sin\theta_{\mathbf{in}} \pm j\frac{2\pi}{d_g} \\ j = 0, 1, 2, \ldots \end{cases} . \quad (2.146)$$

where $\boldsymbol{k}_{\text{in},x}$ represents again the projected-in wave vector that is smaller than the vector of SPPs in a same dielectric, but on gratings, it can be shifted to $\boldsymbol{k}_{j+1,x}$ along x-axis to acquire an extra distance of $\boldsymbol{jk_g}$ (Fig. 2.18b). This wave vector-shifting technology is intrinsically different from that of ATR. After shifting, the diffraction vector can meet with $\boldsymbol{k}_{\text{SPP}}$ at the cross point (the small red circle in Fig. 2.18b) to resonate with each other or to obey Eq. 2.146. At the resonance point, we have $\boldsymbol{k}_{j,x} = \boldsymbol{k}_{\text{SP}}$ and hence following formulae:

$$\begin{cases} \theta_{\text{SP}} = \sin^{-1}\left(\frac{1}{\sqrt{\varepsilon_{\text{in}}}}\sqrt{\frac{\varepsilon_1\varepsilon_2}{\varepsilon_1+\varepsilon_2}} \pm j\frac{2\pi\lambda_{0,\text{in}}}{\sqrt{\varepsilon_{\text{in}}}d_g}\right) \\ \lambda_{\text{SP}} = \lambda_{0,\text{in}}\left(\sin\theta_{\text{in}} \pm j\frac{2\pi}{d_g}\right)\sqrt{\frac{\varepsilon_1+\varepsilon_2}{\varepsilon_1\varepsilon_2}} \end{cases} . \quad (2.147)$$

Clearly at a given order of diffraction, the optical incident angle or wavelength varies with ε_1 (and/or ε_2), which is similar to ATR configurations. While at a given θ_{in} or λ_{in}, the diffraction order j can also be a function of ε_1, which is different from ATR couplers.

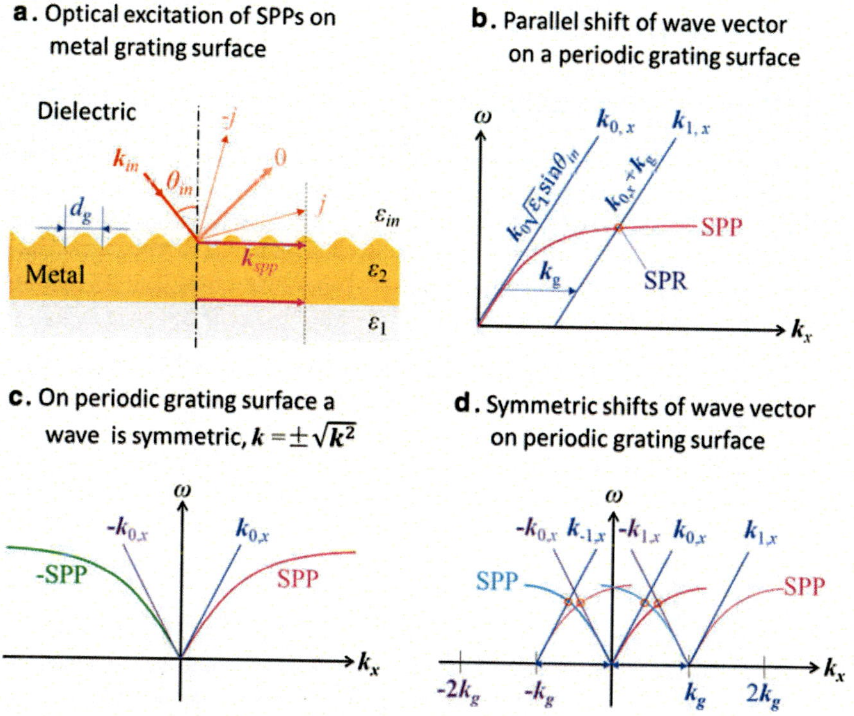

Fig. 2.18 Photon-excitation of SPPs on a metal grating with a grating constant at d_g. **a** Rotating the incident vector at different diffraction direction. **b** Shift the diffracted wave vector along $+x$-axis. **c** Symmetric diffraction. **d** Shifting diffraction vector periodically at jk_g

Furthermore, on the periodic grating surface, propagation dispersion symmetrizes k_g (Fig. 2.18c), which can also be inferred form Eq. 2.112 that indicates that all wave vectors have actually two roots, positive and negative. The periodic structure can fold up the dispersion, making each shifting point become a new origin of SPPs, which is equivalent to horizontal shift of SPP dispersion curve. This and the shift of $k_{j,x}$ make grating-based SPPs acquired multiple resonant points (Fig. 2.18d), much different from the ATR couplers that have only one. It is also due to these increased resonance points to distribute the energy that a grating-based SPR devices lower their sensitivity.

It is worth of mentioning that all the discussed optical resonance coupling configurations can be used reciprocally to make use of SPPs leaking optical emissions.

2.4.2 Resonance with Electrons

It is reasonable that "dropping" of electrons into the electron "sea" seems more easily to induce electronic wave than that of photons. In fact, electrons are more powerful than photons to excite SPPs attributed to the collective electron oscillations [8, 9]. The earliest demonstration of electron-excited SPPs may be traced back to 1957 when Powell and Swan [10] found the energy losses of electrons after their passing through an Al membrane. The earliest purposive excitation of SPPs by electrons was reported in 1977 by Lencante and Ballu [11] who used parabolic trajectory electrons to excite SPPs. The basic principles for electrons to excite SPPs are as follows:

(i) Electrons are easily scattered into various directions as they are moving through a metal, which makes their surface momentum sufficiently available to excite SPPs.
(ii) Electrons with much larger particle mass than photons can have much higher energy and larger momentum in vacuum. They only need to convey a small part of their energy and momentum to polarize SPs. Momentum matching is not more an issue so that vertical projection of electrons (Fig. 2.19a) becomes implementable in practice.

Clearly, the electrons are thus allowed to be projected onto a metal at any an angle, and the most convenient angle is naturally vertical [12] while the most ideal direction to project electrons is along the surface [13]. In the case that an electron beam skims the metal surface at a velocity of $\boldsymbol{v}_{e,x}$, it creates various evanescent fields ready for excitation of and resonance with SPPs at $\boldsymbol{k}_{e,x}= \omega_e/\boldsymbol{v}_{e,x}= \boldsymbol{k}_{\text{spp}}$:

$$\frac{\omega_e}{v_{e,x}} = \frac{\omega_{\text{spp}}}{c}\sqrt{\frac{\varepsilon_2}{1+\varepsilon_2}}. \tag{2.148}$$

For electrons vertically projected on the surface, only those scattered equivalently at an angle of θ_{sc} to have an x-axis component of $\boldsymbol{k}_{e,x}$ can excite SPPs. According to Fig. 2.19a, the related excitation condition is $\boldsymbol{k}_{\text{spp}}= \boldsymbol{k}_{e,x}= \boldsymbol{k}_{e,\text{sc}}\sin\theta_{\text{sc}}= -\boldsymbol{k}_{e,r}\sin\theta_r$, where $\boldsymbol{k}_{e,r}$ is the minimum vector of electrons recorded from the other side of SPPs at an angle θ_r which corresponds to resonant angle. The following equation can thus be obtained:

$$\left|\frac{\omega_e}{v_{e,-x}}\sin\theta_r\right| = \frac{\omega_{\text{spp}}}{c}\sqrt{\frac{\varepsilon_2}{1+\varepsilon_2}}. \tag{2.149}$$

Except for the difference of energy, the parallel excitation yields coherent SPPs while the vertical not. Nevertheless, the parallel excitation has till now much less studied than the vertical excitation due to the manipulation problem in early studies that used the electron microscopes (i.e., scanning electron microscope).

Some recent studies have reduced this issue to some degrees. In 2006, Bashevory et al. [8] have illustrated a layout to excite SPPs-based emission by direct injection of electron beam into nano-structured metal surface (Fig. 2.19b). In 2011, Wang et al.

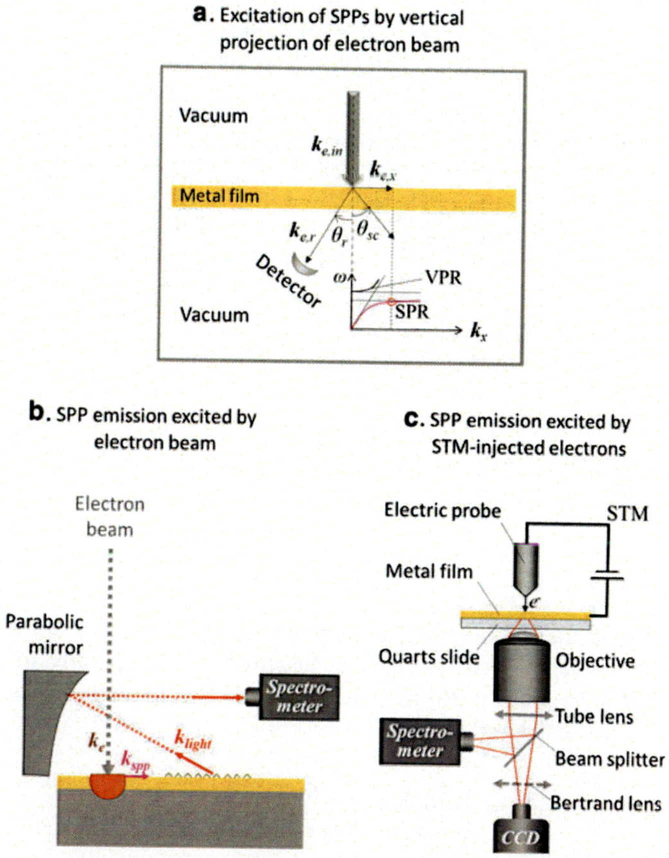

Fig. 2.19 Configurations for electron excitation of SPPs and their emissions by **a**, **b** direct electron bombardment or **c** STM tip injection of electrons

[14] reported a bit easier technique to excite SPP emission by injection of electrons directly into a smooth metal surface from an STM probe tip (Fig. 2.19c).

In theory, electronic excitation is advantageous over photons. As mentioned, it does not need to match their momenta for an electron that has a very large momentum. This simplify the excitation coupling structure. However, it is not easy and convenient to manage electrons compared with photons because electrons should move inside a high vacuum environment. They are still under studies, far from real applications, and will not be discussed further in this book.

2.5 Surface Plasmon Polarizations Scattered by Discrete Molecules and/or Particles

This topic has been lightly touched in Sect. 2.4.1.4 where the metal surface is embedded with metal NPs. It can be generalized as discrete particles or molecules on a metal membrane surface. These discrete objects will affect the propagation of SPPs by scattering mechanism (Fig. 2.20), which is different from the continuous layer of molecules or particles that join in the resonance with SPPs (through affecting the effective value of ε_1). The scattered wave ($\boldsymbol{E}_{sc}$) will further interfere with the propagating SPPs, yielding unique interference fringe pattern with a parabolic tail along both sides of propagation. A back-reflecting tail may also appear if back-reflection SPPs occurs.

A scattered SPP (Fig. 2.20) may couple with another SPP, which in turn changes the electric current density in Maxwell's equation as follows [15]:

$$\boldsymbol{J}_f = -i\omega \sum_{n=1}^{N} \boldsymbol{p}_n \delta(\boldsymbol{r} - \boldsymbol{r}_n), \quad (2.150)$$

where $\boldsymbol{p}_n$ is the dipolar moment of the nth dipole, $\boldsymbol{r}_n$ denotes the adsorption site of the nth dipole, and δ ($\boldsymbol{r}$) is the delta function. Considering free of magnetic effect ($\nabla \times \boldsymbol{H} = 0$) and $\boldsymbol{E} = \boldsymbol{E}(\boldsymbol{r})e^{i\omega t}$, the displacement current density associated with a single particle is given by

$$\begin{cases} \boldsymbol{J}_D = \frac{\partial}{\partial t} \boldsymbol{p}_p \delta(\boldsymbol{r}) = -iJ\boldsymbol{E}_{ex}\delta(\boldsymbol{r}) \text{ or} \\ \frac{\partial}{\partial t} \boldsymbol{J} = -i\omega J \boldsymbol{E}_{ex}\delta(\boldsymbol{r}) \end{cases}, \quad (2.151)$$

where $\boldsymbol{p}_p$ is the effective dipole moment of a single particle at the adsorption site, and J is the effective magnitude of the polarization current excited by an effective electric field $\boldsymbol{E}_{\text{eff}}$. From Eqs. 2.12, 2.14 and 2.16, we have

$$\nabla \times (\nabla \times \boldsymbol{E}) = \nabla \cdot (\nabla \cdot \boldsymbol{E}) - \nabla^2 \boldsymbol{E} = i\omega\mu \boldsymbol{J} + \omega^2 \mu\varepsilon \boldsymbol{E} \quad (2.152)$$

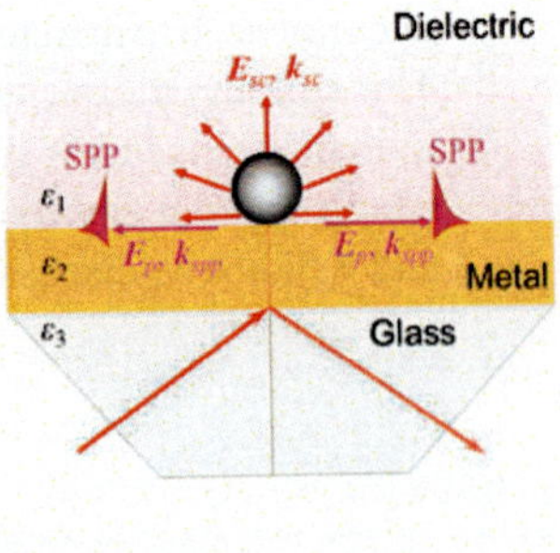

Fig. 2.20 Propagating and scattering of SPPs on a metal surface with a particle

or

$$\begin{cases} (\nabla^2 + k^2)\boldsymbol{E} = -Q\boldsymbol{E}_{\text{eff}}\delta(\mathbf{r}) \\ Q = \omega\mu J \end{cases}, \tag{2.153}$$

where Q denotes the source term. Let $g\boldsymbol{E}_{\text{eff}} = \boldsymbol{E}$, the solution to Eq. 2.153 can be obtained via Green's equation

$$(\nabla^2 + k^2)g = -\delta(\mathbf{r}) \tag{2.154}$$

and multiplying Q. By focusing on a planar wave that propagates at the surface of $z = 0$ free of angle dependency, the solution will concern only with $\boldsymbol{r}$ component that is basically the reflection of the surface

$$\boldsymbol{E}_r = \boldsymbol{E}_{\text{eff}}\left| Q \int\limits_{v} g(\boldsymbol{r}')\delta(\boldsymbol{r} - \boldsymbol{r}')\mathrm{d}v \right| = \boldsymbol{E}_{\text{eff}}|Qg(\boldsymbol{r})|. \tag{2.155}$$

The reflection wave consists of multiple contributions such as, particle-surface interaction, SPP propagation, light polarization. Let $\boldsymbol{E}_{\text{spp}}$ be the solution for the field of propagating SPPs on a particle-free layering system, and $\boldsymbol{E}_{\text{sc}}$ be the solution for the field scattered by the discrete objects deposited on the sensor surface, the following equation can be obtained by neglecting other contributions:

$$\begin{cases} \boldsymbol{E}_r = \boldsymbol{E}_{\text{eff}}\left| Q'_{\text{spp}}g_{\text{spp}} + Q'_{\text{sc}}g_{\text{sc}} \right| = \boldsymbol{E}_{\text{spp}} + \boldsymbol{E}_{\text{sc}} \\ \boldsymbol{E}_{\text{spp}} = -i\boldsymbol{E}_{\text{eff}}Q'_{\text{spp}}\sqrt{\frac{\pi}{2k^2}}e^{-ik|x-x_0|} \\ \boldsymbol{E}_{\text{sc}} = \boldsymbol{E}_{\text{eff}}\frac{Q'_{sc}}{4\pi^2}\int\limits_{R^2}\frac{e^{-i\omega r}}{(\omega^2-k^2)}\mathrm{d}\omega^2 \end{cases}, \tag{2.156}$$

where Q' denotes the source factor and g denotes Green's function. With the aid of Bessel functions J_0

$$\begin{cases} \frac{1}{\pi}\int\limits_0^{\pi} e^{\pm i\omega_r r\cos\alpha}\mathrm{d}\alpha = J_0(\omega_r r) \\ \int\limits_0^{\infty}\frac{J_0(\omega_r r)}{\omega_r^2+(ik)^2}\omega_r\mathrm{d}\omega_r = K_0(kr) = -\frac{i\pi}{2}H_0^{(2)}(kr) \end{cases}. \tag{2.157}$$

$\boldsymbol{E}_{\text{sc}}$ can be simplified in form as

$$\boldsymbol{E}_{\text{sc}} = \boldsymbol{E}_{\text{eff}}\frac{Q'_{\text{sc}}}{2\pi}K_0(ikr) = -\boldsymbol{E}_{\text{eff}}\frac{iQ'_{\text{sc}}}{4}H_0^{(2)}(kr), \tag{2.158}$$

where $H_0{}^{(2)}$ is the zeroth order Hankel function of second kind [16]. The full solution in Cartesian coordinates can be obtained by use of $r^2 = x^2 + y^2$:

$$
\begin{aligned}
E_r =& E_{\text{eff}}\left| i Q'_{\text{spp}}\sqrt{\frac{\pi}{2k^2}} e^{-ik|x-x_0|} + \frac{i Q'_{\text{sc}}}{4} H_0^{(2)}\left(k\sqrt{(x-x_0)^2+(y-y_0)^2}\right)\right| \\
=& E_{\text{eff}}\left| i \frac{Q_{\text{spp}}}{k} e^{-ik|x-x_0|} + i Q_{sc} H_0^{(2)}\left(k\sqrt{(x-x_0)^2+(y-y_0)^2}\right)\right|,
\end{aligned}
\tag{2.159}
$$

where Q_1 and Q_2 are field enhancing factors.

Here, Eq. 2.158 remains complex for calculation. To simplify it, $\boldsymbol{E}_{\text{sc}}$ can be considered to be proportional to $\boldsymbol{E}_{\text{spp}}$ according to elastic scattering theory for a discrete object much smaller than SPP wavelength [17–21]:

$$
\boldsymbol{E}_{\text{sc}}(\boldsymbol{r}) = \gamma \boldsymbol{E}_{\text{spp}}(\boldsymbol{r}_0) e^{-i\boldsymbol{k}\cdot(\boldsymbol{r}-\boldsymbol{r}_0)} e^{-\boldsymbol{\kappa}\cdot(\boldsymbol{r}-\boldsymbol{r}_0)},
\tag{2.160}
$$

where γ is the fraction of $\boldsymbol{E}_{\text{spp}}$ being scattered, $\boldsymbol{r}_0$ is the location of the scattering object and $\boldsymbol{\kappa}$ is the decaying constant of SPP that needs to make the decay within a short distance. Thus, Eq. 2.156 can be written as

$$
\boldsymbol{E}_r(\boldsymbol{r}) = \boldsymbol{E}_{\text{spp}}(\boldsymbol{r}) + \gamma \boldsymbol{E}_{\text{spp}}(\boldsymbol{r}_0) e^{-i\boldsymbol{k}\cdot(\boldsymbol{r}-\boldsymbol{r}_0)-\boldsymbol{\kappa}\cdot(\boldsymbol{r}-\boldsymbol{r}_0)}.
\tag{2.161}
$$

In theoretical calculations, $\boldsymbol{E}_{\text{spp}}(\boldsymbol{r}_0)$ may be approximated as $\boldsymbol{E}_{\text{spp}}$® by neglecting its higher order of variations. In case of multiple discrete objects, the scattering term should be added up to give

$$
\boldsymbol{E}_r(\boldsymbol{r}) = \boldsymbol{E}_{\text{spp}}(\boldsymbol{r}) + \sum_{j=1} \gamma_j \boldsymbol{E}_{\text{spp}}(\boldsymbol{r}_{j,0}) e^{-i\boldsymbol{k}\cdot(\boldsymbol{r}-\boldsymbol{r}_{j,0})-\boldsymbol{\kappa}\cdot(\boldsymbol{r}-\boldsymbol{r}_{j,0})}.
\tag{2.162}
$$

We have to end this chapter here to save space but hope the above-discussed theory is sufficient to understand SPRi. It should however be noted that more theoretical concepts and principles are still required to develop practical SPRi assays. It is reasonable to discuss them in relative chapters whenever required.

References

1. Drude P (1900) Zur Elektronentheorie der metalle. Ann Phys 306(3):566–613. https://doi.org/10.1002/andp.19003060312
2. Mie G (1908) Beiträge zur Optik trüber Medien, speziell kolloidaler Metallösungen. Ann Phys 25:377–455
3. Link S, Mohamed MB, El-Sayed MA (1999) Simulation of the optical absorption spectra of gold nanorods as a function of their aspect ratio and the effect of the medium dielectric constant. J Phys Chem B 103:3073–3077
4. Homola J, Ctyroky J, Skalsky M, Hradilova J, Kolaravo P (1997) A surface plasmon resonance based integrated optical sensor. Sens Actuat B 38–39:286–290
5. Homola J, Yee SS, Gauglitz G (1999) Surface plasmon resonance sensors: reviews. Sens Actuat B Chem 54:3–15
6. Homola J (2003) Present and future of surface plasmon resonance biosensors. Anal Bioanal Chem 377:528–539

7. Marazuela MD, Moreno-Bondi MC (2002) Fiber-optic biosensors: an overview. Anal Bioanal Chem 372:664–682
8. Bashevoy MV, Jonsson F, Krasavin AV, Zheludev NI, Chen Y, Stockman MI (2006) Generation of travelling surface plasmon waves by free-electron impact. Nano Lett 6:1113–1115
9. Cai W, Sainidou R, Xu JJ, Polman A, de Abajo FJG (2009) Efficient generation of propagating plasmons by electron beams. Nano Lett 9:1176–1181
10. Powell CJ, Swan JB (1959) Origin of the characteristic electron energy losses in aluminum. Phys Rev 115:869–875
11. Lencante J, Ballu Y (1977) Electron-surface-plasmon scattering using a parabolic nontouching trajectory. Phys Rev Lett 38:36–40
12. Raether H (1988) Surface plasmons on smooth and rough surfaces and on gratings. Springer, New York
13. García de Abajo FJ (2010) Optical excitations in electron microscopy. Rev Mod Phys 82:209–275
14. Wang H, Zhang Y, Yuan X, Chen Y, Yan M (2011) A universal protocol for photochemical covalent immobilization of intact carbohydrates for the preparation of carbohydrate microarrays. Bioconjug Chem 22:26–32
15. Novotny L, Hecht B, Pohl DW (1997) Interference of locally excited surface plasmons. J Appl Phys 81:1798–1804
16. Abramowitz M, Stegun IA (1965) Handbook of mathematical functions. National Bureau of Standards. In: Applied Mathematical Series #55. Dover Publications, New York
17. Hecht B, Bielefeldt H, Novotny L, Inouye Y, Pohl DW (1996) Local excitation, scattering, and interference of surface plasmons. Phys Rev Lett 77:1889–1892
18. Shchegrov AV, Novikov IV, Maradudin AA (1997) Scattering of surface plasmon polaritons by a circularly cymmetric surface defect. Phys Rev Lett 78:4269
19. Yu H, Shan X, Wang S, Chen H, Tao N (2014) Molecular scale origin of surface plasmon resonance biosensors. Anal Chem 86:8992–8997
20. Bozhevolnyi SI (1996) Localization phenomena in elastic surface-polariton scattering caused by surface roughness. Phys Rev B 54:8177–8185
21. Bozhevolnyi SI, Coello V (1998) Elastic scattering of surface plasmon polaritons: Modeling and experiment. Phys Rev B Condens Matter Mater Phys 58:10899–10910

Chapter 3
Instrumentation

SPRi has now a wide spectrum of instruments and devices to suit for various methods. Based on the key parameters of measurement, SPRi instruments can be classified into four categories: angle- and wavelength-interrogation imagers, and optical intensity- or phase-based imagers. Among them, the most often adopted design is a combination of angle-interrogation with optical intensity and the corresponding SPRi devices and instruments output usually monochromatic or grayscale images. Another optional SPRi instruments are fabricated based on wavelength interrogation able to produce real color pictures, and they are hence called color SPRi instruments. Presently, color SPRi is suffering from some loss of sensitivity due to the reduced sensitivity of present commercial color video detectors. For better sensitivity, phase-based SPRi imagers are preferred, but they are seemingly not commercially available yet and are not very easy to fabricate at low cost.

SPRi can also be sorted by its data presenting format such as 3D or 2D SPRi. Aiming at the applications, SPRi can also be sorted into membrane- and non-membrane-measuring classes. Importantly, no matter how many classes of SPRi instruments may have, their structure can be generalized and can thus be discussed in a more reasonable and clearer way as is given in this chapter.

3.1 Sensitivity

For an instrument, the core feature is its sensitivity. By sensitivity in general is meant two concepts: (i) Inherent hardware sensitivity—it is the response of SPRi devices to the variation per refractive index unit (RIU) caused by net variation of analytes in the systems, which depends on the plasmonic excitation modality, signal measuring geometry and noises. Targets-dependent sensitivity—it varies with analytes due to their variable reaction activity, affinity, adsorbability, diffusivity and their interaction ability with photons. The first sensitivity should well be considered during design and selection of SPRi instruments, which is also cost-dependent decision.

Y. Chen, *Surface Plasmon Resonance Imaging*, Lecture Notes in Chemistry 95,
https://doi.org/10.1007/978-981-99-3118-7_3

Particularly in SPRi and SPR as well, the hardware "sensitivity" is often ambiguously referred to the response magnitude of signals (e.g., intensity, resonant angle or wavelength, phase or polarization) due to a unit variation of refractive index normally caused by analytes. This type of sensitivity is first influenced by the exciting light source (e.g., wavelength, monochromaticity and stability) and camera (e.g., resolution, photoelectric conversion efficiency). Thus, the comparisons between or among the different SPRi instruments fabricated based on variable interrogating principles, a term of limit of detection (LOD) is also coined and used more often in SPRS and SPRi. The so-called LOD is meant herein the minimal measurable refractive index of a testing analyte. LOD is dependent not only on instruments but also on environments and especially analytes. It should be noted that the refractive index is temperature-dependent; for example, water varies its refractive index at a level of 10^{-4} RIU/°C. Thus to reach LOD of 10^{-8} RIU, the temperature variation should be kept within 10^{-4} °C [1].

3.2 Basic Instrumental Blocks

The most often used SPRi device is in fact a 2D imager. It aims at taking surface pictures from the reflected light, somehow similar to an optical camera that also records the reflected lights from the surfaces of an object. The difference lies in that SPRi records the reflected light containing evanescent field information rather than far field that is the case of common optical reflection. The common camera can take a picture of an object at any an angle while SPRi not.

In general, an SPRi instrument includes three major units of hardware: optical, sampling and recording units. The optical unit offers and conducts a light beam to a resonant coupler where the light beam excites and interacts with SPPs and analytes via a metal film sensor at an appropriate incident angle, and finally, the optical unit guides the light to the recording unit. The optical unit usually includes four basic parts: a light source, light path, coupler and mechanic system to adjust the angle.

The sampling unit commonly includes a flow cell and a liquid delivery system. In order to record a 2D image of analytes, we need to put the analytes onto a flat sensor surface often sealed inside a flow cell and assembled on a coupler. The liquid delivery system brings target analytes to the flow cell to make them interact with the detection light via the coupler. It should be noted that the flow cell is better thermostatically controlled to have repeatable and reproducible pictures.

The recording unit is mainly composed of an image recorder such as CCD or complementary metal oxide semiconductor (CMOS). It may also include an image displayer and data treating and saving blocks. In a more common case, the recorded pictures are sent to a computer so that we can use computer technology to analyze the data and withdraw the required analytical information.

In fact, a more complete SPRi instrument consists of seven basic units. They are the light source, optical path, SPR coupler, liquid delivery system, image detection

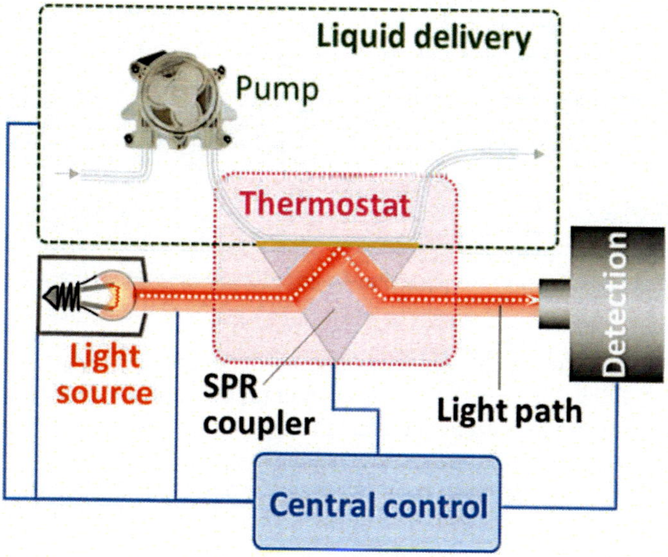

Fig. 3.1 Block diagram of an abstract SPRi instrument

part, thermostat and central controller, as schematically shown in Fig. 3.1. They are to be discussed as follows.

3.2.1 Core Coupler

SPRi needs a coupler that is the core to fabricate an SPRi instrument. It brings the incident light and analytes together on the sensor surface to interact with each other through SPR mechanism. This coupler distinguishes the SPRi instruments from each other and from other optical instruments, especially from the ordinary optical camera.

According to the presented theory, SPPs can be excited by EMWs ranging from X-ray up to terahertz. However, except for X-ray, most photons are unable to penetrate through metals or metal foils and thus cannot directly excite and then couple with SPPs. Their wave vector has to be enlarged through a near field coupler.

As discussed in Chap. 2, there are many different theoretical coupling configurations available, but in this chapter, we discuss only the ATR-based couplers. The very easy version is to use a bunch of fiber-end couplers, aligned tightly in parallel as shown in Fig. 3.2. The significant advantage of a fiber-based SPRi (f-SPRi) is that it produces deformation-free pictures, while its obvious disadvantages are that it greatly lowers imaging intensity and spatial resolution (at least to the size of the outer diameter of a fiber).

In order to improve the lateral resolution to at least a level of micrometers, the fibers should be narrowed to a same size. A simplified alternation to improve the

Fig. 3.2 Core device of an optical fiber-based SPRi with the fibers as pixels

resolution is to "melt" the fibers into an optical prism. Figure 3.3a illustrates an Otto configuration-based imaging coupler [2] assembled with the flow cell. Its major different form SPP excitation configuration shown in the Chap. 2 is the expansion of collimated light beams to cover all the imaging surface. Similarly, Kretschmann configuration-based SPRi coupler [3] can be designed and fabricated schematically shown in Fig. 3.3b. These two couplers are both based in ATR technique and all make use of the evanescent composition of an incident light. The major advantage of Otto configuration is that the metal sensor can easily be fabricated because there is no thickness limitation; nevertheless, the gap thickness of the flow cell has to be finely adjusted to within the range of optical quantum tunneling thickness. The adjustment is not very convenient in practice.

Differently, the Kretschmann configuration does not need to finely control the thickness of the flow cell but needs to precisely control the thickness of the metal film, for example, 50 nm Au. The metal sensing film can directly be deposited on the prism bottom or pre-coated on glass slides in batch to facilitate practical usage. Clearly, the Kretschmann configuration is of easy fabrication and manipulation. Therefore, most of the practical SPRi systems are built up based on Kretschmann configuration and used frequently in many common laboratories. In our studies, Kretschmann configuration is almost the dominated structure in designing various experimental systems and prototypes. It is the focus in this book.

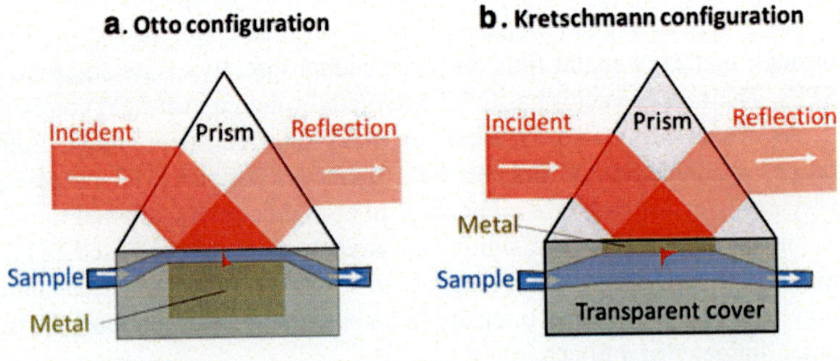

Fig. 3.3 Practical SPR imagers explored from, **a** Otto and **b**Kretschmann configurations

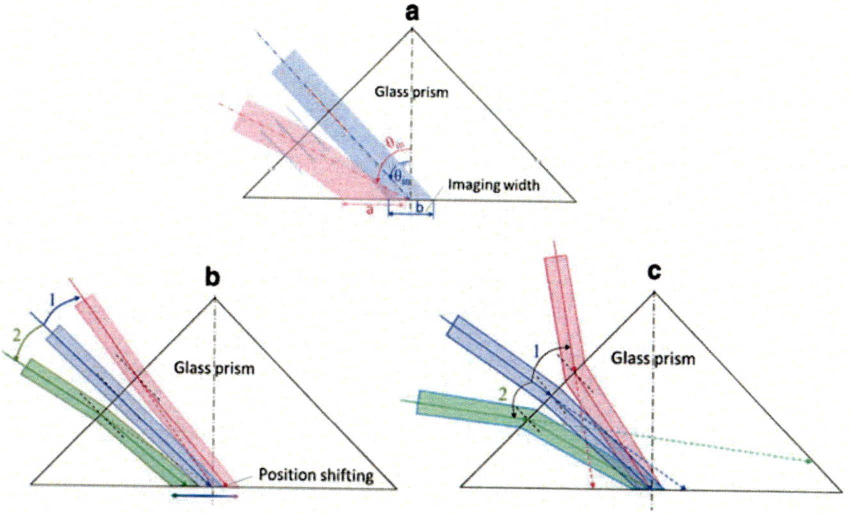

Fig. 3.4 Prism surface refraction-caused undesired shift of excitation position. **a** change of imaging width with incident angle; **b** shift of imaging center with a constant regulation of the incident angle; **c** need of variable adjustment of the incident angle to keep the imaging center unchanged

There is an inherent problem in use of Kretschmann and Otto configurations: They shrink unequally the acquired pictures along the incident direction due to TIR mechanism. Furthermore, the prism surface refraction will also lead to not linear variation of imaging width with the incident angle as shown in Fig. 3.4a, and the incident angle adjustment is also not uniform. Figure 3.4b illustrates that the incident light beam will reach different locations on the prism bottom if we equally regulate the incident angle without considering the surface effect, while Fig. 3.4c demonstrates that the incident angle must be adjusted nonlinearly to keep a same projection location, which is hardly realized in practice.

Through a prism, the optical propagation path for ATR cannot be straightened even within a same beam since it is not possible to make all the light rays penetrate into the prism in parallel to each other. The only possibility is to make the incident interface always perpendicular to the incident light rays. This can hardly be realized in practice by utilization of a hard glass prism unless we use numerous prisms with different angles and replace the prisms once an incident angle is selected. This will not only cost high but be inconvenient in manipulation.

In order to remove these issues more effectively, we have explored a total solution, that is, by use of "soft" prisms made of polydimethylsiloxane (PDMS). However, PDMS deforms light path as it is twisted. To avoid the twisting effect, liquid prisms were tried and proven to work well [4]. There are various optically transparent liquids and solutions with high relative reflection indexes. Examples are the aqueous solutions of glycerol, cellulose and ion liquids. Some organic solvents and/or solutions also possess high refraction indexes. Figure 3.5 illustrates a type of liquid-based ATR coupler where the incident interface (glass slide) is fixed always at 90° to

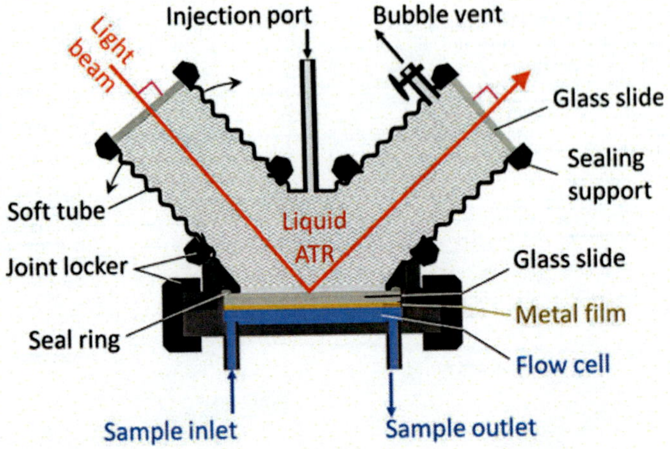

Fig. 3.5 Setup for photon-excitation of SPPs with a liquid prism free of light refraction caused by the prism surface to overcome the glass prism-caused imaging distortions

the light beam and the same is the reflected light beam. It is worth of mentioning that the liquid-based prism couplers may have the problem of bubbles. To eliminate this interference, the coupler needs to have a design of a bubble vent. In order to make the inner bubbles move up and escape out of the liquid, the vent must be located at a sufficiently high and convenient position. It is highly proposed to have an injection port design for filling and later supplying transparent liquid into the coupling area. The injection port is better located at a low position. In short, such an ATR coupler allows the light beam always hitting onto the correct position, without shifting along the incident panel. It should however be noted that this liquid prism remains unable to overcome the image deformation along the incident direction due to the angle-dependent reflection from the prism bottom.

The working principle by all the illustrated ATR couplers is that the incident light beam is guided to the location of metal sensor film attached or deposited on the bottom of the prism or fiber ends to induce evanescent field for exciting and resonating with SPPs, which are affected by the adjacent substances like analytes as they are flowing across the sensor surface or just settle on the surface of the metal, especially through probe capture. The so-called adjacent is meant within a distance shorter than the propagation depth of SPPs, normally less than 200 nm.

Herein, we have shown that a coupler actually consists of three parts, an ATR structure, a metal sensor and a sealable flow cell. The ATR part is made of a transparent substance such as optical glass, plastics like polymethyl methacrylate (PMMA) or various liquids, normally having high refractivity. The metal sensor is actually a thin film deposited directly on the bottom of an ATR structure or first on a glass slide that is then attached to the ATR structure via an index-matching oil or liquid. They can in theory be made of any a metal with negative permittivity. In the range of visible light, gold is presently the most often used metal for it is stable in air and in most of aqueous solutions (except for aqua regia) and solvents (except for ionic liquids and their solutions). Silver is another adoptable metal in the visible light range and shown to be more sensitive than gold, but silver has to be used under protection for anti-oxidization in air and solutions. The oxidization causes SPRi signals to drift with time. Protection of silver is achieved by embedding it between two thin gold films or between gold and polymer films. By protection, some other metals such as Cu and even Al can also be used, with increasing detection sensitivity in most cases.

3.2.2 Light Source and Path

Light source strongly impacts on detection sensitivity. Present SPRi normally uses visible lights. Although lasers are typically monochromatic, spatially coherent and polarized light sources, they cause speckles on rough surfaces and interference patterns from every particle appeared and the edges of optical components in the light path [5]. In addition, the intensity of lasers usually fluctuates relatively large (>1%). Lasers have since been avoided in order to get off obvious interference fringes in the image and intensity fluctuation. Nowadays, a superluminescent light emitting diode (LED) with a life of 10,000 or even 100,000 h is available and can be utilized to overcome the lasers-caused issues. Various LED lights, especially the red lights, are now easily available (with a power up to over 10W) and are cost-effective and compact. To build up reliable SPRi instruments, it is never too late to use a stable and long-life source of lights.

The incident light beam needs to be expanded so as to cover all the imaging area. A light source can either be divergent or be in nature a point, it has to be shrunken or expanded to become a collimated beam. This collimated light is then guided to the sensing part of the coupler. After excitation and resonance with SPPs, the light beam carrying the information of analytes is guided to project onto a camera to record the imaging signals.

Since only the TM mode of light can excite SPPs, a polarizer is required. In principle, it can be inserted in any a position before the light enters into the camera, but in common, it is inserted just after the first lens of collimation to save space. To purify the light source or to narrow the light wavelength, a high quality of bandwidth filter is better used, commonly placed just before the camera to prevent also the environmental lights from entering. The random reflection lights from environments can largely blur the pictures recorded.

In practice, we need to select or adjust the incident angle of the optical beam. Thus, the light path has to be adjustable and commonly is isolated into two parts: The first part corresponds to building up the incident light beam (the left arm in Fig. 3.6a) where all the optical elements are fixed on a rotatable arm so that they can turn simultaneously along a same direction to adjust the incident angle. The second part concerns with the reflected light beam. Similarly, its optical elements are fixed on another rotatable arm to guide the reflected light beam to the camera (the right arm in Fig. 3.6a). According to theoretical study, the two arms must be finely adjustable, at a resolution down to10^{-3} degree or even finer.

An easy-to-think fine angle adjuster is a conjugate precession machinery as shown in Fig. 3.6b where the two loading arms are co-hooked at a rotatable knob (6) through two linking arm. By tuning the frontal rotatable knob right or left, this point will move forward or backward to conjugately widen or narrow the incident and reflect angles of the light beams. The angle resolution is determined by the pitch of the rotatable screw. If you need to adjust the two loading arms independently, the design illustrated in Fig. 3.6c is suggested, where the guideway can also be finely geared.

The advantages of these designs include easy adjustment of the angles and allow intuitive observation of the light rays to turn its direction along with the turning arms, while the disadvantages are that the angle adjusting system occupies a lot of space and the turning points are easily worn to loose. This ruins angle accuracy that is further worsened by the backlash. To increase the adjusting accuracy, the rotary elements and mechanics should be as few as possible. There is in theory a possibility to do so by use of a microscopic objective as an ATR device. Figure 3.7 illustrates a design used in author's laboratory where only the optical holder (9) needs to be adjusted slightly up or down to increase or decrease the incident angle. The microscope objective is commonly selected with high magnification and high refractive index (e.g., 60 ×, with numeric aperture of ca. 1.5). This type of coupler can offer high spatial resolution, down to ca. 50 nm, and is applicable to the study of cellular surface receptors. It is also called SPRM.

Another design is to turn the prism coupler by shaping the prism into a form able to reflect back the light beam as shown in Fig. 3.8a. The optical elements for incident and reflected light paths can thus be fixed or rotation-free as shown in Fig. 3.8b.

How to design the light path depends not only on the coupler but also on the light source. There are several light sources ready for selection. As mentioned, lasers have been gradually abandoned due to the issues of interference fringes and power fluctuation. Incandescent lamps were also tried at the very beginning but not adopted soon due to their short life time (several hundreds of hours). They have been replaced by LED that normally emits quite a thin and parallel light. Usually, an LED light must be expanded first by inserting a convex lens prior to collimation.

It is due to its low cost and compact that LED light sources can easily be arrayed in relatively limited space to vary the incident wavelength or incident angle by electric switches rather than rotation machinery. The cost of using arrayed LEDs is no longer a problem.

The power of a LED is normally selected based on the camera used. A very sensitive CCD detector needs only a low power LED (possibly around 10 mW), while a not very sensitive detection camera asks for above 1 W LED. A common movie camera and even a public monitor video working at natural lights may also become usable with a 2 ~ 10W LED.

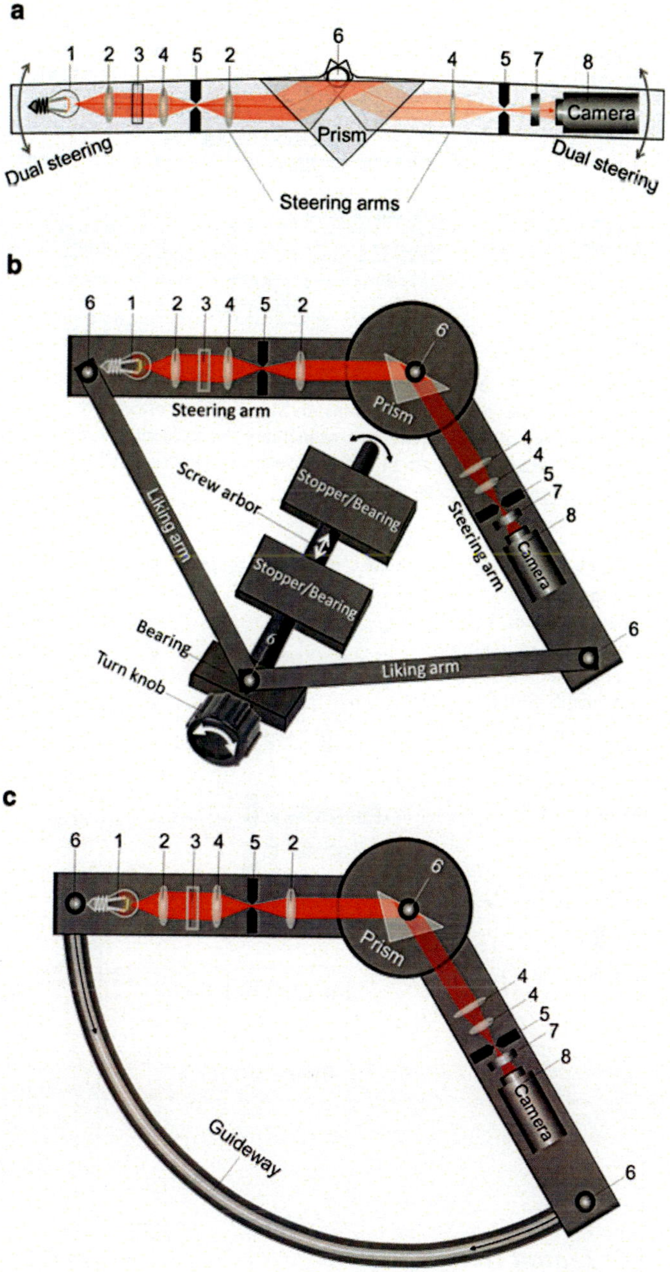

Fig. 3.6 SPRi-used dual steering light path systems with, **a** independent, **b** linked and **c** track-guided moving arms. 1. Light source; 2. Collimating convex lens; 3. Polarizer; 4. Focusing lens; 5. Slit; 6. Turning knob; 7. Bandpass filter; 8. Camera such as industrial CCD or CMOS

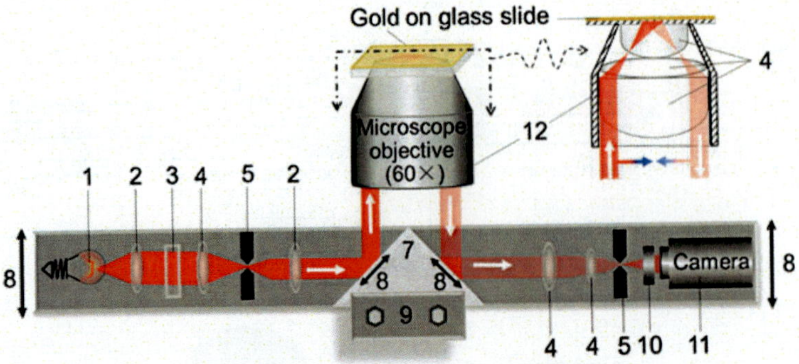

Fig. 3.7 Microscopic objective-based SPR imager or microscope. 1. Light source; 2. Collimating convex lens; 3. Polarizer; 4. Focusing lens; 5. Slit; 7. Total reflector (e.g., 90° prism); 8. Incident angle regulation (up or down to decrease or increase the incident and reflection angles); 9. Holders of reflectors; 10. Bandpass filter; 11. Camera such as industrial CCD or CMOS; 12. Microscopic objective and its sectioned view

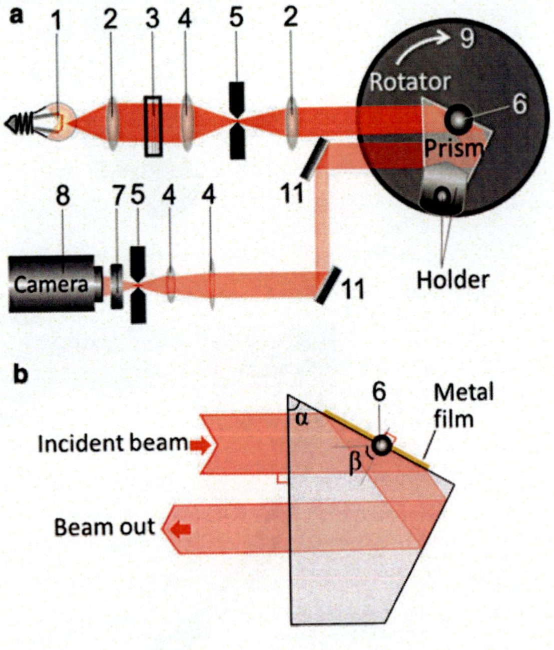

Fig. 3.8 **a** Compact SPR imager by use of, **b** a back-reflection prism. 1. Light source; 2. Collimating convex lens; 3. Polarizer; 4. Focusing lens; 5. Slit; 6. Knob; 7. Bandpass filter; 8. Industrial CCD or CMOS camera; 9. Conjugated angle regulation; 11. Total reflection mirror

3.2.3 *Liquid Delivery Unit*

An analyte has to be brought to the sensor surface for SPRi detection. There are presently at least three ways and/or techniques to make an analyte come down to the surface: The first way commonly adopted is to capture an analyte by its probes previously immobilized on the surface; the second way is to directly make the analyte contact the sensing surface by diffusion through either spotting or solution-flowing technique; and the third way is an *in situ* technique that directly places a hard or soft sample or its sliced sheet on the sensing surface aiming at measuring the surface-absorbed substances or compositions coming adjacent to the surface. They all need a system or means to deliver liquids, aiming at cleaning the sensor surface and/or the flow cell, modifying the sensor surface and bringing target analytes to the measuring area especially the location of their probes.

A liquid delivery system commonly consists of a pump, a pipeline with inlet and outlet, and a flow cell to hold or deposit the analyte solution(s) on the sensor surface. There are two types of pumps often used in SPRi instruments, peristaltic pumps and injection pumps. The former can endlessly pump liquids but delivers pulse flows, while the latter pumps create smooth and stable flows but need to refill the pump after the liquid is used up. To control the flow down to a level of microliters, the injection pumps are better utilized rather than the peristaltics. Injection pumps can nowadays be automatically controlled by a programmable turning mechanism.

The liquid path is guided to the flow cell and out of it by use of inert capillary tubes such as fluoroplastics. Teflon tubes are hence frequently selected but fused silica capillaries are also optional in many cases if their non-specific adsorption effect can be omitted. In a simplified SPRi device built in a laboratory, the liquid path may be value-free, but in many commercial instruments, some values are designed to close or open the flow path. In some cases, reciprocal flow may facilitate to wash off absorbates and/or to speed up the capture of analytes.

The flow cell is aligned on the metal sensor to enclose a space for sample solution to flow through or to stay inside for a while. For easy observation, the flow cell is often made of transparent materials such as glass or organic polymers like PMMA, but to block environmental diffusing reflection lights, the cell is better made of light-proof materials. Furthermore, in order to well control the reaction temperature, the cell can also be made of a metal (at least partially) such as stainless steel to facilitate heat transfer.

The flow cell position and shape influence on the distribution of liquid and especially on the exclusion of air bubbles that easily retain inside a thin and tiny flow cell, no matter where the prim is positioned. A simple way to reduce the bubbles is to erect the flow cell as illustrated in Fig. 3.9a. Modification of the cell surface can further help to suppress the retaining degree of bubbles but not necessarily completely. Optimizing the inner shape of the flow cell to facilitate liquid flowing through is more helpful. One of the effective shapes is streamline with all round corners (Fig. 3.9b, c). Such a complex flow cell can nowadays be fabricated by 3D-printing techniques (further polishing may be required). It may be worth of mentioning that the streamline-shaped flow cell occupies wider space and needs to roughly double the use of liquids but this is worth for having smooth and bubble-free flow.

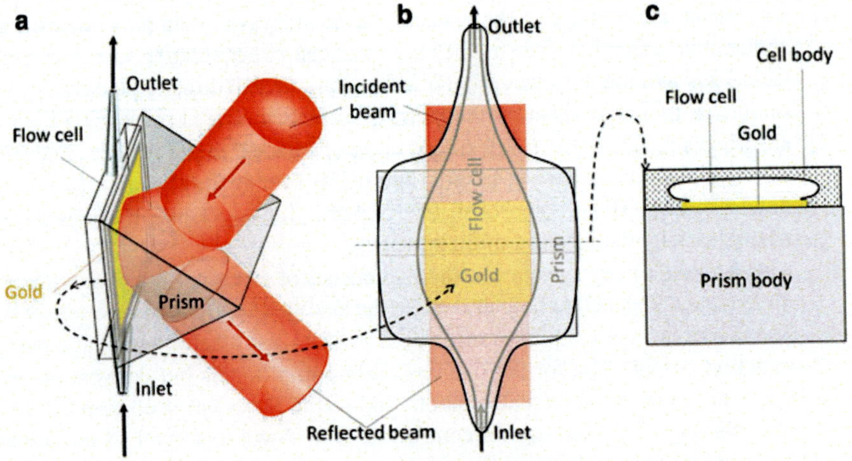

Fig. 3.9 **a** Vertically positioned SPRi configuration assembled with, **b** a streamline-shaped flow cell (top view) having, **c** round edges (cross section) to exclude bubbles

3.2.4 *Thermostat*

It is better to place an SPRi instrument in a clean, air-conditioned laboratory in order to stabilize the imaging signals, but this cannot necessarily ensure its high performance. Some special parts of SPRi have to be thermostatically controlled, with temperature adjustable. This is also important in the investigations of molecular interactions and/or chemical reactions such as nucleic acid hybridization reactions. Temperature control is a basic measure to have repeatable and reproducible determinations.

The most critical part that has to be thermostatically controlled is the flow cell. Two basic techniques may be considered: thermostatically control the liquid (especially sample solutions) just before it flows into the cell or better *in situ* control the whole cell. The former is simply achieved by inserting a section of the delivery pipe (normally a capillary) into a thermostat and insulating the flow cell from environment by a heat-proof sleeve.

The *in situ* thermostatic control can be realized nowadays by use of a semiconductor cooling device. Peltier elements are highly recommended for their very compact and variable size (Fig. 3.10a). With a suitably sized Peltier cooling element, the temperature of a small flow cell can quickly reach a set value and maintained stably. The Peltier element can be placed directly on the flow cell as schematically illustrated in Fig. 3.10b. It is also possible to directly control the temperatures of both the flow cell and the coupler by attachment of two Peltier elements at the non-optical lateral sides of the prism (Fig. 3.10c). This simultaneous control benefits the combination of SPRi with fluorescent and other optical detection techniques and helps to suppress the environmental temperature drifts. The obvious and unwanted disadvantages are that this control takes longer time to stabilize the temperature and

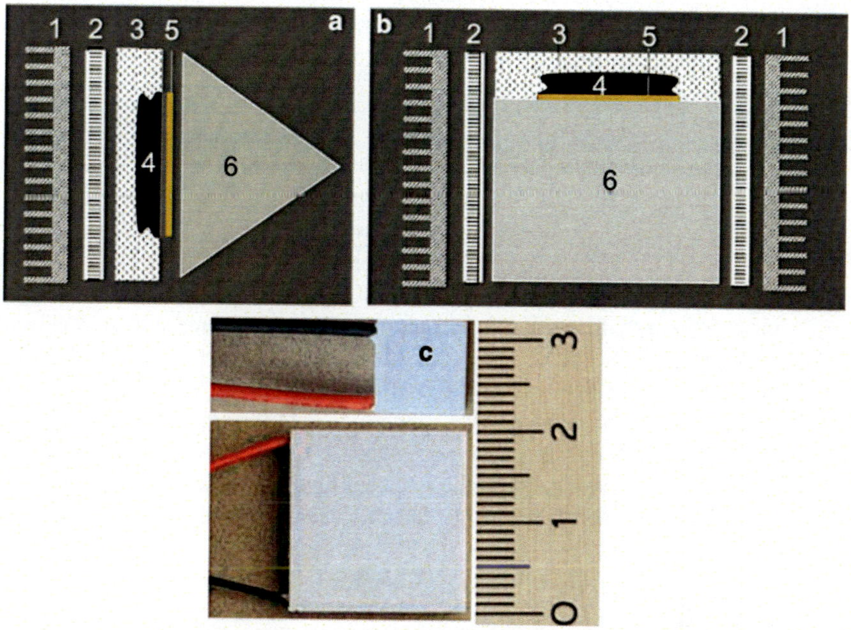

Fig. 3.10 SPRi thermostatic controller. **a** Thermostat placed directly on the flow cell; **b** use of two thermostats to clip the flow cell-prism assembly on the non-optical sides; **c** two adoptable peltier coolers. 1. Heatsink; 2. Peltier element; 3. Flow cell body; 4. Flow cell chamber; 5. Metal sensing film, 6. Prism

a temperature gradient may appear across the flow cell in between the two Peltier elements.

The temperature needed depends on imaging purpose and applications. In most case the temperature of the flow cell maintained at 15–50 °C is sufficient. In some special applications such as to denature double-strand DNA in PCR experiments, the reaction temperature must cover a wide range, up to 95 °C.

The temperature can also be controlled through forced air convection between the thermostatic element(s) and the flow cell/coupler. This design can simplify the structure of SPRi instruments by lowering the control precision somewhat. If the space to be controlled is very wide, the control power is better increased to speed up the process.

3.2.5 Detection Unit

3.2.5.1 Basic Way to Record Imaging Signals

Similar to other imaging techniques, SPRi utilizes a photoelectric converter to detection the imaging signals. The converter can be shaped into a point, a line or a surface. In order to record a surface picture, a point detector needs to scan all the surface points as is commonly adopted in scanning probe microscopy. Due to slow recording rate, this point scanning detection method is not adopted in present SPRi instruments. The linear or linearly arrayed point detectors can largely increase the scanning rate but remain rarely used because nowadays surface array detectors are readily available in the visible light range such as CCD or CMOS cameras. Although well-cooled and extremely sensitive cameras have been produced for scientific researches (e.g., fluorescent imaging), only a common or industrial camera is enough to construct a quite sensitive SPR imager because the reflected light carries sufficient photon flux. This makes the manufacturing of SPRi imagers cheap, simple and easy, thus facilitating the promotion and applications of SPRi methods.

Among various video cameras, CMOS is especially recommended not only for its cost-effectiveness and easy availability but also for its fastness and convenience in use. CCD that is so familiar to everyone nowadays is also often adopted in SPRi devices. In setting up of a very simple SPRi device, a mobile phone can also be used as an image detector, where sophisticated software packages can be used directly and/or subjected to further exploitation for different purposes.

The optimal choice of a camera can be achieved through calculation based on the required sensitivity: For a 2^j-bit camera able to yield a relative signal change of m counts by measuring a testing sample with a refractive index change of n, the corresponding resolution is n/m per count. When taking into account the standard deviation of signal (3σ) during the measurement time (t min), a final sensitivity S_{sen} is given by

$$S_{\text{sen}} = \frac{3\sigma n t}{m} \tag{3.1}$$

Based on this equation, the calculated sensitivity will be 5.5×10^{-7} RIU for a 2^4-bit camera able to yield 1545 counts for $n = 1.2 \times 10^{-4}$ RIU testing sample, with $3\sigma = 2.36$ in 3 min measuring time.

3.2.5.2 Camera Position

According to optical imaging principle, the camera needs to be placed on position 1 as shown in Fig. 3.11a–c. However, this position has never been adopted in practical SPRi instruments because this position needs fine and accurate adjustment of camera orientation for every incident angle (comparing **a** with **b** or **c** in Fig. 3.11). This can hardly be realized or is not necessary in practice. Furthermore, the acquired images are asymmetrically foreshortened along the light path due to the oblique incident angle through the coupling prism. It requires complicate optical setups or computation methods to rescue the distorted images.

The dominant position adopted in present SPRi devices is to position the camera perpendicular to the optical beam (2 in Fig. 3.11a, b), which has also been adopted in SPRS. Except for the optical collection angle, the orientation of the video camera does not require any additional adjustment once the incident angle is changed. This largely facilitates the design and practical control of the instrument. Nevertheless, this position 2 cannot avoid the image foreshortening issue except that the asymmetric distortion is largely reduced. By Fig. 3.11a, b, the foreshortening of an object (ab) can be observed to increase with the incident angle (e.g., from smaller θ_1 to lager θ_2). In case that a liquid ART structure is used to eliminate the prism side surface refraction, the foreshortening rate α of the object ab can be calculated by:

$$\alpha_{x,p_2} = \frac{\text{ab} - (\text{am} + \text{nb})}{\text{ab}} F_{\text{am}} = \frac{\text{ab} - \text{ab}\cos\theta}{\text{ab}} F_{\text{am}} = (1 - \cos\theta) F_{\text{am}}, \quad (3.2)$$

where F_{am} is an amplification factor. For $F_{\text{am}} = 1$, the foreshortening rate is calculated to be 35.7% at an incident angle of 50°. The distortion is significant but can roughly be recovered by computational correction or by a concave cylindrical lens if necessary. Due to the incorrect optical imaging position, the image quality also varies along the x-direction but this variation is negligible compared with the background.

The best but not yet used are the parallel positions, 3 and 4, as indicated in Fig. 3.11c if we do not care about the position that conforms to the principle of optical imaging. The position 4 makes the spatial orientation of images consistent with its object, while the position 3 inverted along the light propagation direction. The obvious advantage of these two positions is that they avoid the foreshortening effect except that the images are zoomed in or out, while the disadvantage compared with the position 2 is that the camera needs to move back or forth horizontally with the incident angle according to following formula:

$$\text{OP} = h \tan\theta_{\text{in}}, \quad (3.3)$$

where OP measures the central distance between the incident beam and the photoelectric element and h is the vertical distance of the camera from the prism bottom (Fig. 3.11c). It should be noted that the camera has to move nonlinearly along the x-direction with the incident angle as shown in Fig. 3.11d. Nevertheless, this is in practice adoptable because real value of the incident angle changes between 40 (glass against air) and 65° (glass against aqueous solutions), where the OP changes from 0.8391 to 2.1445, very close to linearly:

$$\text{OP}/h = 0.0512 \tan\theta_{in} - 1.2986, \quad R^2 = 0.962. \quad (3.4)$$

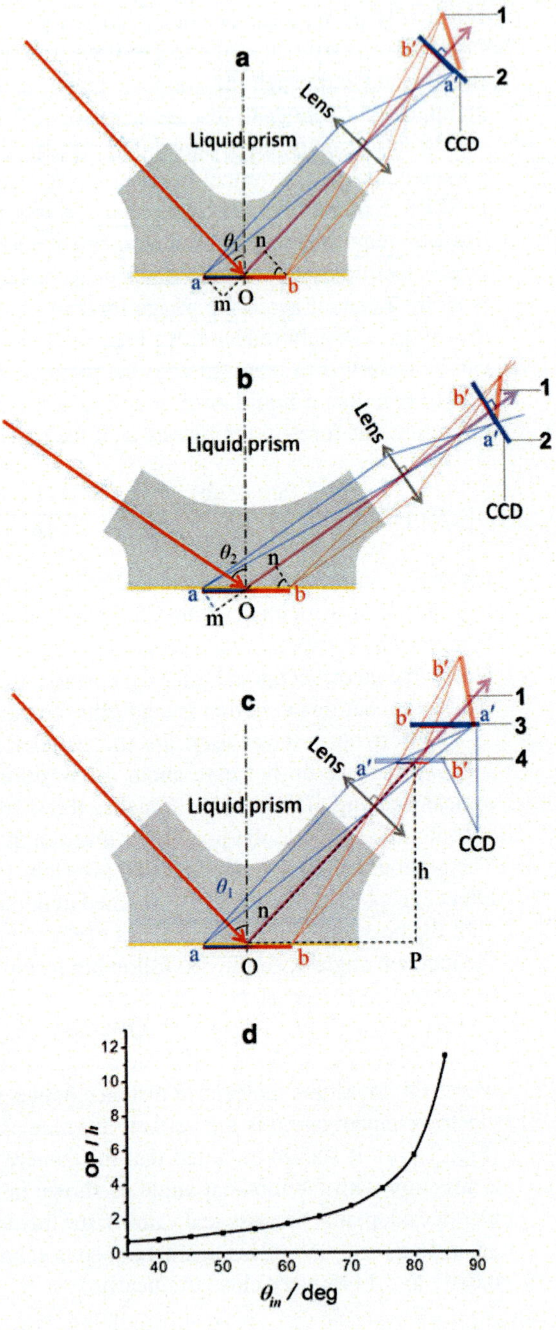

Fig. 3.11 Four potential camera positions for fabrication of an SPR imager 1 **a–c**. Optically correct but hardly adjustable imaging position; 2 **a–c**. Optically incorrect but dominantly used imaging position nowadays; 3 and 4 **c**. Not-yet-reported and optically incorrect but free of image-distortion imaging positions, where the camera needs to slightly shift horizontally according to the curve shown in **(d)**. Note, the position 1 and 2 foreshorten the acquired images along the incident direction

In the case of a prism with $n = 1.50$–1.70 against aqueous solutions with $n = 1.333$, the incident angle will change between 62.8 and 51.64°, or widened to 51–63°, the moving equation will be much more close to linear one:

$$OP/h = 0.06008 \tan \theta_{in} - 1.8622, \quad R^2 = 0.9911. \tag{3.5}$$

For a giving sample, the incident angle will change less than 1°, the linearity will become ideal. Even nonlinear, the horizontal movement can easily be manipulated by sliding, and the correct position can easily be judged or regulated by image position.

3.2.6 Central Control Unit

The central control unit is commonly a computer that has been installed with an SPRi imaging workstation able to control all the actions of machine and to serve as a receiver and processor of imaging data and related signals. The workstation itself can perform post-experimental analysis of the received imaging data. In a better case, it can also act as a virtual SPRi device to conduct pre- and post-experimental studies.

Machine actions are commonly realized by command statements sent through I/O interface board(s). The commands include on/off switches and values, which in turn start some hardware to rotate, move, and to have fluids flow and so forth.

The computers used in SPRi are fairly common such as desktops or a labtops. It is better to have a large memory and storage space because the imaging data are fairly space-occupying. In case to perform numerous reactions, a large memory and fast computing speed are preferred; otherwise, the computer may crash by rushing data.

Compared with the hardware, software is even crucial but you may have not chance to choice if you use a commercial SPRi instrument. All the functions are fixed and you have to bear it until a new version of workstation is issued and you can also afford it. However, in case you can self-design and fabricate SPRi devices, you are able to self-edit the imaging workstation. For control of a machine through Window system, C + + language is proposed to edit the software because it can reach the underlying layer of operation commands. Nowadays, it is highly recommended to edit your workstation via LabVIEW that is easy to learn, possessing high compatibility and powerful edition function. If only imaging itself is concerned, MATLAB is a good choice because it is inherently assigned to treat data matrixes. If you decide to edit yourself a workstation, the following functional blocks are better considered.

(i) Direct control of machine actions such as switch, knob, angle/wavelength selection, pump and so forth;
(ii) Method program including online and offline processes such as imaging, non-imaging, dynamics and/or reactions/interactions;
(iii) Online/offline data processing including imaging and non-imaging, noise suppression, background deduction, format transfer for import/export of data, 2D/3D display, report and print, and so forth;

(iv) SPRS/SPRi-hyphenation related datum-processing block.

The key points to be concerned carefully are: (i) how to take a picture and/or a video film, (ii) how to analyze and convert the image data and (iii) how to sharp the mages through reducing the noise and background. To go a further step, it is also an issue how to display the pictures, for example, by 2D, 3D or even holographic techniques. Noise suppression or filtration of a picture is different from a linear curve.

In practice, there may be various replaceable imaging software available that can directly be adopted in SPRi or adapted to SPRi systems after some modifications. In the easiest case, one can use a mobile phone to record the SPRi signals and then treated by normal picturing software such as Photoshop.

3.3 Summary

In summary, a SPRi device can vary its outer shape but its core structure is more or less unchanged. To demonstrate, Fig. 3.12a, b illustrates two ordinary SPRi systems fabricated in our laboratory for routine analysis. Although both of the models were fabricated based on Kretschmann configuration, the model TX7100 was fabricated according to the design shown in Fig. 3.6b, while the model BX8100 according to Fig. 3.8. Generally, a p-polarized light beam is collimated and directed at an adjustable angle onto the prism bottom where a gold sensing chip is placed via a refraction index-matching oil (i.e., cedar oil) and sealed in a flow cell able to accommodate a certain volume for solution (e.g., 40 μL). The reflected light beam is projected onto a CCD camera (WAT-902B, Watec Co., Ltd., Japan) after filtered through a narrow bandpass filter (normally centered at 645 nm) to record the images in real time. The recording rate can be set at $\leq$ 24 frames per second, then saved and analyzed with a laboratory-edited imaging workstation. The temperature of the flow cell is controlled by a Peltier cooling and heating system, allowing to regulate the temperature from 2 to 95 °C. All the solutions are delivered with a replaceable syringe pump depending on the requirement of flow rate (normally at alevel of μL/min). The imaging size can be adjusted for several times by shortening or lengthening the distance between the CCD camera and the sensor chip.

For exploitation of novel imaging principles and more comprehensive methods, an integrated (up to six in one at present) SPPs-based device was built and used for high resolution (tens of nanometers) imaging analysis, especially for the analysis of discrete macromolecules and/or particles. Figure 3.13 illustrates the major construction consisting of basically four parts: (i) excitation light source that can offer p-light at the desired wavelengths from the expandable or changeable lasers; (ii) inverted microscope that is used for recording the SPR images and for excitation of SPPs; (iii) multi-role microscope that is used to perform bright and/or

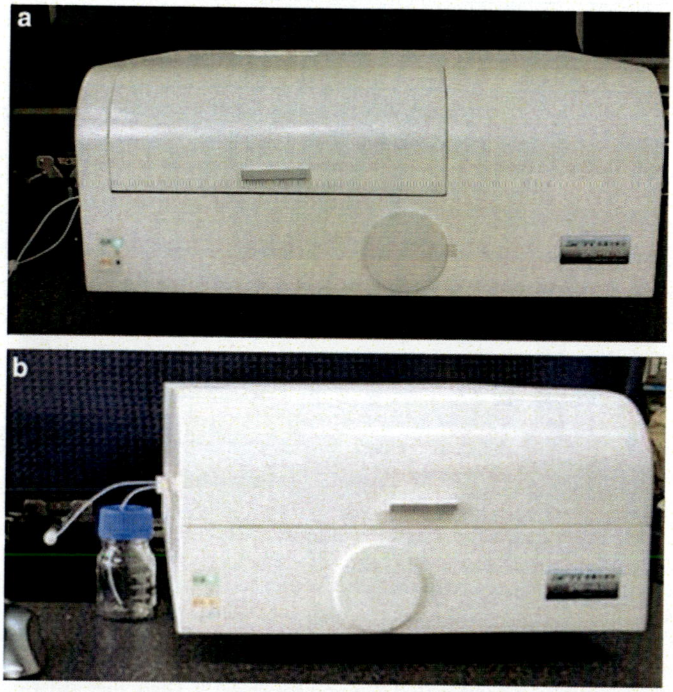

Fig. 3.12 Two models of our laboratory-made wide-angle SPRi imagers. **a** Model SPRi-TX7100 that uses a regular prism; **b** SPRi-PX8100 that utilizes the back-reflection prism

dark field microscopy (normally for comparison or for finding analyte location), to conduct evanescent field-excited or SPP-induced fluorescence microscopy, to carry out evanescent field- or SPPs-scattering microscopy, or to record evanescent field- or SPP-induced Raman spectra and/or images; (iv) motion mechanism to adjust the incident angle or to center and focus the sample. The device is not yet completely fixed, allowing partial modification or even large change of some parts.

It may be worth of mentioning that all the illustrated devices can further be optimized or be reduced to a smaller size because the inner space has not yet been fully occupied. If one likes to known about SPRi systems working on other principles such as angle or phase variation, other books or reviews should be referred.

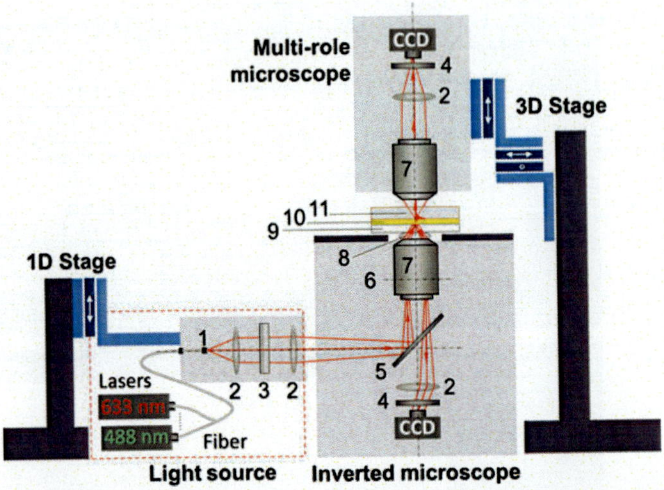

Fig. 3.13 Laboratory-built six-in-one-integrated experimental device for conducting lens-based SPRi, bright and/or dark field microscopy, SPPs-induced fluorescence microscopy, Raman scattering spectroscopy and/or microscopy and SPP scattering microscopy, where the 1D stage is used to adjust incident angle (up for smaller angle and down for larger), the 3D stage to center and focus sample, and lasers are addable or changeable. 1. Fixing collar; 2. Convex lenses; 3. Polarizer; 4. Filter; 5. Beam splitter; 6. Focal plane; 7. Objectives with different numerical apertures (normally NA = 1.49 to conduct oil microscope); 8. Oil or other liquid; 9. Glass slide; 10. Gold film normally at 48–50 nm thickness; 11. Sample (normally solution)

References

1. Huang Y, Ho HP, Kong SK, Kabashin AK (2012) Phase-sensitive surface plasmon resonance biosensors: methodology, instrumentation and applications. Ann Phys (Berlin) 524:637–662. https://doi.org/10.1002/andp.201200203
2. Otto A (1968) Excitation of nonradiative surface plasma waves in silver by the method of frustrated total reflection. Z Phys A Hadrons Nucl 216:398–410
3. Kretschmann E, Raether H (1968) Radiative decay of non-radiative surface plasmons excited by light. Z Naturforsch 23A:2135–2136
4. Xu J, Chen Y (2018) Surface plasmon resonance sensing with adjustable sensitivity based on a flexible liquid core coupling unit. Talanta 184:468–474
5. Hausler P, Jobst S, Fischer J, Roth C, Bierl R (2021) Homogeneous light source for surface plasmon resonance imaging. In: Albella P, Raposo M, Andrews D, Ribeiro P (eds) Proceedings of the 8th international conference on photonics, optics and laser technology (PHOTOPTICS 2020), 1st edn. Scitepress, Sebútal, pp 163–167

Chapter 4
Methodology

This chapter discusses two aspects of methodology, the theoretical and practical. The theoretical aspect includes the general imaging principles based on various types of SPPs, the specific principles for SPRi, imaging signal or data treatment and relative mathematic techniques, which are discussed majorly in Sects. 4.1–4.3.

The practical methods need various executable measures and logical steps that can roughly be divided into four essential phases: sample preparation, sensor chip preparation, surface chemistry, image recording and analysis. They are discussed majorly in Sects. 4.4–4.8.

4.1 Potential Plasmonic Principles of Imaging

The essence of imaging is to vividly illustrate the variations of one or more parameters in a space within a given time. In ordinary life, you often take color photos by using the parameter of optical wavelength or take black-and-white pictures by a parameter of optical intensity, which is convenient since most camera or detectors respond to optical intensity. In sciences, other parameters have also been explored as imaging variable such as molecular mass (in mass spectrometric imaging), absorption coefficients (in X-ray-based imaging) and radio wave absorption or emissions (in NMR imaging). Nowadays, images are usually digitized into pixels in which the signal intensity is averaged and the continuous spatial change of the intensity is conversed into staircase-like change to facilitate the treatment and storage of image data in computers. These are inherited by SPRi.

Theoretically, SPRi can record and output the spatiotemporal information in a form of static pictures or videos that are a series of static images or frames taken at a certain time interval. Clearly, the key lies in how to take the static pictures. This needs at least one parameter to display its spatial variation in respect of the strength (e.g., intensity or concentration) or strength gradient at a time.

Y. Chen, *Surface Plasmon Resonance Imaging*, Lecture Notes in Chemistry 95,
https://doi.org/10.1007/978-981-99-3118-7_4

With SPPs, there can be at least three types of variables for imaging, that is, LSPPs, SPP-scattering-caused optical fluctuation and SPR-caused optical absorption.

4.1.1 Localized Surface Plasmonic Absorption Imaging

According to Eq. 2.141, metal NPs can adsorb excitation light, which is somehow similar to macromolecular absorption and has been largely applied to the sensing of trace substances and even small ions [1–8]. It is thus easy to think of LSPPs-based imaging methods by direct adopting the ways for molecular dying or fluorescent imaging. The prerequisite condition is that the to-be-imaged objects must possess excitable LSPPs or can be physically and/or chemically "dyed" with a type of metal nanoparticles. Once "dyed", the objects can be viewed under a normal optical microscope, and the observed images can also be recorded through the LSPPs-caused adsorption, emission or scattering effect. Clearly, this sort of imaging is applicable to the analysis of either discrete objects or continuous layers, in either solutions or on solid surfaces. This is however not new in respect of imaging principles, except for the utilization of LSPP effect. Therefore, it will not be discussed further unless necessary.

4.1.2 Surface Plasmonic Scattering Imaging

Discrete objects (e.g., molecules and particles) can be imaged according to Eqs. 2.161 or 2.162. It can be understood that the scattering ratio of an object, γ or γ_j, and its corresponding image contrast increase with its size within the SPP propagating depth. By $I = <E^2>$, we can found that (i) the propagation pattern of the discrete object scattered SPPs has a fan-shaped parabolic interference fringes starting from the particle; (ii) if the back-reflection waves are also recorded, the picture becomes a pair of fan-shaped parabolic fringes facing the top and (iii) the imaging contrast decreases fast away the objects.

To image more sensitively, we need high lateral resolution and effective coupling configuration. The most often adopted configuration is based on the microscopic objective with high numerical aperture (Fig. 2.15c). By this configuration, the incident light will transversally transmit for a while to interact with the analytes adjacent to the evanescent field and then leak or be reflected back into the objective. The leaking and reflected beam is recorded by a camera to image the excited area. Because the leaking beam carrying the scattering information of analytes, the recorded microscopic images have contrast differences between analytes and background under a monochromatic light or have color variations from analyte to analyte under a white light. This will be further discussed in Sect. 4.3.

4.1.3 Surface Plasmonic Resonance Imaging

A layer of adjacent objects (e.g., molecules or particles) on a metal surface can be treated as a smooth film, if the distance between the objects is shorter at least than the propagation length (several microns) of SPPs, better than the wavelength of SPPs (hundreds of nanometers). The on-surface-deposited uniform films at a molecular size in thickness can thus be imaged based on Eqs. 2.145 or 2.146. It is understandable that the film will couple with the metal layer and causes the change of either resonance position or absorption intensity. The change is dependent on the permittivity of the objects, as a consequence, different analytes will produce different imaging contrasts and they are recognizable if their position on the metal surface is addressable. This is usually achieved by dividing the surface into indexed areas, for example, spotting analytes or analytes-specific probes on the known areas of the sensing surface.

This is the typical SPRi that works according to the SPR-caused absorption curves (i.e., the negative-peaked reflection spectra). It needs a newly designed device to record the microscopic images. In addition to the microscopic objective configuration (Fig. 2.15c), the prism-based configurations (Fig. 2.15c, b) are usually adopted for its easy construction, especially the Kretschmann one.

The basic analytical feature of this type of SPRi is that it can image on-surface events in batch. The so-called on-surface events include surface-related permittivity or concentration variations caused by flowing, reactions, affinity recognitions, absorptions and so forth. SPRi can also be used as a label-free detector for various chips if they can be prepared on or coupled with metal surfaces. With a lateral resolution of above 10 μm, the SPRi can produce a real image of chips with a density higher than 20,000 dots per square centimeters.

4.2 Imaging of Continuous Adlayer

SPRi typically measures analytes able to vary the permittivity or refractivity on the sensor surface and/or in its vicinity. This is induced by mass transfer processes, including physical and chemical origins: The most often adopted physically caused variations of permittivity is to pump in a substance, normally in liquid state, onto the sensor surface. This allows quantification and/or qualitative analysis based on the fact that the permittivity and hence the intensity or color of recorded images are a specific function of the concentration and the nature of analytes [9].

The change in refractivity on or adjacent to the sensor surface will shift the resonant angle, wavelength and phase of the reflect light. The reflection intensity at a given incident angle or wavelength will also change. These enable at least 4 different ways to have imaging contrasts: resonant angle variation, wavelength or color shift, phase rotation and optical intensity change. They may hence be termed angle, color, phase and intensity SPRi, respectively. A bit pity is that although the phase and angle SPRi are sensitive and quantitative, they are not very convenient in use; The color SPRi is easy to realize, but it is currently less sensitive than the intensity SPRi because the present color cameras are less sensitive than monochromatic cameras. Unless specifically stated, only the intensity SPRi is focused in later discussions. For brevity, the intensity SPRi is often shorted as SPRi.

4.2.1 *Imaging Parameter*

The so-called continuous adlayer can be a complete film or segments like spots, both on the metal sensing surface. In theory the film or spots can all be imaged by SPRi based on their spatial variation features such as resonant angle, resonant wavelength and the resonance-induced variations of optical intensity, phase and polarization. They are hence spatiotemporal function, $f(x, y, z, t)$. In static imaging, the time variable is removed and the function become 3D or stereo form of $f(x, y, z)$, and for a surface, the function is further reduced to $f(x, y)$, leaving z for signal intensity. Once a parameter is selected, its spatial strength function is controlled by optical path that should be pre-adjusted to match the size of a digital detector used (e.g., CCD or CMOS).

In modern scientific imaging by use of CCD, CMOS or similar detection array as a recorder, the digitization of space is automatically achieved while the strength of a parameter like optical intensity is normally digitized between 0 and 255 (or finer, depending on the digital resolution and precision) through a digital I/O interface.

4.2.2 *Monochromatic Imaging*

In common SPRi, the monochromatic reflection intensity and angular variation are the basic contrast parameters. Theory tells that the SPR-caused absorbing negative peaks are fairly wide and the resonant positions increase with ε_1 as a function of analytes (Fig. 4.1a). Although a wide absorbing peak is not preferred to have sharp imaging contrast, it opens a door to construct images in a variable or selectable position along the curve, from the dropping side to the rising side.

Therefore, the imaging contrast is dependent on the curve position. As illustrated in Fig. 4.1b, b′, imaging at the dropping side of the peak will cause an increase of the optical intensity. As a consequence, the recorded images are black-and-white (Fig. 4.1b). While imaging at the rising side of the peak, the intensity decreases, giving white-and-black images (Fig. 4.1b′). Between these two extremes, the image contrast increases with the incident angle and normally pictured in white-and-black forms (Fig. 4.1c). It should be noted that the images recorded from a chip spotted with a same sample are sharpened the most at the two extreme positions but more quantitative with the angle shift. If different analytes are spotted, a compromised position should be searched out that is normally at the dropping side of the peak but not necessarily at the extreme point (or $2I_{\min}$) of the differentiated SPR curve. In

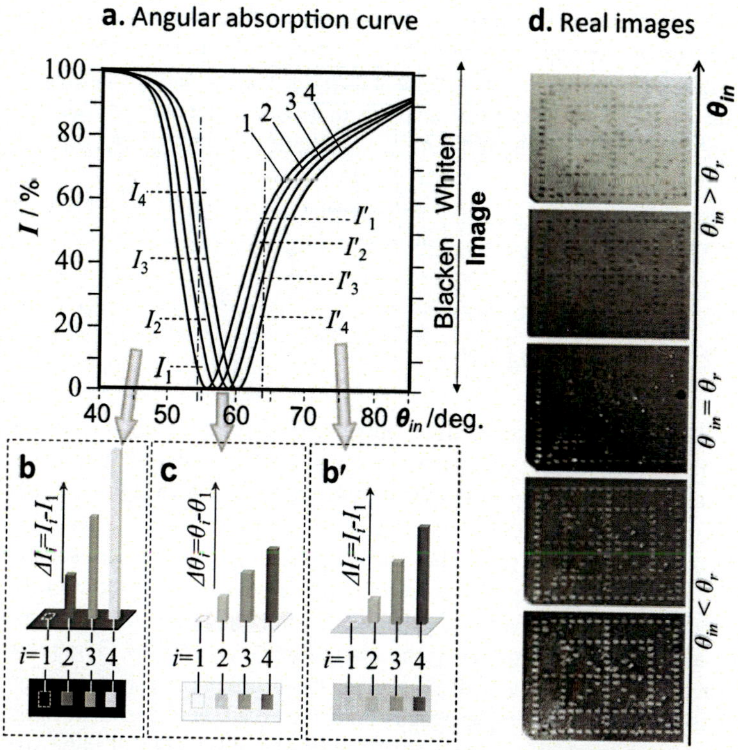

Fig. 4.1 Principle to construct monochromatic images based on the SPR-caused angular absorption curves. **a** Angle-interrogated SPR curves; **b** theoretical black-and-white 3D (upper) and 2D (lower) images pictured from the dropping side of the curves; **b′** white-and-black 3D (upper) and 2D (lower) images pictured from the rising side of the curves; **c** 2D image drawn from the resonant angle shift; **d** intensity images (aligned to a same size) of spotted BSA measured from the dropping side at different incident angles on an imager, model SPRi-PX8100; $\Delta\theta$, incident angle shift used as imaging signal strength (multiplication a factor and normalization are required); 1, background; 2, 3, 4, assumed different analytes

general, different analytes have different imaging position for the highest contrast; therefore, compromised measures are unavoidable to acquire the optimal overall contrast.

The intensity SPRi can produce images in real time but has a certain dynamic range. A sample spotted on a chip far away the optimal imaging position (refer to the vertical lines in Fig. 4.1a) is not preferred since it may produce seriously distorted images. This issue can be removed by use of resonant position shift (e.g., $\Delta\theta_j = \theta_j - \theta_1$, where 1 denotes background or a control) as the imaging contrast (Fig. 4.1c). This angular SPRi can give not only a very wide dynamic imaging range but also reliable quantitative information. Nevertheless, angular SPRi is in general not possible to image in real time because the resonant points can only be calculated after interrogation of all the positions.

Both intensity and angular SPRi provide 2D images as the direct spatial functions of ε_1 or (in the real part) refractive index n_1 that distributes on the sensing surface within a given time and varies once analytes appear in the dielectric 1. Because ε_1 and n_1 are also functions of analytes and their concentration, the imaging signals can be associated with the quantity and identity of analytes and are hence usable for quantitative and qualitative analysis of substances in the dielectric 1. SPRi can sense label-free substances in batch and is suitable for monitoring surface reactions and/ or variations.

4.2.3 Color Imaging

Instead of monochromatic light, polychromatic or white light can be used to excite SPPs that enable the performance of color SPRi at or above the resonant angle based on the wavelength-dependent absorption spectra as shown in Fig. 4.2a. The maximum absorption wavelength of a given spot is in fact the sum of a target analyte and its background beneath (e.g., chip surface including coatings and solvents adsorbed). The recorded color should thus vary with the type and concentration of samples as a function of ε_1 for a given chip. The image color resulted from the resonant absorption at a wavelength is in theory the one complementarity to the absorption. By the wavelength range indicated in Fig. 4.2a, the absorbed color shifts from green yellow to orange, so that, the complementary color observed should be in the range from purple to cyan or green (Fig. 4.2b). This was validated by SPRi of three spotted proteins as illustrated in Fig. 4.2c, where the image color also varies with the incident angle of the excitation light, giving the optimal image contrast at an angle around 46.98°. This is the consequence that the resonance angle and wavelength are dependent with each other as demonstrated in Fig. 6.2d measured from IgG spotted at different concentration with ethanolamine as a control.

The same as monochromatic images, with a polychromatic light to excite SPPs, gray (or color) intensity image can also be obtained at a given incident angle according to:

$$I_{\text{Gray}} = 0.299 I_{\text{Red}} + 0.587 I_{\text{Green}} + 0.114 I_{\text{Blue}}. \tag{4.1}$$

Equation 4.1 is applicable only to RGB color system.

4.2.4 Phase Imaging

Under the conditions of SPR, the TM mode of incident light will vary its phase (φ_p) through energy exchange between photons and surface plasmons, while the TE

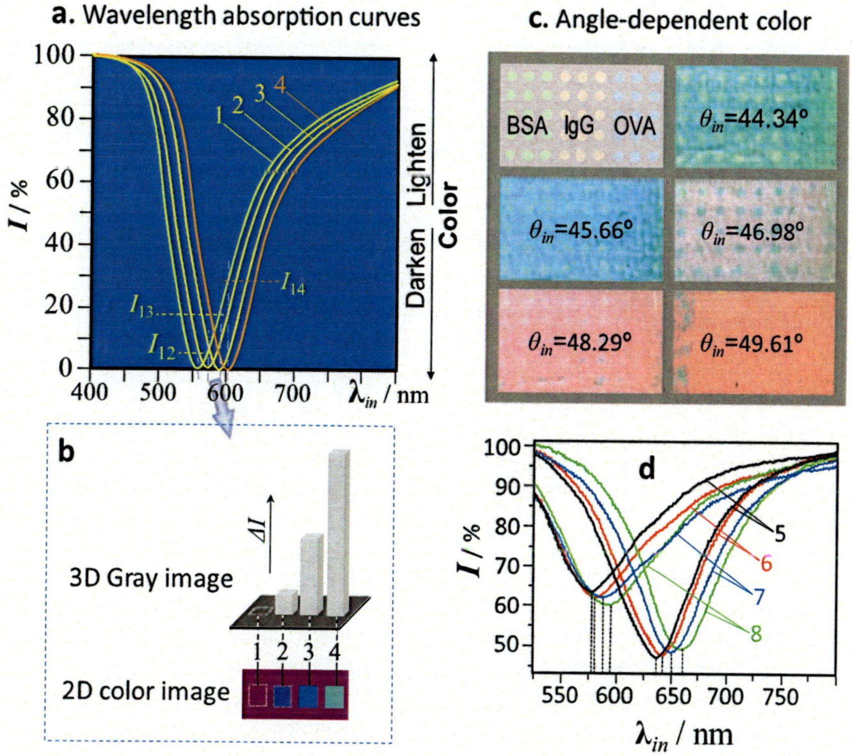

Fig. 4.2 Principles of color SPRi and related positions to construct complementary color images from wavelength-interrogated spectra. **a** Wavelength-interrogated SPR curves; **b** theoretical 3D gray (upper) and 2D color (lower) images drawn from the complementary color of the resonant wavelength at the spots; **c** raw color images (aligned to a same size) of spotted BSA, human IgG and hen OVA measured against air at different incident angles on a laboratory-built device equipped with a color CCD and a white excitation light source; **d** sensorgram of real sample spotted on MUA-modified a gold chip reprinted from [10] with permission; Spotted sample: 1 = background; 2, 3, 4 = assumed different analytes; 5 = MUA-ethanol; 6 = MUA-IgG at 0.05 mg/mL; 7 = MUA-IgG at 0.50 mg/mL; 8 = MUA-IgG at 1.00 mg/mL

mode of light keeps nearly unaffected. The variations of phase and amplitude can be derived according to Fresnel formulae Eq. 2.40 and a coordinate as shown in Fig. 4.3a:

$$\begin{cases} r_{\mathrm{TM}} = |r_{\mathrm{TM}}|e^{i\varphi_{\mathrm{TM}}} = \dfrac{E_r}{E_{\mathrm{in}}} = \dfrac{r_{\mathrm{in2}} + r_{21}e^{i2k_{2z}d}}{1 + r_{\mathrm{in2}}r_{21}e^{i2k_{2z}d}} \\ r_{\mathrm{in2}} = \dfrac{\varepsilon_2 k_{\mathrm{inz}} - \varepsilon_{\mathrm{in}} k_{2z}}{\varepsilon_2 k_{\mathrm{inz}} + \varepsilon_{\mathrm{in}} k_{2z}};\ r_{21} = \dfrac{\varepsilon_1 k_{2z} - \varepsilon_2 k_{1z}}{\varepsilon_1 k_{2z} + \varepsilon_2 k_{1z}} \\ k_{\mathrm{mz}} = \sqrt{k_0^2 \varepsilon_m - k_x^2}, \quad (m = \mathrm{in}, 1, 2) \end{cases} \tag{4.2}$$

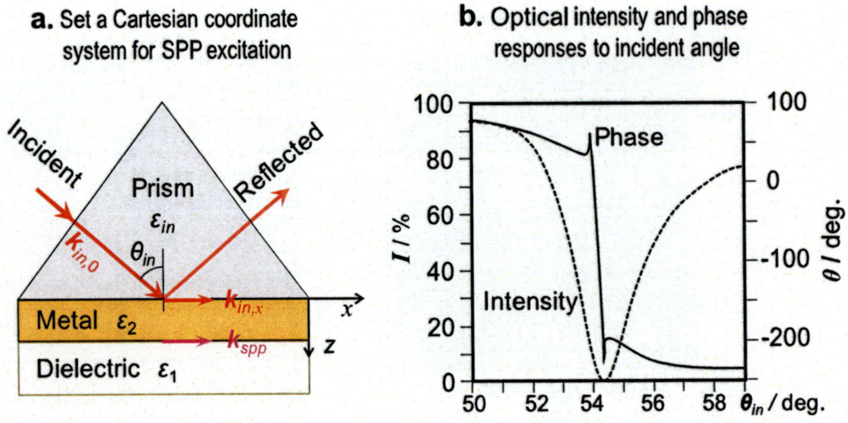

Fig. 4.3 Schematic excitation of an SPP by photons in, **a** a Cartesian coordinate system to produce, **b** a resonant absorbing curve (dash) and a phase jumping curve (solid) versus angle

Equation 4.2 shows that the amplitude and phase are very sensitive to ε_1 or n_1 under the resonant conditions when other parameters are kept unchanged. Different from the variation of reflection coefficient r or r^2 that drops on the resonance slopes, the sharp phase jump takes place only around the minimum of SPP curves (Fig. 4.3).

The phase jump was found, before SPRi, in 1976 when Abelès was doing his ellipsometric characterization of thin films [11]. This sharp phase jump under a certain thickness of environment helps to increase the resolution of the acquired images, reaching $\Delta\varphi/\Delta n \sim 10^5$ degree/RIU that corresponds to an LOD of 10^{-8} RIU [12–14], about 2 orders of magnitude better than other SPRi strategies.

Phase variation can be measured by interferometry, elliptical polarimetry, optical heterodyning and more. SPR interferometry extracts phase information from an interference pattern between the TM mode of light and a reference optical beam such as a coherent TM mode of light without passing through the target area or the unaffected TE mode of light. This approach implies the extraction of "pure" phase information with the reference as a background. The main disadvantage is that the setup and the process are somewhat complicated, needing fine fabrication, careful adjustment and maintenance of SPRi systems.

SPR polarimetry extracts the phase information through analyzing the elliptical light polarization while TM and TE lights present simultaneously in the measurement. The polarization ellipse can be traced by rotating a polarizer (Fig. 4.4), a spatial beam modulator and temporal phase modulation techniques such as liquid crystal, piezoelectric, mechanical [15] and photoelastic [16, 17] modulators. These techniques are more suitable for 2D measurement and allow to modulate the harmonic frequency to reduce noise.

Although the phase SPRi increases sensitivity for about 2 orders of magnitudes, it is much rarely utilized than the current intensity SPRi. The reasonable reasons are due not only to the easy fabrication and manipulation of the intensity SPRi instruments

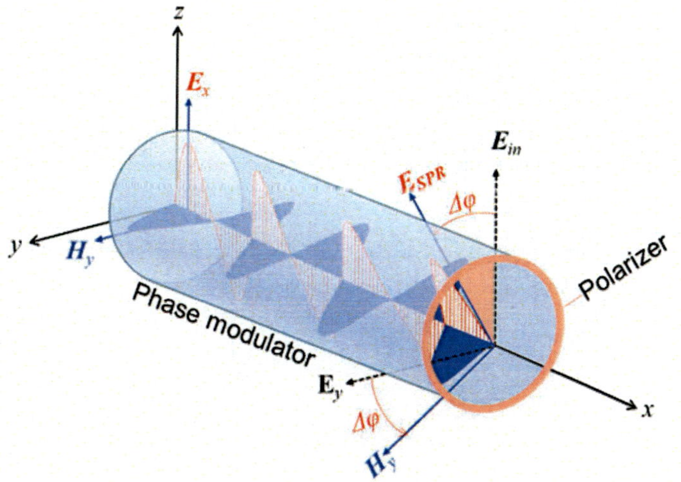

Fig. 4.4 Easy way to extract phase information in SPRi from the optical heterodyning based on the differential frequency between the orthogonally polarized TE and TM components in the reflected light

but also to the possibility to increase the sensitivity by non-hardware techniques such as signal amplification.

4.2.5 Adsorption Isothermal Measurement

To enable above-discussed SPRi measurements, the analytes need to reach the sensor surface or its vicinity so that they can sufficiently affect ε_1. In general, the target analytes are made to adsorb either specifically or non-specifically on the sensor surface, by various interaction mechanisms. In the real world, it is hard to find two neighboring substances that are free of interactions. From the point of view of chemistry, any two adjacent chemicals can have a sort of interaction, from weak to very strong, in a selective or non-selective way. The selective interaction is the origin of molecular recognition. A very strong interaction, once happens, can even result in the formation of new product(s). This is commonly called chemical reactions including also coordination and supra macromolecular complexations.

Molecular adsorption is in general treated based on isotherm equilibrium. Upon an SPRi configuration as shown in Fig. 4.5a, it is easy to derive the on-sensor adsorption isotherm. Consider a measurement starting from the very beginning at $t = 0$ when it is free of adsorption, giving the response of SPRi at a resonant angle $\theta_{r,t=0}$ or a resonant wavelength $\lambda_{r,t=0}$, the timely measured resonant value will be θ_r or λ_r. This resonant value corresponds to the total or effective refractive index, n_{eff}, that is averaged from both the adsorbed layer (n_a) and solution (n_s). It is theoretically expected that the resonant change has a linear relationship with the n_{eff} over an appropriate range:

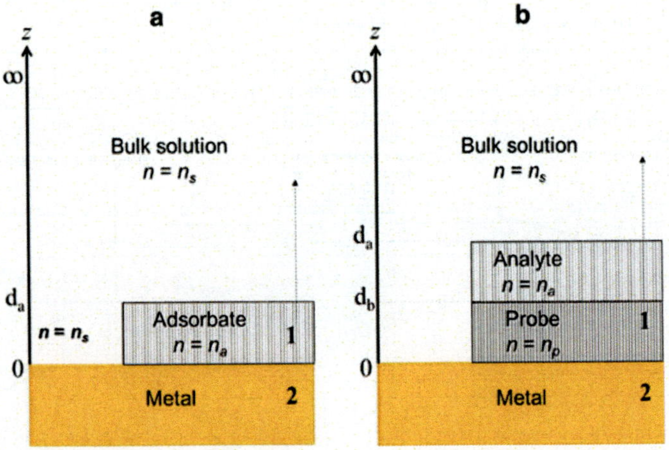

Fig. 4.5 Two typical structure of SPRi sensor chip with **a** one or **b** two adlayers

$$\begin{cases} \theta_r = a n_{\text{eff}} + b \\ \lambda_r = a' n_{\text{eff}} + b' \end{cases}. \tag{4.3}$$

Taken the starting point of measurement $t = 0$ as a reference and assume that the a, a', b and b' keep unchanged in the range (which is not always true), from the Eq. 4.3, we can have

$$\left.\begin{cases} \Delta\theta_r = S_\theta(n_{\text{eff},t} - n_{\text{eff},t=0}) = S_\theta \Delta n_{\text{eff}} \\ \Delta\lambda_r = S_\lambda(n_{\text{eff},t} - n_{\text{eff},t=0}) = S_\lambda \Delta n_{\text{eff}} \end{cases}\right\} \Rightarrow \begin{cases} S_\theta = \dfrac{\Delta\theta_r}{\Delta n_{\text{eff}}} \\ S_\lambda = \dfrac{\Delta\lambda_r}{\Delta n_{\text{eff}}} \end{cases}, \tag{4.4}$$

where S is the slope of the variation curve, denoting the sensitivity of a SPRi system and the subscripts indicate the change of measured angle and wavelength, respectively. In SPRi, optical intensity change, ΔI, is more often adopted and can be expressed as follows according to Eq. 4.4 within its linear range,

$$\begin{cases} \Delta I = S_I \Delta n_{\text{eff}} \\ S_I \propto S_{\theta \text{ or } \lambda} \end{cases}. \tag{4.5}$$

It should be noted that ΔI is better measured at the maximum position, normally on the left side of SPR curve at about $2I_{\text{min}}$ (refer to Fig. 4.1a or Fig. 4.2).

To make the Eq. 4.4 or Eq. 4.5 practical, we have to know the value of n_{eff}. It is easy to fix $n_{\text{eff},t=0} = n_s$ just before the sensor surface starts to adsorb analytes on a bare gold surface (Fig. 4.5a) or on a probe layer (Fig. 4.5b). In this phase, both the senor and the contacting sample solution are not disturbed, or the disturbance is negligible. In practical applications, it is more convenient and more reliable to measure the background optical intensity adjacent to sample spots than the initial. For non-specific adsorption sensor surface, the n_{eff} above the sample-free areas can also be considered to be equal to n_s (left side in Fig. 4.5).

The value of n_{eff} on a sample spotted areas is not directly available but may be calculated after some assumptions. Considering a sample spot of mono-adlayer that is covered under a solution or buffer (right side in Fig. 4.5), the value of n_{eff} would be the average of $n(z)$ along the decaying direction weighted by a factor of e^{-z}. If we take the intensity propagation depth $\delta_{I,1z}$ or amplitude depth $\delta_{A,1z}$ as a unit of variable, the exponential factor can be given by $e^{-2z/\delta_{I,1z}}$ or $e^{-4z/\delta_{A,1z}}$. For a uniform layer parallel to the sensor surface, its $n(z)$ will maintain unchanged at z along the whole layer, and the $n(z)$ can thus be averaged along z by integration [18]:

$$
\begin{aligned}
n_{\text{eff},t} &= \frac{2}{\delta_{I,1z}}\int_0^\infty n(z)e^{-\frac{2z}{\delta_{I,1z}}}\mathrm{d}z = \frac{4}{\delta_{A,1z}}\int_0^\infty n(z)e^{-\frac{4z}{\delta_{A,1z}}}\mathrm{d}z \\
&= -\int_0^{d_a} n(z)\mathrm{d}e^{-2d/\delta_{I,1z}} - \int_{d_a}^\infty n(z)\mathrm{d}e^{-2d/\delta_{I,1z}}.
\end{aligned} \tag{4.6}
$$

If the adlayer is pure or solution-free, with a refractive index at n_a and an averaged thickness equal to d_a, there will be $n(0 < z < d_a) = n_a$ and $n(d_a < z < \infty) = n_s$. The integral of Eq. 4.6 becomes

$$
\begin{aligned}
n_{\text{eff},t} &= n_a(1 - e^{-2d_a/\delta_{I,1z}}) + n_s e^{-2d_a/\delta_{I,1z}} \\
&= n_a - (n_a - n_s)e^{-2d_a/\delta_{I,1z}} \\
&= n_s + (n_a - n_s)[1 - e^{-2d_a/\delta_{I,1z}}].
\end{aligned} \tag{4.7}
$$

Combine Eq. 4.7 with Eqs. 4.4 or 4.5, we get

$$
\Delta y = S_y(n_a - n_s)[1 - e^{-2d_a/\delta_{I,1z}}], \quad (y = I, \theta_r, \lambda_r). \tag{4.8}
$$

Equation 4.8 connects the measured signal change with the adlayer thickness and the propagation depth. Once the adsorption thickens sufficiently after an infinitive time, Eq. 4.8 changes to a limit:

$$
\Delta y_\infty = S_y(n_a - n_s)\left(1 - e^{-2d_{a,\max}/\delta_{I,1z}}\right), \quad (y = I, \theta_r, \lambda_r), \tag{4.9}
$$

where Δy_∞ is directly measurable. Divide Eq. 4.8 over Eq. 4.9, an SPRi-based measurement of adsorption isotherm is obtained as follows:

$$\begin{cases} \dfrac{\Delta y}{\Delta y_\infty} = \gamma_{y\max}\left(1 - e^{-2d_a/\delta_{I,1z}}\right), \quad (y = I, \theta_r, \lambda_r) \\ \gamma_{y\max} = \dfrac{1}{1 - e^{-2d_{a,\max}/\delta_{I,1z}}} = \dfrac{e^{2d_{a,\max}/\delta_{I,1z}}}{e^{2d_{a,\max}/\delta_{I,1z}} - 1} > 1 \end{cases}. \tag{4.10}$$

Equation 4.10 depicts that a mono-adlayer gives an upward asymptotic curve, reaching its limit at $\Delta y/\Delta y_\infty = 1$. The adlayer can also be associated with the adsorption equilibrium with a constant at K_a for a simplified binding reaction:

$$A + B \underset{k_{\text{off}}}{\overset{k_{\text{on}}}{\longleftrightarrow}} AB, \tag{4.11}$$

$$K_a = \frac{k_{\text{on}}}{k_{\text{off}}} = \frac{[AB]}{[A][B]} = \frac{\beta_a}{1 - \beta_a}\frac{1}{c_A}, \tag{4.12}$$

where A denotes the target analyte in sample solution, B is the adsorptive sites on sensor surface that can be expressed as a concentration at c_B, β_a is the ratio of B occupied by A at an equilibrated concentration of $[A]$ or c_A, k_{on} is binding kinetic constant and k_{off} is dissociation kinetic constant. Equation 4.12 can be rearranged as

$$\beta_a = \frac{c_A K_a}{1 + c_A K_a}. \tag{4.13}$$

Combine Eq. 4.13 with Eq. 4.10, we have

$$\frac{\Delta y}{\Delta y_\infty} = \gamma_a \frac{\beta_a}{\beta_{a,\infty}} = \gamma_a \frac{c_A}{c_{A\infty}} \frac{1 + c_{A\infty} K_a}{1 + c_A K_a} = \gamma_a \frac{c_A}{c_B} \frac{1 + c_B K_a}{1 + c_A K_a}, \tag{4.14}$$

where γ_a is a ratio coefficient that may be calculated from experimental data, $c_{A\infty}$ denotes the maximum adsorbed concentration of the analyte. Equation 4.14 can be rewritten into

$$\begin{cases} \dfrac{\Delta y_\infty}{\Delta y} c_A = a c_A + b \\ \begin{Bmatrix} a = \frac{c_{A\infty} K_a}{1 + c_{A\infty} K_a} \frac{1}{\gamma_a} \\ b = \frac{c_{A\infty}}{1 + c_{A\infty} K_a} \frac{1}{\gamma_a} \end{Bmatrix} \Rightarrow \begin{cases} K_a = \frac{a}{b} = \frac{k_a}{k_d} \\ \gamma_a = \frac{c_{A\infty}}{b + a c_{A\infty}} \end{cases} \end{cases} \tag{4.15}$$

Equation 4.15 shows that the curve of $c_A \Delta y_\infty / \Delta y$ against c_A is linear, allowing to calculate the slope a and intercept b, which in turn makes it possible to calculate the adsorption equilibrium constant K_a and the ratio coefficient γ_a.

If a light of TE mode is taken as a reference to reduce noise and background drift, we can have following formula:

$$\frac{R_{\text{TM,ref}}}{R_{\text{TE,ref}}} = \frac{I_{\text{TM}} - I_{\text{Bk}}}{I_{\text{TE}} - I_{\text{Bk}}} \frac{T_{\text{TE,in}}^2}{T_{\text{TM,in}}^2}, \tag{4.16}$$

where R is in fact the reflectance at gold-prism interface, T is the transmittance at the prism entry, I is the pixel density in grey level across a single spot and the subscript Bk denotes the background pixel. With TE as a reference, the value of $R_{\mathrm{TM,ref}}$ is actually the percentage reflectivity averaged cross the spot, the same is T. The apparent density of on-sensor analyte spot can also be given by

$$\begin{cases} \beta_a = \dfrac{n_a - n_s}{S_c} d_a = \dfrac{\Delta R}{S_R S_c} \delta_{I,1z} = \dfrac{\Delta R}{S_{Rc}} \delta_{I,1z} \\ S_c = \dfrac{\delta n}{\delta c_A} \\ S_R = \dfrac{\partial R}{\partial n}, \end{cases} \tag{4.17}$$

where S_c connects the refractive index change with analyte concentration in sample solution and S_R is reflectivity change per unit refractive index. They both can be calibrated by experiments. S_R is at scale of about 2.25×10^{-3} and S_c is $1.9 \times 10^{-10}\mathrm{mm}^3/\mathrm{pg}$ for proteins and nucleic acids.

4.2.6 Adsorption Kinetic Measurement

SPRi is more often performed in a flowing way. The adsorption and desorption never reach the equilibrium state but instead reach a steady state. Thus the adsorption and desorption kinetic constants cannot be detected directly by the steady state unless some assumptions are adopted. In most cases, a quasi-first order of reaction is assumed for convenience. This can be true if the reactions are controlled to have only one chemical variable in its concentration while others fixed. This corresponds to two phases: Phase one exists at the very beginning of adsorption reaction when the products are negligible while the adsorptive sites (i.e., the immobilized probe molecular number) are fixed at a given concentration, and phase two happens at the starting point of dissociation by pumping in zero concentration of analyte. The reactions can be written schematically as follows:

$$\begin{cases} A + B \xrightarrow{k_{\mathrm{on}}} AB \underset{k_{\mathrm{off}}}{\longrightarrow} B + A \\ c_A \; c_B \; c_{AB} \\ \dfrac{\mathrm{d}c_{AB}}{\mathrm{d}t} = k_{\mathrm{on}} c_B c_A - k_{\mathrm{off}} c_{AB} \doteq k_{\mathrm{on}} c_B c_A, \text{ starting to adsorb analyte} \\ \dfrac{\mathrm{d}c_{AB}}{\mathrm{d}t} = k_{\mathrm{on}} c_A - k_{\mathrm{off}} c_{AB} = -k_{\mathrm{off}} c_{AB}, \text{ washing with analyte - free buffer} \end{cases} \tag{4.18}$$

Herein the reaction is different from that shown in Eq. 4.11. At the phase of desorption, after adsorption equilibration, by pumping in a fresh buffer without any analyte, that is $c_A = 0$, Eq. 4.18 becomes $dc_{AB}/dt = -k_{off}\, c_{AB}$, its solution is

$$\begin{cases} \ln\dfrac{c_{AB}}{c_{AB,0}} = -k_{off}t \text{ or} \\ c_{AB} = c_{AB,0}e^{-k_{off}t} = c_{AB,max}e^{-k_{off}t}. \end{cases} \tag{4.19}$$

Suppose that the signal of SPRi varies linearly with c in a certain range, that is, $I = Bc$ where B is constant independent on the value of concentration (which is not always true) and I is the signal intensity of SPRi at time t, Eq. 4.19 is converted to

$$\begin{cases} I = I_{max}e^{-k_{off}t} \text{ or} \\ \ln\dfrac{I}{I_{max}} = -k_{off}t \end{cases}, \tag{4.20}$$

where I_{max} is the maximum signal at the very beginning of desorption or at the end of adsorption. Both Eqs. 4.19 and 4.20 show that the reaction kinetic constants are the slop of $\ln(I/I_{max})$ or $\ln(c/c_0)$ against reaction time t. Specially at the half life time $t_{1/2}$ when either the initial concentration or signal has a change of 50%, there will be

$$k_{off} = \frac{\ln 2}{t_{1/2}}. \tag{4.21}$$

For adsorption, the kinetics differs because the concentration of analyte is in common kept unchanged by pumping in fresh sample solution, that is, c_A = constant. SPRi intensity is generally an upward asymptotic curve, due to the completion of adsorption with desorption reactions at steady state:

$$\begin{cases} A + B \underset{k_{off}}{\overset{k_{on}}{\rightleftharpoons}} AB \\ c_A\ (c_{B,0} - c_{AB})\ c_{AB} \\ \dfrac{dc_{AB}}{dt} = k_{on}c_A(c_{B,0} - c_{AB}) - k_{off}c_{AB} = -(k_{on}c_A + k_{off})c_{AB} + k_{on}c_Ac_{B,0} \end{cases}. \tag{4.22}$$

Let $I = Bc$ and $I_{t=0} = I_0$ as a baseline, that is, $I_0 = 0$, the solution to Eq. 4.22 is

$$\begin{cases} c_{AB} = \dfrac{k_{on}c_A}{k_{on}c_A + k_{off}}\left[1 - e^{-(k_{on}c_A + k_{off})t}\right] = c_{AB,max}\left[1 - e^{-(k_{on}c_A + k_{off})t}\right] \\ I = I_{max}\left[1 - e^{-(k_{on}c_A + k_{off})t}\right] \end{cases}. \tag{4.23}$$

In Eq. 4.23, $c_{AB,max}$ is the maximum adsorbed concentration of the analyte that is just equal to the balanced concentration of AB: $c_{AB} = K_ac_A/(K_ac_A + 1)$, we can thus have balanced or steady state measuring formula at $c_A/c_{AB,\max} \ll 1$:

$$\begin{cases} \dfrac{c_{AB}}{c_{AB,\max} - c_{AB}} = \dfrac{c_A(K_a c_{A,\max} + 1)}{(c_{A,\max} - c_A)} \approx K_a c_A \\ \dfrac{I}{I_{\max} - I} \approx K_a c_A \end{cases}. \tag{4.24}$$

It should be noted that all the value of I in Eq. 4.24 must be measured after the adsorption reaches equilibration or steady state. On other conditions, by data fitting after measurement according to Eq. 4.23, both k_{on} and k_{off} can be obtained, or by use of the k_{off} measured according to Eq. 4.20 or Eq. 4.21, k_{on} can be calculated from Eq. 4.23. For convenience, Eq. 4.23 can be rearranged into following two ways to plot linear kinetic curves, $I \sim t$ at a given c_A or $I \sim c_A$ at a given t:

$$\begin{cases} \ln \dfrac{I_{\max}}{I_{\max} - I} = (k_{\text{on}} c_A + k_{\text{off}})t \\ \ln \dfrac{I_{\max}}{I_{\max} - I} = k_{\text{off}} t + k_{\text{on}} t c_A \end{cases}. \tag{4.25}$$

Equation 4.25 allows to measure k_{on} and k_{off} from the slope and/or the intercept. To plot the curve, a serials of samples at an increasing concentration have to prepared and measured at different times. Note that the value of $I_{\max}$ can be measured by adsorption of a sufficiently concentrated sample.

4.2.7 *Estimation of Propagation Depth and Adsorbed Parameters*

Equation 4.10 connects the measured optical intensity change with the adlayer thickness and the propagation depth and hence enables the estimation of these parameters after rearrange:

$$\begin{cases} \delta_{I,1z} = -2d_a / \ln(1 - \Delta y/\Delta y_\infty) \\ d_a = -\dfrac{\delta_{I,1z}}{2} \ln(1 - \Delta y/\Delta y_\infty) \\ y = I, \theta_r, \lambda_r \end{cases}. \tag{4.26}$$

The value of $\delta_{I,1z}$ that is theoretically predicted in Eq. 2.128 is now measureable by mathematic fitting if d_a is given. Note that propagation depth varies weakly with n_{eff}. By calculation with a gold sensor film at 665 nm, $\delta_{I,1z}$ can decrease by a factor of 30% as n_{eff} increases from 1.33 (pure water) to 1.57 (pure protein). The decrease becomes only 7% after a protein film is formed at 22 nm thickness. Although weak variation, $\delta_{I,1z}$ depends on local refractive indices and the use of unchanged data implies an approximation that is allowed to calculate the adsorbed amounts of analytes, giving an error below 15%. If the refractive indices in the double adlayers of the sample are sufficiently different and their thickness is > 0.2 $\delta_{I,1z}$, more rigorous treatment is required.

For $d_a << \delta_{I,1z}$, Eq. 4.26 can be simplified to

$$d_a = \frac{\delta_{I,1z}}{2}\frac{\Delta y}{\Delta y_\infty} = \frac{\delta_{I,1z}}{2S_y(n_a - n_s)}\Delta y. \tag{4.27}$$

Again a linear relationship is found between d_a and the response signal for a given SPRi system. This makes it probable to directly determine the thickness of adlayer by some calibration approaches. The absolute accuracy can reach below a few percent after calibration but may be worsened to about ± 35% without any calibration. From Eq. 4.27, the detection sensitivity S_y and refractive index of the adlayer n_a also contribute to the accuracy especially in case of a thin adlayer. The relative error grows rapidly as the true thickness increases to near a half of $\delta_{I,1z}$, for example, thickness error is 35% at $d = 0.35\delta_{I,1z}$.

The above discussions are based on the layered system where a metal film is absorbed by one adlayer beneath an infinitively thick solution or liquid (Fig. 4.5a). For a trilayers formed with several different substances (Fig. 4.5b), we can have η_{eff} by extension of Eq. 4.6:

$$\begin{aligned} n_{\text{eff}} &= -\int_0^{d_b} n_b \mathrm{d}e^{-\frac{2z}{\delta_{I,1z}}} - \int_{d_b}^{d_b+d_a} n_a \mathrm{d}e^{-\frac{2z}{\delta_{I,1z}}} - \int_{d_b+d_a}^{\infty} n_s \mathrm{d}e^{-\frac{2z}{\delta_{I,1z}}} \\ &= n_b\left[1 - e^{-2d_b/\delta_{I,1z}}\right] + n_a\left[e^{-2d_b/\delta_{I,1z}} - e^{-2(d_b+d_a)/\delta_{I,1z}}\right] + n_s e^{-2(d_b+d_a)/\delta_{I,1z}}. \end{aligned} \tag{4.28}$$

Equation 4.8 can thus be changed to

$$\begin{cases} \Delta I = S_I(n_a - n_s)[1 - e^{-2d_a/\delta_{I,1z}}]e^{-2d_b/\delta_{I,1z}} \\ \quad\; = S_I^b(n_a - n_s)[1 - e^{-2d_a/\delta_{I,1z}}] \\ S_I^b = S_I e^{-2d_b/\delta_{I,1z}} \end{cases}. \tag{4.29}$$

This is only for the optical intensity change upon the adsorption of a target analyte after the presence of probe b as in ordinary SPRi measurements. S_I is again the slope of the calibration curve in the range of n_{eff} while $S_I{}^b$ makes the Eq. 4.29 be in the same form of Eq. 4.8. Their only difference lies in a factor of $e^{-2d_b/\delta_{I,1z}}$ that approaches 1 as d_b becomes 0. The adlayer b acts as a magnitude or sensitivity reducing factor in SPRi response and is negligible at $d_b << \delta_{I,1z}$. This makes it possible to have a probe adlayer to facilitate a selective capture of target analytes and to assist qualitative analysis as well. Equation 4.29 also reveals that the probe interlayer can severely decrease the sensitivity if it becomes too thick. At $d_b = \delta_{I,1z}$, the sensitivity decreases by a factor of ca. 7 if S_I is kept unchanged. Thus S_I is dependent on the thickness of b. In above discussion, we considered $\delta_{I,1z}$ as a constant. This is incorrect but it is the most reasonable approximation we can adopt.

From the estimated average thickness, d_a, of a uniformly absorbed adlayer, the corresponding surface molecular concentration, C_{ss} (molecular number per cm^2), or the volume molecular number density, C_{sv} (molecules per cm^3 on the sensor surface), can be calculated:

$$C_{ss} = d_a C_{sv} = d_a N_A \frac{\rho}{M}, \tag{4.30}$$

where ρ is its volume density of the adsorbate in unit of g/cm^3.

It is also important to estimate n_a, which is not necessarily achieved by SPRi or SPRS. Reasonably, the on-sensor molecules are assumed not to be obviously perturbed even after their forming bounds on the metal surface, thus the value of n_a must be very close to that of their pure and condensed state. This has been validated to be correct for macromolecules whose size is so big that their binding fraction is negligible. An example is the absorbed proteins with $n_a \sim 1.6$ [19] that is very close to their pure value.

For molecules whose refractive index is unknown or unmeasurable, the value of n_a can also be estimated from other molecules with similar structure. For example, alkanethiolates anchored on gold surface had n_a very close to their free thiols because their lost hydrogen atoms after adsorption are negligibly tiny compared with the whole molecular volume [18]. Similarly, the refractive indices of the bounded alkyls, carboxylates and ammonium cations can be estimated from the value of their corresponding alkanes, carboxylic acids and amines, respectively.

A further method to estimate the refractive indices is based on the Clausius–Mossotti Equation for a pure compound j with a refractive index of n_j

$$\frac{n_j^2 - 1}{n_j^2 + 2} = C_{v\text{tol},j} \frac{\chi_j}{3\varepsilon_0}, \tag{4.31}$$

where ε_0 is the permittivity in free space, $C_{v\text{tol},j}$ the total molecular number of j per unit volume in pure state, and χ_j the frequency-dependent polarizability of the molecule. In case of a solution mixed from substances j and k, at a number density of $C_{v,j}$ and $C_{v,k}$, the mixed refractive index, n_{mix}, for an ideal solution, can be estimated by a similar formula called Lorenz–Lorentz equation [20]:

$$\begin{cases} \dfrac{n_{\text{mix}}^2 - 1}{n_{\text{mix}}^2 + 2} = C_{v,j}\chi_j + C_{v,k}\chi_k = \gamma_j \dfrac{n_j^2 - 1}{n_j^2 + 2} + \gamma_k \dfrac{n_k^2 - 1}{n_k^2 + 2} . \\ \gamma_j = C_{v,j}/C_{v\text{tol},j}; \quad \gamma_k = C_k/C_{v\text{tol},k} \end{cases} \tag{4.32}$$

Equation 4.32 shows that the mixed refractive index n_{mix} is averaged from each composition with a weighting factor of γ. A close linear relationship exists between n_{mix} of the mixed solution and the volume fraction, within a few percent where n_j and n_k do not surpass SPR measuring range (e.g., between 1.33 and 1.60):

$$n_{mix} = \gamma_j n_j + \gamma_k n_k = \gamma_j (n_j - n_k) + n_k. \tag{4.33}$$

Both Eqs. 4.32 and 4.33 are applicable to the mixed bulk solution. For example, addition of 1.0 g/L protein into an aqueous buffer solution containing 0.3 M NaCl will cause n_{mix} to increase by1.8 × 10^{-4} RIU·L/g (a bit lower for glycol- and lipo-proteins). The known specific volume of proteins in aqueous buffer is ca. 0.77 mL/g [21]. From Eq. 4.33, we have

$$\because n_{mix} - n_{buf} = \gamma_p (n_p - n_{buf}) = 1.80 \times 10^{-4}$$

$$n_p - n_{buf} = \frac{1.80 \times 10^{-4}}{0.770 \times 10^{-3}}$$

$$\therefore n_p = \frac{1.80 \times 10^{-4}}{0.770 \times 10^{-3}} + \eta_{buf} = \frac{1.80 \times 10^{-4}}{0.770 \times 10^{-3}} + 1.336 = 1.57,$$

where n_p = 1.57 RIU denotes the refractive index of water-free protein. Nearly, a same value can be calculated from Eq. 4.32. They are close to 1.60 RIU for protein crystals but greater than the data estimated by ellipsometry for "adsorbed protein films" including a great deal of water [22]. These imply that we are better to take the adsorbed protein on SPRi sensor as "pure protein film" and neglect the appearance of water.

Equation 4.32 and Eq. 4.33 are also applicable to the estimation of refractive index of an adsorbed single molecule from its fragments. By dividing a molecule into small fragments similar to small molecules with known refractive indices, its n can be estimated by weighted sum of the refractive indices over all the fragments, with the fragmental weight factor equal to the fragmental volume over the whole molecular volume. It is known that alkyl thiols, alcohols and carboxylic acids will all approach a limit n of the longest alkane [23]. By assumption of a same group volume of CH_3 and CH_2, with refractive indices of 1.183 and 1.471, respectively, the n value of $CH_3(CH_2)_4CH_3$ calculated is equal to (2 × 1.183 + 4 × 1.471)/6 = 1.375 RIU, very close to 1.3751 RIU. In many cases, the volume of a fragment or group is not known but can be geometrically estimated through bond length, angle and van der Waals radii of atoms. It can also be obtained by fitting the equation with the known molecular refractive indices.

4.2.8 Determination of Analytes in a Sample Solution

It is generally more interested in determining the concentration, c, of analytes in a real sample solution contacting with the sensor. There are two obvious SPRi ways: The first is to correlate ΔI directly with c or x that is a function of n and d; and the second is to correlate them with adsorption equilibration.

4.2.8.1 Correlation of Imaging Signal with the Solution Concentration of Analytes

Based on the relationship of the refractive index with the molecular number of an analyte, SPRi-measured ΔI can be proportional directly to the adsorbed surface concentration or indirectly to the real concentration of analytes in samples. We can assume that the following formula holds

$$n_s = ax + n_{x=0}, \tag{4.34}$$

where x is the mass fraction ratio of solute (g/100 g), $n_{x=0}$ is the refractive index of analyte-free solution such as pure water ($n_x = n_{-a} = 1.333$), and a is a slope for a given solute, for example, $a = 1.84 \times 10^{-3}$ for NaCl or sucrose, and $a = 3.30 \times 10^{-4}$ for ethanol. From Eq. 4.8 at $d_a << \delta_{I,1z}$, we have

$$\Delta I = \frac{2S_I^b}{\delta_{I,1z}}(n_a - n_s)d_a. \tag{4.35}$$

Here d_a is the averaged adsorption thickness of the target analyte. It can be proportional to the covering rate β_a of the adsorptive area or sites available on the metal sensor. According to the Langmuir's adsorption isotherm for mono-molecular layer, β_a cannot vary linearly with the solute concentration c in bulk sample solution unless the sample solution is highly diluted. In this case, we can assume $d_a \propto \beta_a = k_a c + b$. Combination of Eq. 4.34 with Eq. 4.35 gives

$$\begin{cases} \Delta I = S_1 c^2 + S_2 c + S_3 \\ S_1 = -\dfrac{2S_I^b}{\delta_{I,1z}} k_a a \\ S_1 = -\dfrac{2S_I^b}{\delta_{I,1z}} [(n_a - n_{-a})k_a - ab] \\ S_3 = -(n_a - n_{-a})b \end{cases}. \tag{4.36}$$

It is not a linear function. If the concentration becomes so high that the adsorption become saturated, β_a becomes constant for a monolayer of molecules or further changes nonlinearly for multi-layer adsorption. In general, the optical intensity change can be an exponential function of analyte concentration in the bulk sample solution.

4.2.8.2 Analyte Concentration at Adsorption Equilibration

By adsorption isotherm, the absorbed quantity w of an analyte is normally an exponential function of its bulk concentration c:

$$w = ac^b, \tag{4.37}$$

where a and b are empirical constants, respectively. This applies to many cases. Similarly, Eq. 4.13 can be rearranged as

$$\beta_a = \frac{c_A K_a}{1 + c_A K_a} = \sum\nolimits_n (-1)^j (c_A K_a)^{j+1}. \tag{4.38}$$

Consider $\beta_a \propto I$, from Eq. 4.38, we have

$$\Delta I = \gamma(\beta_a - \beta_{c=0}) = \gamma \frac{c_A K_a}{1 + c_A K_a} = \gamma \sum_n (-1)^j (K_a)^{j+1} (c_A)^{j+1}. \tag{4.39}$$

Equations 4.36, 4.37 and 4.39 all show the nonlinear function between the SPRi signal and the bulk concentration of an analyte in a sample solution. They are all complicated exponential functions, giving often a very narrow "linear window" for quantification. To widen the linear window, we suggest to plot double logarithmic curve:

$$\ln(\Delta I) = A \ln x + B. \tag{4.40}$$

By Eq. 4.40, the linear range can cover more than 2 orders of magnitude but the calibration should always be performed for each analyte with a series of its standard solutions.

We have studied the linear window between the ΔI (gray value after deduction of background) and c. As expected, no linear curve was measured, only quasi-linearity was found within one order of magnitude with c. The linear window was expanded to at least two order of magnitudes by plotting $\log(\Delta I)$ against $\log c$ (Fig. 4.6). A problem was further encountered, that is, the plots were easily accompanied with great errors that seriously damage the linear relationship. More experiments revealed that the linearity was much dependent on the background and noises. In the worst case, the linearity disappeared or became extremely poor, with a linear coefficient approaching ca. 0.5 as illustrated in Fig. 4.7a where the linear correlation coefficient is only at $R^2 = 0.54$. However, the linearity was improved after deduction of the deformed background (Fig. 4.7b), with R^2 improved to 0.8858 (Fig. 4.7c). The linear correlation coefficient was further improved up to $R^2 = 0.9401$ (Fig. 4.7d) after filtration of noise by wavelet transformation. Deduction of the uneven background (Fig. 4.7e) and filtration of noise for a second time, the linearity can even be improved to $R^2 = 0.9982$ (Fig. 4.7f), with errors kept at < 10% [24]. It meets now with the criteria for quantitative analysis.

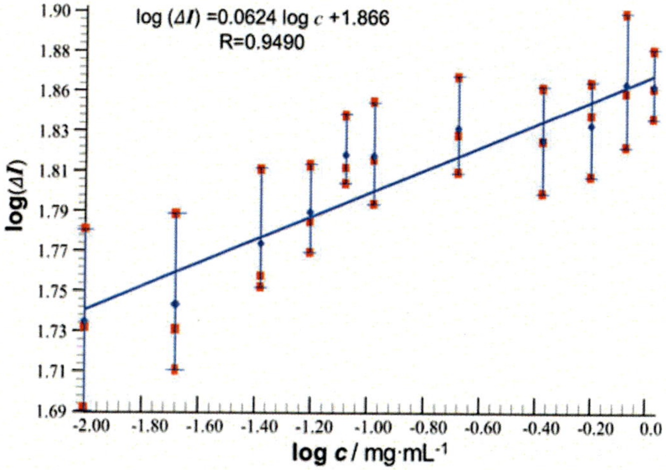

Fig. 4.6 Plot of the logarithm of net gray value, lg Δ*I*, of bovine serum albumin spots against the logarithm of its concentration, lg *c*, in bulk sample solution measured on an imager model SPRi-TX7100 at 25 °C and averaged over 6 spots

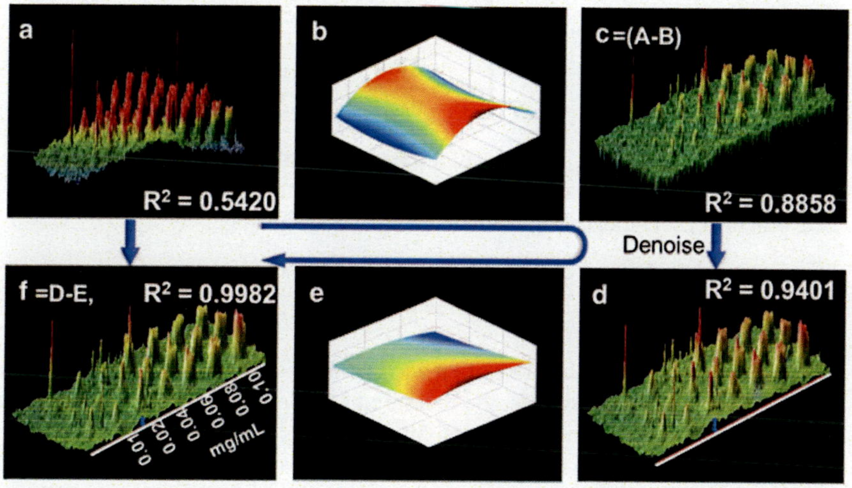

Fig. 4.7 Improvement of quantification performance through, **a–c**background deduction and **d–f** denoise of an image measured on the same conditions as in Fig. 4.6. Reconstructed from Ref. [24] with permission

4.2.9 *Limit of Detection*

The performance of SPRi can also be qualified by LOD corresponding to the bulk refractive index n, bulk concentration c, surface concentration C_{ss} and/or adlayer thickness d_a.

4.2.9.1 Statistic Definition of LOD

In analytical chemistry, the minimum measurable signal $y_{\min}$ is defined by blank (target analyte-free) signal y_{blank} plus several number of its standard deviation σ_y:

$$\begin{cases} y_{\min} = \langle y_{\text{blank}} \rangle + j\sigma_y \\ \sigma_y = \sqrt{\dfrac{1}{i-1}\sum\limits_{i=1}^{i=N}\left(y_{\text{blank},i} - \langle y_{\text{blank}} \rangle\right)^2} \\ \langle y_{\text{blank}} \rangle = \dfrac{1}{N}\sum\limits_{i=1}^{i=N} y_{\text{blank},i} \end{cases}, \tag{4.41}$$

where j is an optional number depending on the degree of confidence, usually going to 3 for 99.7% confidence and i denotes the number of measurements, normally at $N = 20$–30 to have statistical reliability. Statistically, LOD is defined by

$$\begin{cases} x_{\text{LOD}} = \dfrac{y_{\min} - \langle y_{\text{blank}} \rangle}{S} \\ S = \dfrac{\mathrm{d}y}{\mathrm{d}x} \end{cases}, \tag{4.42}$$

where x can be n, c, d_a, or C_{ss} of a measured analyte, and S is the sensitivity of net SPRi response, for example, $\mathrm{d}(\Delta I)/\mathrm{d}c$. Substitute Eq. 4.41 into Eq. 4.42, we have practical LOD measuring formula for 99.7% confidence:

$$x_{\text{LOD}} = \frac{3\sigma_y}{S}. \tag{4.43}$$

Equation 4.43 is the most often adopted measuring formula to estimate the limit of detection. In our studies, the measured LOD of SPRi did not show a significant variation with methods in use of 50-nm-thick gold sensors. Nevertheless, it was largely dependent on noise and background drift or image deformation as discussed in Sect. 4.2.8.2. As a cost for increasing throughput and easing the measurement of optical intensity, SPRi loses its LOD by a factor of ca. 5–10 folds compared with SPRS, ca. 1×10^{-5} RIU bulk solution *vs.* ca. 1×10^{-6} RIU. The LOD of SPRi can be improved obviously by denoise and subtraction of the deformed background from the original signals. An easy denoise technique is to average the signals over a short distance, but the better way is based on mathematic (such as Fourier or wavelet) transform to filter out the non-signal frequencies. Again the deformed image background can be flattened by subtraction of a fitting surface from the original (Fig. 4.7). A further technique to suppress the background drift is to control the temperature of the flowing cell. Finally, mechanical and chemical stability of the index-matching fluid between the sensor and prism should be well considered. Our suggestion is to seal up the interface into a specially designed cell.

4.2.9.2 Estimation of LOD

LOD needs to measure rigorously for a specific SPRi assay because it depends not only on device but also on methodology. However, it is helpful to estimate the value of limit of quantification (LOQ) for some important parameters such as n, c, d_a, or C_{ss}. From Eq. 4.35, we have

$$x_{\mathrm{LOD}} \approx \frac{\Delta I_{\min}}{S_{I,\theta}} = \frac{2d}{\delta_{I,1z}}(n_a - n_s). \tag{4.44}$$

Equation 4.44 allows to estimate different types of LOD from $n_{\mathrm{LOD}} \sim 1 \times 10^{-5}$, for example, the minimum averaged thickness of a uniform single adlayer can be estimated by taking a typical value of $\delta_{I,1z} \approx 180$ nm from a 50 nm gold film against aqueous solution at $(\eta_a - \eta_s) = 0.1$

$$d_{\mathrm{LOD}} = d_{a,\min} \approx \frac{1 \times 10^{-5} \times 180}{2 \times 0.1} = 0.009\,\mathrm{nm} \approx 0.1\,\text{Å}$$

The minimum adlayer is only about 1/15 of a carbon atomic (ca. 1.5 Å in diameter) layer. In case of propylamine, its $(n_a - n_s)$ can be lowered to 0.026 RIU [18], its d_{LOD} can reach 0.35 Å. The minimum averaged thickness can also be transferred to the lowest occupied surface concentration of the adsorptive sites on the gold surface according to Eq. 4.30:

$$\begin{aligned} C_{\mathrm{ss,LOD}} &= \frac{\rho}{M} d_{\mathrm{LOD}}\ (\mathrm{mol/cm^2}) \\ &= \rho d_{\mathrm{LOD}}\ (\mathrm{g/cm^2}). \end{aligned} \tag{4.45}$$

For a protein adsorbate with $(n_a - n_s) \sim 0.24$ RIU and $\rho \sim 1$ g/cm^3, we have $d_{\mathrm{LOD}} \sim 0.04$ Å or $c_{s,\mathrm{LOD}} \sim 0.4$ ng/cm^2. With surface concentration and film thickness, the LOD of analyte concentration in bulk sample solution c_{LOD} can further be estimated by

$$c_{\mathrm{LOD}} = \frac{C_{\mathrm{ss,LOD}} d_{\mathrm{LOD}} A_{\mathrm{eff}}}{V_{\mathrm{cell}}}, \tag{4.46}$$

where A_{eff} denotes the effective sensor surface area (really exposed to sample solution) in the flowing cell with a volume at V_{cell}. The estimation with Eq. 4.46 can have a large deviation due to the not sufficient adsorption of analyte from the solution. A more reliable estimation can be derived from Eq. 4.13. In the case of LOD, β_a will approach a negligible level so that $(1 - \beta_a) \approx 1$, Eq. 4.13 can be simplified into

$$c_{\text{LOD}} = \frac{\beta_{a,\text{LOD}}}{K_a} = \frac{C_{\text{ss,LOD}}}{K_a C_{\text{ss,max}}}, \tag{4.47}$$

where the potential binding surface concentration $C_{\text{ss,max}}$ can be measured by saturated adsorption of the analyte on the sensor. It should be aware that, in general SPRi, the total adsorption sites are determined by the effective number of immobilized probes on the sensor surface. This opens another way to estimate $C_{\text{ss,max}}$. In this case, K_a may change and is better measured *in situ* according to Eq. 4.13. In these measurements, the factor of $e^{-2d_b/\delta_{I,1z}}$ is included in the results. However, this is in common not obvious since $e^{-2d_b/\delta_{I,1z}}$ is very close to unity. LOD may also vary as molecules cluster unevenly on the surface, causing deviation of the experimentally measured data from theoretic calculations.

4.3 Imaging of Discrete Objects

4.3.1 Basic Principle

When the imaging objects like molecules or particles are extremely diluted, the imaging principle for the continuous analyte films has to be modified. Although the common SPRi principle derived from a model of multi-layer reflection shows the relationship of resonance angle with a couple of optical properties as indicated in Eqs. 2.145, 2.146 and 4.2, it does not provide any spatial information on the local reflectivity caused by a reflective object such as NPs. The physics of the reflection-caused point source function (PSF) needs other theoretical models to treat the interaction of an isolated particle with the propagating SPPs [25–27]. There are three components involving in the interference of the propagating SPPs, that is, the partially reflected wave and the particle-scattering wave of SPP that is a decaying cylindrical plasmonic wave. The SPRi intensity is at least the sum of the SPP intensity of a layered sensing system and the object-scattered SPP intensity. Considering the scattered SPP occupies a ratio of γ over the total SPP, from Eq. 2.161, we have

$$I(\boldsymbol{r}) = \left|\boldsymbol{E}_{\text{spp}} + \gamma \boldsymbol{E}_{\text{spp}}\right|^2 = \left|\boldsymbol{E}_{\text{spp}}(\boldsymbol{r}) + \gamma \boldsymbol{E}_{\text{spp}}(\boldsymbol{r}_0) e^{-(i\mathbf{k}+\boldsymbol{\kappa})(\boldsymbol{r}-\boldsymbol{r}_0)}\right|^2. \tag{4.48}$$

This equation counts two terms of wave, a planar wave of SPP (the first term in the right side of Eq. 4.48) and a cylindrical wave of scattered SPP (the last term in Eq. 4.48). Figure 4.8 schematically illustrates the physical interpretation of an SPP

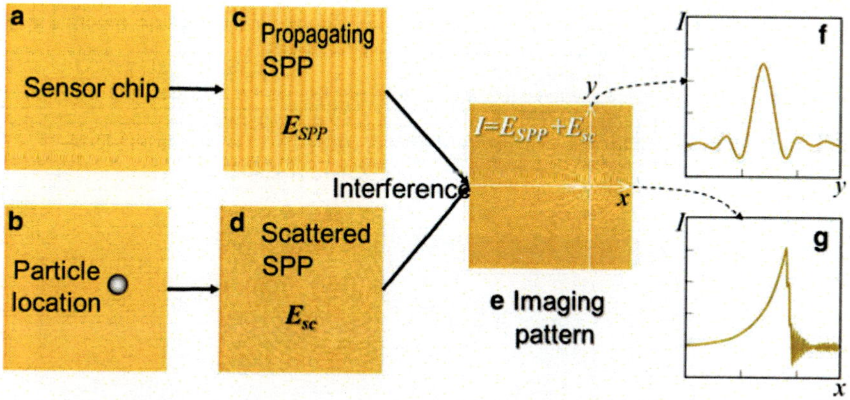

Fig. 4.8 Schematic illustration of the measuring principle in SPRi of a particle absorbed on the sensor surface excited at an incident angle above the critical resonance one. **a** Gold sensor chip; **b** location of a particle on the chip surface; **c** electric field distribution along the propagating direction of an SPP; **d** electric field distribution of the particle-scattered SPP; **e** interfered image of the two fields; **f** imaging intensity across the particle along *y* direction perpendicular to the propagating SPP; **g** imaging intensity across the particle along *x* direction parallel to the propagating SPP

excited on a metal film (Fig. 4.8a, c) and the scattered field $\boldsymbol{E}_{sc}$ on the same surface by a scattering object (Fig. 4.8b, d). They will interfere with each other to produce a unique interference fringe pattern as schematically shown in Fig. 4.8e, featuring C2 symmetry across the propagation direction (Fig. 4.8f) and slow decaying against the propagation direction (Fig. 4.8g).

Similar to the continuous molecular layers, the image contrast of a discrete analyte is also sensitive to the incident angle (Eqs. 2.157–2.159). Figure 4.9 schematically illustrates that, at an incident angle smaller than the resonant, the image contrast is positive, and it becomes negative as the incident angle shifts to above the resonant. The contrast inversion of the image around the resonant angle is due to the phase inversion of the reflected field. It should be noted that, at the resonant angle, the image contrast becomes the smallest (the middle in Fig. 4.9), which is somewhat comparable with the continuous SPRi (Fig. 4.1b).

4.3.2 Image Enhancement

Images with low contrast are hard to recognize. This happens especially in SPRi of discrete objects due to the weak particle-surface interaction to produce low scattering electric field of $\boldsymbol{E}_{sc}$. In addition to the interference with $\boldsymbol{E}_{SPP}$, the weak $\boldsymbol{E}_{sc}$ will further be disturbed by the shot and mechanical noises, making the measured images very blurry. Enhancement of the images are highly necessary, which can reasonably be achieved by theoretical techniques [27]. There are numerous theoretical techniques available such as background deduction, denoise and/or image

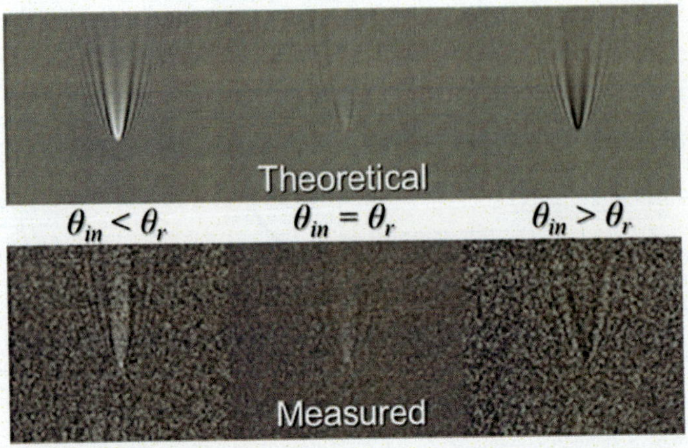

Fig. 4.9 Theoretical (upper) and experimental (lower) SPRi of a single object due to the interference of a planar SPP with its scattered cylindrical wave by the object to produce positive, faint and negative images at an incident angle, θ_{in}, smaller, equal to and larger than the resonance angle, θ_R, respectively. Reconstructed from Ref. [26] with permission

reconstruction by Fourier or wavelet transform, arithmetic or geometric smoothness, contrast adjusting (e.g., via gamma function), differential technology and so forth. The following discussed are several practically useful techniques.

4.3.2.1 Background Deduction

Background deduction is constantly used in SPRi to construct qualified images, not just limited to the discrete objects. The deduction can reduce systematic deviations and long-range fluctuation of signals caused by various sources like surface variations, temperature fluctuation, solution gradient change and time-dependent factors. As in SPRi of continuous adlayers, there are two techniques to deduct the background in SPRi of discrete substances: temporal and spatial deductions. Temporal deduction is performed by taking off the image signal measured at a reference time t_{ref} when target analytes are out of the vision field,

$$I(x, y, t) = I_{\text{raw}}(x, y, t) - I_{\text{raw}}(x, y, t_{\text{ref}}), \tag{4.49}$$

where I_{raw} is the raw or original imaging signal and t_{ref} can be earlier or later than t. It is batter to select t_{ref} at the very beginning of recording (e.g., $t_{\text{ref}} = t_0 = 0$) for real-time analysis. This technique is suitable for treating images independent on time but dependent on space. It is normally used to "flatten" the background of a same chip.

Reversely, for temporally dependent but spatially independent imaging signals, the spatial deduction technique is suggested,

$$I(x, y, t) = I_{\text{raw}}(x, y, t) - I_{\text{raw}}(x', y', t), \tag{4.50}$$

where (x', y') indicates a location free of analytes. This technique is more often used than the temporal one in taking off strong background and some noises to highlight the net imaging signals.

4.3.2.2 Difference Signal

For images digitized into pixels, the difference signal $I(x, y, t)$ can be calculated by following definition:

$$I(x, y, t_{j+1}) = \beta \frac{I_{\text{raw}}(x, y, t_{j+1}) - I_{\text{raw}}(x, y, t_j)}{I_{\text{raw}}(x, y, t_j)}, \tag{4.51}$$

where β is an artificial amplification factor (e.g., 100), $I_{\text{raw}}(x, y, t)$ denotes an original signal for a pixel (x, y) taken at time t. The time interval of $(t_{j+1} - t_j)$ to take the imaging signals can be within several seconds up to even minutes depending on how fast the measuring system changes. Suppose that an object appearing in an original image at time t has a timely proportion of γ, and the original images are mainly determined by an SPP and its scattered fields, the raw intensity will be

$$\begin{aligned} I_{\text{raw}}(x, y, t) &\propto \boldsymbol{E} \bullet \boldsymbol{E}^* = \boldsymbol{E}(x, y)e^{i\omega t} \bullet \boldsymbol{E}^*(x, y)e^{-i\omega t} = \boldsymbol{E}(x, y) \bullet \boldsymbol{E}^*(x, y) \\ &= (\boldsymbol{E}_{\text{spp}} + \gamma \boldsymbol{E}_{\text{sc}})(\boldsymbol{E}_{\text{spp}} + \gamma \boldsymbol{E}_{\text{sc}})^* \\ &= \boldsymbol{E}_{\text{spp}} \bullet \boldsymbol{E}^*_{\text{spp}} + \gamma^2 \boldsymbol{E}_{\text{sc}} \bullet \boldsymbol{E}^*_{\text{sc}} + \gamma \text{Re}(\boldsymbol{E}^*_{\text{spp}} \bullet \boldsymbol{E}_{\text{sc}}) + \gamma \text{Re}(\boldsymbol{E}_{\text{spp}} \bullet \boldsymbol{E}^*_{sc}) \\ &= E^2_{\text{spp}}(x, y) + \gamma^2 E^2_{\text{sc}}(x, y) + 2\gamma \text{Re}[E^*_{\text{spp}}(x, y) E_{\text{sc}}(x, y) \cos\theta] \\ &= E^2_{\text{spp}} + \gamma^2 E^2_{\text{sc}} + 2\gamma \text{Re}[E^*_{\text{spp}} E_{\text{sc}} \cos\theta], \end{aligned} \tag{4.52}$$

where $\boldsymbol{E}^*$ denotes the complex conjugate of $\boldsymbol{E}$ and θ is the angle between two vectors. The time variable can thus be removed from I_{raw}. Combine Eq. 4.52 with Eq. 4.51, we can have the difference intensity as follows:

$$I(x, y, t) \propto \frac{(\gamma^2_{j+1} - \gamma^2_j) E^2_{\text{sc}} + 2(\gamma_{j+1} - \gamma_j) \text{Re}(E^*_{\text{spp}} E_{\text{sc}} \cos\theta)}{E^2_{\text{spp}} + \gamma^2_j E^2_{\text{sc}} + 2\gamma_j \text{Re}(E^*_{\text{spp}} E_{\text{sc}} \cos\theta)}. \tag{4.53}$$

Equation 4.53 shows that the absolute variance of SPR events is dependent much more on angle than on time in the common case of $\boldsymbol{E}_{\text{spp}} >> \boldsymbol{E}_{\text{sc}}$. Note, Eq. 4.53 may over parametrize. To reduce this risk, $\boldsymbol{E}_{\text{sc}}$ is assumed to have only contribution to the latter image while negligible to the former, that is, $\gamma_{j+1} \approx 1 >> \gamma_j \approx 0$, and thus Eq. 4.53 can be simplified into

$$I(x, y, t) \propto \frac{E^2_{\text{sc}} + 2\text{Re}(E^*_{\text{spp}} E_{\text{sc}} \cos\theta)}{E^2_{\text{spp}}}. \tag{4.54}$$

Equation 4.54 is suitable for determining the relative intensity change in a fixed time interval with more than one indistinguishable interaction events that blend to a statistical distribution. It must be noted that Eq. 4.53 determines the absolute value of the interaction events with time as a variable, while Eq. 4.54 determines only the relative intensity change in a fixed time span. Because more than one indistinguishable interaction events can occur in a giving time interval, and these events will blend to a statistical distribution, it is more preferable to use Eq. 4.54 rather than Eq. 4.53 to reduce the over-parametrization issue.

The noises in the difference images can be filtered by fast Fourier transform (FFT) technique as illustrated in Fig. 4.10. First an image is pictured according to Eq. 4.51 as shown in Fig. 4.10a. It is then transformed by FFF into a picture with double spheres in frequency space (Fig. 4.10b). The double spheres are filtered by masking its inside and outside areas (Fig. 4.10c), with a radius determined by the critical resonance condition $k^2 = {\omega_x}^2 + {\omega_y}^2$. Finally, the filtered double spheres are reversely Fourier-transformed to produce an image (Fig. 4.10d) purer than its original (Fig. 4.10a). With the improved image, the maximum intensity at the object point becomes clearer and more suitable for counting the imaged objects. Thus, if only the point intensity $I_{\max}$ of each object is stored or displayed, the storing space can be largely saved and the object sites will be highlighted.

In order to dig up more information from the treated images, their intensity varying tendency must be built up, which can be obtained by fitting technique in association with some SPR-related parameters. Considering $I_{\text{fit}} = (I + I_0) \propto \boldsymbol{E}^2$, where I_0 is background intensity, a fitting equation can be constructed by simulation of Eqs. 2.156, 2.158 and 4.51,

$$\begin{cases} I_{\text{fit}} = I_0 + \dfrac{q^2\left|\boldsymbol{H}_0^{(2)}(\boldsymbol{k}\bullet\boldsymbol{r})\right|^2 + 2q\text{Re}\left[-i\boldsymbol{H}_0^{(2)}(\boldsymbol{k}\bullet\boldsymbol{r})\bullet\frac{-i}{\boldsymbol{k}}e^{-i\boldsymbol{k}\bullet\boldsymbol{x}}\cos\theta\right]}{\left|-\frac{i}{\boldsymbol{k}}e^{-i\boldsymbol{k}\bullet\boldsymbol{x}}\right|^2} \\ q = \dfrac{Q_2}{Q_1};\quad r = \sqrt{x^2+y^2} \end{cases}, \quad (4.55)$$

where q is the field enhancing parameter specific to an imaging object. Note that $\boldsymbol{k}$ is a complex vector. The fitted function contains, in fact, only the primary propagation direction, along the x axis herein.

Simulation from Eqs. 2.162 and 4.51, another fitting equation can be derived,

$$I = \frac{\left|\boldsymbol{E}_{\text{spp}}(\boldsymbol{r}) + \sum\limits_{n=1}\alpha_n\boldsymbol{E}_{\text{spp}}(\boldsymbol{r}_{n,0})e^{-i\boldsymbol{k}\bullet(\boldsymbol{r}-\boldsymbol{r}_{n,0})-\kappa(\boldsymbol{r}-\boldsymbol{r}_{n,0})}\right|^2 - \left|\boldsymbol{E}_{\text{spp}}(\boldsymbol{r})\right|^2}{\left|\boldsymbol{E}_{\text{spp}}(\boldsymbol{r})\right|^2}. \quad (4.56)$$

For $n = 1$, it gives

$$I = 1 + \frac{\alpha}{e^{\kappa(\boldsymbol{r}-\boldsymbol{r}_0)}}\left(\frac{\alpha}{e^{\kappa(\boldsymbol{r}-\boldsymbol{r}_0)}} + \frac{1+e^{i2k(\boldsymbol{r}-\boldsymbol{r}_0)}}{e^{ik(\boldsymbol{r}-\boldsymbol{r}_0)}}\right). \quad (4.57)$$

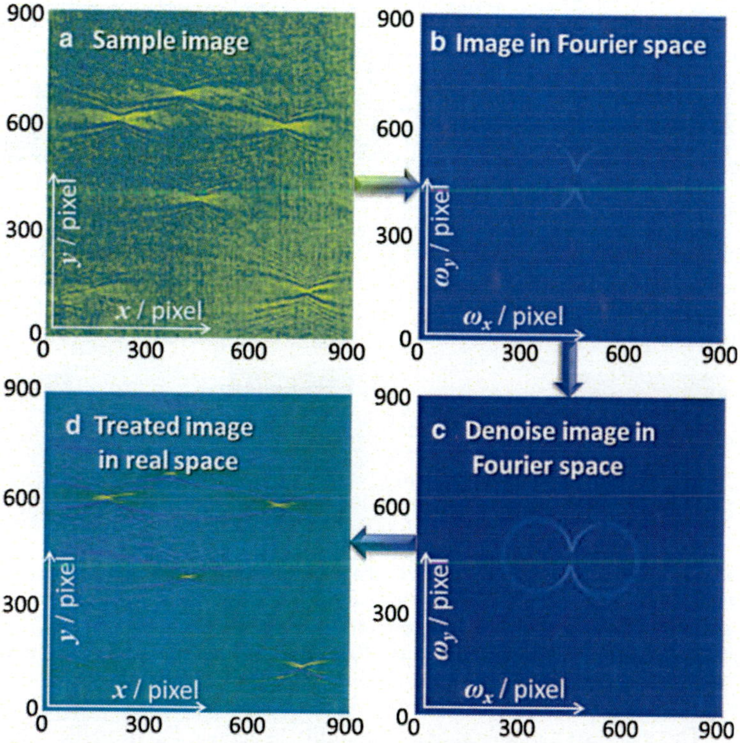

Fig. 4.10 Fast Fourier transform image denoising pictured with false color, where the dark circle in the center of Fourier-transferred images is used to highlight the images. **a** A sample image pictured after Eq. 4.51 (with $\beta = 100$ and $t_{j+1} - t_j = 3$ s); **b** Fourier-transformed frequency image; **c** noise-filtered image in Fourier space; **d** recovered image by reverse FFT. Redrawn from Ref. [27] with permission

4.3.2.3 Image Reconstruction

Due to the interference of $\boldsymbol{E}_{\rm sc}$ with $\boldsymbol{E}_{\rm spp}$, the measured images of strong scattering objects like nano and microparticles are accompanied by parabolic tails (Fig. 4.9) that leads to poor lateral resolution and disturbs the recognition of imaging objects (Fig. 4.10a). Unfortunately, this interfering issue cannot be removed by noise filtering and/or signal amplification measures. It needs some special mathematic techniques to suppress or even clear off the tails. An often adopted technique is to reconstruct the image with the aid of Fourier transform as schematically illustrated in Fig. 4.11.

According to Eq. 2.156, we have

$$I = \left|\boldsymbol{E}_{\rm spp} + \boldsymbol{E}_{\rm sc}\right|^2. \tag{4.58}$$

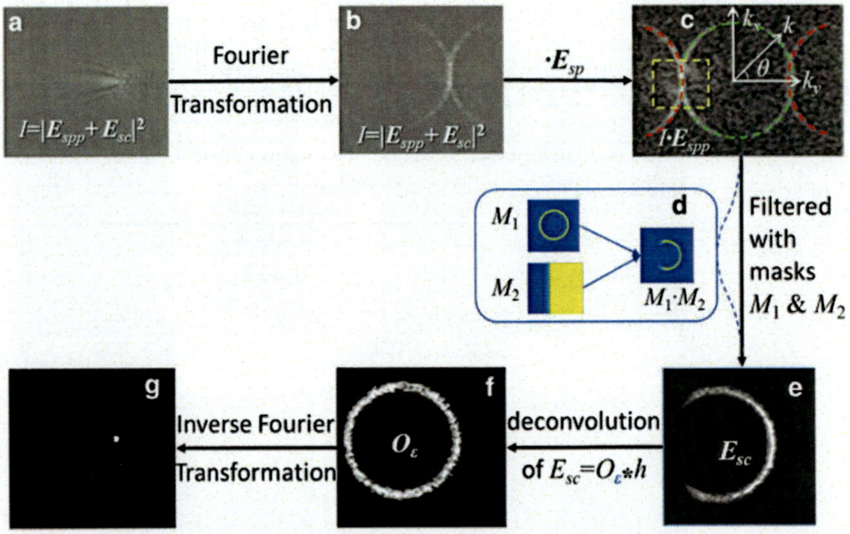

Fig. 4.11 Route to reconstruct an SPR image via Fourier transform. **a** Measured image of a scattering object (e.g., a particle); **b** image in Fourier space; **c** Convolution with an SPP; **d** two masks, M_1 and M_2, for filtering noise; **e** image filtered with the masks; **f** deconvolved image; **g** reconstructed images by inverse fast Fourier transform. Reconstructed from Ref. [29] with permission

This suggests a mathematical possibility to deduce $\boldsymbol{E}_{\text{sc}}$ from the detected I; therefore, it possible to reconstruct the image with the derived $\boldsymbol{E}_{\text{sc}}$. To simplify the treatment, a single analyte is considered [28]. In mathematical physics, $\boldsymbol{E}_{\text{sc}}$ is the convolution (*) of the spatial distribution of the objective permittivity, O_ε, and the point spread function, h, of SPPs [26, 29]

$$E_{\text{sc}} = O_\varepsilon * h, \tag{4.59}$$

where h can be expressed as a cylindrical function with amplitude decaying along the propagation direction

$$h = \chi e^{-\boldsymbol{\kappa}\bullet\boldsymbol{r}-i\boldsymbol{k}\bullet\boldsymbol{r}}, \tag{4.60}$$

where χ is the polarizability of the scattering object (e.g., NPs) and $\boldsymbol{r}$ is radius to the scattering center. To extract $\boldsymbol{E}_{\text{sc}}$, we have to multiply I with $\boldsymbol{E}_{\text{spp}}$,

$$I \bullet \boldsymbol{E}_{\text{spp}} = \left(\left|\boldsymbol{E}_{\text{spp}}\right|^2 + |\boldsymbol{E}_{\text{sc}}|^2\right) \bullet \boldsymbol{E}_{\text{spp}} + \boldsymbol{E}_{\text{spp}}^2 \bullet \boldsymbol{E}_{\text{sc}}^* + \left|\boldsymbol{E}_{\text{spp}}\right|^2 \bullet \boldsymbol{E}_{\text{sc}}, \tag{4.61}$$

where the first term on the right side of Eq. 4.61determines a planar wave, the second determines the conjugate of the scattered wave, and the third determines the scattered field that is in theory a sphere in Fourier space (indicated by green dashed circle in

Fig. 4.11c). They three also overlap partially in Fourier space (dashed square in Fig. 4.11c).

It can be imagined that the Fourier-transformed sphere of $\boldsymbol{E}_{sc}$ determines the image contrast of the scattering object in real space. To isolate it from the other two terms of Eq. 4.61, a circular filtering mask M_1 at $z = 0$ (Fig. 4.11d) is designed in Fourier space:

$$M_1(k_x, k_y) = e^{-\frac{\left(\pm\sqrt{k_x^2+k_y^2}-k_0\right)^2}{k_{s1}^2}}. \tag{4.62}$$

M_1 equals 1 at k_0 and k_{s1} is a decaying constant that determines the thickness of the circle. M_2 is designed to reduce the noises away the circular line as follows:

$$M_2(k_x, k_y) = \begin{cases} 1 & k > k_1 \\ e^{-\frac{(k_x \cos\theta + k_y \sin\theta - k_1)^2}{k_{s2}^2}} & k < k_1 \end{cases}, \tag{4.63}$$

where θ determines the propagation direction of an SPP (Fig. 4.11c) while k_1 and k_{s2} must be chosen according to the signal-to-noise ratio around k_0.

With these two masks, $\boldsymbol{E}_{sc}$ is isolated through

$$\mathscr{F}[\boldsymbol{E}_{sc}] \sim \mathscr{F}\left[I \bullet \boldsymbol{E}_{sp}\right] \bullet M_1 \bullet M_2 \tag{4.64}$$

where $\mathscr{F}$ is Fourier transform operator. $\boldsymbol{E}_{sc}$ can further be treated as follows to withdraw O_ε:

$$\because \mathscr{F}[\boldsymbol{E}_{sc}] = \mathscr{F}[O_\varepsilon * h] = \mathscr{F}[O_\varepsilon] * \mathscr{F}[h]$$

$$\therefore \begin{cases} \mathscr{F}[O_\varepsilon] = \dfrac{\mathscr{F}[\boldsymbol{E}_{sc}]}{\mathscr{F}[h]} \\ O_\varepsilon = \left|\mathscr{F}^{-1}\{\mathscr{F}[O_\varepsilon] \bullet M_1\}\right| \end{cases} \tag{4.65}$$

where M_1 is applied once again to reduce the potential noise amplified by the denominator $\mathscr{F}[h]$. In theory, O_ε draws a bright spot with the parabolic tail removed (Fig. 4.11g). Clearly, this technique is applicable to any one discrete object and is expandable to more discrete objects.

4.4 Preparation of Imaging Samples

Preparation of imaging samples is meant in this book not only the pretreatment of target analytes but also the transfer of the analytes onto the sensor surface. Similar to other instrumental analysis, SPRi may need a long term to prepare the imaging samples depending on analytes, measuring purpose and approaches to be used. Because SPRi can measure a broad spectrum of samples, including adsorptive, transferable and/or soluble components in solids, soft matter and other matrices (including gases), the ways to prepare imaging sample are also variable. Following discussed are several typical techniques.

4.4.1 Pretreatment of Imaging Samples

SPRi usually measures pure analytes that are separately spotted in addressable locations on the sensor chip. Therefore, the imaging samples need to be pre-extracted and/or purified from different sources before spotting. Various techniques may thus be concerned such as extraction, filtration, separation and concentration or dilution.

In common SPRi measurements, it is rare to directly spot the pure analytes on the sensor chips but instead, the target analytes are usually captured by specific probes pre-spotted on the sensor surface. In this case, mixed analytes become measurable by directly application or flow of a sample solution to the imaging surface. Sample preparation requirements can thus be greatly reduced to save samples, time and cost, and in turn to improve analytical precision. With highly specific probes, even natural samples such as biological fluids can also be analyzed directly with SPRi.

Particularly for quantification, a series of standard solutions have to be prepared at different concentrations; while for identification, blank, positive and/or negative controls have to be prepared. The bank control is always required in SPRi to have background for deduction, while the positive and negative controls are critical for the analysis of biological events.

4.4.2 Contact Transfer Technology

Analytes may be on/in insoluble substrates (especially soft matter like tissues) or in substances unable to be seriously treated or needed to keep for further inspections. In these cases, the analytes may be imaged after direct or indirect transfer. Direct transfer is performed by application of a target sample onto the sensor surface while indirect transfer may be conducted by various ways, for example, through a proper membrane for spatial fidelity. Most often, the sample is sliced or cut to have a suitable surface for contacting with the sensor surface. Flat sections are more easily imaged on a SPRi sensor surface than other shapes of samples. To have a correct contact of the cut or sectioned surface with the sensor for better signal, the sampling surface may need to be gently pressed onto the sensor surface. In practice, SPRi of tissue sections can be conducted, where direct contact of the tissue section with the sensor chip surface is applicable. However, it should be noted that inappropriate contact transfer technology will lose image fidelity, e.g., distortion or reduction of image contrast. The better solution is to develop fidelity transfer technology that can be achieved, according to our experience, by insertion of a layer of buffer (<1 mm in thickness) or better a layer of buffer-wetted nylon membrane with 0.4 μm pore size in between the section and sensor surface. Note, the transfer assembly should be kept silently to avoid blurring effect caused by lateral mass transfer due to vibration, diffusion, thermal fluctuation and other heterogeneous factors. The lateral blurring effect can be suppressed by fast chemical fixation techniques, for example, fast capture of target analytes with specific probes pre-immobilized on the sensor surface. To this end, the sensor chip must be pre-modified, normally via self-assembly of a monolayer of mercaptan at different length of alkyl. To adjust polarity, further deposition of hydrophilic (e.g., cellulose or nitrocellulose) or hydrophobic (e.g., Teflon or CYTOP that is a trade name of perfluoro(1-butenyl vinyl ether) polymer) may be required, which will be discussed later.

Polarity matching technology is another solution to the lateral blurring effect. On a metal sensor surface, a sample solution just spotted or transferred from a tissue section can either spread out or shrink in along the surface due to mismatching of polarity between the sample solution and sensor surface, giving tortured or even wrong pictures in respect of their location, distribution and size. To restrict these unwanted position variations for higher spatial fidelity, the sensor surface has to be modified so as to have opposite polar toward the sample liquid. In case of transferring hydrophilic compositions via aqueous buffer from a section, it is better inserting a hydrophobic dense mesh between the section and sensor to confine the liquid path, as has been mentioned above.

In transfer of some macromolecules such as proteins from tissue sections, cleaning and chemical fixation of the sections are required. Usually, the sections are subjected to de-lipids by quick (ca. 30 s) washing in about 70 ~ 90% isopropanol for 2–3 times and then dried in a desiccator. The dried sections can be stored in cryogenic refrigerators for later use.

In imaging skins, direct contact or touching transfer is optional. This can be exemplified by SPRi of a fingerprint transferred by direct and gentle touching the sensor surface with your finger.

4.4.3 Mechanical Spotting Technology

Addressable spotting of chemical solutions (not just limited to samples) onto a sensor surface is more or less a routine task in SPRi since the dotted chips are much more often used than other types of chips. In theory, different substances can be arrayed on a same sensor chip by manual or automatic mechanical spotting technology. Manual spotting can simply be performed with a microsyringe that allows to inject a given volume of sample from the needle tip onto the desired location of the chip. The syringe can be replaced by a capillary under pressure to force the sample solution to flow (Fig. 4.12). The pressure can be formed by compressed gas or a pump. The desired position to be deposited is adjusted manually or by program. Figure 4.12a schematically illustrates the designed structure and Fig. 4.12b its real system fabricated in our

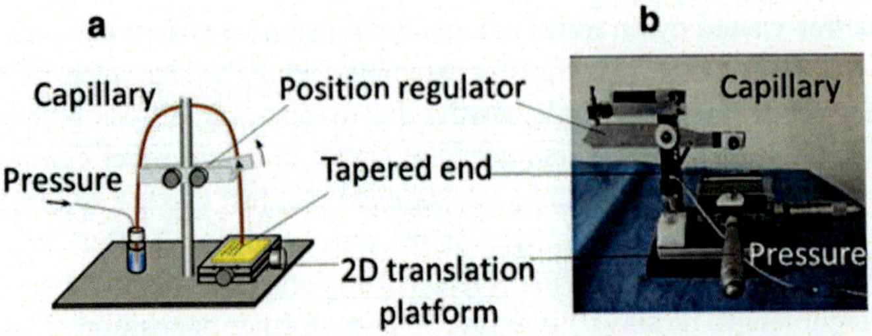

Fig. 4.12 Laboratory-fabricated microspotting device. **a** Designed structure; **b** photo of a practical device

laboratory. Fully automatic spotting is usually performed on a commercial spotting machine.

Given following are two practical approaches what we developed:

Approach 4.1, For direct spotting thiol-modified probes onto a gold sensor chip,

Step 1, clean the gold surface (normally on glass chips) thoroughly with ethanol and ultrapure water, dry it under a nitrogen gas stream and activate the surface in a plasma cleaner for 1 min.
Step 2, lay a gold chip on the 2D stage (Fig. 4.12) and spot thiol-modified samples or probes by pressing the solution vial to make a solution flow through the capillary and drop in the desired location on the gold surface one after another, and from solution to solution, where the location is controlled by manually or automatically adjusting the 2D stage, with a space between two spots > 40 μm.
Step 3, incubate the spotted sensor chip(s), overnight, in a wet chamber at 80% humidity and room temperature.
Step 4, soak the chip(s) in 1 ~ 10 mg/mL bovine serum albumin (BSA) solution for 30 min to block the remaining active and non-specific adsorption sites.
Step 5, wash the chip(s) thoroughly with ultrapure water for three times and dry under a nitrogen gas stream.

This approach is often applied to the immobilization of thiol-modified nucleic acids and their fragments. The resulted chips can be used directly or stored in dark at 4 °C for about a week.

Approach 4.2, For spotting analytes on amine-modified chip surface,

Step 1, clean and dry a gold sensing chip as in Approach 4.1.
Step 2, modify the bare gold surface with a monolayer of 11-mercaptoundecylamine ($MUNH_2$) by immersing the bare gold chip(s) into 1 ~ 5 mM $MUNH_2$ in ethanol overnight, rinse the surface with propanol, water and acetone, and dry it under a nitrogen gas stream.
Step 3, activate the $MUNH_2$-modified surface in (400 ~ 500) mM N-ethyl-N'-(3-dimethylaminopropyl)carbodiimide (EDC) and (100 ~ 200) mM N-hydroxysuccinimide (NHS) for 20–40 min at room temperature, rinse the chip

with running buffer (normally, phosphate buffered saline or PBS) for > 5 min, and dry it under nitrogen gas stream.

Step 4, array-spot analyte solutions (several μmol/L analyte in 10 mmol/L PBS at pH7.0 or 10 mmol/L acetate at pH4.5) onto the chip and incubate it in a wet chamber at 80% humidified and room temperature for ca.2 h.

Step 5, freeze the chip by nitrogen liquid and immediately transfer it into 1–10 mg/mL BSA or 1 mol/L ethanol-HCl (pH8.0) solution to block the unreacted sites and non-specific adsorption sites for 30 min.

Step 6, wash the chip with water, dry it under nitrogen gas stream, and mount it onto SPRi system for measurements.

This approach is often used to spot protein and peptides through reaction of amino group with carboxylic terminal. The resulted chis are better used as soon as possible but can be stored at 4 °C for about a week.

4.4.4 Photochemical Spotting Technology

Instead of mechanical technology, photochemical reactions can also be explored as addressable sporting technology. This technology needs to combine with finely designed masks to light up the target locations. The photochemical reactions are flexible but may be limited by the unavailability of appropriate specific photoresponsive reagents. To reduce the dependence on the availability of photoresponsive reagents, it can be merged with lithographic technology. The excellent feature of photochemical spotting technology is that it enables the preparation of high-density sample-dotted chips, with a lateral resolution at least finer than 40 μm (Fig. 4.13) that allows the dot density up to near 2.0×10^4 dots/cm^2. According to Fig. 4.13c, there is a possibility to reach a lateral resolution down to ca. 4 μm, comparable with the reported data in literature [30–36]; thus, the spotting density can reach near 0.20×10^6 dots/dcm^2. The measured lateral spatial resolution of a real chip depends on several parameters including the inherent resolution of SPRi, the mask spatial resolution, molecular diffusion effect and light diffraction. In theory, the inherent resolution is determined by the lateral propagation length of SPPs that is at the level of about micrometers depending on the incident light wavelength. Molecular diffusion appears wherever a concentration gradient creates, which raises lateral broadening effects. The diffusion can be suppressed by increase of solution viscosity to be spotted or by polarity matching technology.

It is fortunate that the practically required dotting density is often at about dozens of dots per square centimeter, far below a thousand (Fig. 4.13b). The chips of very high density are used only in rare case.

Approach 4.3, For patterning a gold and/or silver surface with photochemical spotting technology that is often combined not only with photolithography but also with

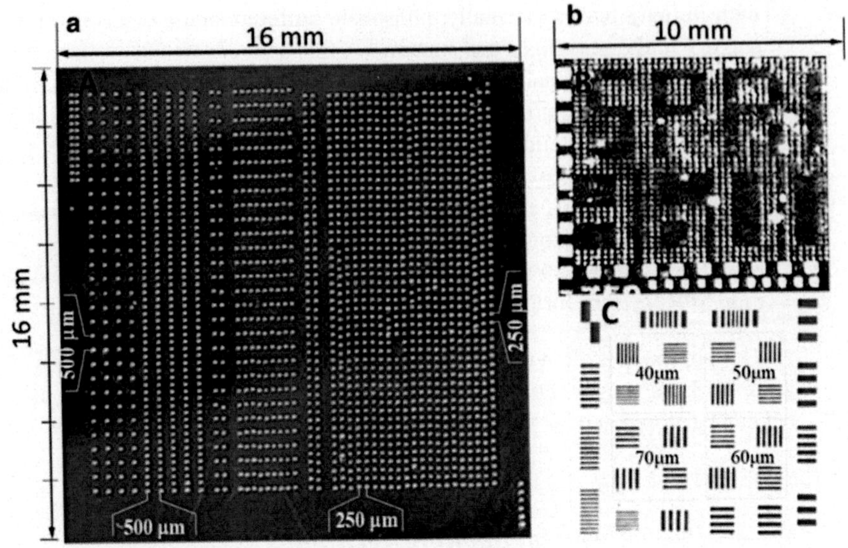

Fig. 4.13 SPR images of BSA spots measured on SPRi-PX8100. **a** BSA spotted at different intervals; **b** patterned BSA with spot interval at 40 μm; **c** patterned BSA lines at different interval as indicated to clarify usable SPRi resolution

surface chemistry and mechanical spotting technology, including following critical steps:

Step 1, functionalize the gold sensing surface with photosensitive mercapto-alcohols or amines.
Step 2, pattern the functioned gold surface under a mask by UV exposure.
Step 3, drop target probe or sample solutions and controls onto their set position.

It should be noted that lithographic techniques themselves are not addressable and the samples should be addressed by localized mechanical sporting or specific photochemical, biochemical and/or chemical reactions, much the same as in the preparation DNA chips.

In common, the chemicals can be anchored on gold or Ag surfaces through physical adsorption, chemical and/or biochemical immobilization. Physical adsorption is simple but normally used in the production of disposable chips. Chemical or biochemical immobilization is a way to prepare stable and re-usable chips. Such chips may be stored and used for a long time. They can thus serve as standard chips for calibration of instruments in qualitative and/or quantitative analysis.

To spot samples, the sensor chips have to be pretreated by rinse and possibly chemical modifications. For detailed chip preparation including spotting, please refer to next section.

4.5 Preparation of Sensor Chips

Currently, there are commercial chips available. Nevertheless, the costed commercial chips may not necessarily meet with your needs. It remains critical to master the technology for the fabrication of your won chips. In the previous section, the preparation of sensor chips has been discussed in part because it can hardly be distinguished from sample pretreatments. As a vital step in the development of SPRi methods, chip fabrication involves not only the sample preparation but also many other aspects such as deposition of sensing metal on a support (commonly a glass slide), surface reformation of the deposited metal, surface modification, probe arraying and immobilization, further clean and assembly of a chip onto SPRi device for further experiment.

4.5.1 Deposition and Reformation of a Metal Film on Glass Slides

There are several techniques to deposit metal films on a glass surface, such as vacumed evaporation deposition, ion sputtering (e.g., magnetron sputtering), pulse laser deposition, chemical vapor deposition, wet chemical deposition (e.g., sol-gel deposition), electroplating and its combined technique [37]. Among these techniques, ion sputtering produces usually the best gold sensor chips but it costs somehow to equip in a common laboratory. Therefore, vacuumed vapor deposition is often the choice for it is relatively cheap and easy to equip for the deposition of gold, silver and other substances on glass slides. The issue is that the resulted metal chips may vary in thickness that depends very much on their location in the chamber. These chips are suggested to be checked in respect of smoothness and thickness. The advanced technique is to observe the chip by AFM but more often we measure their transmission spectra at 532 nm or their SPR absorbance. Only the chips that have a correct optical absorbance are selected and used.

To make a noble metal strongly adhere on a glass surface, a thin ($\leq$ 2 nm) layer of adhesive mater is required. The most often used adhesive layer is chromium or titanium. According to their permittivity, titanium is bit better than chromium but chromium is more often used in practice due to its mature deposition technology and convenience. However, both chromium and titanium are opaque, easily affecting the SPR strength. This can be reduced by use of transparent adhesive matter. This can be exemplified by ZnS that is transparent at wavelength above 380 nm and can be evaporated at 1200 °C and 10^{-2} Pa. ZnS also features a high refractivity at 2.3–2.6 in the range of visible light.

Approach 4.4, A general program for deposition of gold film on glass slides:

Step 1, select glass slides in batch to save time and cost.
Step 2, brush the slides in water to remove possible particles and wash them in ethanol, acetone, isopropanol and water again under sonication to remove

organic and inorganic substances. Note, for used or dirty slides, serious oxidative cleaning should be performed in piranha solution (prepared by slow addition of 30% hydrogen peroxide into concentrated sulfuric acid at a volume ratio between 1:3 and 1:7 at around 70 °C) for 30 min. Caution: Piranha solution is corrosive and must be used in a fume hood under protection!

Step 3, blow dry the cleaned slides with compressed nitrogen gas or on a heating plate at 120 °C for 30 min (in a clean room).

Step 4, place the cleaned and dried glass slides, Cr (2.2 mg per batch) and Au (88 mg per batch, both in a tungsten wire basket) into the vacuum chamber of a vacuum evaporation coating machine, seal the chamber by a bell-shaped cover, turn on first the mechanic vacuum pump under water-thermostat and then the oil pump to lower the inner pressure from about 3 to 10^{-3} Pa.

Step 5, evaporate deposition of first ca. 2 nm Cr at 9–15A and 3–4 × 10^{-4} Pa for 10 min, then 48 nm Au at 7–13 A for 30 min. Note, the deposition current must be pre-selected that depends on the melting point of metal, for example, Au at 1063 °C, Ag at 960 °C, Al at 660 °C, Cu at 1083 °C, Cr at 1857 °C and Ti at 1678 °C, respectively.

Step 6, de-vacuum after the system is cooled to room temperature. Note, the quality of newly deposited Au films can be improved to some degree by annealing and reforming at a certain temperature.

Step 7, check all the gold-coated slides under microscope and kick off chips with observable defects

Step 8, measure the optical absorbance of the checked chips at 532 nm on a UV spectrometer or measure the θ_r or λ_r on an SPRi instrument (against air) to screen out usable chips for later use.

If there is not an expensive coating equipment for the time being, wet chemistry or electroplating can also be used to prepare gold film. Following introduced is an electroless coating approach for plating gold on glass slides [37] that was developed by our laboratory to lower the technical threshold of preparing gold sensor chips.

Approach 4.5, For rapid plating of a limited number of gold sensor chips:

Step 1, preparation of ~ 2.5 nmgold seeds: add 1 mL of 1% aqueous $HAuCl_4 \cdot 3H_2O$ into 100 mL H_2O under vigorous stirring at room temperature, add 1 mL 1% aqueous trisodium citrate in 1 min and mix the solution for another 1 min; further add 1 mL 0.075% $NaBH_4$ (in 1% trisodium citrate), let the solution to react for 5 min and cool the reacted solution at 4 °C until use.

Step 2, microwave-accelerated coating of glass slides with APTMS: immerse a well cleaned glass slides in 10% APTMS in methanol and accelerate the coating reaction by placing them in a microwave oven at the center position and irradiation at 118 W for 1 min; clean the coated slides with copious amount of methanol and water and use them immediately.

Step 3, microwave-assisted assembly of a monolayer of gold seeds on glass slides: immerse cleaned glass slides, with or without a fresh APTMS coating, into the gold seed solution and irradiate them in microwave oven at 118 W for 6 and 2 min

for the bare and coated glass slides, respectively. Note, the seeded glass slides are better subjected to direct electroplating rather than storage.
Step 4, electroless plating of gold film on the seeded glass slides: immerse the freshly seeded glass slides in an aqueous solution of 0.1% $HAuCl_4 \cdot 3H_2O$ and 1 mM hydroxylamine hydrochloride under agitation for about 10 min, thoroughly rinse them with water and blow them dry under a nitrogen gas stream. Note, the gold film forming on the substrates will change its color from pink to purple, to blue and finally to golden as its thickness increased, which helps the judgment of coating time; to have gold film at a required thickness (e.g., 50 nm), the coating time is better re-optimized for a particular laboratory.

4.5.2 Preparation of Gold Microarray

Although the sensor chips usually used in SPRi are fully covered with continuous gold film, it will facilitate the preparation of patterned spots if the whole film can be segmented into microarray with boundary having desired polarity. This is possible by use of photolithography as shown in Fig. 4.13 and Approach 4.6 that is adjustable developed in our laboratory.

Approach 4.6, For photolithographic preparation of on-glass gold microarray with hydrophobic boundary:

Step 1, heat a cleaned and dried gold-coated chip at 80 °C for 1 min.
Step 2, spin-coat the gold surface with a positive photoresist and dry the coating at 80 °C for 10 min.
Step 3, expose the photoresist coating, under a mask, to a UV light (normally for 20 s but dependent on the type and thickness of the photoresist coated).
Step 4, develop the exposed chip in a solution (e.g., 0.4% (w/v) NaOH for ca. 90 s) to remove the unwanted photoresist coating lines.
Step 5, etch the exposed gold surfaces in an aqueous solution of 2.5% (w/v) I_2 and 4.0% (w/v) KI for ca. 7 min at room temperature, clean and dry the chip at 80 °C for 10 min.
Step 6, spin-coat the chip with a polymerizing or curing solution, for example, hydrophobic CYTOP that can be formed from a solution consisting of an initiator and a monomer at a volume ratio of 1:10 to fill the boundary lines.
Step 7, lay the chip in open air for 30 min, heat it at 90 °C for 2 h, and immerse it, after cooling, in acetone for 1 min, sonicate for seconds, wash it with water and blow dry with nitrogen gas.

The critical step D shown in Fig. 4.14 is variable depending on what polar material is to be coated. To prevent a hydrophilic solution from spreading laterally, a hydrophobic material should be used to form hydrophobic boundary, and oppositely, to restrict hydrophobic solutions, a hydrophilic material needs to be coated.

Alternatively, the continuous gold surface can also be segmented into pseudo microarray by photochemical technology as is exemplified by Approach 4.7.

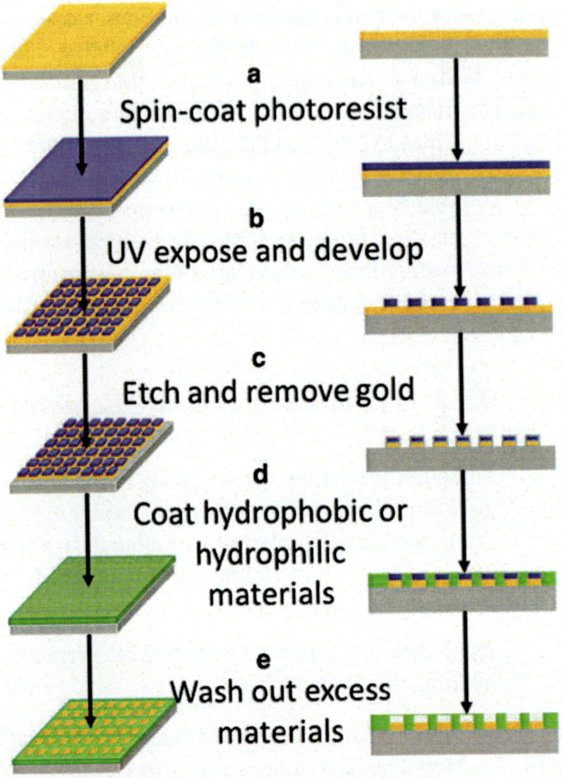

Fig. 4.14 General process to prepare gold flake array for SPRi use

Approach 4.7, For photochemical deposition of hydrophobic gridlines on gold surface:

Step 1, insert a gold sensing chip into 1 mM $MUNH_2$ at 4 °C for 12 h (to assemble a monolayer of $MUNH_2$), rinse the chip with propanol, water and acetone, and blow dry with nitrogen gas.

Step 2, modify the amino terminal with hydrophobic fluorene methoxycarbonyl (Fmoc).

Step 3, expose the chip, under a mask with transparent squares (e.g., $200 \times 200 \mu m^2$), to a strong UV (e.g., mercury xenon arc lamp at 400 W) for 60 min to cleave Au–S bounds, rinse the chip thoroughly with ethanol and blow dry by nitrogen gas, which leaves hydrophobic boundary around all the exposed gold surfaces.

Step 4, functionalize the exposed gold surfaces with desired mercato-containing substances.

4.5.3 Chemical Modification of Chip Surface

Gold-coated chips usually need modifications for variable purposes. Although gold is stable in solutions except for aqua regia and many ion liquids, many metals like Ag are instable and need protection to avoid corrosion. Nearly, all metal surface tends to adsorb substances non-specifically, which has to be blocked in SPRi of many biological macromolecules like proteins that tend to non-specifically adsorb on various surface including gold film. Surface coating is the most often used modification technology, and the frequently adopted coatings are hydrophilic polymers such as cellulose, chitosan, starch, polyethylene glycol (PEG) and polyvinyl alcohol (PVA). Chip surface also needs to be activated or functionalized to enable further modifications or to spot samples or probes.

The modification can normally be realized within one or more steps by use of surface chemistry to be discussed next section. For coatings with either polyhydroxyl or amino groups, a bridging reagent of cyanuric chloride (CC) is suggested. Following is an approach to chemically activate gold microarray surfaces aiming at spotting DNA probes [38, 39].

Approach 4.8, For spotting DNA probes able to suppress cross-contamination:

Step 1, modify chips prepared in Approach 4.7 by self-assembly of mercaptoundecanoic acid (MUA) on the exposed gold microarray surfaces.
Step 2, modify the COOH terminal with hydrophilic maleimide by covalently anchoring a bifunctional linker of sulfosuccinimidyl-4-(N-maleimidomethyl)cyclohexane-1-carboxylate (Sulfo-SMCC).
Step 3, drop 1 mM thiol-derived DNA probes to the target gold flakes using a microsyringe or spotter.
Step 4, remove the reversible protecting group Fmoc around the flakes in a mild base to reveal the $MUNH_2$-modified surrounding.
Step 5, seal the amine-terminal with PEG-NHS to block its non-specific adsorption and to suppress the cross-contamination between the flakes.

The chips are ready for spotting of DNA or DNA-labeled analytes. This can simply be achieved by hybridizing the DNA probes with their complementary DNA strands at 27 °C in a buffer of 20 mM phosphate at pH 7.7 added with 300 mM NaCl, 1 mM EDTA and 100 mM urea.

4.5.4 Preparation of Analytical Sensor Chip

To prepare an analytical SPRi sensor chip, we need surface chemistry and sample spotting or transferring technology. The former is to be discussed in Sect. 4.6, while the latter have been delivered in Sects. 4.4.2–4.4.4.

It is not rare that the prepared analytical chips cannot offer clear images. Further online or offline treatments of the chip surface may rescue the chip. The online treatments are performed by directly pumping in some rescuing solutions, one after another, into the flow cell to perform *in situ* surface modification. This method is usable only in the case that the target spots are stable enough or you just want to take the pictures simply against air to increase the contrast. Otherwise, it is better to dismount the chip and treat it in required solutions.

There are several ways to improve the image contrast besides the measuring against air. The first way is to make the probe-immobilized chip react with samples for a longer time or repeatedly. This is effective if a sample is too diluted and the probes have sufficient capturing capacity. The second way is to make the captured analytes react with other substances in order to enlarge their size for signal amplification. This concerns with many principles and operation techniques. They will be specifically discussed later in the relative chapters. The final way is to regenerate the chip.

4.5.5 Regeneration of Chips

In many cases, a sensor chip is deposed after use. However, reuse of a chip can benefit to many researches, at least some tedious manipulations can be avoided or the investigations can be conducted under more homogeneous conditions. In theory, all chips have the potential to be regenerated. For example, you can remove all the organic compositions on the metal surface by oxidations or photo-cleavage and then re-treat the chip as a new one. More often, by chip regeneration is meant to recover a probe-immobilized chip through some mild conditions. This include immune-chips, affinity chips and reversible reaction-based chips.

The regeneration techniques are hence dependent on the probes used. In regeneration of an immunological chip, normally a high salted solution is sufficient to remove the captured antibodies or antigens. For example, a 2-M NaCl solution is commonly used to deplete the captured proteins from their probes by keeping the chip in the NaCl solution for several minutes. Similarly, for other chips, the regeneration procedure is created by de-affinity or de-recognition means.

4.6 Surface Chemistry

Surface chemistry is another key in developing practical SPRi methodology. It involves various aspects in SPRi such as metal deposition, chip surface modification, sample transfer and capture, probe immobilization, surface denaturing and recognizing events, chip regeneration, etc. There are presently numerous useful chemical reactions available for SPRi use. Here we introduce several simple and practical surface reactions for SPRi.

4.6.1 *Chemistry for Bare Metal Surface*

4.6.1.1 Thiol Chemistry

Chemical modification of metal (and glass as well) surfaces is unavoidable in SPRi of real samples. Since current SPRi majorly uses an on-glass-deposited Au and/or Ag films for excitation of SPPs, thiol-based chemistry is on top of selection. Accordingly, mercaptan chemicals are highlighted that are a type of weak acids (pKa≈10.6), a bit stronger than alcohol. They can serve as nucleophilic substitution reagents and can be either self-oxidized into disulfide bond in diluted H_2O_2 or I_2, or violently oxidized by strong oxidants like HIO_4, HNO_3 or $KMnO_4$ to produce organic sulfonic acids. The well-known thiol reactions are the formation of insoluble salts with metal ions such as Hg^{2+}, Cu^{2+}, Ag^{+} and Pb^{2+}. It is helpful to explore and utilize the free radical-mediated thiol-ene click reactions.

However, the most useful thiol chemistry for SPRi is to assemble mercapto molecules on metal surface under mild conditions. In SPRi, the most frequently used is the direct thiol-metal reactions, of which S–Au reaction is a basic technology to prepare highly ordered monolayer of self-assembly membrane (SAM) on a gold sensor surface. The basic process is that, in a solution, a thiol molecules are first tiled on the gold surface until full coverage, and then stand up (molecular chain at about 30° tilt angle) through lateral pressure induction [40–42]. The advantage is that the assembled molecular layer can be cleaved under UV light irradiation in combination with the cleaning of organic solvents (Fig. 4.15).

The possible assembling reaction steps are as follows [43]:

$$R - SH + Au(0)_n \rightarrow R - S - Au(I)Au(0)_{n-1} + 1/2H_2. \quad (4.r1)$$

$$2R - SH + 2Au \rightarrow 2R - S - Au + H_2. \quad (4.r2)$$

$$2R - SH + 2Au + O_2 \rightarrow 2R - S - Au + H_2O_2. \quad (4.r3)$$

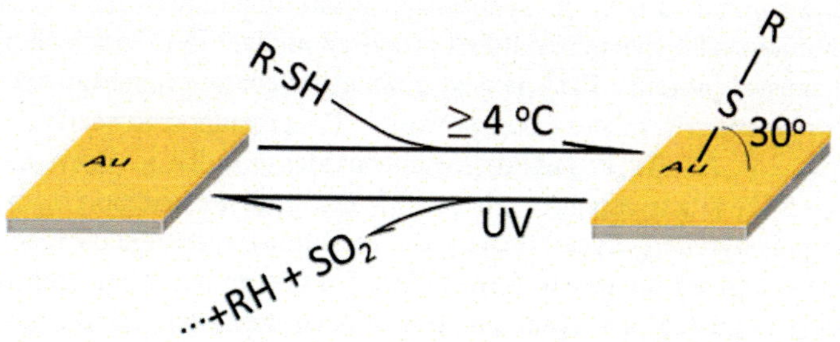

Fig. 4.15 Assembly of an alkanethiol on gold surface and its cleavage under UV light

$$R - S - Au \xrightarrow{UV} Au(0)_n + RH + SO_2 + \cdots . \quad (4.r4)$$

The reactions can generally be carried out at 4–25 °C for 12 h. The cleavage of the formed SAM on Au (Eq. 4.r4) is generally realized by UV irradiation from a mercury xenon arc lamp at 400 W for 1–2 h, and the cleaved residues can be washed off with ethanol.

Besides Au, thiol substances can also form SAMs on some other solid metals such as silver through S–Ag bound at 8–12° chain tilt angle [44, 45], copper through S–Cu bound at 12° chain tilt angle [46–48], and even iron possibly through S–Fe bound [49–51].

Very recent study with STM suggested that the thiol-based SAMs on gold would be formed from physical rather than chemical adsorption because the thiol hydrogen atoms retained on the molecules [52]. Fortunately, this retaining does not affect the well-established procedure to modify SPRi sensor chips.

Based on thiol-metal reaction, various organic molecules can be assembled on the sensor surfaces with required terminal such as hydroxyl, carboxyl and amino groups. In practice, the modification is simply achieved by immersing the metal articles into a solution of mercaptoundecanol (MUOH), $MUNH_2$, MUA, or their mixtures, depending on the need of terminals. These groups enable the grafting of more substances on the sensor surface. The advantage in use of the mixed modifiers is that the types and their ratios of their terminals can be quantitatively regulated through the change of their concentrations.

Approach 4.9, A general procedure to assemble mercapto compound on gold film:

Step 1, insert cleaned bare gold chips into 1 mM mercapto compound(s) in ethanol and keep reaction at 4 °C for 12 h or at room temperature for > 4 h.
Step 2, wash the chips sequentially with ethanol and water and blow dry with nitrogen gas.

4.6.1.2 Polydopamine

Polydopamine (PDA) is synthesized from dopamine. It is a mussel-inspired biomimetic adhesive material first proposed in 2007 [53], much similar to the adhesive mussel proteins. PDA is now a universal coating material, specifically in wet environments. Freshly or *in situ* prepared PDA can adhere to nearly all solid including metals (e.g., Au, Ag, Pt and Pd), metallic oxides (e.g., TiO_2, Al_2O_3 and others), amorphous SiO_2, silica, ceramics, glass and hydroxyapatite, and polymers (e.g., polyethylene, polystyrene, polycarbonate and even Teflon). Although the reaction mechanism remains unfixed on how to form PDA from dopamine, it has been well agreed that PDA is in general not a pure compound but a mixture [54] that vary with synthetic and post-synthetic reaction conditions such as solution pH, initial concentration of dopamine [55] and reaction temperature [56]. Many characterizations revealed that PDA contains covalent rings and non-bound assembled structures [57, 58]. Although the adhesive matter may reasonably be called dopamine-melanin [59, 60], PDA has been widely accepted and hence is adopted in this book.

In spite that the reaction mechanism is not really clear yet, the oxidation-induced polarization and assembly or aggregation can be conceived as illustrated in Fig. 4.16, where the covalent cross-linking between or among the benzene rings may be suspected [61] and the co-existence of physical stacking with covalent bounding may also be possible [57].

PDA has a pI at about 4.0 and hence is a type of amphoteric compound(s). Its coating will be charged negatively at pH > pI, for example, $\zeta_{PDA} = -40$ mV at pH 8.5 [62]. It is hence hydrophilic, with a contact angle at 50–60° that is independent on the coated materials [63]. PDA coating is in general advantageous over alkanethiol coatings. However, PDA is not really stable in strong acid and base solutions, being easily detached. In addition, the coating thickness and deposition rate depend very much on oxidation conditions and the concentration of reactants. Yang et al. have reported a quite thorough study on the impact of reaction conditions and how to stabilize the coated PDA on gold surface [56]. The following presented are 3 approaches to coat different types of PDA on a gold sensor chip.

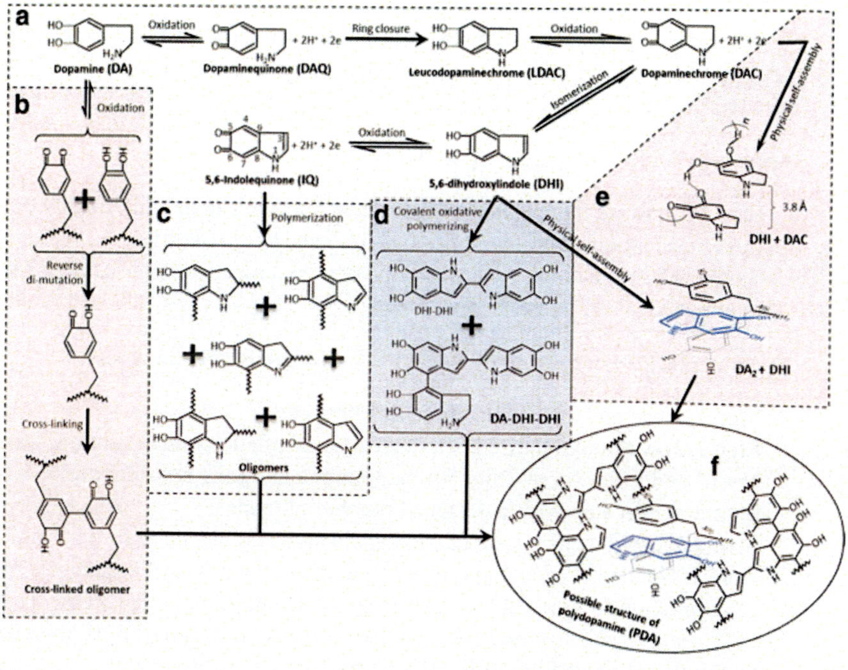

Fig. 4.16 Potential routes to form polydopamines induced by oxidation of dopamine. **a** Oxidation and ring-closure of dopamine; **b** cross-linkage of dopamine oligomers; **c** formation of oligomers from the oxidized products; **d** covalent oxidative polymerization; **e** physical assemble of oxidized dopamine monomer through either hydrogen bond or π-π stacking; **f** potential structure of polydopamine

Approach 4.10, A normal procedure for coating PDA:

Step 1, immerse cleaned gold sensor chip(s) in 2 mg/mL freshly prepared dopamine (in 10 mM Tris–HCl at pH8.5) at 25 °C for 15–60 min.
Step 2, wash the chip(s) with water and blow dry with nitrogen gas for later use.
Note 1: These chips are embedded with weak physical aggregates that can be washed off or be further stabilized through "aging" techniques. To wash off the aggregates, the chips must subject to NaOH washing; and for aging, they need to stay in Tris–HCl buffer at pH8.5 for at least 30 h.
Note 2: Tris–HCl buffers have a faster coting rate than phosphate buffers but slower than Carbonate buffers. Borate buffers prohibit the formation of PDA and boron containing reagents are better removed from coating solutions.
Note 3: The oxidation of dopamine in solutions is realized normally by the dissolved oxygen (dis-O_2). The coating reaction rate can hence be varied by replacing the dis-O_2 with other oxidants like Cu^{2+}, H_2O_2, ammonium persulfate (APS), $NaIO_4$, or polyethyleneimine (PEI). A coating rate order was found as follows:

$$CuSO_4/H_2O_2 > NaIO_4 > APS > disO_2 > PEI > CuSO_4.$$

Whenever the oxidant is change, the coating condition should be re-optimized since the reaction conditions are different, including dopamine concentration, reaction temperature and buffer pH.

Approach 4.11, For coating tight PDA:

Step 1, immerse the cleaned gold chips in 5 mg/mL dopamine freshly dissolved in 50 mM carbonate buffer at pH 10.0–11.5 and keep reaction at 70 °C for 60 min.
Step 2, clean the chips with water and dry the chips for latter used or for aging treatments.

Approach 4.12 For coating of stable PDA with adjustable thickness:

Step 1, clean and array gold chips in a flow cell.
Step 2, flow through the cell in a sequence of fresh dopamine solution, water, 1 M NaOH and water for 30 min, 5 min, 30 min and 5 min, respectively.
Step 3, repeat step 2 for a designed number of cycles.
Step 4, clean the chips with ethanol and water, and dry them for later use.
Note 1: NaOH solution can effectively remove the physical aggregates from the coating, leaving normally a monolayer of strongly adsorbed PDA. This fresh PDA remains reactive and adhesive for further deposition of next layer of PDA; therefore, cyclic coating of PDA layer by layer is enabled.
Note 2: This coating technique can linearly thicken the coating for the first 10 cycles, with a linear correlation coefficient > 0.999.

In conclusion, PDA contains abundant functional terminals such as catechol hydroxyl and amino groups, able to further react with amino-, carboxyl- and/or mercapto groups. Therefore, PDA can also be used as a universal linking agent to graft functional molecules onto the sensing surfaces.

4.6.2 Linking Chemistry

In addition to surface coating, linking chemistry is unavoidable to bridge different coating layers or molecules. Herein introduced are more or less universally applicable linking reagents and some practical linking approaches.

4.6.2.1 Cyanuric Chloride Chemistry

Cyanuric chloride or CC, also called chlorotriazine or tricyanogen chloride, is formally termed 2,4,6-trichloro-1,3,5-triazine (TCTA) in organic chemistry. It is a cheap chemical of white crystalline powder, soluble in chloroform, carbon tetrachloride, ethanol, ether, acetone, dioxane, benzene and acetonitrile, but insoluble in cold water. CC can keep stable, in water at 0 °C, for about 12 h and then gradually hydrolyzes into cyanuric acid (Fig. 4.17). CC is unstable, volatile and irritating in air; while in aqueous alkali, CC soon decomposes into trichlorocyanic acid and release smoke-like hydrochloric acid gas. CC thus needs moisture-proof storage, and its solutions are better freshly prepared just prior to use.

CC is widely used as a bridging reagent because it can link primary and secondary amines, hydrazine, azide, Grignard reagents and compounds containing OH, SH and even CO_2H groups [64, 65]. The unique structure of CC endows different reaction

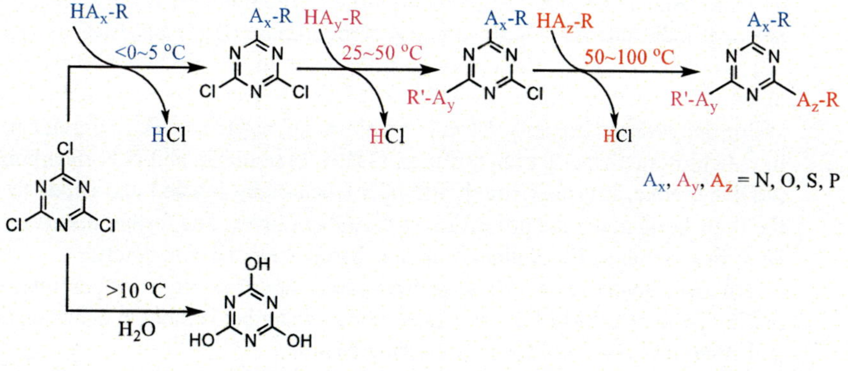

Fig. 4.17 Temperature-regulated stepwise substitution of the chloro groups in cyanuric chloride at the optimized temperature (for steps 2 and 3, respectively) dependent on the nucleophilic reagent used

activities to the three chlorine atoms on the triazine ring. As a consequence, the chlorine atoms can be replaced in step by a nucleophile at different reaction temperatures to produce mono-, di- and tri-substitutes [66]. Normally, the first chlorine atom is replaceable at 0–5 °C or a bit below, the second at room temperature or above and the third at above 50 °C [67]. It should be noted that the practical reaction temperature needs to be screened out from the type and steric hindrance of a nucleophile to be used, the existing substituents on the triazine ring, the alkalinity of reaction environment and the polarity of solvent. Nucleophilic groups with strong alkalinity and low steric hindrance are more reactive than others [68]. Polar solvents (such as water, acetone and their mixtures) generally facilitate the substitution. It is also performable to stepwise substitute the chlorine atoms in one pot by programmed control of the reaction temperature and time, in combination with a proper selection of the reaction solvent and acid absorbent, and timely addition of the nucleophilic reagents [69]. In all the controls, the key is to absorb the released HCl with a strong base. At pH7.0–8.0, the reactions take place for about 2 h.

Approach 4.13, For CC reaction with phenol:

Step 1, dissolve 5 mmol (1.01 g) CC and 5 mmol phenol (e.g., 2-hydroxyl-3-naphthoic acid) in 20 mL acetone at 0 °C.
Step 2, stir the solution while dropping 20 mL of 5 mmol (0.20 g) aqueous NaOH, and keep reaction for 3 h.
Step 3, stop the reaction by addition of ice-cooled water.
Step 4, harvest the powder product by vacuum drying, giving a yield of 77%.
Step 5, store the final product by re-dissolving it in dichloromethane.

4.6.2.2 EDC/NHS Chemistry

This chemistry makes a carboxylic acid be amidated or esterificated with amines or alcohols. The coupling reagents are alkyl carbodiimides, for example, dicyclohexylcarbodiimide (DCC), diisopropyl carbodiimide (DIC) and EDC as well [70–72], which can activate the carboxyl group to facilitate amidation or esterification under mild reaction conditions. DCC is frequently used because it is cheap but has strong dehydration capacity. DCC is a white solid, easily soluble in organic solvents like dichloromethane, tetrahydrofuran (THF), acetonitrile and N,N-dicarbonamide (NDBA). Note, it causes cough and rash when being inhaled and needs to use in the fume hood under the protection with rubber gloves. The further disadvantage of DCC lies in that its by-product of urea is hard to remove completely.

DIC is a colorless or yellowish transparent liquid at room temperature, with a boiling point at 145–148 °C. It is insoluble in water but soluble in benzene, ethanol and ether. It is also used for its cost-effectiveness.

EDC is thus more often used for it is soluble in water (> 0.2 g/mL) and in ethanol, able to activate carboxyl and phosphate. It is commonly used in pharmaceutical synthesis and biological researches, but it is more expensive than DCC and DIC. EDC is one of the zero length cross-linking agents because it directly forms amide bond(s). Its by-product of urea derivative(s) is in most cases water-soluble and is easy to wash off from chips and organic systems.

In general, this carbodiimide-based acylation can have variable reaction pathways. Figure 4.18 illustrates that an acid first reacts with the carbodiimide to form a reactive but unstable intermediate of O-acylisourea ester. This ester can either change back to the original carboxylic acid through hydrolysis (Path *a*) or rearrange into stable by-product(s) of N-acylurea amide derivative(s) if without the presence of a stronger nucleophilic reagent (Path *b*). The reaction can also stop at the stage of the active ester if there is a large hindrance or a strong electron-accepting group at the α-site of the acid, giving a strong MS signal [73]. Once an amine is added, the ester will change into amide (Path *c*) but at low rate and yield. To improve the reaction rate and specificity, an acylation catalyst is added (Path *d*) such as NHS, 4-N,N-dimethylpyridine (DMAP), 1-hydroxybenzotriazole (HOBt), and so forth, as shown in Fig. 4.19.

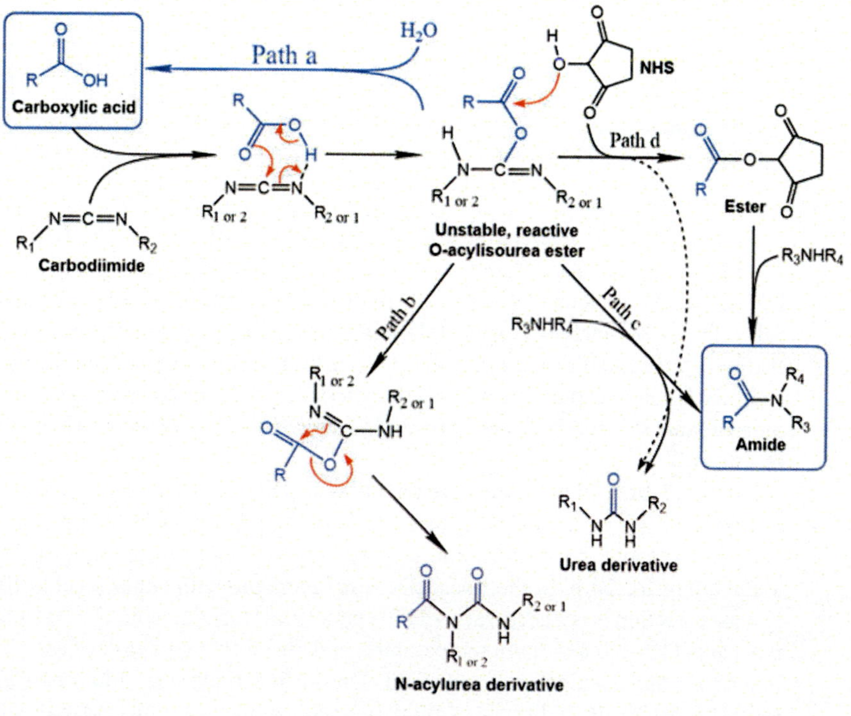

Fig. 4.18 Reaction paths of carbodiimide-based acylation of carboxylic acid. The carbodiimide-carboxylic addition produces reactive and unstable O-acylisourea ester that can change back to the original acid through hydrolysis (Path *a*). It will rearrange into one (for $R_1 = R_2$) or two (for $R_1 \neq R_2$) acyl urea derivatives if there is no nucleophilic reagent (Path *b*). Once a primary or secondary amine is added, the O-acylisourea ester can be substituted to form amide (Path *c*) but at a low yield. The yield and reaction rate can largely be increased by addition of NHS to form a new ester that can easily be replaced by amines (Path *d*)

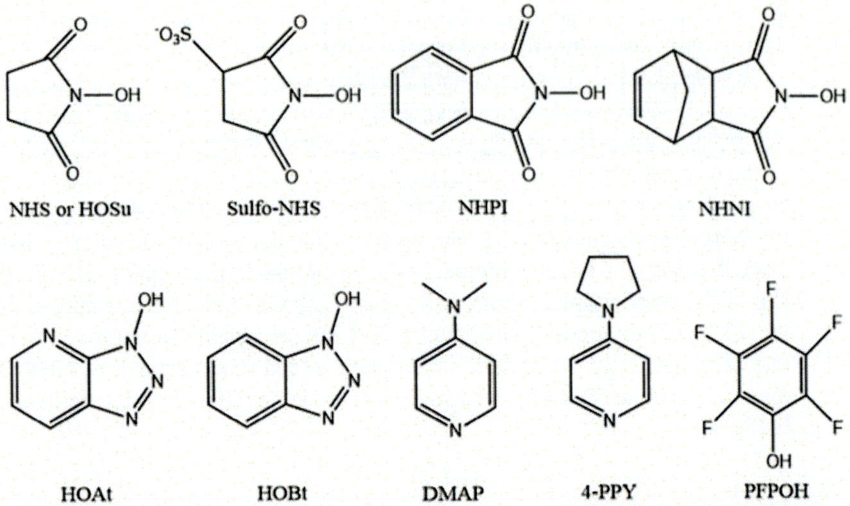

Fig. 4.19 Activation or catalytic reagents useable in aqueous and organic solutions. NHS = HOSu: N-Hydroxysuccinimide, Sulfo-NHS: N-Hydroxysulfosuccinimide (normally sodium salt), NHPI: N-Hydroxyphthalimide, HINI: N-Hydroxy-5-norbornene-2,4-dicarboximide, HOAt: 1-Hydroxy-7-azabenzotriazole, HOBt: 1-Hydroxybenzotriazole, DMAP: 4-dimethylaminopyridine, 4-PPY: 4-Pyrrolidinopyridine, PFPOH: Pentafluorophenol

Although all the nucleophiles illustrated in Fig. 4.19 are usable, DCC is often paired with DMAP to catalyze the esterification of the acid with an alcohol or phenol. DMAP can be replaced by 4-PPY to improve the reaction activity for about 1000 folds. The disadvantage of using DMAP is its producing dicyclohexylurea by-product that is slightly soluble in organic phase and is difficulty to remove completely. In this case, DCC can be replaced with DIC that produces diisopropyl urea highly soluble in most organic solvents for easy removal. In fact, DIC is more often used in solid-phase reactions.

For amidation, EDC is often combined with NHS or N-hydroxysulfosuccinimide (Sulfo-NHS) for better solubility in water. In the linkage of proteins, EDC needs only very mild reaction conditions, normally stirring for 1–2 h and keeping still at room temperature for 24 h. Sometimes EDC may combine with water-soluble HOBt. If aqueous reaction is not necessary, EDC is better used together with DMAP, which can react at pH4.5–6.0 and room temperature, and the addition order does not obviously impact on the reaction rate and yield. During reactions, the solutions should be neutralized with alkali. In case of amidation with amine or amino acid hydrochloride, N-methylmorphorphine or diisopropyl ethylamine (DIEA) is added for 2–3 folds excess in mole. In peptide treatment, NHS, HOBt (water-free) and organic acids are normally dissolved in CH_2Cl_2, chloroform, THF, or dimethylformamide (DMF) that should be used with carefulness due to its high boiling point. In case that a large excess of EDC [74] is added in aqueous solution at pH4.5 and reacts at 35 °C, the

O-acylurea intermediate product will rearrange through O → N migration via cyclic electron transfer (Path *b* in Fig. 4.18) to form a stable N-acylurea product [75–77].

Specifically, HOBt can resist racemization and is often used in a water-free state. In the presence of a carbodiimide (e.g., EDC), NHS or sulfo-NHS will react with a carboxyl group to form an active or transition ester. The formed ester can easily change into amides by reaction with a primary or secondary amine. NHS or sulfo-NHS itself is not necessary but can divide the reaction into two steps to largely improve the yield. The transition ester is in common stable for several minutes up to several hours, depending on solution pH, for example, its half-life ($t_{1/2}$) is 4–5 h, 1 h and 10 min at pH 7.0, 8.0 and 8.6, respectively. It is thus better to perform amination reaction immediately after esterification or *in situ*. The optimal reaction pH for EDC/sulfo-NHS pair is between 4.5 and 7.2. Note, EDC reaction is better conducted in 2-(N-morpholino)ethanesulfonic (MES) buffers at pH 4.7–6.0, while the amination of sulfo-NHS esters better in PBS at pH = 7.2–7.5. All the buffers used must be free of amino and carboxyl groups.

NHS dissolves in water, easily in acetone, alcohol and ethyl acetate, but slightly in chlorinated hydrocarbons, ethers, toluene and benzene. Its solutions have strong UV absorption at 280 nm. It is interesting that NHS-based active ester can be very stable and stored for a long time, for example, N,N′-1,3-phenylene bismaleimide (that can be used to directly link amino group(s) at extremely high yield). It has since been often applied to the linkage of enzyme, antigen and/or anti-body. Four approaches are given bellow for reference.

Approach 4.14, For linking carboxyl with amino in aqueous EDC/NHS:

> **Step 1**, insert carboxyl-terminated sensor chips in 0.1 M MES buffer (pH6.0) containing 0.5 M NaCl, 2 mM (10 × carboxyl) EDC and 5 mM NHS and shake the solution with the chips for 15–30 min.
> **Step 2**, quench the reaction by transferring the chips into 20 mM mercaptoentanol.
> **Step 3**, insert the chips into an amine (10 × excess) solution prepared in 0.1 M sodium phosphate and 0.15 M NaCl at pH7.2–7.5 or in 0.1 M $NaHCO_3$ and keep reaction at room temperature for 2 h.
> **Step 4**, wash the chips with water and blow dry with nitrogen gas.

Approach 4.15, For liking amino with carboxyl via EDC/HOBt in dichloroform:

> **Step 1**, insert an amino-terminated (0.1 mmol in total) gold chip into 0.1 mmol carboxyl acid in CH_2Cl_2 pre-cooled to 0 °C.
> **Step 2**, add 0.1 mmol (1.5 mg/mL) HOBt and 0.1 mmol (2.11 mg/mL) EDC into the solution and allow to react under stirring at room temperature for 10 h.
> **Step 3**, wash the chip in sequence with 5% aqueous HCl, 5% aqueous $NaHCO_3$, water, brine and water again and blow dry with nitrogen gas.

Approach 4.16, For linking between carboxyl and amino via EDC/HOBt in DMF:

> **Step 1**, insert carboxyl-terminated (0.076 mmol intotal) gold chips in 1 mL DMF solution of 76 mM (10.3 mg/mL) HOBt and 100 mM (19.2 mg/mL) EDC under stirring at room temperature for 20 h.

Step 2, add 76 mM amines and 0.17 mL of 0.15 M 4-methylmorpholine in DMF to above solution and keep stirring at room temperature for 4 h.
Step 3, wash the chips with ethyl acetate, propanol and water and blow dry with nitrogen gas.

Approach 4.17, Linking chemistry via DCC/HOBt:

Step 1, insert gold sensor chips with carboxyl terminal in 2.5 mL of anhydrous DMF containing 0.28 M HOBt and 0.28 M DCC under shaking at room temperature for l h.
Step 2, add 1 mL of 0.5–0.7 M amine (in DMF) to the solution and keep reaction under shaking at 25 °C for 23 h.
Step 3, wash the chips with DMF, isopropanol, 10% HCl, $NaHCO_3$ and water and blow dry with nitrogen gas.

Approach 4.18, For linking amine with phenol via DIC/HOBt:

Step 1, insert amino-terminated (0.3 mmol in total) gold chips in 2 mL of anhydrous DMF solution containing 116 mg (0.3 mmol or 0.15 M) Fmoc-Phe-OH and 44.8 mg (0.33 mmol or 0.15 M) HOBt.
Step 2, add in an aliquot of 56 μL DIC (0.36 mmol or 0.18 M) under stirring at room temperature for 12 h.
Step 3, wash the chips with DMF, and blow dry with nitrogen gas.
Step 4, insert the chips in 2.0 mL piperidine under stirring at room temperature for 1 h to remove the Fmoc protection group.
Step 5, clean the chips with ethanol and water and blow dry with nitrogen gas.

4.6.2.3 Carbonyl Diimidazole Chemistry

N,N′-carbonyldiimidazole (CDI) features high activity, low cost and easy post-treatment and is hence much superior to DCC and EDC in practical applications. It is due to the activation of its carbonyl group by the imidazole that becomes a highly active formylation reagent. The two imidazole rings can be replaced stepwise or in "one pot", depending on the reaction conditions and manipulations. The replacement happens once CDI encounters active hydrogen reagent such as hydroxyl or amino groups, which leads to the formation of carbonyl derivatives. In case of stepwise reaction, single substituted N-imidazole formyl ester or N-imidazole formamide will be formed in the first step of reaction, with a loss of one imidazole ring, while the second imidazole can further be replaced to form a carbonyl derivative with asymmetric structure as shown in the route ①; in Fig. 4.20 [78, 79]. When a substrate with two active hydrogen functional groups reacts with CDI, intramolecular carboacylation occurs to produce a cyclic product referring to the route ②; in Fig. 4.20 [80–82]. If the functional group with the active hydrogen is on a carboxylic acid, its corresponding acyl imidazole is generated that activates the acyl of the acid (route ③; in Fig. 4.20), making it reactive with either amino or hydroxyl group to form corresponding amides (route ④; in Fig. 4.20) or carboxylate esters (route ⑤; in Fig. 4.20)

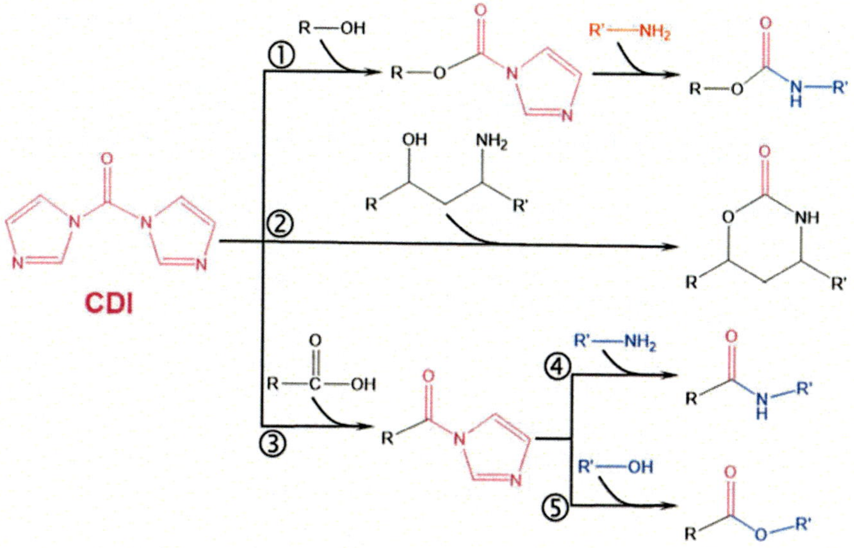

Fig. 4.20 Liking routes of N,N′-carbonyldiimidazole or CDI

under very mild conditions [83–85]. This type of two-step reaction process is generally conducted in "one pot". The amidation reaction may be accelerated by CO_2 [86, 87].

CDI is often applied to the linkage of proteins through their N-terminal (α-amino) and/or lysine side chains (ε-amino), which can form uncharged urethane-like derivatives with excellent chemical stability.

4.6.2.4 Succinic Anhydride Chemistry

Anhydrides can be esterified and amidated with hydroxyl and amino groups, respectively, which is an easy chemistry to change OH terminal to carboxylic group for further linkage, especially useful to terminate saccharides with carboxylic groups as shown in Fig. 4.21 and Approach 4.19.

Approach 4.19, For terminating saccharide-coating with carboxyl:

Step 1, insert saccharide-coated chips into a DMF solution containing 10 mg/mL succinic anhydride and 15 mg/mL DMAP at room temperature for 16 h [88].

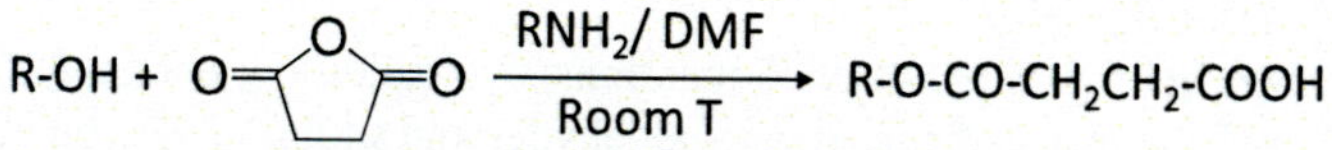

Fig. 4.21 Change of hydroxyl terminal to carboxyl via succinic anhydride

Step 2, wash the chips sequentially with DMF, isopropanol and water and blow dry with nitrogen gas.

In anhydrous pyridine, the hydroxyl group can react with succinic anhydride to obtain an intermediate with carboxyl group, succinic acid semester. It can further react with amino group in the presence of carbodiimide or another anhydride, resulting in the insertion of a succinyl group between the hydroxyl and amino groups.

4.6.2.5 Glutaraldehyde Chemistry

Glutaraldehyde can add a 5-carbon bridge between two amino groups through Schiff bonds, often used to link proteins under mild reaction conditions. The linking reaction can be performed in buffers at pH 6.0–8.0 and 4 ~ 40 °C. Glutaraldehyde-based linking can be performed in one pot reaction, but at the cost of low cross-linking efficiency (1–5%). The cross-linking efficiency can be increased to 5–25% through stepwise reaction.

Note, glutaraldehyde is affected by light, temperature and alkalinity. Its linking ability may be weaken by its self-polymerization. It is better used freshly.

Approach 4.20, For linking between amino groups:

Step 1, insert amino-terminated sensor chips in a 1.25% glutaraldehyde freshly diluted from 25% glutaraldehyde with 0.01 M phosphate buffer (PB) at pH 6.8 at 20 °C for 18 h.
Step 2, clean the chips and blow dry with nitrogen gas.
Step 3, immerse the glutaraldehyde-treated chips in an amine solution at pH 9.0–9.6 (e.g., 5 mg/mL protein in 0.15 M NaCl and 1 M carbonate buffer) under stirring for 24 h.
Step 4, wash the chips with water and add BSA solution to block the remaining aldehyde groups at 4 °C for 2 h.
Step 5, wash the chips with water and blow dry with nitrogen gas.

4.6.2.6 Click Chemistry

Click chemistry was coined in 2001 by Sharpless [89] who shared the 2022 Nobel Prize in Chemistry. It can quickly and reliably synthesize various molecules through the splicing of small units via carbon heteroatomic bonds (C–X–C) to widen the molecular diversity. The chemistry features high yield and high steric selectivity under mild, water-insensitive reaction conditions. It has also tolerance to the variation of functional groups.

There are four types of click reactions: cycloaddition, nucleophilic ring opening, carbonyl chemistry of non-alcohol aldehydes and addition reaction of carbon–carbon multi-bond. It can also be classified as metal-catalyzed and non-metal-catalyzed reactions. The classic click reaction is metal-catalyzed cycloaddition reaction of alkyne with azide, normally producing region-selective 1,4-disubstituted-1,2,3-triazole. The non-metal-catalyzed click reactions are majorly based on Michael addition of thiol with terminal enes.

Azide-alkyne cycloaddition is originally catalyzed with Cu(I) [90] as illustrated in Fig. 4.22. The reactants can be mono-substituted alkynes and organic azides, while the active Cu(I) catalyst can be formed from a commercial salt or complex, or *in situ* oxidation of copper metal or reduction of Cu(II). Besides Cu(I), other metal or metal complexes have also been explored, for example, Ru(I) in pentamethylcyclopentadiene ruthenium chloride or [Cp·RuCl] that can catalyze the cycloaddition of azide to either terminal or non-terminal alkynes to produce a single stereoselective 1,5-disubstituted or fully substituted 1,2,3-triazole [91]. This reaction system can thus be conducted in aqueous solutions but the yield depends very much on the catalyst applied.

Thiol-ene click reaction is a type of representative non-metal-catalyzed click reactions. Different from azide-alkyne reactions, thiols can react with alkene, terminal alkyne or even bromine substitutes mediated by free radical or catalyzed by nucleophilic substances such as tertiary amine as summarized in Fig. 4.23 [92–96]. The radical-mediated click reaction can be initiated by either UV irradiation or heating, and it is often fast and convenient in applications. The reaction of thiols with terminal alkyne will finally produce branched molecules. By use of nucleophilic catalyst, Michael addition of thiols needs electron deficient double bounds such as α,β-unsaturated ketone, butyraldehyde, maleic anhydride, maleic acid ester, acrylonitrile and cinnamic acid.

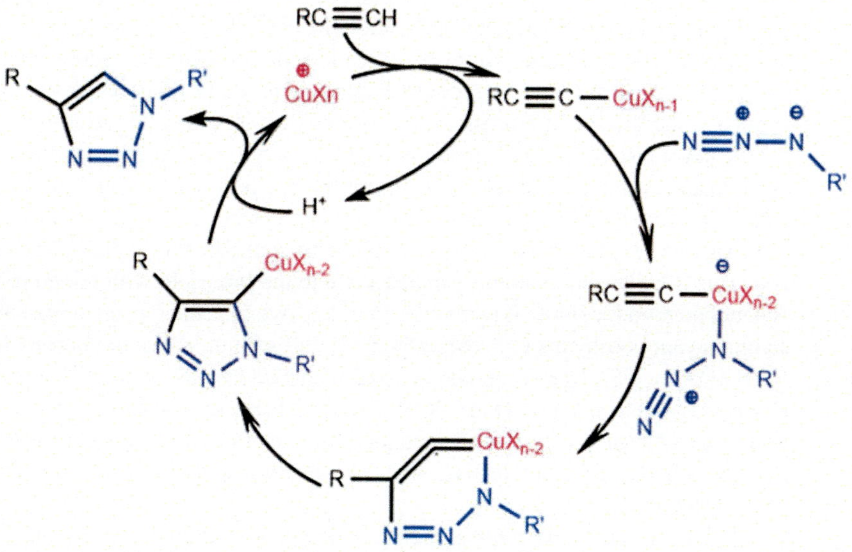

Fig. 4.22 Cu(I)-catalyzed click reaction of azide with alkyne

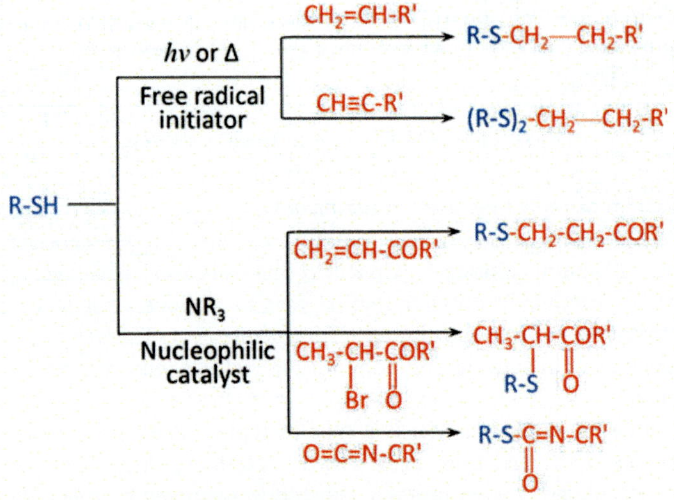

Fig. 4.23 Linkage through thiol-ene click reaction

4.6.2.7 Reductive Amination Linking Chemistry

Reductive amination makes aldehydes or ketones link with ammonia, primary amines or secondary amines in the presence of reductant to yield primary, secondary or tertiary amines. It is also applicable to the modification of carbohydrates with amines in combination with $NaIO_4$ to oxidize and break orthohydroxyl carbon bonds. The oxidation has commonly around 70% yield. The overall reactions include (i) production of aldehyde or ketone groups by oxidation, (ii) formation of Schiff's base and (iii) reduction of the base to reduced amines (Fig. 4.24). This chemistry is often applied to the linkage of enzyme without an evident loss of catalytic activity.

The key in use of this combined linking chemistry is the selection of the reducing agent that must selectively reduce the iminium ions or imines rather than aldehydes and/or ketones. In the case of large scale organic synthesis, catalytic hydrogenation can easily be adopted in the presence of platinum, palladium or nickel metallic catalyst. This is however not suitable in SPRi due to the low yield with mixed products. A better method for solution reaction is to utilize hybrid reducing agent, like $NaBH_4$, sodium cyanoborohydride ($NaBH_3CN$) [97] and sodium triacetoxyborohydride or $NaBH(OAc)_3$ [98]. $NaBH_4$ should be used with care because it can reduce not only C = N to C–N but also C = O to C–OH. To suppress the second reduction, the formation of Schiff base must be conducted prior to reduction. It can also be replaced by $ZnBH_4$, $NaBH_4$-$ZnCl_2$ or $NaBH_4$-NiCl. The catalytic activity increases with Lewis or Brönsted acids.

$NaCNBH_3$ and $NaBH(OAc)_3$, being hydrogen-substituted $NaBH_4$, have better selectivity than metallic salts of BH_4^-, allowing to conduct "one pot" amination with nearly free of C–OH by-products. $NaCNBH_3$ is soluble in hydroxyl solvents like methanol and is stable in acidic (ca. pH3.0) environments, featuring pH-dependent

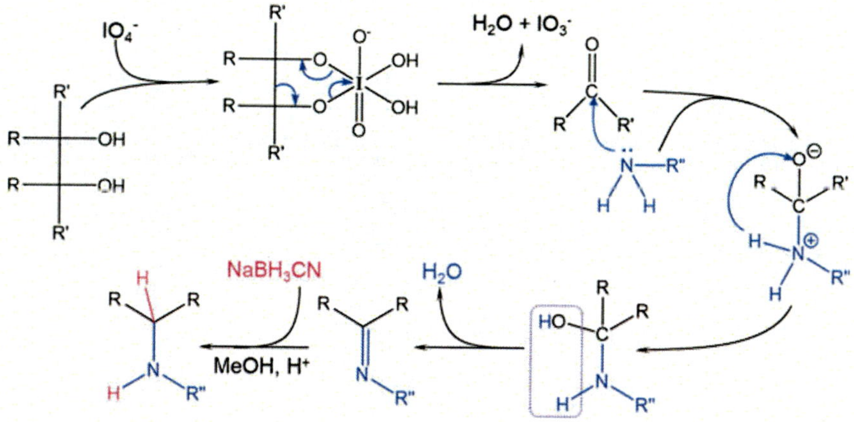

Fig. 4.24 Oxidation-coupled reductive amination reaction route

selectivity [99]. It can effectively reduce C = O to C–OH at pH3.0–4.0 [100] but preferentially reduces C = N to C–N at pH6.0–8.0. The shortages of $NaCNBH_3$ are that it needs > 5-folds excess of amines, itself is high toxic and can further produce toxic by-products of HCN and/or NaCN. Differently, $NaBH(OAc)_3$ is much less toxic than $NaCNBH_3$ but it is instable or decomposes in alcohol and water. It should be solved in non-hydroxylic solvents (e.g., CH_2Cl_2, THF, acetonitrile, etc.) and is now more reactive than $NaCNBH_3$, able to reduce the Schiff bases formed from ketones and weak bases of aromatic amines, with high yields. The reaction can further be accelerated by microwave, and similar to $NaBH_4$, $NaCNBH_3$ and $NaBH(OAc)_3$ can both be enhanced with Lewis acids.

The reduction can also be achieved with boranes such as BH_3-pyridine or BH_3-picoline, $B_{10}HH_{14}$. Borane-2-picoline is usable in aqueous reaction. Following are approaches for reference.

Approach 4.21, For oxidation-combined reductive amination:

Step 1, add freshly prepared 10 mM $NaIO_4$ into 5 mg protein or saccharide in 0.1 M $NaHCO_3$ at an equal volume, mix and keep reaction in dark for 2 h; alternatively, mix fresh 0.1 M $NaIO_4$ with the sample at a volume ratio of 1:3 under stirring in dark at room temperature for 20 min, or let the reaction proceed at room temperature for 3 h or 2 h for sample prepared in PBS at pH7.0 or in 0.01 M $NaHCO_3$.

Step 2, add freshly prepared 5 mg/mL $NaBH_4$ into the reacted solution at a volume ratio of 1:20, mix the solution thoroughly and keep reaction at room temperature for 30 min.

Step 3, add another portion of 5 mg/mL $NaBH_4$ at volume ratio of 2:20, mix and keep reaction at room temperature for 60 min or at 4 °C for 2 h.

Approach 4.22, For stepwise amination:

Step 1, dissolve an aldehyde or ketone in methanol, add a target amine and heat the mixture at about 40 °C for ca. 5 h to completely form Schiff base that can be tested by thin layer chromatography (TLC): no aldehyde or ketone band observed.
Step 2, remove the methanol, adjust the solution to pH6.0 or above with NaOH or $NaHCO_3$ and cool the reacted solution in ice-salt bath.
Step 3, add $NaCNBH_3$, rise to room temperature and keep reaction overnight.

4.6.3 Group Protection Chemistry

Many reactive or instable groups such as amino must be protected in many chemical procedures. The often used protective groups are alkoxycarbonyl, acyl and alkyl, where the alkoxycarbonyl includes benzyloxycarbonyl (Bzoc or Cbz), t-butyloxy carbonyl (Boc), Fmoc, allyloxycarbonyl (Alloc), trimethylsilylethoxycarbonyl (Teoc), methoxycarbonyl (Moc) and ethoxycarbonyl (Eoc); the acyl includes phthaloyl (Pth), p-toluenesulfonyl (Tos), trifluoroacetyl (Tfa), o- or p-nitrobenzene sulfonyl (Ns), tervaleryl (Tv) and benzoyl (Bz); and the alkyl includes trityl (Trt), 2,4-dimethoxybenzyl (Dmb), p-methoxybenzyl (Pmb) and benzyl (Bz). Alkyl protection usually needs complicated reaction manipulation, and it is since rarely used in SPR and will not be discussed furthermore.

4.6.3.1 Alkoxycarbonyl Chemistry

Alkoxycarbonyl reagents, especially Boc and Fmoc, are often used to protect amines in SPRi. Boc can easily be removed by acidolysis in trifluoroacetic acid (TFA) or 1:1 (v/v) TFA/CH_2Cl_2. The produced tert-butyl cation will further decompose into evaporative isobutene and CO_2, leaving no by-product. Differently, Boc-amines are stable during alkali hydrolysis, hydrazinolysis or catalytic hydrogenolysis, able to resist nucleophilic attack and suitable for long-term storage. The protection reaction is commonly conducted by mixing amine(s) with Boc_2O in dioxane/water mixed solvent added with $NaHCO_3$. In case that an amino compound is sensitive to water, the protection reaction can be performed with Boc_2O in triethanolamine (TEA)/MeOH (methanol) or DMF at 40–50 °C. This type of reactions normally produces easily removed byproducts. Although the Boc-protective groups can be removed by TFA, it is more often acidolyzed with 1–2 M HCl in dioxane. For washing, methanol is commonly used, nevertheless, in case of tert-butyl esters, especially existing together with free carboxyl groups, HCl/MeOH should be avoided because they accelerates the formation of methyl esters. HCl in ethyl acetate can maintain the washing stability of unstable Boc-amines such as tert-butyl and non-phenolic esters. Under neutral anhydrous condition, trimethylsilane iodide can remove Boc group and break carbamate, esters, ethers and ketals. Some chemicals (e.g., thioalcohol, thiophenol and thioether) and electron-rich aromatic compounds (e.g., indole, thiophene, pyrazole and furan polyphenol hydroxyl substituted benzenes) may react with the by-product of tert-butyl carbon ion to form alkylated products. This can be avoided by use of phenylthiophenol as an additive or be scavenged by addition of anisole, phenylthiomethyl ether, benzothiophenol, cresol and dimethyl sulfide. Boc protective reaction and cleavage are schematically shown in Fig. 4.25.

Fmoc-protected amines are highly stable in acidic environments, which facilitates the deprotection of Boc and benzyl. Fmoc-protective group can be added onto amino with Fmoc-Cl agent (Fig. 4.26). The reaction is normally performed in dioxane or CH_2Cl_2 saturated with $NaHCO_3$ and kept at 0 °C—room temperature for 1–2 h. Instead of Fmoc-Cl, Fmoc-OSu (O-linked succinimide) is now more often used, which needs mixed solvents of acetonitrile and water. Its advantage is that it produces few oligopeptides in the protection of amino acids.

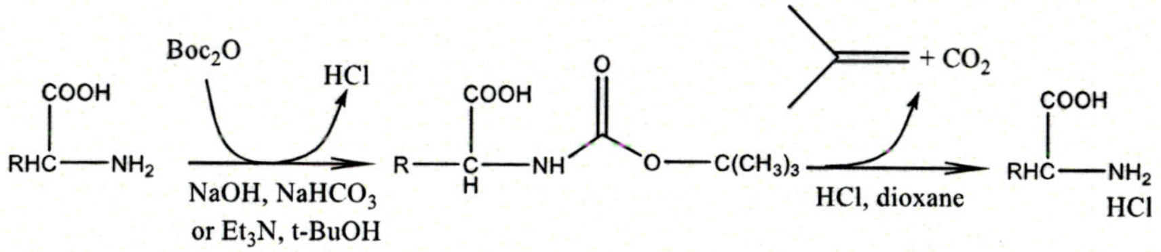

Fig. 4.25 Boc-protective and removing reaction route

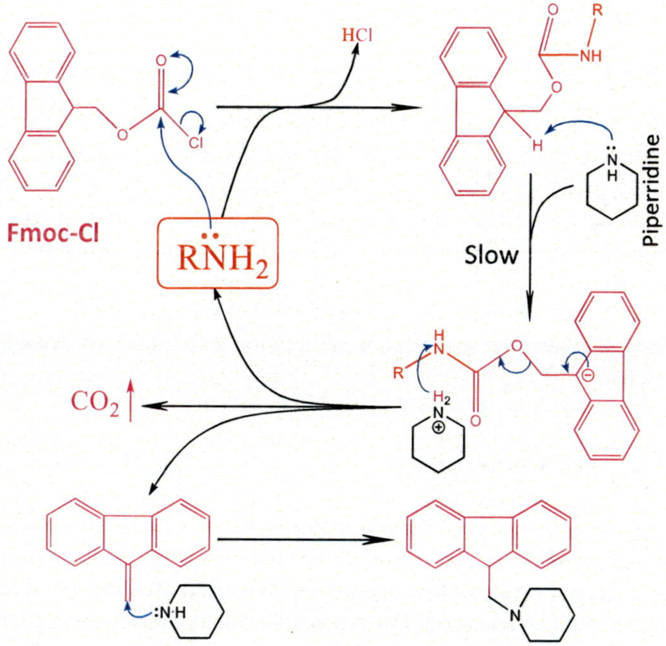

Fig. 4.26 Protection of amines by reaction with Fmoc-Cl and its removal

The added Fmoc group can simply be removed by addition of amines, without hydrolysis (Fig. 4.26). The deprotection reaction rate depends on the base used. In DMF, 20% piperidine is quite fast, about decade minutes. Fmoc protective group can also be removed with 30:9:1 (v/v/v) concentrated ammonia, dioxane and 4 M NaOH, or with 50% CH_2Cl_2 solution of piperidine, ethanolamine, cyclohexylamine, morphorphine, pyrrolidone or other amines. In common, a tertiary amine (e.g., triethylamine) has only poor ability to remove Fmoc group, while the amines with steric hindrance is the worst. In addition, $Bu_4N^+F^-$/DMF can be used at room temperature.

In addition to amino groups, Fmoc can also react with sulfhydryl and hydroxyl groups. The unreacted Fmoc and its hydrolysates can be clear off through pentane extraction.

Approach 4.23, For the protection of an amino compound:

Step 1, dropwise add 1 mL ethanol solution of 0.337 g/mL Fmoc-OSu into 1 mL solution of amino compound at pH 9.0 and keep reaction for 1–2 h, during this period check pH and add $NaHCO_3$ if pH < 8.0.
Step 2, remove the ethanol by distillation after adjusting the solution pH to ca. 1.5 with HCl.
Step 3, add 1 mL water into the residue and mix with 3 × 0.5 mL ethyl acetate to extract the product.
Step 4, wash the extract twice with saturated salt, dry the product and recrystallize the protected amine in 1 mL ethyl acetate.

Approach 4.24 For the deprotection of an Fmoc-amino compound:

Step 1, dissolve 1.0 mmol Fmoc-amino compound in 10 mL DMF and add 1.0 mL diethylamine (DEA) and keep reaction at room temperature for 1.5 h; alternatively, add 1.0 mL DMF/piperidine to react for 10 min, 1.0 mL acetic anhydride/piperidine for 10–20 min, or 1.0 mL 30% DEA/dichloromethane for 3 h.
Step 2, remove the diethylamine and the solvent in vacuum at a thermostatic temperature < 30 °C or by rotary evaporation.
Step 3, triturate the residue with 25 mL petroleum ether.
Step 4, collect the sediment, wash it twice with 20 mL ether and finally dry it in vacuum.

4.6.3.2 Acyl Chemistry

Among the named acyls, Pth is normally used to selectively protect primary amino, in combination with the prevention of racemization (Fig. 4.27). This protection chemistry can resist catalytic hydrogenolysis, HBr/HAc treatment and reduction by Na/NH_3. It should be noted that Pth-protected primary aminos are instable in basic environments. For example, the phthalimide ring will beak in alkali saponification to yield O-carboxybenzoyl derivatives.

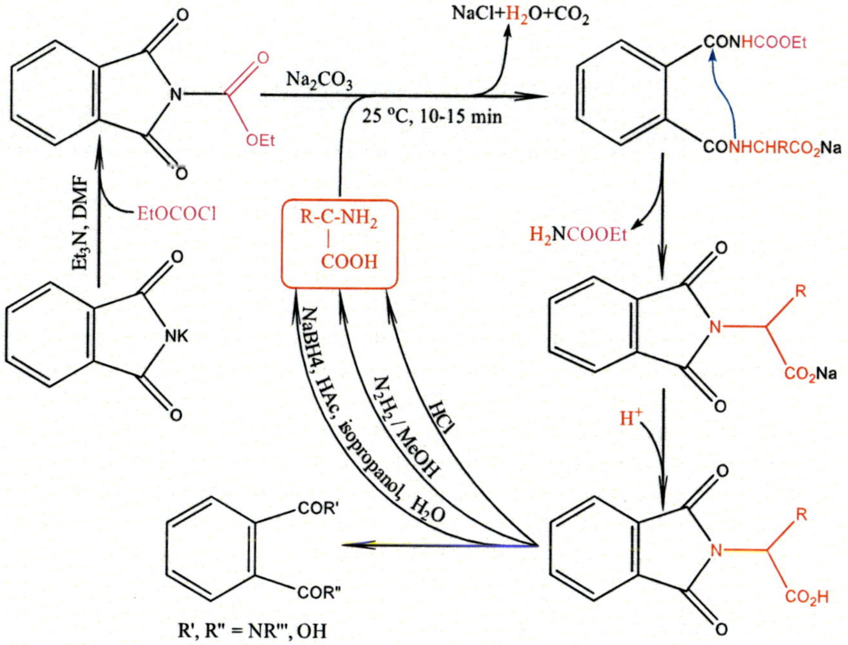

Fig. 4.27 Selective protection and deprotection of primary amine in secondary amines with an *in situ* synthesized N-ethoxyphthalimide

The protected amino compounds can easily be released by removal of Pth group with aqueous or hydroxyl solution of hydrazine, which takes 1–2 days at room temperature. The deprotection can be accelerated in methanol solution of hydrazine hydrate by heating reflux for 2 h. If the hydrazine does not work ideally, the distillation can be altered in 6:1(v/v) isopropanol-water solution of $NaBH_4$ and acetic acid at 80 °C for 5–8 h. The distillation can also be performed in concentrated HCl. In common, the deprotection of Pth does not affect the Boc-protected groups, allowing multiple protection operations.

Another often used reagent is Tos that can react with amino substances and p-toluenesulfonyl chloride to yield p-toluenesulfonamide in pyridine or aqueous alkaline (e.g., NaOH or $NaHCO_3$) solution. This protection chemistry is often used to protect amino acids, pyrrole and indole. Tos is the most stable amino protection reagent, against catalytic reduction, alkaline or acidic (e.g., TFA or HCl) hydrolysis. Nevertheless, weakly alkaline amine such as pyrrole- or indole-formed p-toluenesulfonamide can indeed be deprotected by alkaline hydrolysis. The deprotection is conducted with sodium naphthalene, Na/liquid NH_3 (which may cause breakage of some peptides), Li/liquid NH_3, HBr/phenol, or Mg/MeOH, of which the last two are used more often. Tos can also be removed by reflux in HF/CH_3CN.

Ns is comparatively easy to react with amino compounds under mild conditions. The protected compounds are normally stable in strong acids and/or alkali. Deprotection can also be realized in mild conditions, such as in $PhSH/K_2CO_3/DMF$ or in $HSCH_2CO_2H/LiOH/DMF$, both at 23 °C, where the second condition will yield $O_2NC_6H_4SCH_2CO_2H$ that can be washed away by aqueous alkaline solutions.

4.7 Imaging Data Recording and Treatment

Once a sensor chip is mounted on SPRi system, the image can immediately be recorded and displayed on a computer screen in real time. To have interaction of the probes dotted on the chip surface with their recognizable targets, they have to be made contact with each other by either drop-addition of the target solutions on their probe spots or pumping their solutions into the flow cell. If you like, you can record all the variation by common video in real time or in a longer or even shorter time interval. To have higher image contrast, the reaction solutions are evacuated and record the pictures against air. Also to clearly reveal the increment of imaging signals, the images acquired after interaction are subtracted by the images taken before interaction, which is called time-difference recoding technique.

The interaction or recognition events can be performed by online or offline format. The online interaction is accomplished by dynamically pumping reaction solution through the flow cell. To have sufficient interaction time, stop-flow technique may be adopted. In fact, the continuous flow-through technique waste a large amount of target analytes because the time for flowing through the flow cell is too short to let the molecules on the upper side of the flowing stream diffuse down to the sensing surface. This is especially an issue in capturing the macromolecules such as proteins. The macromolecules have quite a small diffusion coefficients and will take quite a long time to diffuse from up down to the sensing surface. For example, a protein has a diffusion coefficient at about 2×10^{-5} mm^2/s, thus inside a 1 mm thick flow cell, the top molecule will take $t_D = 1^2/(2 \times 2 \times 10^{-5}) = 2.5 \times 10^4$ s ≈ 6.94 h to diffuse down to the interaction surface. However, the flow-through time is only about minutes. Even use of the stop-flow technique, it remains not sufficient to have all target analytes come down to interact with their probes. We have tried to apply an electric field between the top and bottom surfaces of the flow cell during pumping in the samples solution, the effect remained not ideal. A better way is to agitate the flow to produce turbulent flow or even counter flow. Reciprocating flow is also a solution in practice.

Another effective way is to conduct the interaction reaction by offline method, that is, by direct insertion of sensor chip into a reaction solution under sonication or stirring for a certain time. The chip is then cleaned and re-assembled onto the SPRi system to take sharper pictures. The advantage of the offline approach is that it makes full use of space and agitation to speed up the recognition reaction by stirring or sonication. While its main shortage is that the take-off and re-mounting are not very easy to manipulate. To overcome the issues, the flow cell must be so designed that it can be opened and closed to facilitate loading and unloading solutions and even the chip. The key shortage of offline is that it sacrifices the real-time imaging function.

4.7.1 *Recording of Images*

There are different ways to collect the SPRi signals depending on the working principle used to construct the instrument. The easiest way is to record the intensity of reflection light with a black-and-white or a color video camera, and then is to detect the resonant angle or wavelength change. Phase change is better recorded but needs to use a specially designed instrument.

As known, SPRi can be performed statically or dynamically. Static SPRi is not a timely mode, and it records only stable pictures normally after a reaction completes or reaches its equilibrium. This mode reduces data volume and allows you to leisurely control a slow reaction event and/or to go less detours because you can conduct next step of experiment after you well analyze the current imaging data. The disadvantages are that you lose the real-time information and also cannot perform post-experimental dynamic analysis.

Dynamic SPRi is the most common mode that records the real-time information. This method can acquire full information of all reactions, including the statics, and allows to do post-experimental analysis vividly at any a time you are convenient. The disadvantages of this method are that it occupies a huge memory space and have potential to congest a computer. This may become a big challenge in a common laboratory if we think that SPRi can perform thousands or over ten thousands of reactions all at once.

4.7.2 *Analysis of Imaging Data*

4.7.2.1 Image Comparison

SPRi data can be expressed in three forms: 2D and 3D images or sectioned spectra as illustrated in Fig. 4.28, which offers the area (Fig. 4.28a), volume (Fig. 4.28b) and relative height (Fig. 4.28c) of a spot. Based on these data, SPRi analysis ordinarily starts from reading and comparisons of images and/or imaging data recorded under same or different conditions, aiming at the withdrawing of qualitative and/or quantitative information.

Image comparison aims at finding the spatial variation, the similarity and/or difference of the compared images. In the analysis of a spatial addressable chip, the pattern of spots can directly reveal the existence of target analytes on a chip with the spot brightness as illustrated in Fig. 4.1b and Fig. 4.28a, b. The pattern itself can offer the

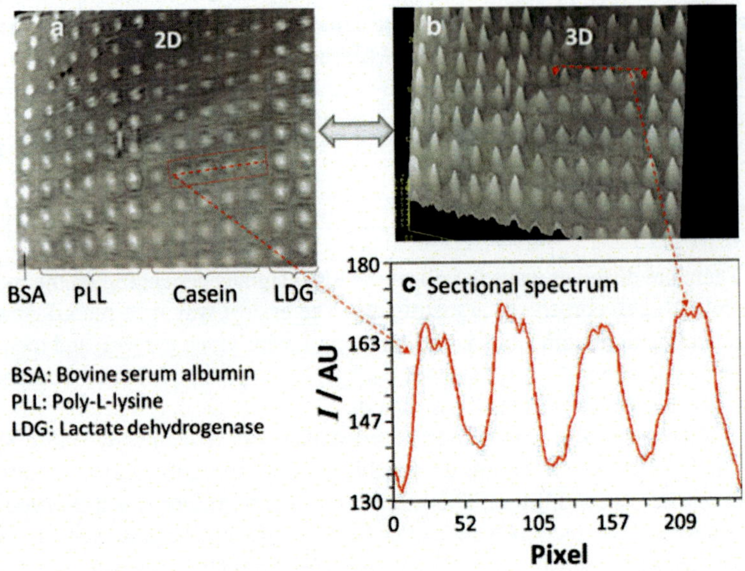

Fig. 4.28 SPRi of spotted proteins acquired on a laboratory-built SPR imager model TX 710. **a**2D image; **b** 3D image; **c** sectional spectrum along the indicated dot line

vivid analytical information, especially the spatial connections or correlations. This may neatly be called pattern recognition.

In order to have time-dependent information, you now need to compare images recorded in a time sequence. It is possible to pick up one or more spots and observe their variation with time. More informative comparison is conducted by pattern recognition and differentiation among images. The comparison is somewhat easy based on the image intensity extracted from a same-sized local surfaces. More complete comparison should be conducted by use of all possible information including the maximum intensity, location, spot size and volume. This needs to have a subprogram able to recognize the spot border, integrate spot area and total volume (by further integration of intensity).

Image pattern comparison does not necessarily play along a time axis but along spatial axes (as mentioned in the first paragraph) or cross-images not in same coordinate system, for example between or among different experiments with variable conditions and even analytes. This help to reveal some unexpected analytical information.

Image pattern analysis has a unique advantage, that is, it offers spatiotemporal connections or correlations. This is not possible by other analytical methods based on a single datum or more scattered data. The obvious disadvantage lies in that image analysis is presently not very familiar to normal users. Fortunately, this issue can soon be get over by transplanting the artificial intelligence (AI) concepts and related programs in SPRi.

4.7.2.2 Qualitative Analysis

No matter the image is composed of discrete spots or continuous film, SPRi utilizes signal patterns or surface address to recognize and/or identify target analytes.

Similar to SPRS, SPRi is not good at qualitative analysis in spite that its signals are dependent on the physical features of analytes such as their relative permittivity ε_1 or refractive index n_1. As known the ε_1 or n_1 is also a function of concentration of a related analyte, it is thus not sufficient to identify a substance.

In order to perform reliable identification of a target analyte by SPRi, additional means should be coupled with. An easy way is to combine with probe recognition. It is known that many analytes can be recognized by their specific probes. The problem lies in that the recognition is often interfered by cross-reactions, leading to unreliable identification. This issue can be removed in many circumstances by use of multiple probes in combination with image patterns.

SPRi is easy to integrate with other optical spectral methods such as fluorescence and Raman spectroscopy to help its qualitative analysis, while the most powerful alternative is to couple SPRi with some highly recognized identification tools like MS, especially MS imaging or MSi. Although there are several possibilities to hyphenate SPRi with MS and/or MSi such as DART (direct analysis in real time), DESI (desorption electrospray ionization) and many ambient MS formats, the most often coupled format is SPRi-MALDI (matrix-assisted laser desorption/ionization) TOF MS or MSi. By coupling with MSi, an SPR image will produce many *m/z*-based MS images with ionic intensity or concentration as contrast. Each of the MS images may correspond to a special identity of an ion species.

4.7.2.3 Quantitative Analysis

SPRi features high-throughput quantitative analysis based on the measured changes of resonant angle $\Delta\theta_r$, resonant wavelength $\Delta\lambda_r$, reflected optical intensity ΔI and/or their relationship with ε_1 or n_1 and in turn with the concentration of an analyte in bulk solution c_b or on the sensing surface c_s.

In theory, quantitative correlation of SPRi signals with the absolute coverages of adsorbed analytes or adsorbates can be obtained from SPR absorption spectra that are plots of the fully reflected light intensity versus the incident angle or wavelength. It is also possible to quantify the adsorbates by fitting a full SPR curve based on the Fresnel equations that give in fact the average thickness of the probed area [33, 101, 102]. On the adsorbed spots, there are not only the target analyte but also the molecules from the contact buffer solution, and the measured refractive index should be a volume average of the adsorbates and buffer [18]. Thus, if their refractive indices are known, the number of a target analyte can be calculated through multiplying the volume fraction of the analyte by the spot thickness, and the molecular mass of the analyte can further be calculated if the density is known.

The target analyte can also be determined according to the resonant shift that can be converted to the volume of adsorbed analyte per unit area. This method requires to calibrate the response of the resonant shift per unit change in bulk refractive index, which should be a constant over the quantitative range. The mass of adsorbed molecules per unit area is calculated by multiplying this with the density of the pure analyte. Through dividing this by the partial molar volume of the analyte, the number of adsorbed molecules per unit area can be calculated.

Because the interrogation of resonant data takes time, real-time tracking the resonant shift is in general not possible. Before a time-independent interrogation technique is developed, the quantification of analyte coverage will have a time delay.

To obtain real-time information about adsorption and desorption kinetics with SPRi, a spatially resolved measurement of the reflectivity at a fixed angle would provide much better time resolution [31, 32, 36, 103–105]. However, the relationship between the reflected light intensity and absolute adsorbate coverage has not yet been addressed.

In the study of DNA binding events, Nelson et al. indirectly calibrated the saturated intensity changes of SPRi with the hybridized target on a DNA-functionalized spot [39]. To measure, fluorescently labeled DNA is captured on the sensor surface with a sufficiently large area and quantified by fluorescence spectrophotometry after releasing the DNA in HCl solution. The results showed that the saturated adsorption needed ca. 1×10^{12} DNA/cm^2 and had a 0.6% reflectivity change in SPRi. There are also other techniques to conduct DNA or DNA hybridization-based quantification that was either reported [106] or is waiting for exploration.

4.7.2.4 Calibration of Instrument

It is critical to calibrate SPRi instruments for reliable SPRi determination. In principle, all the related devices especially those for quantitative analysis must be calibrated to eliminate systematic errors. Particularly, the SPRi instrument needs the calibration with one or two standards to define its response to the variation of refractive index n or concentration c. The easily available calibration standards include neutral sucrose, glucose, ethanol and electrolytes like NaCl. Their calibration equations can be measured from their aqueous solutions by Abbe refractometer. Following refractive index equations were measured at 20 °C and 589 nm for reference:

$$
\begin{aligned}
n_{\mathrm{NaCl}} &= 1.33329 + 1.84 \times 10^{-3}x + 2.91822 \times 10^{-7}x^2 \\
&\quad (x = 0 \sim 25\%\ \mathrm{w/v},\ r^2 = 0.9998) \\
n_{\mathrm{sucrose}} &= 1.33312 + 1.41 \times 10^{-3}x + 8.78638 \times 10^{-6}x^2 \\
&\quad - 9.9172 \times 10^{-8}x^3 + 1.08797 \times 10^{-9}x^4, \\
&\quad (x = 0 \sim 60\%\ \mathrm{w/v},\ r^2 = 0.9998) \\
n_{\mathrm{ethanol}} &= 1.33333 + 3.59591 \times 10^{-4}x + 1.38977 \times 10^{-5}x^2 \\
&\quad - 3.08117 \times 10^{-7}x^3 + 1.68846 \times 10^{-9}x^4 \\
&\quad (x = 0 \sim 50\%\ \mathrm{w/v},\ r^2 = 0.9999);
\end{aligned}
$$

$$= 1.33333 + 1.76 \times 10^{-3}c + 7.62802 \times 10^{-4}c^2 - 1.25642 \times 10^{-4}c^3$$
$$+ 7.81249 \times 10^{-6}c^4 - 1.80533 \times 10^{-7}c^5,$$
$$(c = 0 \sim 15 \text{ mol/L}, \ r^2 = 0.9999). \tag{4.66}$$

An SPRi device is better calibrated each day before starting quantification but after a chip is mounted, cleaned and well equilibrated. Following illustrated is a common protocol we used.

Approach 4.25 for calibration of an SPRi device:

Step 1, mount a chip to be used on the SPRi device.
Step 2, switch on power and temperature control.
Step 3, clean the system with required solutions, normally water followed by running buffer.
Step 4, equilibrate the chip with running buffer until baseline is stable.
Step 5, inject a sufficiently long plug of refractive index calibration solutions from low to high concentration.
Step 6, record the images and/or sensorgrams.
Step 7, adjust the gain and related parameters of the SPRi device to the target RIU value.

4.8 General Program for Method Development

There are some key points involving the general program to develop SPRi methods, such as theoretical limitations or intrinsic possibility of SPRi, analytical tasks and realistic conditions (including equipment level and cost affordability). The intrinsic possibility of SPRi is the decisive factor that is dependent on the working principles as have been discussed and on the relationships of SPRi with neighboring techniques (e.g., potential hyphenation or collaboration). The core of developing more powerful SPRi methodology lies in the hyphenation and/or integration of current SPRi technology with more sophisticated technologies. Figure 4.29 illustrates four categories to empower SPRi. Among the indicated fields, SPRS and SPRi are a pair of fusible methods because they measure the same parameters of resonant angle, resonant wavelength, intensity change and/or phase variation. An ideal SPRi instrument should be flexible, with optional ability in angle- and/or wavelength interrogations and able to give corresponding absorption spectra. In addition to SPRS, other types of sensors may also have the potential to couple to or integrate with SPRi. A successful example is the integration of electrochemical sensors with SPRi.

Separation analysis such as chromatography, capillary electrophoresis and their miniaturization formats like micro/nanofluidics are a vast field that has been touched in SPRi development but is still waiting for effective exploration. In fact, SPRi is a novel type of label-free chip technology, therefore, it can easily collaborate with the

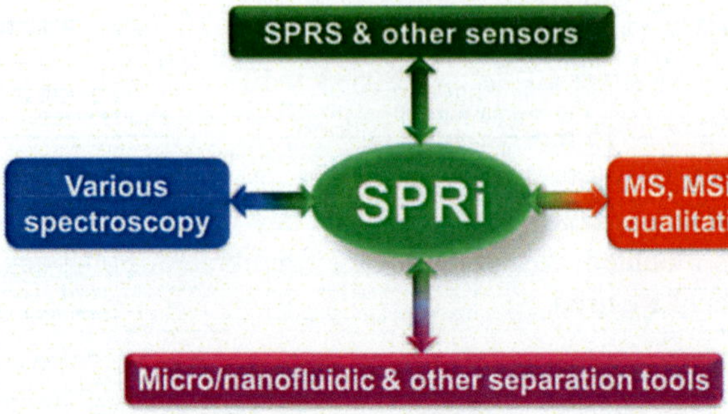

Fig. 4.29 Four categories for development of novel SPRi assays through the coupling or integration of SPRi with other instrumental methods

micro/nanofluidics. As an imaging means, SPRi can also serve as a novel type of cost-effective multi-channel detector for simultaneous detecting the multiply separated bands.

As an absorption optical technology, SPRi is in nature suitable for coupling or integrating with other optical spectroscopic techniques such as *in situ* fluorescence, Raman spectroscopy, infrared spectroscopy and their imaging technology.

Unfortunately, SPRi itself is short of identification ability. This has been largely overcome by probing technology. Further solution is *in situ* or offline coupling of it with powerful identification technologies such as MS especially MSi. This is the current effort in the development and utilization of novel SPRi assays for precise identification of on-chip-captured unknown substances.

A more universal program or thinking to develop a practical SPRi method is illustrated in Fig. 4.30 for reference. There are at least three major goals that have to be carefully considered in the design and exploitation process: (i) investigation of reactions and kinetics including weak and strong molecular events such as recognition, specific interaction and chemical or biochemical reactions, (ii) quantitative analysis including the continuous adlayers and discrete state (counting in this latter case), and (iii) qualitative analysis including identification of unknown substances and screening of specific targets. These goals may lead to two different final measures, utilization of multi-channel or spotting technology. The solution may be achieved by simply use of SPRi alone or by easy or complicated coupling with other analytical tools.

It is worth of further saying several words on the workflow why we tart to develop a method from analytical purpose: It was efficient according to our experience. The factor that may promote or hinder the development of a practical SPRi method lies in that if you can reliably identify some unknown substances in your samples or analytes without proper probes. In this case, it is necessary to combine the use of SPRi with MS or MSi. Other confining factors may include the research environments and laboratory conditions that are out of the scope of this book.

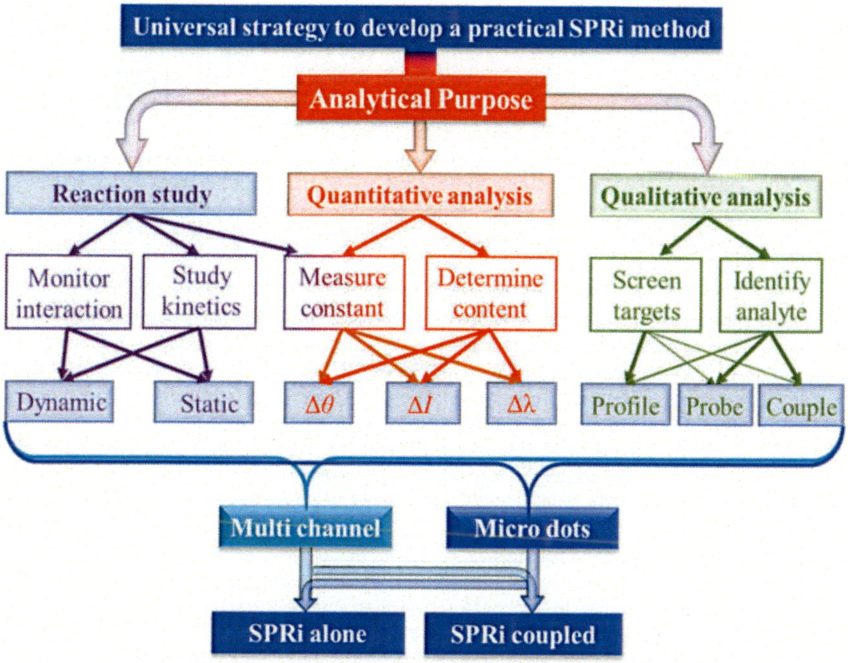

Fig. 4.30 Workflow for developing a practical SPRi method adopted in our laboratory

References

1. Wang Z, Chen Y (2001) Detection of metal ions using wavelength interrogation surface plasmon resonance spectroscopy with calixarane derivatives as sensing films. Anal Lett 34:2609–2619
2. Wang Z, Chen Y (2001) Analysis of mono- and oligo-saccharides by multi-wavelength surface plasmon resonance (SPR) spectroscopy. Carbohydr Res 332:209–213
3. Han Z, Qi L, Shen G, Liu W, Chen Y (2007) Determination of chromium(VI) by surface Plasmon field-enhanced resonance light scattering. Anal Chem 79:5862–5866
4. Huang H, Huang S, Liu X, Zeng Y, Yu X, Liao B, Chen Y (2009) Label-free optical biosensors based on Au_2S-coated gold nanorods. Biosens Bioelectron 24:3025–3029
5. Chen S, Zhao Q, Liu F, Huang H, Wang L, Yi S, Zeng Y, Chen Y (2013) Ultrasensitive determination of copper in complex biological media based on modulation of plasmonic properties of gold nanorods. Anal Chem 85:9142–9147
6. Zhao Q, Huang H, Zhang L, Wang L, Zeng Y, Xia X, Liu F, Chen Y (2016) Strategy to fabricate naked-eye readout ultrasensitive plasmonic nanosensor based on enzyme mimetic gold nanoclusters. Anal Chem 88:1412–1418
7. Wang X, Xu J, Wang Y, Wang F, Chen Y (2016) A universal strategy for direct immobilization of intact bioactivity-conserved carbohydrates on gold nanoparticles. RSC Adv 6:85333–85339
8. Yuan X, Chen Y (2012) Visual determination of Cu^{2+} through copper-catalysed in-situ formation of Ag nanoparticles. Analyst 137:4516–4523
9. Chen J, Xu J, Chen Y (2013) Interaction of straight chain alcohol vapors with self-assembled MOF film by surface plasmon resonance sensing and imaging. Chin Chem Lett 24:651–653
10. Shen G, Han Z, Liu W, Chen Y (2007) Color surface plasmon resonance imaging of protein microdots arrays. Chem Lett 36:926–927

11. Abelès F (1976) Surface electromagnetic waves ellipsometry. Surf Sci 56:237–251
12. Kabashin AV, Patskovsky S, Grigorenko AN (2009) Phase and amplitude sensitivities in surface plasmon resonance bio and chemical sensing. Opt Express 17:21191–21204
13. Kabashin AV, Nikitin PI (1997) Interferometer based on a surface-plasmon resonance for sensor applications. Quantum Electron 27:653–654
14. Kabashin AV, Nikitin PI (1998) Surface plasmon resonance interferometer for bio- and chemical-sensors. Opt Commun 150:5–8
15. Patskovsky S, Maisonneuve M, Meunier M, Kabashin AV (2008) Mechanical modulation method for ultra-sensitive phase measurements in photonics biosensing. Opt Express 16:21305–21314
16. Law W-C, Markowicz P, Yong K-T, Roy I, Baev A, Patskovsky S, Kabashin AV, Ho H-P, Prasad PN (2007) Wide dynamic range phase-sensitive surface plasmon resonance biosensor based on measuring the modulation harmonics. Biosens Bioelectron 23:627–632
17. Markowicz PP, Law WC, Baev A, Prasad PN, Patskovsky S, Kabashin AV (2007) Phase-sensitive time-modulated surface plasmon resonance polarimetry for wide dynamic range biosensing. Opt Express 15:1745–1754
18. Jung LS, Campbell CT, Chinowsky TM, Mar M, Yee SS (1998) Quantitative interpretation of the response of surface plasmon resonance sensors to adsorbed films. Langmuir 14:5636–5648
19. Armstrong Jr SH, Budka MJE, Morrison KC, Hasson M (1947) Preparation and properties of serum and plasma proteins. XII. The refractive properties of the proteins of human plasma and certain purified fractions. J Am Chem Soc 69:1747–1753
20. McMeekin TL, Groves ML, Hipp NJ (1964) Refractive indices of amino acids, proteins, and related substances. In: Stekol JA (ed) Amino acids and serum proteins. Adv Chem 44:54–56. https://doi.org/10.1021/ba-1964-0044.ch004
21. Darnell JE, Lodish H, Baltimore D (1990) Molecular cell biology. Scientific American Books, New York
22. Gölander C-G, Kiss E (1988) Protein adsorption on functionalized and ESCA-characterized polymer films studied by ellipsometry. J Colloid Interface Sci 121:240–253
23. Lide DR (ed) (1990) Handbook of chemistry physics, 71st edn. CRC Press, Boston
24. Chen J, Chen Y, Xu J, Zhang Y, Liao T (2012) Post-experimental denoising and background subtraction of surface plasmon resonance images for better quantification. Chemom Intell Lab Syst 114:56–63
25. Pitarke JM, Silkin VM, Chulkov EV, Echenique PM (2007) Theory of surface plasmons and surface-plasmon polaritons. Rep Prog Phys 70:1–87
26. Yu H, Shan X, Wang S, Chen H, Tao N (2014) Molecular scale origin of surface plasmon resonance biosensors. Anal Chem 86:8992–8997
27. Viitala L, Maley AM, Fung HWM, Corn RM, Viitala T, Murtomäki L (2016) Surface plasmon resonance imaging microscopy of liposomes and liposome-encapsulated gold nanoparticles. J Phys Chem C 120:25958–25966
28. Bozhevolnyi SI, Coello V (1998) Elastic scattering of surface plasmon polaritons: modeling and experiment. Phys Rev B: Condens Matter Mater Phys 58:10899–10910
29. Yu H, Shan X, Wang S, Wang S, Tao N (2017) Achieving high spatial resolution surface plasmon resonance microscopy with image reconstruction. Anal Chem 89:2704–2707
30. Aust EF, Sawodny M, Ito S, Knoll W (1994) Surface plasmon and guided optical wave microscopies. Scanning 16:353–361
31. Thiel AJ, Frutos AG, Jordan CE, Corn RM, Smith LM (1997) In situ surface plasmon resonance imaging detection of DNA hybridization to oligonucleotide arrays on gold surfaces. Anal Chem 69:4948–4956
32. Berger CEH, Beumer TAM, Kooyman RPH, Greve J (1998) Surface plasmon resonance multisensing. Anal Chem 70:703–706
33. Knoll W (1998) Interfaces and thin films as seen by bound electromagnetic waves. Annu Rev Phys Chem 49:569–638
34. Lyon LA, Holliway WD, Natan MJ (1999) An improved surface plasmon resonance imaging apparatus. Rev Sci Instrum 70:2076–2081

35. Rothenhäuslar B, Knoll W (1988) Surface plasmon microscopy. Nature 332:615–617
36. Zizlsperger M, Knoll W (1998) Multispot parallel on-line monitoring of interfacial binding reactions by surface plasmon microscopy. Prog Colloid Polym Sci 109:244–253
37. Huang H, Zhang S, Qi L, Yu X, Chen Y (2006) Microwave-assisted deposition of uniform thin gold film on glass surface. Surf Coat Technol 200:4389–4396
38. Brockman JM, Frutos AG, Corn RM (1999) A multistep chemical modification procedure to create DNA arrays on gold surfaces for the study of protein−DNA interactions with surface plasmon resonance imaging. J Am Chem Soc 121:8044–8051
39. Nelson BP, Grimsrud TE, Liles MR, Goodman RM, Corn RM (2001) Surface plasmon resonance imaging measurements of DNA and RNA hybridization adsorption onto DNA microarrays. Anal Chem 73:1–7
40. Camilone N, Chidsey CED, Liu G-Y, Scoles G (1993) Substrate dependence of the surface structure and chain packing of docosyl mercaptan self-assembled on the (111), (110) and (100) faces of single crystal gold. J Chem Phys 98:4234–4245
41. Jimenez A, Sarsa A, Blazquez M, Pineda T (2010) A molecular dynamics study of surfactant surface density of alkanethiol sefl-assembled monolayers on gold nanoparticles as a function of the radius. J Phys Chem C 114:21309–21314
42. Vericat C, Vela ME, Benitez G, Carro P, Salvarezza RC (2010) Sel-assembled monolayers of thiols and dithiols on gold: new challenges of a well known system. Chem Soc Rev 39:1805–1834
43. Ulman A (1999) Formation and structure of self-assembled monolayers. Chem Rev 96:1533–1554
44. Bryant MA, Pemberton JE (1991) Surface raman scattering of self-assembled monolayers form from 1-alkanethiols: behavior of films at gold and comparison to film at silver. J Am Chem Soc 113:8284–8293
45. Schoenfish MH, Pemberton JE (1998) Air stability of alkanethiol self-assmbled monolayers on silver and gold surfaces. J Am Chem Soc 120:4502–4513
46. Laibinis PE, Whitesides GM, Allara DL, Tao YT, Parikh AN, Nuzzo RG (1991) Comparison of the structures and wetting properties of self-assembled monolayes of n-alkanethiols on the coinage metal surfaces: copper, silver and gold. J Am Chem Soc 113:7152–7167
47. Campos MAC, Trilling AK, Yang M, Giesbers M, Beekwilder J, Paulusse JMJ, Zuihof H (2011) Self-assembled functional organic monolayers on oxide-free copper. Langmuir 27:8126–8133
48. Laibinis PE, Whitesides GM (1992) Self-assembled monolayers of n-alkanethiols on copper are barrier films that protect the metal against oxidation by air. J Am Chem Soc 114:9022–9028
49. Volmer M, Stralmann M, Viefhaus H (1990) Electrochemical and electron spectroscopic investigation of iron surface modified with thiols. Surf Interface Anal 16:278–282
50. Nozawa K, Nishihara H, Aramaki K (1997) Chemical modification of alkanethiol monolayers for protecting iron against corrosion. Corros Sci 39:1625–1639
51. Nozawa K, Aramaki K (1999) One- and two-dimensional polymer films of modified alkanethiol monolayers for preveting iron from corrosion. Corros Sci 41:57–73
52. Inkpen MS, Liu Z-F, Li H, Campos LM, Neaton JB, Venkataraman L (2019) Non-chemisorbed gold–sulfur binding prevails in self-assembled monolayers. Nat Chem 11:351–358. https://doi.org/10.1038/s41557-019-0216-y
53. Lee H, Dellatore SM, Miller WM, Messersmith PB (2007) Mussel-inspired surface chemistry for multifunctional coatings. Science 318:426–430
54. d'Ischia M, Napolitano A, Pezzella A, Meredith P, Sarna T (2009) Chemical and structural diversity in eumelanins: unexplored bio-optoelectronic materials. Angew Chem Int Ed 48:3914–3921
55. Della Vecchia NF, Avolio R, Alfè M, Errico ME, Napolitano A, d'Ischia M (2013) Building-block diversity in polydopamine underpins a multifunctional eumelanin-type platform tunable through a quinone control point. Adv Funct Mater 23:1331–1340
56. Yang W, Liu CJ, Chen Y (2018) Stability of polydopamine coatings on gold substrates inspected by surface plasmon resonance imaging. Langmuir 34:3565–3571

57. Hong S, Na YS, Choi S, Song IT, Kim WY, Lee H (2012) Non-covalent self-assembly and covalent polymerization co-contribute to polydopamine formation. Adv Funct Mater 22:4711–4717
58. Liebscher J, Mrówczyński R, Scheidt HA, Filip C, Hădade ND, Turcu R, Bende A, Beck S (2013) Structure of polydopamine: a never-ending story? Langmuir 29:10539–10548
59. Bernsmann F, Ball V, Addiego F, Ponche A, Michel M, Gracio JJ, Toniazzo V, Ruch D (2011) Dopamine-melanin film deposition depends on the used oxidant and buffer solution. Langmuir 27:2819–2825
60. d'Ischia M, Napolitano A, Ball V, Chen C-T, Buehler MJ (2014) Polydopamine and eumelanin: from structure-property relationships to a unified tailoring strategy. Acc Chem Res 47:3541–3550
61. Dreyer DR, Miller DJ, Freeman BD, Paul DR, Bielawski CW (2012) Elucidating the structure of poly(dopamine). Langmuir 28:6428–6435
62. Ball V (2010) Impedance spectroscopy and zeta potential titration of dopa-melanin films produced by oxidation of dopamine. Colloids Surf A 363:92–97
63. Kim BH, Lee DH, Kim JY, Shin DO, Jeong HY, Hong S, Yun JM, Koo CM, Lee H, Kim SO (2011) Mussel-inspired block copolymer lithography for low surface energy materials of teflon, graphene, and gold. Adv Mater 23:5618–5622
64. Smolin EM, Tapoport L (1959) The chemistry of heterocyclic compounds, s-triazine and derivatives. Itersciences, New York
65. Bruckner H, Strecker B (1992) Various concepts for toppingsteam plants with gas turbines. J Chromatogr 627:97–105
66. Thurstojanme JT, Dudleyd JR, Kaiser DW, Hechenbleikner I, Schaefer FC, Holm-Hansen D (1951) Cyanuric chloride derivatives. I. Aminochloro-s-triazines. J Am Chem Soc 73:2981–2983
67. Blotny G (2006) Recent applications of 2,4,6-trichloro-1,3,5-triazine and its derivatives in organic synthesis. Tetrahedron 62:9507–9522
68. Jan JZ, Huang BH, Lin JJ (2003) Facile preparation of amphiphilic oxyethylene–oxypropylene block copolymers by selective triazine coupling. Polymer 44:1003–1011
69. Steffensen MB, Simanek EE (2003) Chemoselective building blocks for dendrimers from relative reactivity data. Org Lett 5:2359–2361
70. Palazon F, Benavides CM, Leonard D, Souteyrand E, Chevolot Y, Cloarec JP (2014) Carbodiimide/NHS derivatization of COOH-terminated SAMs: activation or byproduct formation? Langmuir 30:4545–4550
71. Montalbetti CAGN, Falque V (2005) Amide bond formation and peptide coupling. Tetrahedron 61:10827–10852
72. Valeur E, Bradley M (2009) Amide bond formation: beyond the myth of coupling reagents. Chem Soc Rev 38:606–631
73. Li D, Guo Z, Liu C, Li J, Xu W, Chen Y (2017) Quantification of near-attomole gibberellins in floral organs dissected from a single Arabidopsis thaliana flower. Plant J 91:547–557
74. Nakajima N, Ikada Y (1995) Mechanism of amide formation by carbodiimide for bioconjugation in aquaus-media. Bioconjug Chem 6:123–130
75. Kuo JW, Swann DA, Prestwich GD (1991) Chemical modification of hyaluronic acid by carbodiimides. Bioconjug Chem 2:232–241
76. Pouyani T, Kuo JW, Harbison GS, Prestwich GD (1992) Solid-state NMR of N-acylureas derived from the reaction of hyaluronic acid with isotopically-labeled carbodiimides. J Am Chem Soc 114:5972–5976
77. Kurzer F, Douraghi K (1967) Advances in chemistry of carbodiimides. Chem Rev 67:107–152
78. Kishikawa K, Nakahara S, Nishikawa Y, Kohmoto S, Yamamoto M (2005) A ferroelectrically switchable columnar liquid crystal phase with achiral molecules: superstructures and properties of liquid crystalline ureas. J Am Chem Soc 127:2565–2571
79. Nyangulu JM, Galka MM, Jadhav A, Gai Y, Graham CM, Nelson KM, Cutler AJ, Taylor DC, Banowetz GM, Abrams SR (2005) An affinity probe for isolation of abscisic acid-binding proteins. J Am Chem Soc 127:1662–1664

80. Grant EB, Weiss JM, Branum S, Hayden S, Johnson S, Guiadeen D, Murray WV, Macielag MJ (2005) The synthesis of (9S)-9-alkyl-9-hydroxyerythromycin A derivatives and their ketolides. Tetrahedron Lett 46:2731–2735
81. Davis FA, Deng J (2004) Asymmetric synthesis of syn-(2R,3S)- and anti-(2S,3S)-ethyl diamino-3-phenylpropanoates from N-(benzylidene)-p-toluenesulfinamide and glycine enolates. Org Lett 6:2789–2792
82. White JD, Hansen JD (2005) Total synthesis of (-)-7-epicylindrospermopsin, a toxic metabolite of the freshwater cyanobacterium aphanizomenon ovalisporum, and assignment of its absolute configuration. J Org Chem 70:1963–1977
83. Iliev B, Linden A, Heimgartner H (2003) An unexpected formation of a 14-membered cyclodepsipeptide. Helv Chim Acta 86:3215–3234
84. Ella-Menye J-R, Sharma V, Wang G (2005) New synthesis of chiral 1,3-oxazinan-2-ones from carbohydrate derivatives. J Org Chem 70:463–469
85. Dandapani S, Curran DP (2004) Second generation fluorous DEAD reagents have expanded scope in the Mitsunobu reaction and retain convenient separation features. J Org Chem 69:8751–8757
86. Pedras MSC, Chumala PB, Quail JW (2004) Chemical mediators: the remarkable structure and host-selectivity of depsilairdin, a sesquiterpenic depsipeptide containing a new amino acid. Org Lett 6:4615–4617
87. Vaidyanathan R, Kalthod VG, Ngo DP, Manley JM, Lapekas SP (2004) Amidations using N, N′-carbonyldiimidazole: remarkable rate enhancement by carbon dioxide. J Org Chem 69:2565–2568
88. Li MH, Choi SK, Leroueil PR, Baker JR Jr (2014) Evaluating binding avidities of populations of heterogeneous multivalent ligand-functionalized nanoparticles. ACS Nano 8:5600–5609
89. Kolb HC, Finn MG, Sharpless KB (2001) Click chemistry: diverse chemical function from a few good reactions. Angew Chem Int Ed 40:2004–2021
90. Himo F, Lovell T, Hilgraf R, Rostovtsev VV, Noodleman L, Sharpless KB, Fokin VV (2005) Copper(I)-catalyzed synthesis of azoles. DFT study predicts unprecedented reactivity and intermediates. J Am Chem Soc 127:210–216
91. Boren BC, Narayan S, Rasmussen LK, Zhang L, Zhao H, Lin Z, Jia G, Fokin VV (2008) Ruthenium-catalyzed azide–alkyne cycloaddition: scope and mechanism. J Am Chem Soc 130:8923–8930
92. Hoyle CE, Lee TY, Roper T (2004) Thiol–enes: Chemistry of the past with promise for the future. J Polym Sci Part A: Polym Chem 42:5301–5338
93. Becer CR, Hoogenboom R, Schubert US (2009) Klick-Chemie jenseits von metallkatalysierten Cycloadditionen. Angew Chem 121:4998–5006
94. Kade MJ, Burke DJ, Hawker CJ (2010) The power of thiol-ene chemistry. J Polym Sci Part A: Polym Chem 48:743–750
95. Ke Z, Melisa AL, Ying W, Gregory NT (2011) Universal cyclic polymer templates. J Am Chem Soc 133:6906–6909
96. Hoyle CE, Browman CN (2010) Thiol–ene click chemistry. Angew Chem Int Ed 49:1540–1573
97. Hutchins RO, Hutchins MK (1991) Reduction of C=O to CHNH by metal hybrids. In: Trost BN, Fleming I (eds) Comprehensive organic synthesis, vol 8. Pergamon Press, New York
98. Albdel-Magid AF, Carson KG, Harris BD, Maryanoff CA, Shah RD (1996) Reductive amination of aldehydes and ketones with sodium triacetoxyborohydride. Studies on direct and indirect reductive amination procedures. J Org Chem 61:3849–3862
99. Borch RF, Bernstein MD, Durst HD (1971) Cyanohydridoborate anion as a selective reducing agent. J Am Chem Soc 93:2897–2904
100. Borch RF, Durst HD (1969) Lithium cyanohydridoborate, a versatile new reagent. J Am Chem Soc 91:3996–3997
101. Homola J, Yee SS, Gauglitz G (1999) Surface plasmon resonance sensors: reviews. Sens Actuators B Chem 54:3–15

102. Brockman JM, Nelson BP, Corn RM (2000) Surface plasmon resonance imaging measurements of ultrathin organic films. Annu Rev Phys Chem 51:41–63
103. Frey BL, Jordan CE, Kornguth S, Corn RM (1995) Control of the specific adsorption of proteins onto gold surfaces with poly(L-lysine) monolayers. Anal Chem 67:4452–4457
104. Jordan CE, Corn RM (1997) Surface plasmon resonance imaging measurements of electrostatic biopolymer adsorption onto chemically modified gold surfaces. Anal Chem 69:1449–1456
105. Nelson BP, Grimsrud TE, Liles MR, Goodman RM, Corn RM (2001) Surface plasmon resonance imaging measurements of DNA and RNA hybridization adsorption onto DNA microarrays. Anal Chem 73:1–7
106. Shumaker-Parry JS, Campbell CT (2004) Quantitative methods for spatially resolved adsorption/desorption measurements in real time by surface plasmon resonance microscopy. Anal Chem 76:907–917

Chapter 5
Interaction and Reaction

Interactions and reactions cover a series of molecular events from weak to strong that universally exist in the world, especially in biological processes. As revealed in Chaps. 2 and 4, SPRi needs to make use of various interactions and reactions to work and therefore is able to study them in turn. This also needs to thank to the invention of non-destructive light source such as LED and semiconductor lasers and the fast response of light, which make the theoretical possibility of SPRi become true. In fact, SPRi is now able to do researches with native or intact substances in various natural and non-natural environments wherever you prefer, suitable for the direct and indirect studies of dynamics and thermodynamics of various interactions or reactions under designable and/or variable conditions, including physiological conditions that arc critical to have insights into the realistic biometric events. Extending form this, SPRi is applicable to the quantitative and qualitative analysis of various substances involving various processes that will be discussed in Chaps. 6, 7 and 8.

In this chapter, we will concisely discuss dynamic and/or thermodynamic measurements as the initiation of SPRi application to the solution of real measuring problems based on majorly the methodology delivered in Chap. 4. Three basic measurement-related topics are designed for the discussion: general methodology to measure the reaction constants, simulation of the recognition of membrane receptors with their ligands and maintenance of SPRi selectivity and sensitivity in the study of reactions. They are the essential in later chapters or normally involve in various real applications of SPRi assays.

5.1 General Methodology

By spot array technology, SPRi can simultaneously observe many binding and recognition events between different molecules such as DNA–DNA [1–5], DNA–RNA [6], DNA–protein [7–9], protein–protein, protein–carbohydrate, carbohydrate–carbohydrate and so forth, which may be endless. The general measuring approach includes:

Y. Chen, *Surface Plasmon Resonance Imaging*, Lecture Notes in Chemistry 95,
https://doi.org/10.1007/978-981-99-3118-7_5

(i) preparation of sensing chip, (ii) preparation and loading of sample (reaction) solution and (iii) real-time measurement that includes record, display, treatment and output of image signals. If gold-coated chips are at hand, the first two steps can be merged together and simplified into surface modification and sample spotting. In dynamic study, at least one of reactants need to be introduced dynamically. The most often adopted technology is to continuously flow the reactant (usually at constant concentration) across the chip surface to initiate and/or maintain the reaction (e.g., capture or adsorption through chemical reaction, specific recognition or affinity interaction and/or desorption by designed conditions) by pumping technique. In some cases, manually addition is performable. Currently, the imaging signals are recorded frame by frame via a video camera. The time resolution is dependent on how fast is the camera, while the practical lateral spatial resolution depends on the spot size and spot-to-spot distance, down to the limit of lateral propagation length of SPPs at about several tens of micrometers.

Based on these three major steps, SPRi of interactions or reactions can be performed either statically or dynamically. Static SPRi measures the images under a reaction-equilibrated or steady state, which is suitable for thermodynamic study, while dynamic SPRi measures the imaging signals continuously under flowing and reacting states, which is suitable for kinetic study. According to the technique used to add reaction solution, we can have four operation combinations:

(i) Stop-flow-based static measurement, which measures only the steady imaging signals after the solution flow stops, when the reactions are normally reach their equilibrium state;
(ii) Solution-addition-based static measurement, which measures the steady images before and after addition of reactive solution(s);
(iii) Continuous measurement, which measures imaging signals continuously at a temporal resolution depending on the set frequency within the allowable range of the equipped video camera;
(iv) Combined measurement, which measures only the required imaging signals at variable frequencies, for example, stop recording during wash or useless reactions or recording at a low frequency for slow variation of signals but at high frequency for fast variation of signals, which is normally adopted in probe-based capturing reactions involving affinity recognition, immunoreaction and other specific interactions.

5.2 Kinetic and Thermodynamic Measurements

As discussed in Chap. 4, SPRi can study the kinetic and/or dynamic events in batch, which opens a new way to inspect molecular interactions that are the starting phase of all reactions. Nevertheless, the term of interaction is more often meant some reversible binding process, which does not necessarily include the formation of covalent bonds. The highly attractive molecular interactions are affinity, supramolecular and host–guest recognitions and immune reactions. A molecule can in theory interact with same and/or different molecules, which will cause a shift of SPRi signals once the interaction events appear on the sensor surface or its vicinity. By analysis of the measured imaging signals, the following information can be obtained:

Monitoring or studying the reaction procedure in real time;
Calculation of relative constants;
Ranking the interaction order by sensitivity or selectivity;
Screening out some specific substances;
The latter two will be delivered in later chapters, while the first two are essential in SPRi of kinetics or thermodynamics.

5.2.1 *Measurement of Constants*

In an equilibrium state, Langmuir adsorption isothermal can be measured. As an example, the immune interaction of human IgG with its antibody has been performed in our laboratory by assumption that the following reaction happens:

$$\text{IgG} + \text{anti-IgG} = \text{immune complex} \tag{5.1}$$

Antibodies are spotted on a SPRi sensor surface in array. The sensor was first modified with a layer of MUA. After the carboxylic terminal was activated in 400 mmol/L EDC and 100 mmol/L NHS for 30 min at room temperature, the chip was soaked in 1.00 mg/mL biotinylated BSA for 30–40 min to form amide bonds with the lysine residue on BSA to create a layer of biotin terminal. After washed with PBS, the chip was inserted in a PBS solution of 0.40–1.00 mg/mL avidin that can bridge any biotinylated substances including proteins, nucleic acids and glycoconjugates. The chip with avidin terminal was spotted with biotinylated antibodies and reacted for 30–60 min in a 70% humidity chamber. The chip was then inserted again in 1.00 mg/mL biotinylated BSA to block the remaining reaction sites. After sufficiently washed with PBS, the chip was mounted in the flow cell and IgG was pumped in at a designed concentration from low to high, step by step, after the imaging signals of spots became stable. Linear curves corresponding to the immune reaction of different antibodies with IgG were plotted according to Eq. (4.24), with R^2 all > 0.99 (Fig. 5.1). From the slope of each curve, the binding or dissociation constants were calculated:

$$\begin{cases} K_{a,\,\text{antibody II}} = 3.32 \times 10^{7}\,\text{M}^{-1} \\ K_{a,\,\text{antibody II}} = 1.89 \times 10^{7}\,\text{M}^{-1} \\ K_{a,\,\text{antibody III}} = 1.65 \times 10^{6}\,\text{M}^{-1} \end{cases} \xrightarrow{K_d = 1/K_a} \begin{cases} K_{d,\,\text{antibody,I}} = 30.1\,\text{nM} \\ K_{d,\,\text{antibody II}} = 52.9\,\text{nM} \\ K_{d,\,\text{antibody III}} = 606\,\text{nM} \end{cases}$$

These data agree well with the reported value [10].

To measure kinetic constants away the equilibrium states, fitting techniques are often used with the real-time recorded data, which is suitable for the study of much complicated systems. Global fitting is independent on the surface coverage and can

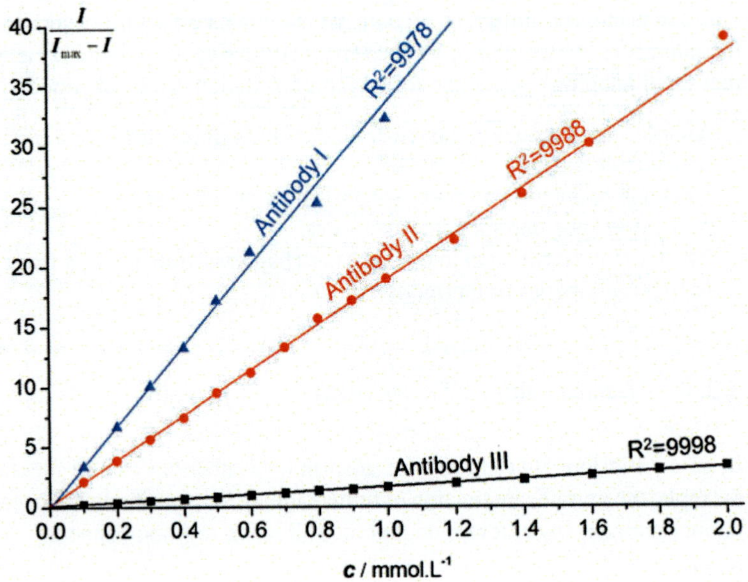

Fig. 5.1 Relative SPRi intensity of three anti-human-IgG antibodies against the molar concentration of human IgG

simultaneously analyze numerous datasets that are acquired from different concentrations of analyte(s) by sharing the same parameters to have the best-fit value [11, 12]. This technique has been applied to the study of interaction between IgE and aptamers by channelized SPRi [13]. An aptamer (i.e., Aptamer I) was checked that can selectively interact with IgE. The study was carried out by immobilization of different aptamers/cysteamine on the sensor surface through Au–S chemistry. The image data (Fig. 5.2) showed that the SPRi intensity increased linearly with the concentration of IgE between 10 and 90 nmol/L, with a limit of detection down to 2 nmol/L. By assumption of a reaction as

$$\text{IgE} + \text{Aptamer I} = \text{IgE} \cdot \text{Aptamer I} \tag{5.2}$$

and the use of nonlinear least-square global fitting technique, they calculated the binding reaction kinetic constant, k_{on}, equal to 3.4×10^4 mol^{-1} L s^{-1} and dissociation kinetic constant k_{off} being 9.2×10^{-3} s^{-1}. The dissociation constant K_d should thus be $k_{off}/k_{on} = 2.7 \times 10^{-7}$ mol L^{-1}.

A similar but dot-based SPRi approach has explored to study the interaction of the spotted dsDNA with proteins (Figs. 5.2 and 5.3 as well). dsDNA probes can easily be immobilized via either Au–S chemistry or biotin–avidin affinity reaction. The dissociation constant, K_d, can be determined by fitting each part of the interaction with a mono-exponential curve, without any problem.

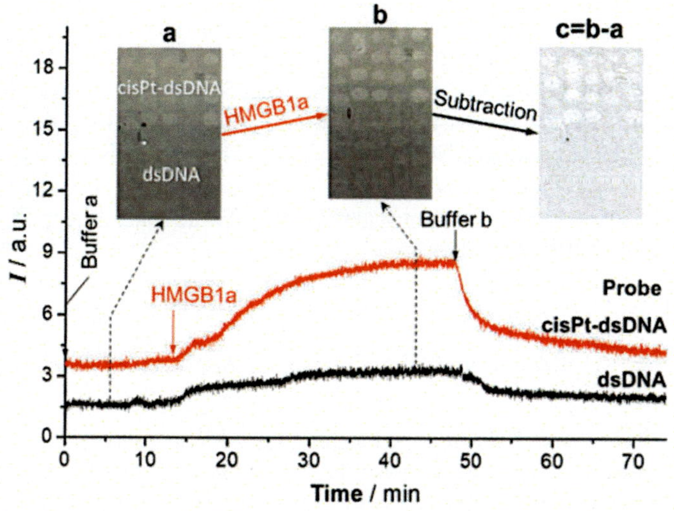

Fig. 5.2 SPRi sensorgram (lower) and images (upper) of protein interaction with spotted dsDNA probes. **a** Image recorded from a chip immobilized with dsDNA and its cisplatin (cisPt) complex as probes after balanced by buffer a; **b** Image recorded after the chip is balanced with a solution of protein HMGB1a (a type of nuclear protein). **c** Image after subtraction; buffer a: 10 mM Tris–HCl at pH7.8; buffer b: 10 mM NaCl and 10 mM Tris–HCl at pH7.8; dsDNA: double-strand DNA fragment; cisPt-dsDNA: cisplatin-bound or damaged dsDNA (refer to Sect. 5.5 for further information). The dot lines with arrow indicate the time point to take images

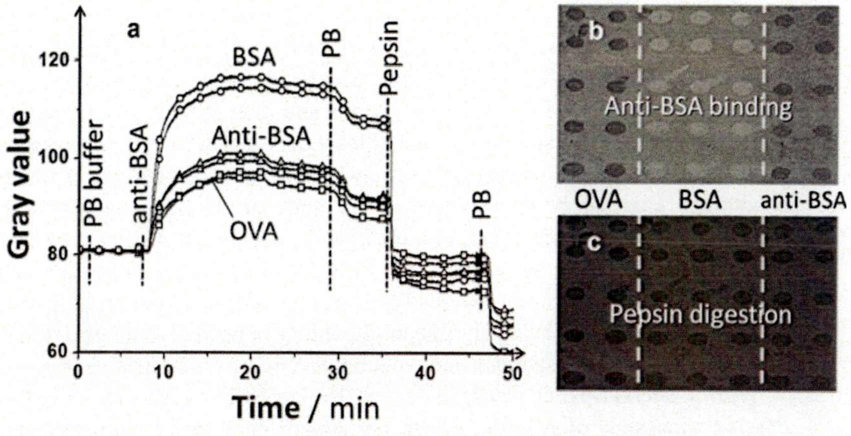

Fig. 5.3 Monitoring of immunoreaction between BSA and anti-BSA, and pepsin digestion of BSA, BSA–anti-BSA complex and hen egg OVA measured on SPRi-TX7100. **a** Sensorgram. **b** Image taken at 20 min after formation of BSA complex with anti-BSA. **c** Image taken at 50 min after pepsin digestion of all the proteins cleaned with PB buffer

The fitting techniques are more or less complicated in data analysis. To simplify, all the interactions can be studied separately, namely either in binding phase or in the dissociation phase. It is well known that reactions have different orders that complicate the study. To remove this difficulty, the relative reactions are often simplified into first-order or pseudo first-order reaction by fixing some parameters. According to the measuring principles given in Sect. 4.2, the initial stage of binding and the starting point of dissociation are the triggers to simplify the study. The relative examples can be found in later chapters.

5.2.2 *Simultaneous Monitoring of Molecular Reactions*

By either channelized or dotted technology, SPRi is capable of simultaneous monitoring a bunch of molecular reactions. The channelized SPRi is preferred to monitor different reactions under variable conditions. In principle, the reactions can be conducted by two categories: (i) add or pump in reaction mixture to the sensor part or flow cell and (ii) immobilize one of reactants in the channels and pump in other reactants (including catalysts if needed) to initiate the reactions. The first category is simple and easy in control and comparison, but it suits only for quite slow reactions or at least slow starting reactions. The obvious disadvantage is its limited throughput. It is unable to build up numerous channels on the limit sensor surface of about 1 cm^2.

The second category is generally adopted, especially in the study of fast and/or reversible reactions and/or interaction. To simultaneous monitor numerous reactions, dot-based SPRi has to be explored and adopted because the spotting technology can largely increase the number of reactions and interactions. The prerequisites to perform spots-based reactions are that at least one of reactants in each reaction can be immobilized on the sensor surface and all the reactions can occur under same conditions. This can be encountered in the study of chemical stability and in the screening of drugs, biomarkers, affinity ligands, specific aptamers, chiral selectors and so forth. Therefore, the spots-based SPRi has a hug space of applications, with an attractive ability to study various types of interactions (from weak to strong) as illustrated in Fig. 5.3, having significant flexibility in respect of the reaction number (≥ 1), consumption of samples and reagents and being cost-effectiveness.

With a lateral spatial resolution down to μm level, SPRi can simultaneously observe thousands of reaction points per square centimeter, with the up limit at around 0.20×10^6 dots/cm^2 that is also confined by computer quality. It has the unique ability to serve as a harmless high-throughput tool for batch monitoring of parallel reaction events or for screening of special targets. We have since tried to monitor 2000 parallel reactions and we got successfulness. Fortunately, the practical reactions that have to be monitored are in common within ca. 20. In this case, the reaction can be regulated finely and compared timely as shown in Fig. 5.4.

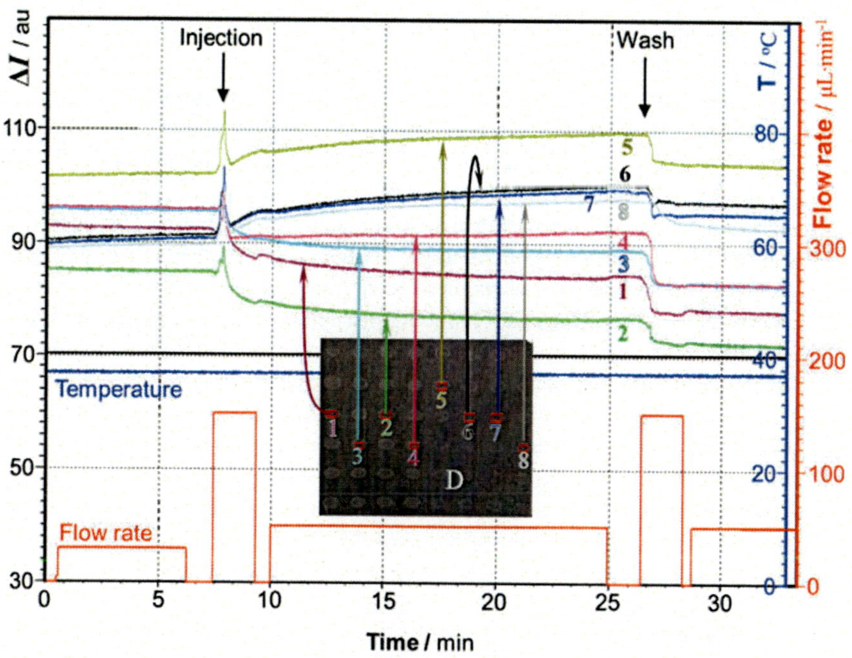

Fig. 5.4 Monitoring the hydrolysis of spotted proteins (6 × 8 = 48 spots) at 37 °C catalyzed by injection of pepsin solution, measured on SPRi-TX7100

5.3 Simulating the Recognition of Membrane Receptors

Biological recognition, which exists universally in living systems, makes the organization of various living systems possible. In this world, we have got abundant biological molecules with various recognition abilities, of which membrane protein receptors play many key roles in various biological processes, e.g., transport of ions and molecules, control of transmembrane potential, generation and transduction of energy, signal recognition and transduction, catalysis of biochemical reactions. These receptor proteins are also associated with numerous diseases that have induced and still need the development of medicines. Driven by membrane molecules, cells tend to adhere to each other to form aggregates once they are out of the physiological conditions. It is hence of great interest to know how these membrane molecules aggregate, but there is less than 1% of the proteins with known structure and binding level. This is however an opportunity for SPRi to exert its ability to simultaneously study the various interactions of membrane proteins with other molecules.

As usual, in order to image such cross-correlated membrane receptor interactions, different potential probes are required and have to be immobilized on a same sensor chip at the addressable sites. A sample that potentially contains some target biomolecules is then poured on or pumped across the chip surface to make the complicated cross-recognitions happen. Timely image signals and/or patterns are then recorded for kinetic and/or thermodynamic studies, including the measurement of related constants, as discussed in Sect. 5.2. To save time and space, we are not going to repeat the same determination in this section. Instead, we are discussing a bit more root issues, namely, immobilization of receptors and/or membrane-associated probes.

Firstly, to study membrane-involved recognitions, the probes must be very specific, which is better confirmed prior to immobilization. The specificity of a probe is ideally tested with its target analytes. If the information on the analytes is poor or unavailable, it is required to estimate their polarity and to perform pre-screening. A general way is to immobilize a potential class of molecules with pairing polarity, for example, hydrophobic probes must be terminated with various alkenyls, while hydrophilic probes are terminated with polar group(s) such as NH_2, COOH, OH and SH. The optimization can be achieved quantitatively by determination of related binding or dissociation kinetic constants. Zhang et al. found that in the study of adsorption behavior of extracellular polymeric substances on model organic surfaces, their adsorption decreased dramatically with pH values but enhanced by CH_3-terminal and further with the addition of Fe^{3+} [14]. Differently, Na^+ promoted them to adsorb on COOH-terminated surface that attracted also Ca^{2+}-mediated complexes, while NH_2-terminal repelled the complexes.

Secondly, the strategy to perform SPRi of membrane-based interactions must be carefully considered. SPRi can investigate the interactions by immobilizing either the probes or reversely the membrane receptors. The former strategy is usually adopted in the capture or screening of biological particles with known probes, which will be discussed in Chap. 7, while the latter way is suitable for observing the interaction of membrane receptors with known and especially unknown molecules involving usually signal transduction. The current technique is to immobilize the target receptors on the gold sensor surface surrounded with a lipid bilayer to correctly fold their functional structure, which is unique compared with the normal probe–molecule interactions.

To demonstrate, a G protein-coupled receptor (GPCR) of rhodopsin is taken as an example. The functional GPCR has seven transmembrane sections with α helix structure and has the binding sites in the inner membrane side for G protein (i.e., guanylic acid binding protein) on the C-terminal and the peptide chain linking the intracellular ring of the fifth and sixth transmembrane helices (Fig. 5.5). In order to maintain the functional structure, a GPCR has to be immobilized on a surface together with lipid bilayer so that it can respond to interacting molecules or transducing like G proteins on the bilayer. Figure 5.6 illustrates a strategy to immobilize rhodopsin through biotin–avidin affinity reaction to enable the observation of receptorligand binding events activated by light. The lipid membrane on the gold chip surface is patterned into micrometer-sized regions of phospholipid bilayers separated by monolayers of MUA fabricated by photochemical technique (refer to Approach 5.1 that is modified from Approach 4.6). The lipid bilayers enable the reconstitution of the transmembrane receptor into functional structure, and the monolayer boundary should prevent the receptor from insertion into it. The lipid layers are also necessary to adsorb G proteins (normally by partition). If the reconstitution is correct, the G protein

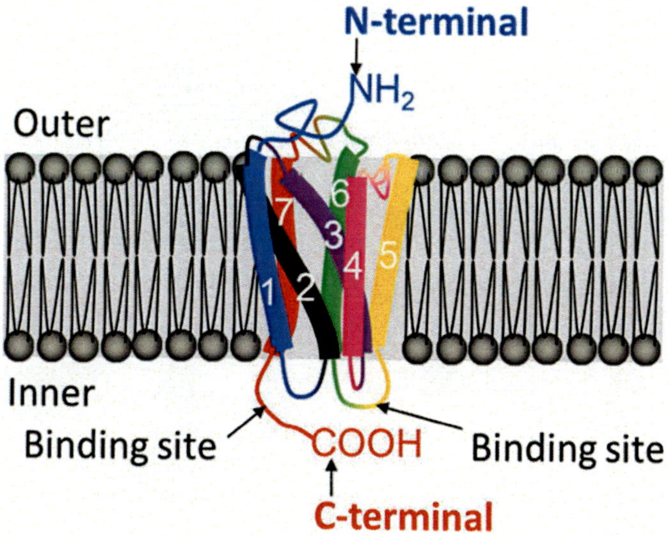

Fig. 5.5 Schematic picture of a G protein-coupled receptor in lipid bilayer, with seven transmembrane α helix peptide segments

will interact with rhodopsin after optical activation, causing expected dissociation and release of the active GTP-bound subunit from G protein and retinal from receptor (Fig. 5.6c, d). The release of mass enables SPRi measurement.

Approach 5.1 For SPRi of interaction events between immobilized and reconstituted membrane receptor (i.e., rhodopsin) and its ligand (i.e., G protein) modified from Approach 4.6 and referred also to literature [15, 16]:

Step 1, micropatterning the gold sensor chip: assemble a monolayer of MUA (at 20 mM in ethanol (EtOH)) on the chip, expose the surface to UV light through a mask, wash off the cleaved residues first with EtOH then by sonication in water for 1 min and blow dry with nitrogen gas. **Note**, microcontact printing can be used: place a PDMS stamp first on a filter paper pre-wetted with 20 mM MUA for 30 s, then on the gold chip for about 10 min, all without additional pressure except for the stamp weight.

Step 2, immobilization of biotin-thiol on the bare gold surface: Insert the patterned chip in 0.5 mM biotin-thiol and 5 mM MUA in EtOH for 1–2 h, rinse the chip in 5 mM MUA for five times to clear off the potential biotin-thiols in the MUA regions.

Step 3, blocking the non-specific adsorption sites: Insert the chip in 10 mg/ml BSA in PBS buffer (0.15 M NaCl, 1 mM $MgCl_2$, 10 mM sodium phosphate at pH 7.0) for at least 30 min, wash it with 1 mg/ml BSA and PBS buffer 3 times each and blow dry with nitrogen gas.

Step 4, binding of biotin with streptavidin: Insert the chip in a solution of streptavidin and biotinylated rhodopsin with washed membrane in a PBS buffer for

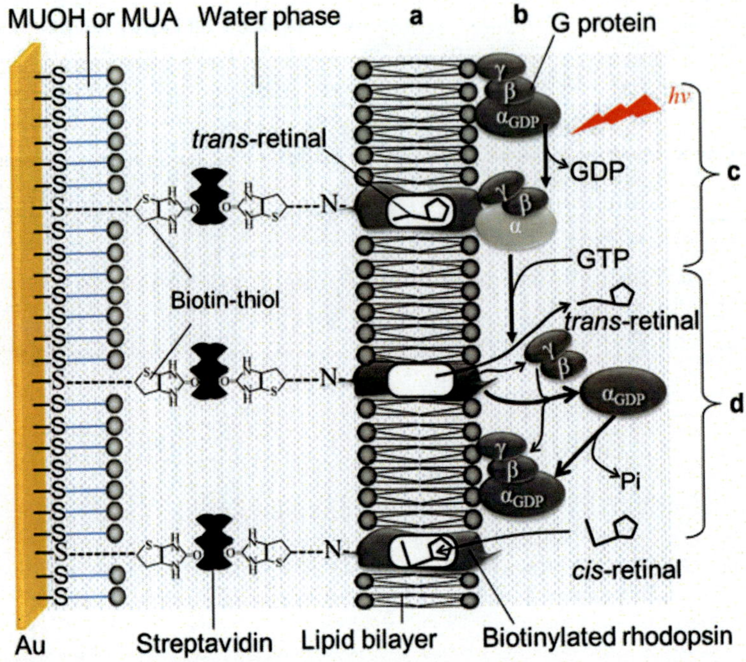

Fig. 5.6 Schematic strategy for **a** on-gold immobilization of rhodopsin (a well-studied membrane receptor) through biotin–avidin affinity reaction with reconstituted functional structure in the self-assembled lipid bilayer and **b** adsorption of G protein on the bilayer, followed by **c** optically induced conformational activation of the receptor and G protein binding to liberate GDP from the α subunit and **d** G protein disassembly and recovery on the consumption of GTP and the release and recombination of retinal isomers via conformational variation. Note, photoisomerization of 11-*cis*- to all-*trans*-retinal triggers the activation of rhodopsin and the mass variation on the membrane and in turn on the chip is detectable by SPRi

1.5 h, and wash it with the buffer. **Note**, (i) to avoid the excitation of rhodopsin, the preparation, from now on, should be conducted under dim red light at about 665 nm; (ii) to correctly orient receptors on a gold surface for attainment of an ideal sensor, the receptors are better biotinylated at the N-terminal, while the water phase spreading at the both sides of lipid bilayer helps to reconstruct the receptor as in a cell environment and (iii) the biotinylated receptor can be prepared by incubation of 50 μM rhodopsin (in washed membranes) with 20 mM $NaIO_4$ and 7.5 mM biotin-hydrazide at room temperature for 1 h, by dialysis (10 k cutoff) at 4 °C against PBS buffer, and by dilution, just before use, to 1/5 (to have ca. 1 μM rhodopsin) with the PBS buffer added with 50 mM octylglucoside.

Step 5, online formation of lipid bilayer around the receptors: Mount the chip in the flow cell on SPRi instrument, wash the cell with MetOH and water, fill in 320 μM egg phosphatidylcholine in the PBS buffer and keep for 30–60 min and slowly pump in PBS first at a flow rate of 40 μL/min to self-assemble the lipid bilayer around the capture rhodopsin, then at 200 μL/min after the formation of

lipid bilayer is completed, and wash the chip with a diluted PBS buffer added with octylglucoside to remove the potential vesicles adsorbed on the membrane.
Step 6, adsorption of G protein on the membrane: Pump 2.5 μM G protein solution into the cell to fast add the protein along the lipid bilayer (k_{on}=0.17 s^{-1}).
Step 7, conformational inspection: Flash a light to the chip surface to induce photoisomerization of 11-*cis* to all-*trans* retinal (active conformation), which binds G protein but releases its GDP (Fig. 5.6c). Once GTP is added, G protein desorbs from the receptor, the activated receptor decays spontaneously to opsin (apoprotein) and G protein binds again to the membrane surface following hydrolysis of GTP (Fig. 5.6d), which causes a reduction of resonant angle. The kinetics can thus be studied. Once *cis*-retinal binds again to opsin, the receptor recovers its photoactivatable rhodopsin, which recovers the resonant angle.

We have measured the release kinetics with SPRi, giving k_{off} at about 0.20 μM^{-1} s^{-1} that is close to 0.24 μM^{-1} s^{-1} [15]. Bieri et al. suggested that the initial slope of the observed desorption signal could be considered a good approximation for the receptor intensity per unit area of surface, giving $k_{off} = 0.06 \pm 0.015°$ per s [16]. The activation caused cleavage of the Schiff's base between receptor and chromophore led to the decay of rhodopsin to opsin and G protein relaxes to its resting state by fast hydrolysis of the bound GTP ($1/k$ = 20 s). This relaxation slowly increases the resonance angle at k_{on} = 0.003–0.006/s, finally close to the starting angle.

This strategy is extendable to the study of other types of membrane-relative interactions, especially for the screening of cell signaling molecules artificially synthesized in laboratory, which is one of the focuses in the development of chemical biology.

5.4 Measuring Selectivity and Sensitivity

It is critical to keep or even improve the selectivity and sensitivity of SPRi assays. As theory and methodology tell (Chaps. 2 and 4), SPRi selectivity and sensitivity depend on numerous factors, while in practical experiments, the fatal control factor is always associated with the immunoreaction of probes. As already discussed in previous section, the membrane receptor has to work in the lipid bilayer environments. A desirable technology to immobilize the probes is that it can maintain and perform their correct or expected function(s). This concerns with not only what kind of binding chemistry should be used but also how to adjust the orientation of the probes during and after immobilization. In general, the binding sites of immobilized probes are easily blocked or shielded by some structure or groups, which makes the anchored probes lose their functions partially or even totally, which leads to at least obvious reduction of the selectivity and sensitivity of the probes and in turn SPRi assays.

By taking SPRi of immunoreactions as an example, the proper immobilization of antigen or antibody remains a pre-requisite in the development of a selective and sensitive SPRi assay. Although there are different chemical approaches suitable for the immobilization of either antigens or antibodies on the gold sensor surface (Sect. 4.6), we usually prefer to immobilize the antibodies, rather than antigens, on gold sensor chips. The reason lies in that the structure of an antigen is much more changeable than that of an antibody. Antibodies also vary their structure to adjust their recognition selectivity, but their basic structure is always the same, i.e., a Y shape that provides some definite sites for functional immobilization.

In the study of biological event, biogenic or biocompatible reactions are more often adopted than the pure chemical reactions. Herein discussed is a biotin-associated recognition reaction as illustrated in the left part in Fig. 5.6. It may worth of mentioning that biotin–avidin-based immobilization technology may not be cost-effective. To reduce cost, some other coupling reagents can be used, instead, such as CC that allows an easy rotation of the immobilized molecules.

Even the avidin–biotin affinity reaction is utilized, it remains a serious challenge to correctly immobilize an antibody because a Y-shaped antibody can stand up or lie down. There are in total four normal orientations on a gold sensor surface (Fig. 5.7a): (i) upright standing with Fc or C-terminal closer to chip surface, (ii) upside down standing with Fabs or N-terminal closer to the surface, (iii) side lying with Fc and either one of the Fabs closer to the surface and (iv) flat lying with Fc and 2 Fabs closer to the surface. Clearly, only the first orientation can make the variable recognition region of V on Fabs become free and accessible. The upside down orientation is not expected to have a chance of interaction with its antigens. The side and flat lying orientations may have some chance to realize immune reactions but are not ideal and should be avoided if possible. Norde and co-workers have measured a simple relationship of the absorbed surface concentration c_s with the molecular orientation: the standing orientation could have a value of c_s at 370–550 ng/cm^2 that was dependent on the angle of the two Fab arms, while the side and flat lying orientations at below 300 ng/cm^2 [17]. This provides us a basic confidence to immobilize an antibody on gold sensor surface.

In fact, there is a practically usable way to make an antibody molecule stand upright after immobilization. It is to anchor it on G protein (at ca. 10 μg/mL in PBS) that binds specifically to the Fc region [18]. Chen et al. have shown that a SAM of calix[4]crown ether (Fig. 5.4b) can also immobilize the antibody on its C-terminal [19], being similar to G protein. The process is exemplified in Approach 5.2.

Approach 5.2 For immobilization of antibodies standing upright on gold surface:

Step 1, wash a gold sensor chip with 1% (v/v) chloroform in methanol, then with methanol for 15 min each, and blow dry with nitrogen gas.
Step 2, incubate a gold sensor chip in 1 mM calix[4]crown ether and 1% (v/v) chloroform in methanol for at least 30 min, wash the chip with methanol, water and PBS buffer in sequence.
Step 3, incubate the chip in 2–100 μg/mL antibody in PBS for 30 min.
Step 4, wash the chip with PBS buffer for 3 time and mounted it in the flow cell for imaging experiments. Note, protein immobilized gold sensor chips are better used tight after immobilization and cleaning; otherwise, the chips will lose their

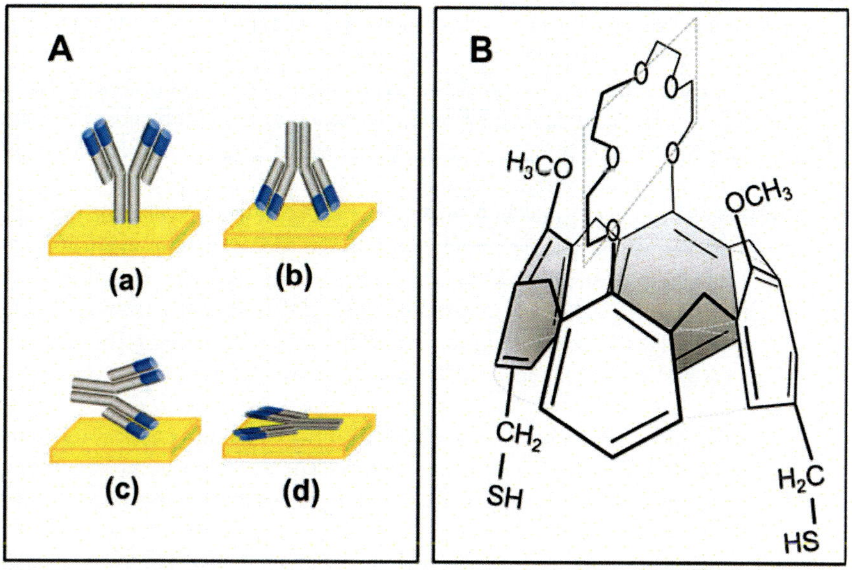

Fig. 5.7 **a** Idealized orientations of an antibody molecule on a gold sensor surface and **b** the chemical structure of a calix[4]crown ether that can form a SAM on gold for erect immobilization of antibody molecules on the SMA surface

selectivity and sensitivity gradually until invalidity due to the degradation of the immobilized proteins.

This Approach 5.2 can also be conducted online by pumping in relative solutions in sequence until SPRi signals surpass the turning corner of plateau. The successful immobilization of an antibody on the calix[4]crown ether SAM can have an increase of 138 Å in thickness and a surface coverage of 428 ng/cm^2, which agrees with the molecular size of the antibody at 142 × 85 × 38 Å^3 and falls in the range of surface concentration for the immobilized molecules of the antibody to stand vertically.

In short, SPRi enables the timely study of kinetic and dynamics concerning with various interactions and reactions at the molecular level. Based on this, SPRi has found a vast door to expand its ability and applications that are going to be discussed in later chapters.

5.5 Mechanism Study of Anti-cancer Medicine

Many medicines function through special recognition mechanisms, therefore, study of these recognition mechanisms can help understand and increase the efficacy of medicines, and it is also a key step to develop new drugs. Currently, SPRi is a powerful tool able to perform high throughput screening of specific mechanisms and chemical candidates. To demonstrate, platinum-based anti-cancer drugs are considered and tried in this laboratory [9].

It is known that cisPt (shorted from cisplatin that is also abbreviated from a platinum complex of *cis*-diamminedichloroplatinum (II)) remains one of the most effective anti-cancer drugs in clinical treatment of various human malignancies. It plays the therapeutic role by binding to dsDNA and forming cisPt-DNA adduct(s). This causes damage to dsDNA and further induces the cancer cells into apoptosis [20–23]. The interaction is associated with the recognition of cisPt-DNA by high mobility group proteins (HMG). Usually the target dsDNA, having a dodecamer duplex containing a d(GpG) intrastrand cross-link, is altered after binding with cisPt to facilitate HMG recognition, for example, a minor groove of 16-base-pair DNA duplex is widened after coordination with *cis*-$[Pt(NH_3)_2]$ to promote HMG1 binding [22].

While the cisPt is used as a clinically effective anti-cancer drug, transplatin has long term shown no anti-cancer effect. The situation is, however, changing: Several transplatin analogues that contain a planar amine ligand have shown their cytostatic activity against some cisPt-resistant tumor cells. An example is *trans*-$[PtCl_2 \cdot NH_3 \cdot$thiazole$]$ or transPtTz that can bind and cause damage to dsDNA. The damaged dsDNA will then be selectively recognized and bound by PC4, a human nuclear protein-positive cofactor 4 [24].

Similar to HMG such as HMGB1 that is able to activate the binding of p53 to the homologous sites by providing curved DNA for p53 recruitment and p53-dependent transcription, and can inhibit the repair of cisplatin-DNA *in vitro* and *in vivo*, PC4 in the same group of p53-interacting proteins can enhance the sequence specific DNA binding and transactivation of p53 that originally maintains the genomic integrity and cellular homeostasis [25]. It has been reported that the transPtTz-DNA contains about an equal proportion of mono- and bifunctional intra and interstranded adducts [26]. However, further information remains poor. It should be valuable to explore SPRi approaches to study the recognition of the damaged dsDNA by the nuclear proteins and to distinguish the protein PC4 from HMG proteins in the recognition of the damaged dsDNA.

5.5.1 Basic Considerations and Approach

It is clear that the binding mechanism and biological effect cannot be exactly the same between *cis*- and *trans*platins, for example, their DNA adducts recognize different nuclear proteins as mentioned above. To demonstrate the applicability of SPRi to the study and differentiation of their recognition mechanisms, the pair of transPtTz and PC4 was assayed together with the pair of cisPt and HMGB1 as models.

For fair comparison, SPRi must be conducted under same conditions. In order to utilize the matured immobilization technology, dsDNA fragments are selected as probes since they can be anchored on a sensor chip on its one end, which is usually not possible for proteins. The immobilized probes are arranged normally in

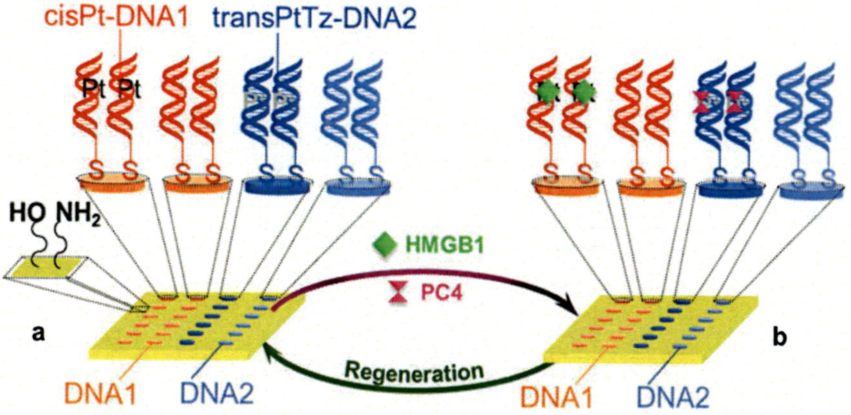

Fig. 5.8 DNA probe arrangement and regeneration of a gold sensor chip for SPRi of cisplatin and transplatin recognition of nuclear proteins. **a** Location of the spotted DNA probes. **b** Theoretical cartoon for the selective capture of nuclear proteins by the DNA probes. DNA1: synthetic thiol dsDNA fragment 1; DNA2: synthetic thiol dsDNA fragment 2; cisPt-DNA1: adduct of DNA1 and cisplatin; transPtTz-DNA2: adduct of transplatin and DNA2

parallel as shown in Fig. 5.8, including the synthetic thiol dsDNA fragments of DNA1 and DNA2 as controls and their respective platinum adducts of cisPt-DNA1 and transPtTz-DNA2 for recognitions. The chemical information of these key materials are listed in Table 5.1.

Once the target DNA probes are immobilized, the gold chip is mounted in the flow cell of SPRi instrument, for example, SPRi-TX7100. The interactions and related kinetics of the platinated DNA with the nuclear protein HMGB1 or PC4 are then monitored by pumping the related protein solution(s) under simulated physiological conditions (Fig. 5.8). The relative operation is collectively delivered in Approach 5.3.

Approach 5.3 For the study of recognition mechanism of platinum-based anticancer drugs:

Step 1, preparation of microarrayed gold sensor chip: Activate a cleaned gold sensor chip by plasma for 3 min, spot it with the four types of thiol DNA probes separately dissolved in a Tris buffer at pH 7.8 (10 mM Tris, 100 mM NaCl, 10 mM $MgCl_2$ and 1 mM EDTA) and incubate it in a humid chamber at room temperature for 2 h; after freezing the chip in liquid nitrogen for 3 min, insert it immediately into a solution of 20 mM cysteamine hydrochloride and 1 mM MCH prepared in the pH 7.8 Tris buffer and keep reaction for 2 h to block the reactive regions and the non-specific adsorption sites (refer also to Sect. 5.5.2.2); wash the chip thoroughly first with the Tris buffer, then with pure water for at least three times, and blow dry it with nitrogen gas. Note, the dried chip can be used directly or stored in dark at −20 °C for at least 1.5 months.

Table 5.1 Key chemicals used in SPRi of cisplatin and transplatin recognition mechanisms

Chemical	Structure or sequence	Abbreviation
Cisplatin	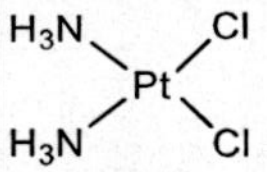	cisPt
Tansplatin	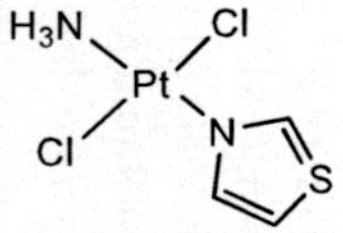	transPtTz
Thiol dsDNA Fragment 1[a]	5′-CCTCTCT-**GG**-ACCTTCC-3′ 3′-GGAGAGA-CC-TGGAAGG-SH-5′	DNA1
Thiol dsDNA Fragment 2[a]	5′-CTCTT-**GTG**-TTCTTCT-3′ 3′-GAGAA-CAC-AAGAAGA-SH-5′	DNA2
Nuclear protein 1	NH_2-MGKGDPKKPRGKMSSYAFFVQTCREEHKK-KHPDASVNFSEFSK KCSERWKTMSAKEKG KFED-MAKADKARYEREMKTYIPPKGE-COOH	HGMB1
Nuclear protein 2	NH_2-MPKSKELVSSSSSGSDSDSEVDKKLKRKKQV-APEKPVKKQKTGETSRALSSSKQSSSSRD DNMFQ-IGKMRYVSVRDFKGKVLIDIREYWMDPEGEMKP-GRKGISLNPEQWSQLKEQISDIDDAVRKL-COOH	PC4

[a] The bold letters indicate the platinum-binding sites

Step 2, SPRi instrument settings: Select an SPR imager (e.g., the model TX7100) and set the related parameters, for example, temperature (e.g., at 20 °C), flow rate, image recording rate (e.g., 5 frames per second to save space) and so forth.
Step 3, SPRi monitoring the recognition of platinum-based drugs with nuclear proteins though DNA adducts: Mount the spotted chip in the flow cell of the SPR imager, model TX7100, equilibrate the system by pumping in the pH7.8 Tris buffer further added with 0.005% Tween 20 (v/v) at a flow rate of 30 μL/min until the signals become stable, pump in a protein solution dissolved in the Tris-Tween buffer at a flow rate of 10 μL/min: After incubation at 20 °C (or other required value) for 10 min, pump in again the Tris-Tween buffer to clean the chip surface.
Step 4, regeneration of the reacted chip: Pump in first 2 M NaCl at 30 μL/min for 3 min and then the Tris-Tween buffer until the recorded signals return to the baseline and are stable (refer to Sect. 5.5.2.3).

5.5.2 *Some Special Conditions*

There are many parameters conditions that may impact on the precision and reliability of SPRi measurements, such as the immobilization content of DNA probes, the concentration of the nuclear proteins pumped in, the reaction time and temperature, chip stability. Among them, the concentration of probes and proteins and the blocking and regeneration of sensor chip are somewhat special in dsDNAprotein interaction and are hence discussed here in particular.

5.5.2.1 Concentration Impact

To perform sufficiently sensitive SPRi study of the interaction of proteins with the platinated DNA probes, we need to know the optimal and lowest concentration for both the proteins and probes. These values are critical for the preparation of a sensor chip and to conduct reliable SPRi measurements. In common, high density of immobilized probes can capture more target proteins, which in turn increase SPRi intensity. Figure 5.9 shows that the monitored SPRi intensity of both HGMB1 and PC4 recognitions increases not only with their own concentration but also with the spotted density of their respective DNA adducts, rising from 1.0 μM up to 10 μM (Fig. 5.9a). The increases are however not linear (Fig. 5.9b), with a turning corner at about 5.0 μM DNA adducts. It is interesting that both HMGB1 and PC4 show their ability to recognize their respective DNA adducts at a concentration down to 100 nM while the PC4 yields much higher SPRi response than HMGB1. Within the tested concentrations, the upper limits of the spotted content of DNA adducts and the recognition concentration of proteins are better adopted to record clear images.

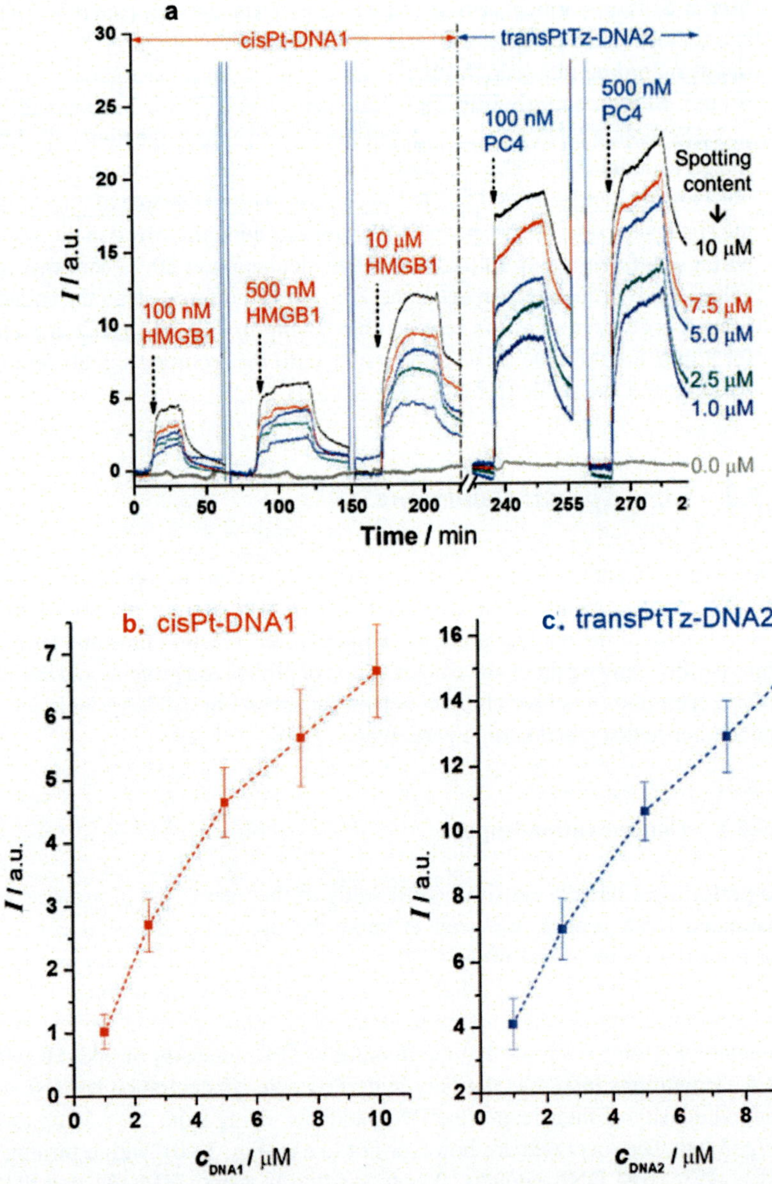

Fig. 5.9 Dependence of SPRi intensity on the spotting content of DNA probes and the concentration of nuclear proteins. **a** Sensorgram measured by pumping in HMGB1 followed by PC4 at the indicated concentration to recognize the DNA probes spotted at the indicated content. **b** SPRi response of 10 μM HMGB1 reacts with 5.0 μM cisPt-DNA1. **c** SPRi response of 500 nM PC4 reacts with 5.0 μM transPtTz-DNA2. Reconstructed from Ref. [9] with permission

5.5.2.2 Chip Blocking Technique

Since proteins are involved in the interaction, the non-specific adsorption effect needs special attention and has to be blocked effectively. A blocking solution of 20 mM cysteamine hydrochloride and 1 mM MCH in pH7.8 Tris buffer was shown to be more effective than the commonly used blocking solution of 0.1% (w/v) BSA (also in the Tris buffer) or 1 M EOA. This is reasonable because, at pH 7.8, BSA (pI = 4.85) is negatively charged and is thus opposite to the positively charged HMGB1 (pI = 10.1). In addition, BSA molecule has an ellipsoidal shape, with hydrodynamic dimensions of $14 \times 4 \times 4\ nm^3$ [27] that is almost the same size as the DNA probes used (16 bases per single strand, giving about 5 nm chain length). Therefore, BSA has the potential to enclose the proteins-binding sites. Differently, the small hydrophilic cysteamine does not have such issues, furthermore, its molecules have sulfhydryl groups to promote their reaction with gold atoms on the surface and have amino groups to repel direct contact between HMGB1 and the gold surface. MCH is better added together with cysteamine to prevent the DNA molecules from falling down on the chip surface or to make the molecular chains stand as vertically as possible to increase their accessibility for proteins to realize specific recognition.

5.5.2.3 Cyclic Use of Probe-Immobilized Chip

It is desirable to cyclically use a sensor chip, which is usually achieved by regeneration technology. When the regenerated chip can maintain its performance, the measuring precision and reliability can become high because we can use the same chip to study the influence of different parameters, especially facilitating the comparative study of the effects of probe and protein concentration changes. The experimental process (particularly chip preparation) and experimental cost will hence be reduced largely.

Obviously, unstable chips cannot be regenerated. This depends not only on the stability of gold costing but more on the immobilized probes. Many macromolecular probes like proteins are easily degraded in contact with solutions, even covalently anchored on the chip. This is another important reason that we use and immobilize DNA probes, in addition to the end-immobilization advantage. To make the cisPt and transPtTz correctly bind to the core DNA structure, precisely synthesized dsDNA fragments are used instead of ssDNA and are made to react with the platinum complexes prior to immobilization.

After these measures, the sensor chip with probes is validated to be sufficiently stable and can be regenerated with 2 M NaCl. Figure 5.10 illustrates the regeneration of a chip immobilized with transPtTz-DNA2 probe tested by its capture and dissociation of PC4. At the fifth regeneration, the net signal reduced from 21 to about 19 (equal to a reduction of 8.5%), including the loss of signal and increase of baseline. This remains suitable for quantification. For qualitative analysis, the chip can be regenerated for 10–20 times.

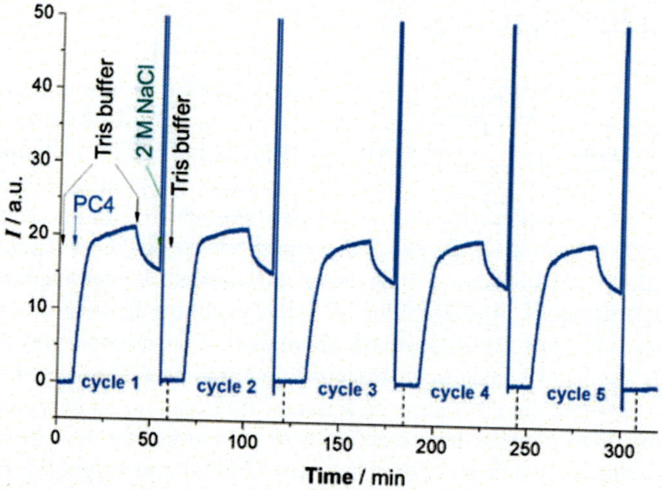

Fig. 5.10 Regeneration of a transPtTz-DNA2 probe-immobilized SPRi chip. Reconstructed from Ref. [9] with permission

5.5.2.4 Protein Interference

Medicines must work in a real body, wherein there are various complex components that can interfere with their function. It is SPRi that can simulate the physiological environment to check the extent of this adverse effect. We did not conduct comprehensive research on this topic because our goal here is to develop and test SPRi methods rather than to develop drugs, but we did check the issue by using several typical proteins as examples such as BSA (pI ~ 4.85), pepsin (pI ~ 8.1), hemoglobin (pI ~ 7.23) and cytochrome c (pI ~ 10.2). From their pI values, it can be estimated, in theory, that BSA and hemoglobin will be charged negatively in a pH7.8 Tris buffer and hence can bind non-specifically to all the DNA probes due to electrostatic repulsion. While in practice, they produce nearly negligible SPRi signals except for hemoglobin that yields obvious SPRi signal increment. Figure 5.11 demonstrates that all these signals can be washed off with the Tris buffer except for the hemoglobin that takes a longer time to completely remove. These interference testings suggest no selective recognition of these irrelative proteins to all the DNA probes immobilized. They additionally confirm the specific recognition of PC4 and HMGB1 with the platinated DNA probes from the other angle.

5.5.3 *Recognition Kinetics and Thermodynamics*

There are basically two strategies to study the reaction kinetics in SPRi. A universal strategy is mathematical fitting that can be applied both linearly and nonlinearly, while a more preferable or convenient way is try to find a linear or quasilinear working

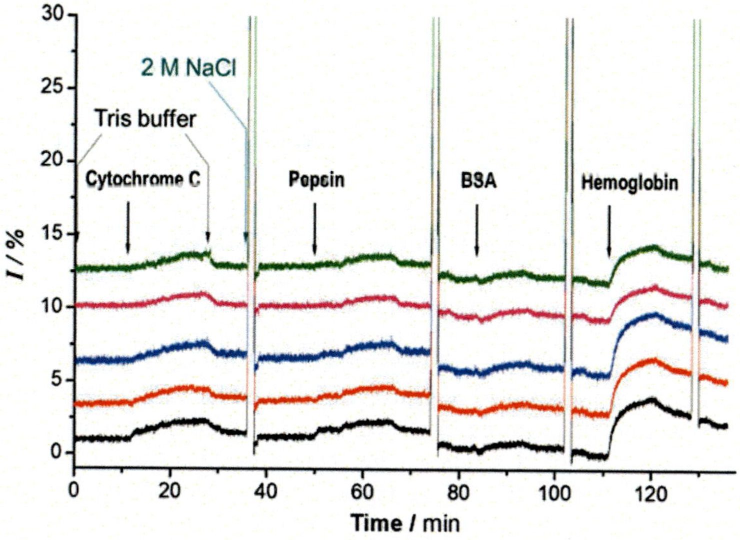

Fig. 5.11 SPRi sensorgram of several irrelative proteins (10 μM each) separately reacting with the platinated DNA probes immobilized on a same sensor chip

range to simplify the mathematical treatments. In general, the kinetics of both binding and dissociating reactions are not linear or exactly, they change exponentially as indicated in Eqs. (4.22) and (4.23). We are discussing first the linear strategy in respect of binding reaction and then fitting strategy for the dissociation reaction.

5.5.3.1 Determination of Binding Rate Constant

To simplify the DNA–protein binding reaction, a type of 1:1 stoichiometry is always suggested under certain conditions [28–31], and thus, the k_{on} and k_{off} in Eq. (4.11) now represent the apparent association and dissociation rate constants, respectively. To have linear mathematics, we can use Eq. (4.20) or Eq. (4.25) in the treatment of SPRi data. However, in order to introduce as many techniques as possible, we discuss an approximation processing approach here. The idea is to make use of the initial stage of reaction. At the very beginning of reaction, an exponential reaction kinetic function (Eqs. (4.22) and (4.23)) can be approximated into quasilinear function, which was also termed "initial rate analysis" in literatures [32, 33]. It can be understood that the binding dominates the reaction in the initial reaction phase while the dissociation is negligible, that is, $c_{AB} \approx 0$; therefore, Eq. (4.22) is reduced to

$$\frac{\mathrm{d}c_{AB}}{\mathrm{d}t} \cong k_{\text{on}} c_{B,0} c_A \doteq k_{\text{on}} c_{AB,\max} c_A, \tag{5.3}$$

where $c_{B,0}$ or $c_{AB,\max}$ is the maximum capacity of the adsorptions sites or the surface density of a specific probe. Let $I = Bc$, Eq. (5.3) changes to

$$\frac{\mathrm{d}I}{\mathrm{d}t} \cong I_{\max} k_{\mathrm{on}} c_A, \tag{5.4}$$

where $I_{\max}$ can be measured at a high concentration of a target analyte able to sufficiently saturate the immobilized probes. Equation (5.4) can be integrated into

$$I \cong I_{\max} k_{\mathrm{on}} c_A t + I', \tag{5.5}$$

where I' is an undetermined constant. Equation (5.5) reveals that, at the initial binding phase, the measured SPRi signal is a quasilinear function of either t at a given c_A or c_A at a given t. This can easily be demonstrated by the imaging intensity recorded at the initial phase of PC4 biding to the immobilized transPtTz-DNA2. Figure 5.12a, b show the plots of difference imaging intensity against reaction time for the first 40 s and 60 s, respectively. Their fitting curves give excellent linear correlation coefficients as collected in Table 5.2. Clearly, the linearity measured for the first 40 s is better than that for the first 60 s. Nevertheless, the two sets of k_{on} do not have significant difference after T-test at a confidence above 95%. An issue has been encountered that needs to be paid a great attention: The linearity is dependent on the LOD of SPRi or the concentration of the reactant, especially at the lower end. In the recognition of PC4 with transPtTz-DNA2, the lowest concentration of PC4 is better added at above 1 μM. It should be noted that this low limit of reaction concentration is also determined by the type of reaction, reactants and especially their SPR response. It is reasonable that highly responsive substances can heave a lower reaction concentration.

Similar to the time plots, the linearity can also be found between the imaging intensity and the reaction concentration for the initial reaction phase as illustrated in Fig. 5.12c, d. Due possibly to the data conversion processing, the fitting curves for the concentration have less linearity than the time curves, giving also worse linear correlation coefficients as shown in Table 5.3. To have $R^2 \geq 0.99$, the reaction time must not be shorter than 30 s. More tests suggest that the so-called initial reaction phase would be better found between 30 and 120 s, which is also dependent on the SPRi instruments, especially on the length of flow path, flow resistance, mass transfer speed, etc. It is better to select or optimized again for a new reaction and/or in use of a different instrument.

It should also be noted that the k_{on} values determined based on concentration are in general a bit larger than the values determined based on time variable. However, the biggest variation (484 and 434) is smaller than 11% that is within acceptable tolerances.

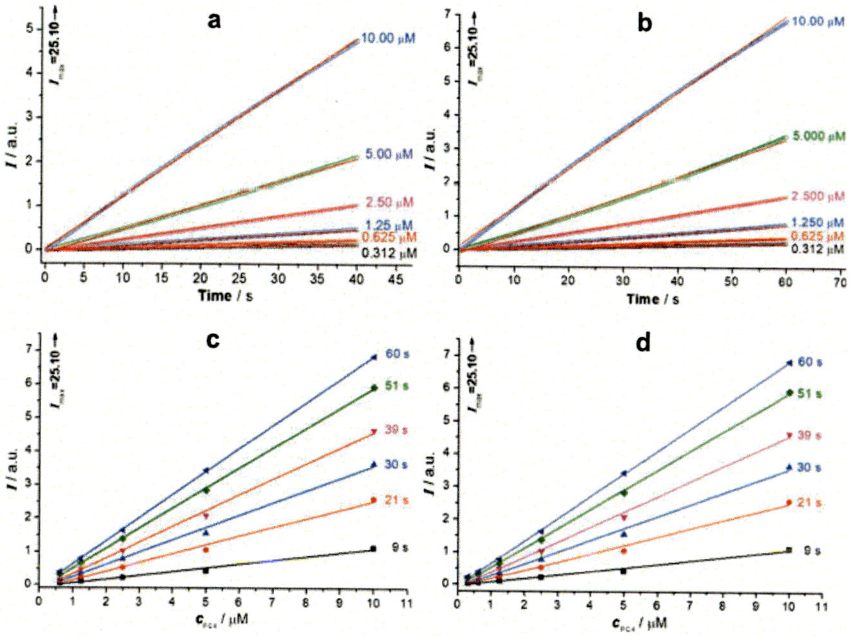

Fig. 5.12 SPRi determination of reaction rate constant at 25 °C in the initial phase according to Eq. (5.4). Dots: SPRi determined net imaging intensity of after removal of background; Lines: Least square fitting curve; Immobilized probe: 10 μM transPtTz-DNA2; Tested protein: PC4 in pH7.8 Tris buffer

Table 5.2 SPRi determination of binding rate constant at 25 °C by fitting the curves of imaging intensity versus time for the initial phase of reaction between PC4 and transPtTz-DNA probe immobilized at 10 μM

Time range	c_{PC4} /μM	Slope/s^{-1}	$k_{on}/10^2\ M^{-1}\ s^{-1}$	R^2
0–60 s	0.312	0.00396	5.057	0.9982
	0.625	0.00656	4.182	0.9998
	1.250	0.01318	4.201	0.9983
	2.500	0.02736	4.360	0.9995
	5.000	0.05727	4.563	0.9980
	10.00	0.11463	4.567	0.9988
Average			4.49 ± 0.32	
0–40 s	0.312	0.00374	4.776	0.9991
	0.625	0.00643	4.099	0.9999
	1.250	0.01249	3.981	0.9992
	2.500	0.02657	4.234	0.9998
	5.000	0.05394	4.298	0.9990
	10.00	0.11978	4.776	0.9995
Average			4.36 ± 0.31	

Table 5.3 SPRi determination of the binding rate constant at 25 °C by fitting the curves of imaging intensity versus the concentration of PC4 measured at different times from the initial of its reaction with transPtTz-DNA2 immobilized at 10 μM

c_{PC4} range	Reaction time/s	Slope/s^{-1}	k_{on}/(10^2 M^{-1} s^{-1})	R^2
0.625–10.00 μM	9.00	0.11391	5.042	0.9815
	21.00	0.26256	4.981	0.9887
	30.00	0.36898	4.900	0.9931
	39.00	0.47106	4.812	0.9966
	51.00	0.60042	4.690	0.9994
	60.00	0.69237	4.597	0.9999
Average			4.84 ± 0.17	
0.312–10.00 μM	9.00	0.11176	4.947	0.9823
	21.00	0.25837	4.902	0.9890
	30.00	0.36394	4.833	0.9931
	39.00	0.46575	4.758	0.9963
	51.00	0.59564	4.653	0.9990
	60.00	0.68867	4.573	0.9997
Average			4.78 ± 0.14	

5.5.3.2 Determination of Dissociation Rate Constant

Dissociation kinetics can be processed as the binding kinetics. To be more comprehensive, this section introduces the process of the dissociation based on Eq. (4.20). Although both linear and nonlinear fittings are applicable, we are discussing only nonlinear fitting technique here. Equation (4.20) depicts the exponential drop of the measured imaging intensity against dissociation time. Figure 5.13 shows the SPRi sensorgram for the dissociation of PC4 from the immobilized transPtTz-DNA2 after saturated adsorption. By fitting based on $\exp(-k_{off} \cdot t)$, the dissociation rat constant is directly available, $k_{off} = 0.11188\ min^{-1} = 1.86 \times 10^{-3}\ s^{-1}$, with $R^2 = 0.9995$. Similar to biding, this fitting is applicable to all dissociation curves above the detection limit.

5.5.3.3 Equilibrium Constant and Gibbs Free Energy Change

Once k_{off} and k_{on} are available, the binding equilibrium constant K_a or dissociation equilibrium constant K_d can easily be calculated by k_{on}/k_{off} or k_{off}/k_{on}. They are reciprocal to each other. Corresponding Gibbs free energy change (ΔG) can be calculated from the equilibrium constants to judge the reaction direction:

$$\Delta G = -RT_e \ln K_a \quad (5.6)$$

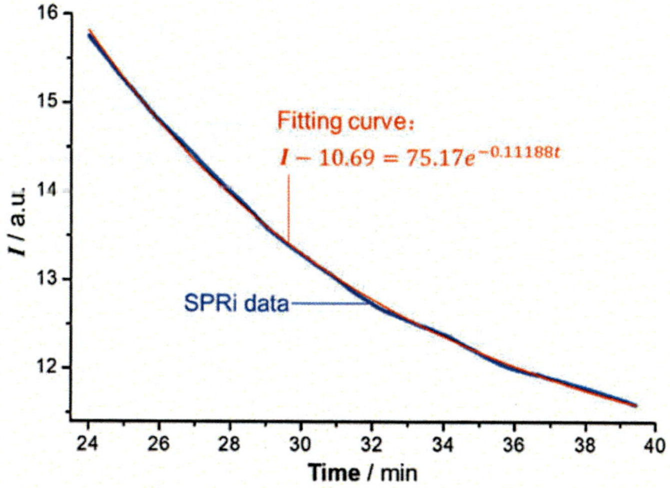

Fig. 5.13 SPRi recorded intensity drop at 25 °C for the dissociation of PC4 from its probe of transPtTz-DNA2 immobilized at 10 μM

By use of above kinetic constants, we are now able to calculate the equilibrium constants. The problem is now that which biding rate constant must be used among different values measured by different techniques and/or parameter ranges. This problem can appear as well for the selection of a dissociation rate constant since its variable values will certainly show up with different determining techniques. The best recommendation is to use a same technique to measure the rate constants. If the such technology is unavailable, the optimized value should be selected. For example, in the determination of binding rate constant, four different values of 4.36×10^2, 4.49×10^2, 4.84×10^2 and 4.78×10^2 M^{-1} s^{-1} have been measured. According to above discussions, the value measured in a shorter allowance time with higher R^2 is on top of selection, that is, 4.36×10^2 M^{-1} s^{-1} is better adopted for calculation. In this case, the calculated K_a (25 °C) is 2.34×10^5 M^{-1} or K_d (25 °C) = 4.27×10^{-6} M, and the corresponding ΔG (25 °C) is − 30.6 kJ/mol that suggests an automatic binding of PC4 onto the probe of transPtTz-DNA2. By using a same approach, all the equilibrium constants for the complexes of PC4 and HMGB1 formed with the four DNA probes (cisPt-DNA1, DNA1, transPtTz-DNA2 and DNA2) can be measured.

5.5.4 Key Differences Between Cisplatin and Transplatin

The known differences between cisPt and transPtTz are that they recognize different dsDNA structures and their damaged DNA can be recognized by different nuclear proteins. However, further information is quite nebulous. Fortunately, we can now use SPRi to qualitatively and/or quantitatively reveal some of their new differences.

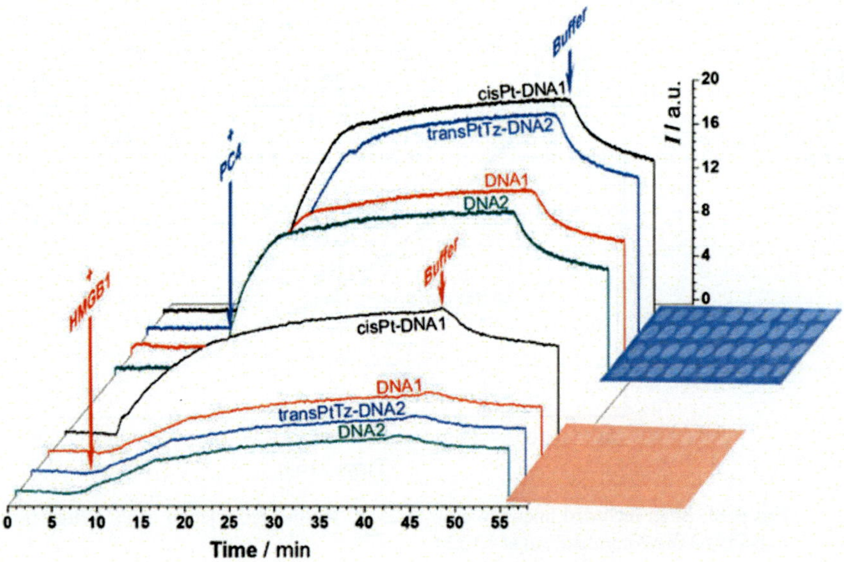

Fig. 5.14 Typical SPRi sensorgrams (left) and images (right) measured from the interactions of HMGB1 (front) and PC4 (rear) with the platinated DNA probes immobilized on a same sensor chip, with platinum-free DNA probes as their controls, respectively, where the spotted DNA probes and the pumped proteins are all at 10 μM. Reconstructed from Ref. [9] with permission

Under simulated physiological conditions, the interactions of the proteins with the platinated DNA probes can be measured in real time, with their respective platinum-free DNA probes as the controls. Figure 5.14 shows that HMGB1 has in general lower SPRi responses than PC4 but has a higher selectivity or affinity to cisPt-DNA1 than to DNA1, DNA2 and transPtTz-DNA2, while PC4 shows binary recognitions with transPtTz-DNA2 and cisPt-DNA1 compared with the controls of DNA2 and DNA1, respectively.

Quantitatively, HMGB1 has higher affinity to cisPt-DNA1 (Fig. 5.15a) than PC4 to transPtTz-DNA2 and cisPt-DNA1 (Fig. 5.15b). In a bit details, HMGB1 shows nearly a same binding strength to both transPtTz-DNA2 and DNA2 along the tested concentration range, especially at its high concentration (above 9 μM), but at its concentration below 1.5 μM, HMGB1 has somewhat weaker binding strength to DNA1 than to DNA2 and transPtTz-DNA2, while at the concentration between 2.0 and 10 μM, HMGB1 binds more strongly with DNA1 than with DNA2 or transPtTz-DNA2.

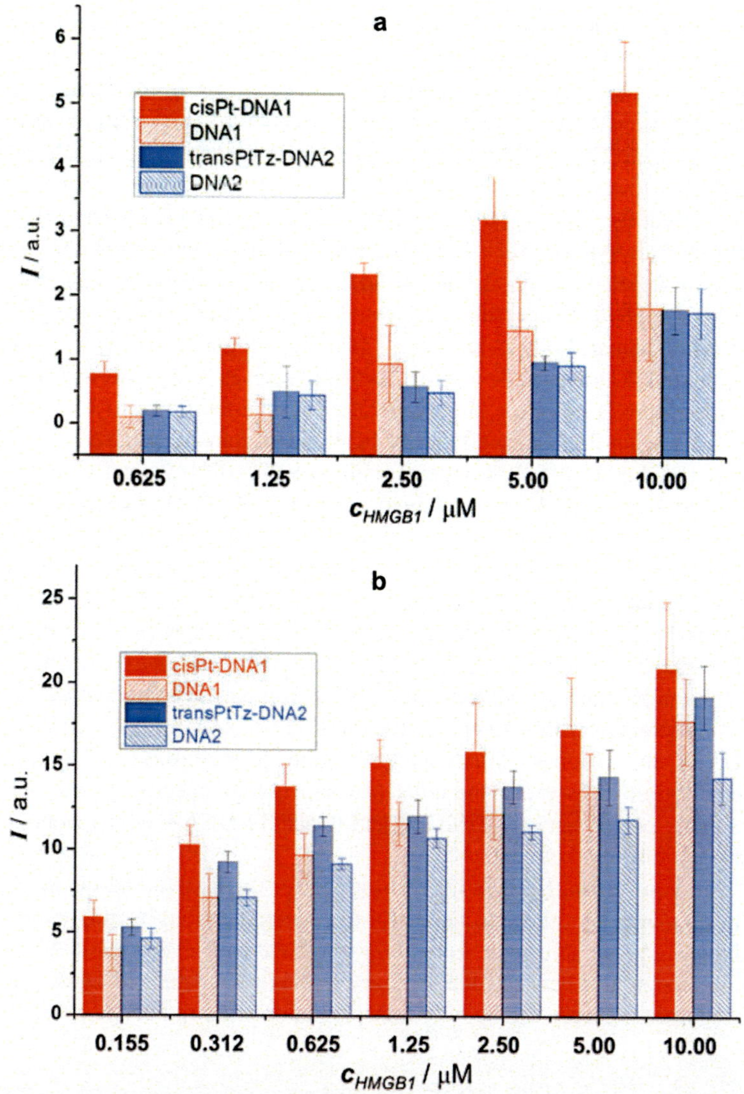

Fig. 5.15 SPRi monitoring the interaction of nuclear proteins with the immobilized platinated DNA probes with their platinum-free DNA probes as control. **a** Recognition intensity of different concentrations of HMGB1 with the indicated DNA probes immobilized on a same chip. **b** Recognition intensity of different concentrations of PC4 with the same DNA probes immobilized on the same sensor chip. Concentrations of spotted DNA probes: 10 μM

Figure 5.15b also reveals that PC4 can bind with both transPtTz-DNA2 and cisPt-DNA1 compared with their respective platinum-free DNA controls, which shows the effectiveness of the platinum-based drugs. However, PC4 has a similar affinity toward the cisPt-DNA1 within its tested concentration range but increases its affinity to the transPtTz-DNA2 slowly with its concentration because the difference imaging intensity toward DNA2 becomes obvious at 10 μM PC4 and above. It should be noted that the binding strength of both HMGB1 and PC4 with their own controls increase always with their concentrations. All the variations are nonlinear.

The differences can also be found by use of the measured equilibrium constant around our body temperature as shown in Fig. 5.16. The normal physiological temperature of human body is around 37 °C, and the maximum temperature can reach around 42 °C under pathological conditions such as fever. Although the lowest body temperature is not clear, the temperature was tested down to 15 °C. Figure 5.16a reveals again the binary affinity of PC4 toward transPtTz-DNA2 and cisPt-DNA1. Stronger binding of PC4 appears with transPtTz-DNA2 than with cisPt-DNA1 at the temperature of 37 °C and below, until 15 °C, however, the binding strength with the transPtTz-DNA2 rapidly decreases to lower than with cisPt-DNA1 at a temperature above 37 °C. This may give an explanation why the transPtTz has not yet been used clinically because the cancer patients may have a higher body temperature than the normal, which may provide a clue for further improving the design of a novel transplatin. Anti-cancer drugs (and maybe other medicines) may need to undergo high-temperature testing rather than just developing at room temperature as usual.

Different from the pair of PC4 and transPtTz-DNA2, the pair of HMGB1 and cisPt-DNA1 shows high specificity with each other. Similar to Fig. 5.15a, the dissociation equilibrium constant of the latter pair shows high specificity, and only negligible binding occurs between HMGB1 and transPtTz-DNA2, which is at or even below the binding level of controls.

In conclusion, SPRi is highly suitable for qualitative and quantitative study of molecular recognition and applicable to screening various ligands or drug candidates. Except for the screening technology, we have well discussed in SPRi analysis of the recognition between the nuclear proteins and platinated DNA probes. All the discussed methods can be fully transplanted to other molecular recognition studies.

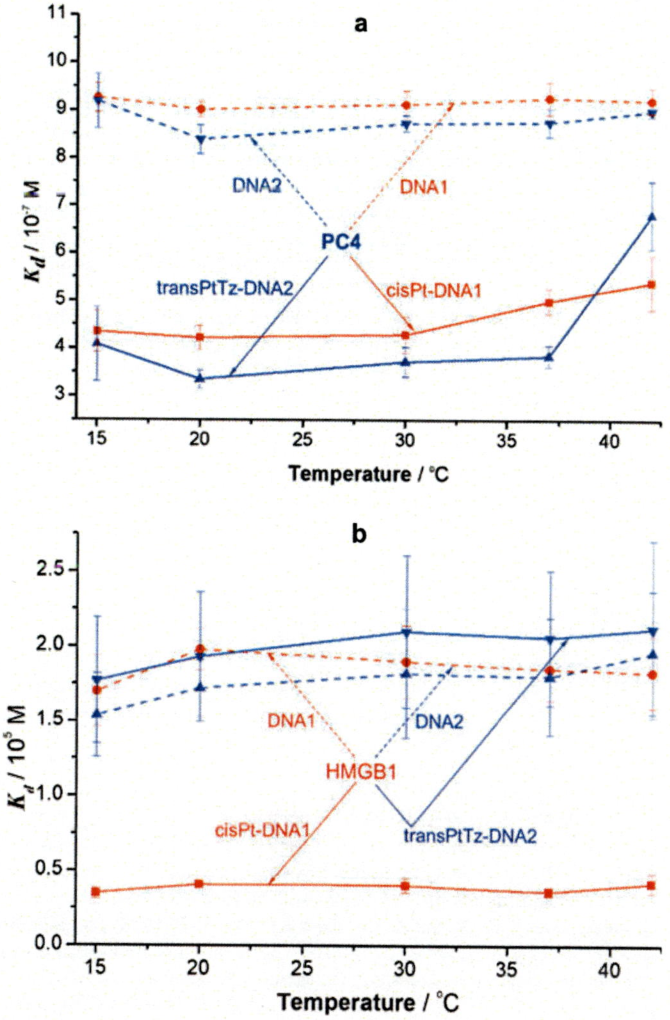

Fig. 5.16 Temperature dependence of the equilibrium constant K_d for the dissociation of **a** PC4 and **b** HMGB1 from their platinated and platinum-free DNA probes as indicated. Solid lines: platinated DNA probes; dot lines: DNA controls; error bar: standard deviation calculated from three sensor chips. Reconstructed from Ref. [9] with permission

References

1. Jordan CE, Corn RM (1997) Surface plasmon resonance imaging measurements of electrostatic biopolymer adsorption onto chemically modified gold surfaces. Anal Chem 69:1449–1456
2. Thiel AJ, Frutos AG, Jordan CE, Corn RM, Smith LM (1997) In situ surface plasmon resonance imaging detection of DNA hybridization to oligonucleotide arrays on gold surfaces. Anal Chem 69:4948–4956

3. Nelson BP, Frutos AG, Brockman JM, Corn RM (1999) Near-infrared surface plasmon resonance measurements of ultrathin films. 1. Angle shift and SPR imaging experiments. Anal Chem 71:3928–3934
4. Guedon P, Livache T, Martin F, Lesbre F, Roget A, Bidan G, Levy Y (2000) Characterization and optimization of a real-time, parallel, label-free, polypyrrole-based DNA sensor by surface plasmon resonance imaging. Anal Chem 72:6003–6009
5. Li M, Lee HJ, Condon AE, Corn RM (2002) DNA word design strategy for creating sets of non-interacting oligonucleotides for DNA microarrays. Langmuir 18:805–812
6. Nelson BP, Grimsrud TE, Liles MR, Goodman RM, Corn RM (2001) Surface plasmon resonance imaging measurements of DNA and RNA hybridization adsorption onto DNA microarrays. Anal Chem 73:1–7
7. Brockman JM, Frutos AG, Corn RM (1999) A multistep chemical modification procedure to create DNA arrays on gold surfaces for the study of protein—DNA interactions with surface plasmon resonance imaging. J Am Chem Soc 121:8044–8051
8. Frutos AG, Brockman JM, Corn RM (2000) Reversible protection and reactive patterning of amine- and hydroxyl-terminated self-assembled monolayers on gold surfaces for the fabrication of biopolymer arrays. Langmuir 16:2192–2197
9. Wang X, Xu J, Liu C, Chen Y (2016) Specific interaction of platinated DNA and proteins by surface plasmon resonance imaging. RSC Adv 6:21900–21906
10. Dong Y, Wilkop T, Xu D, Wang Z, Cheng Q (2008) Microchannel chips for the multiplexed analysis of human immunoglobulin G–antibody interactions by surface plasmon resonance imaging. Anal Bioanal Chem 390:1575–1583
11. Morton TA, Myszka DG, Chaiken IM (1995) Interpreting complex binding kinetics from optical biosensors: a comparison of analysis by linearization, the integrated rate equation, and numerical integration. Anal Biochem 227:176–185
12. Myszka DG, Arulanantham PR, Sana T, Wu ZN, Morton TA, Ciardelli TL (1996) Kinetic analysis of ligand binding to interleukin-2 receptor complexes created on an optical biosensor surface. Protein Sci 5:2468–2478
13. Wang Z, Wilkop T, Xu D, Dong Y, Ma G, Cheng Q (2007) Surface plasmon resonance imaging for affinity analysis of aptamer–protein interactions with PDMS microfluidic chips. Anal Bioanal Chem 389:819–825
14. Zhang P, Chen Y-P, Guo J-S, Shen Y, Yang J-X, Fang F, Li C, Gao X, Wang G-X (2014) Adsorption behavior of tightly bound extracellular polymeric substances on model organic surfaces under different pH and cations with surface plasmon resonance. Water Res 57:31–39
15. Heyse S, Ernst OP, Dienes Z, Hofmann KP, Vogel H (1998) Incorporation of rhodopsin in laterally structured supported membranes: observation of transducin activation with spatially and time-resolved surface plasmon resonance. Biochemistry 37:507–522
16. Bieri C, Ernst OP, Heyse S, Hofmann KP, Vogel H (1999) Micropatterned immobilization of a G protein-coupled receptor and direct detection of G protein activation. Nat Biotechnol 17:1105–1108
17. Buijs J, Lichtenbelt J, Norde W, Lyklema J (1995) Adsorption of monoclonal IgGs and their $F(ab')_2$ fragments onto polymeric surfaces. Colloids Surf B Biointerfaces 5:11–23
18. Neubert H, Jacoby ES, Bansal SS, Iles RK, Cowan DA, Kicman AT (2002) Orientation of an immunoglobulin G using recombinant protein G. Anal Chem 74:3677–3683
19. Chen H, Huang J, Lee J, Hwang S, Koh K (2010) Surface plasmon resonance spectroscopic characterization of antibody orientation and activity on the calixarene monolayer. Sens Actuators B 147:548–553
20. Kašpárková J, Brabec V (1995) Recognition of DNA interstrand cross-links of cis-diamminedichloro-platinum (II) and its trans isomer by DNA-binding proteins. Biochemistry 34:12379–12387
21. Jamieson ER, Lippard SJ (1999) Structure, recognition, and processing of cisplatin—DNA adducts. Chem Rev 99:2467–2498
22. Ohndorf UM, Rould MA, He Q, Pabo CO, Lippard SJ (1999) Basis for recognition of cisplatin-modified DNA by high-mobility-group proteins. Nature 399:708–712

23. Jung Y, Lippard SJ (2007) Direct cellular responses to platinum-induced DNA damage. Chem Rev 107:1387–1407
24. Du Z, Luo Q, Yang L, Bing T, Li X, Guo W, Wu K, Zhao Y, Xiong S, Shangguan D, Wang F (2014) Mass spectrometric proteomics reveals that nuclear protein positive cofactor PC4 selectively binds to cross-linked DNA by a trans-platinum anticancer complex. J Am Chem Soc 136:2948–2951
25. Banerjee S, Kumar BRP, Kundu TK (2004) General transcriptional coactivator PC4 activates p53 function. Mol Cell Biol 24:2052–2062
26. Marini V, Christofis P, Novakova O, Kašpárková J, Farrell N, Brabec V (2005) Conformation, protein recognition and repair of DNA interstrand and intrastrand cross-links of antitumor trans-[$PtCl_2(NH_3)$(thiazole)]. Nucleic Acids Res 33:5819–5828
27. Bolduc OR, Masson J-F (2008) Monolayers of 3-mercaptopropylamino acid to reduce the nonspecific adsorption of serum proteins on the surface of biosensors. Langmuir 24:12085–12091
28. Yamamoto A, Ando Y, Yoshioka K, Saito K, Tanabe T, Shirakawa H, Yoshida M (1997) Difference in affinity for DNA between HMG proteins 1 and 2 determined by surface plasmon resonance measurements. J Biochem 122:586–594
29. Krishnamoorthy G, Beusink JB, Schasfoort RBM (2010) High-throughput surface plasmon resonance imaging-based biomolecular kinetic screening analysis. Anal Methods 2:1020–1025
30. Dey B, Thukral S, Krishnan S, Chakrobarty M, Gupta S, Manghani C, Rani V (2012) DNA–protein interactions: methods for detection and analysis. Mol Cell Biochem 365:279–299
31. Pillet F, Sanchez A, Formosa C, Severac M, Trévisiol E, Bouet JY, Leberre VA (2013) Dendrimer functionalization of gold surface improves the measurement of protein–DNA interactions by surface plasmon resonance imaging. Biosens Bioelectron 43:148–154
32. Edwards PR, Leatherbarrow RJ (1997) Determination of association rate constants by an optical biosensor using initial rate analysis. Anal Biochem 246:1–6
33. Hoebel S, Vornicescu D, Bauer M, Fischer D, Keusgen M, Aigner A (2014) A novel method for the assessment of targeted PEI-based nanoparticle binding based on a static surface plasmon resonance system. Anal Chem 86:6827–6835

Chapter 6
Analysis of Molecules and Biomolecules

This chapter is designed to use SPRi for a comprehensive analysis of molecules that can impact on the gold sensor chip through various interaction effects such as non-specific or specific adsorption or deposition, or just flowing past the vicinity of the sensor surface. Pure molecules can directly be imaged through their impact on the permittivity ε_1, while mixed molecules have to be separately captured by their own probes pre-immobilized on the sensor surface or to be captured non-selectively and then recognized selectively. On a sensor chip chemically spotted with various arrays of probes, SPRi can simultaneously imaging different biologically and/or clinically important substances ranging from small molecules (e.g., metal ions, amino acids, hormones, and microRNA) to large molecules (e.g., DNA and proteins like biomarkers, antibodies) and even particles (e.g., nanoparticles, virus and cells). Here we need to stay awake that, in SPRi of molecules, especially the small molecules, signal amplification techniques must not be forgotten.

Considering that molecular types are too abundant to be covered within a limited chapter, we will first discuss SPRi of three representative biomolecules, that is, nucleic acids, saccharides and proteins, and then introduce SPRi of fingerprints as a representative of comprehensive analysis. Obviously, the key of the discussion lies in improving the detection sensitivity.

6.1 Analysis of Nucleic Acid

Nucleic acids such as DNA and RNA are till now the clearest information biomolecules that play normal and abnormal functions in life systems. Broadly speaking, they are a class of chemicals ranging from monomers (e.g., nucleotides), oligomers (e.g., DNA and RNA fragments like microRNA (miRNA)) and polymers (e.g., DNA and RNA) to complexes (e.g., double- and triple-stranded DNA, RNA–DNA hybrids and other hybrids or embeddings). Their most significant chemical feature is their sequence pairing that makes specific hybridization happen to form

Y. Chen, *Surface Plasmon Resonance Imaging*, Lecture Notes in Chemistry 95,
https://doi.org/10.1007/978-981-99-3118-7_6

duplexes. The hybridization opens unlimited routes for the design and synthesis of probes and selective capture of nucleic acids.

Normally SPRi can directly sense large nucleic acids at a concentration down to μM level. However, it will become a serious challenge as the molecular size become smaller and smaller. To sufficiently increase their SPRi response, signal amplification should be integrated into the measuring process. The present amplification principle is to raise the analytes-associated permittivity ε_1 by:

(i) enriching the analytical molecules on the related probes as more as possible;
(ii) specific addition of external molecules or particles with large mass or high refractive index;
(iii) increasing the apparent size of the analytical molecules; and
(iv) use of the above measures all together.

These measures can be realized chemically and/or physically as well through surface design aiming at signal amplification. Currently, the most useful technology is developed from chemistry. In this section, miRNA will be discussed as a representative.

6.1.1 *Analysis of MicroRNA*

MicroRNAs or miRNAs are a class of non-coding RNAs with 19–23 conserved nucleotide sequence. They are closely associated with the pathogenesis of human diseases, for example, acute kidney injury or AKI [1], polycystic kidney disease or PKD [2] and diabetic nephropathy or DN [3–5]. Some miRNAs (e.g., miRNA-21 and miRNA-192) can affect specific genes associated with diabetic nephropathy [5]. Accurate determination of multiple miRNAs is a key step for early clinical diagnosis, treatment and prevention of diseases. The challenges in SPRi of miRNAs are their low abundance with short and similar sequences [6]. Although there are many methods that are now available, such as northern blotting [7, 8], quantitative reverse-transcription real-time polymerase chain reaction or qRT-PCR [9] and various photo-sensing methods, e.g., electrochemiluminescence [10], colorimetry [11], fluorescence [12], scanometric array profiling chips [13, 14], photonic microring resonators [15, 16] and surface-enhanced Raman spectroscopy [17], SPRS [18, 19] and SPRi [20, 21]. Among them, SPRi are on the top of choice for batch analysis and it can give real-time images and global sensorgrams [22, 23]. Nevertheless, its real applications need to amplify the signals, especially in assaying the trace samples in order to avoid false information such as in the case of single polymorphisms [13].

6.1.1.1 Enrichment-Based Signal Amplification

Enriching miRNA on a sensor surface is the easiest technique to think of and was first explored in early SPRS analysis. Small miRNA molecules were selectively captured and condensed on their specific probes pre-immobilized on the sensor chip. This strategy was shown to be effective in sensing even ions [24], saccharides [25] and chiral amino acids [26]. This technique can clearly be transferred to sensing the miRNAs with SPRi, and the only differences are that multiple probes can be spotted on a same sensor chip, and the optimization for the sensing should be re-conducted to avoid the cross-talk among the immobilized probes. Based on highly specific probes, SPRi can selectively enrich various miRNAs from sample solutions onto their relative probing locations for directly detection of the *in situ* captured and enriched miRNAs. This method can normally detect mg/mL-μg/mL miRNA in sample solutions. It is comparable with normal optical adsorption assays but remains not applicable to detect the trace miRNAs, especially with limited sample resource.

6.1.1.2 Plasmonic Signal Amplification

To further improve SPRi sensitivity, signal amplification other than enrichment should now be considered and explored. There are numerous nucleic acid chemical reactions ready for exploitation, for example, enzyme-catalyzed reactions [27], e.g., PCR [28], strand displacement cyclic reaction (SDCR, [8, 27, 29] and enzyme-free reactions [18, 19], e.g., polymerization [21] and attachment of NPs like metallic [30, 31], non-metallic (e.g., SiO_2, [32]), quantum dots [33] and magnetic nanoparticles [34, 35, 36]. Coupling of nanoparticles, especially those with LSPPs such as AuNPs, into SPRi can often induce high magnification of imaging signals, which can be termed plasmonic amplification of signals. To amplify, specific reactions with or without natural or artificial enzymes are required and have been shown to be applicable, with high effectiveness.

To make effective use of LSPPs, metallic NPs, especially AuNPs are widely used in the signal amplifications of SPRS and SPRi owing to the matured technology to reproducibly synthesize AuNPs with high amplification [30, 35–38]. In addition, AuNPs can conveniently be loaded with high specificity through DNA hybridization reaction by the following Approach 6.1 and as illustrated also in Fig. 6.1.

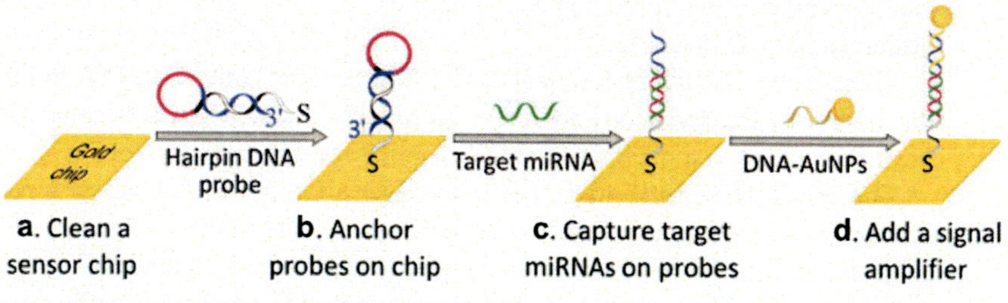

Fig. 6.1 Signal amplification in SPRi of miRNA with AuNPs as amplifiers based on the base-pairing hybridization reactions and Au–S chemistry

Approach 6.1 for plasmonic amplification of miRNA signals with AuNPs as amplifiers:

Step 1 clean an SPRi sensor chip as usual (Fig. 6.1a).
Step 2 on-chip immobilize miRNAs-specific hairpin probes with an end sequence complementary to the single-strand DNA of DNA-AuNPs (Fig. 6.1b).
Step 3 capture the target miRNAs on their specific probes by *in situ* hybridization (Fig. 6.1c).
Step 4 further hybridize the signal amplifiers of DNA-AuNPs with the complementary end of the probe to form a sandwich structure for sensitive SPRi (Fig. 6.1d).

Approach 6.1 can be conducted intermittently outside the SPRi devices or continuously inside the flow cell. The in-flow-cell process is recommended for channelized SPRi systems. Although the in-cell hybridization reduces throughput, it has the advantage to monitor the reactions dynamically or in real time. It must be noted that a functional hairpin DNA probe needs to consist of at least two sections of sequence, one complementary to a target miRNA and the other to DNA-AuNPs. To facilitate its attachment onto the gold sensor surface, a thiol group should be pre-grafted onto the sequence end, normally at the 5′-end. An on-chip immobilized hairpin probe must be stable enough to maintain its hairpin structure before contact with its target miRNA, and sufficiently easy to open, once its target miRNA appears. The sequence pairing with the DNA-AuNP must be located at the 3′-end that can be universalized so that we can use only a common signal amplifier of DNA-AuNP. In practice, the miRNA-specific sequence length has to be optimized in respect of ideal binding and high selectivity. After optimization, the signal gain is < 30 RIU after capture of the target miRNAs and may further increase to 30–60 RIU after the load of AuNPs.

6.1.1.3 Offline Cyclic Reaction-Based Signal Amplification

There are two basic routes to cyclically amplify the SPRi signals in the detection of miRNA. The first one is to increase the copy number of a target miRNA or its related molecules in solution by offline technology, while the second is to on-chip densify and/or enlarge the captured miRNA or its associated molecule(s). In this section, the offline strategy is discussed.

There are at least two theoretical ways to increase the copy of miRNA in solutions, that is, reverse transcription and SDCR. Herein discussed is SDCR because it is more flexible, easier and more cost-effective than the reverse transcription.

The principle to perform SDCR is to make a target miRNA act as a primer to initiate DNA copying reaction from a probing template in the appearance of related polymerases and dNTP (deoxy-ribonucleoside triphosphate, where N is a variable of A, G, T and C). The template is a hairpin-structured DNA that normally consists of three sections of sequences, one complementary to miRNA, one for capture and one for joint and cutting. In a bit detail, the loop at 5′-end has a full section of sequence complementary to the target miRNA, followed by a section of sequence the same as the free end of an on-chip trapper, and a section of sequence the same as the free end of a DNA-AuNP, where the last two sections of sequence, which link with each other, will act as a DNA or molecular amplifier. The copy of the template can be initiated only after its hairpin loop is opened by the target miRNA. This copying reaction can be circulated once the newly elongated DNA sequence can be cut and released from the template, which is implementable with a nicking endonuclease or nickase. This can be exemplified by the nickase of Nt.BstNBI or Nb.BbvCl to specifically recognize and nick the non-palindromic sequence in the following double-strand DNA (dsDNA):

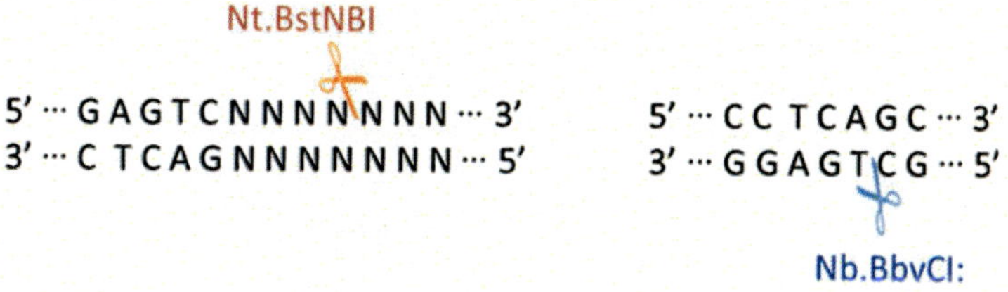

The miRNA-complementary sequence of the template is clearly used to specifically bind the target analytes, while the molecular amplifier, which was termed trigger in literature [8, 27, 29], is used to hybridize with an on-chip immobilized DNA trapper to sensitize SPRi detection. Further increase of the detection sensitivity is achievable by the use of a plasmonic amplifier.

Practical SDCR is usually conducted in solutions by mixing the template with the target miRNA. Template-miRNA hybrids will start its replication from the joint of dsDNA and single-strand DNA (ssDNA) to form a complete duplex in the presence of Kelnow fragment, polymerase and dNTP (Fig. 6.2a and b), where the Kelnow fragment will guide the polymerase to start the replication from the 3′-end of the miRNA. The complete duplexes can be recognized and cut at the joint by the addition of either Nb·BbvCI or Nt.BstNBI (depending on the base sequence at the joint) to initiate a new cycle of replication (Fig. 6.2b). The newly replicated sequence will then displace and release the previous sequence into solution. Numerous DNA amplifiers can thus be synthesized cycle by cycle (Fig. 6.2c). These molecular amplifiers will be captured by another complementary DNA trapper immobilized on a sensor chip (Fig. 6.2d) for sensitive SPRi detection (Fig. 6.2e). If the sensitivity remains insufficient, the plasmonic amplifiers of DNA-AuNPs would further be utilized to form sandwich-structured complexes for more sensitive SPRi detection as shown in Fig. 6.2f. In our experiment, it was found that the signal gain of plasmonic amplification is about 5 folds of that of DNA amplifiers (Fig. 6.3), with a dynamic range between 0.5 and 200 pM miRNA-21 [29] at $R^2 = 0.99$ and LOD at ca. 0.1 pM. It should be noted that these DNA-based SPRi sensor chips can be regenerated (Fig. 6.2f → d) for cyclic uses. A protocol used in this laboratory is shown in Approach 6.2 for reference.

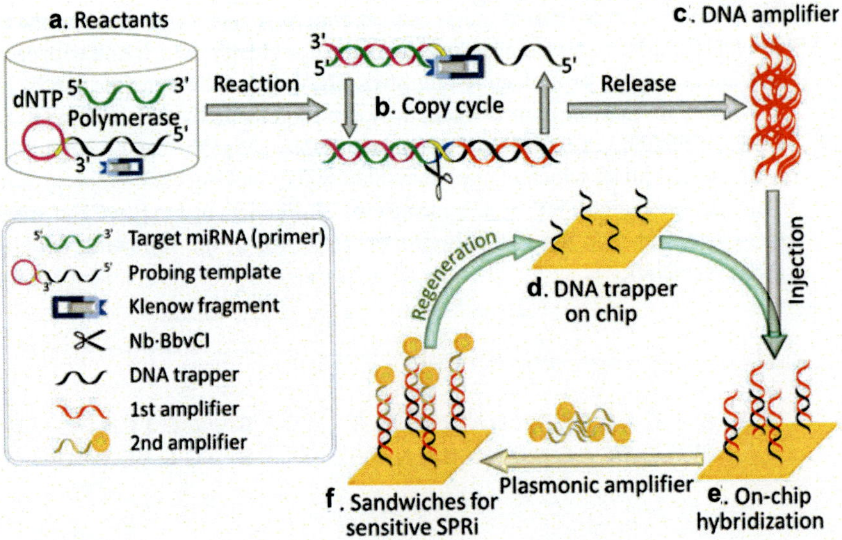

Fig. 6.2 A general process for signal amplification in SPRi of miRNA through DNA synthesis combined with its strand displace amplification and AuNPs-based amplification. The copying reaction cycle is offline-conducted rather than inside the SPRi flow cell

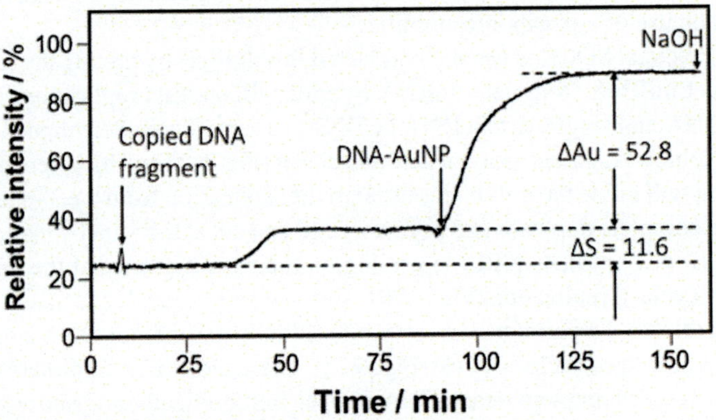

Fig. 6.3 Increment of relative SPRi intensity amplified first by DNA amplifiers (ΔS) replicated through solution-based cyclic reaction of miRNA-21 and then by a plasmonic amplifier of AuNP (ΔAu) measured on SPRi-PX8100

Approach 6.2 for cyclic copy of miRNA-oriented DNA fragments in solution state:

Step 1 mix miRNA solution (at a known or unknown concentration) in a vial with 400 nM hairpin-structured DNA template, 10 U/μL Klenow fragment, 10 U/μL Nb·BbvCI, 250 μM dNTP, 1.2 U/μL RNase inhibitor, 10 × Tris-acetate buffer, 10 × Tris-HCl buffer and diethylpyrocarbonate-treated water at a volume ratio of 2.0:4.0:0.6:0.6:2.0:1.5:2.0:2.0:5.3 μL to form 20 μL reaction solution in total. Note, Tris-acetate buffer is composed of 20 mM Tris-acetate at pH 7.9 added with 500 mM potassium acetate, 10 mM magnesium acetate and 100 μg/mL BSA; while the Tris-HCl buffer consists of 100 mM Tris-HCl at pH 7.5 added with 70 mM $MgCl_2$ and 1 mM dithiothreitol.

Step 2 place the solution at 37 °C to react for 80 min, and terminate the reaction at 80 °C for 10 min.

Step 3 dilute the solution to 400 μL with a buffer composed of 30 mM sodium phosphate, 450 mM NaCl, 3 mM EDTA and 0.25% Triton X-100 at pH 7.4.

Step 4 use the solution for further experiment or store it at – 20 °C for later use.

The diluted solution can directly be injected into parallel flow cells for flow-hybridization at 10 μL/min per channel. After the signals become stable, solutions of DNA-functionalized AuNPs are further pumped in, also at 10 μL/min, to form sandwich assembles that will significantly increase the SPRi signals. After completion of imaging measurements, the chip can be regenerated (Fig. 6.2f → d) by pumping in 50 mM NaOH. All the experiments are usually performed at room temperature (ca.25 °C) unless specific notation.

The on-chip DNA trappers are pre-immobilized on the gold chip surface via S–Au bound. To immobilize, the chip cleaned in a fresh piranha solution and in water is carefully deposited with 1 μM thiolated DNA trappers in 1 M KH_2PO_4 at pH 3.8 (needing ca. 200 μL to fully cover the sensing surface). The chip is incubated at 4 °C overnight in a humid box, washed with water and further blocked by sequential immersion in 1 mM aqueous 6-Mercapto-1-hexanol and in 1% BSA for 1 h each. The chip is getting ready for use after washed with water and blown dry with nitrogen gas stream.

6.1.1.4 On-Chip Cyclic Reaction-Based Signal Amplification

This on-chip amplification of signals is more effective and easier to manipulate than the offline one. There are also many surface reactions ready for exploitation. The examples include the increase of the surface concentration of analytes, utilization of specific feature of the analytes and enlargement of the captured molecules. They can also combine with each other and further with plasmonic amplification.

There are two basic strategies to increase the signals based on surface chemistry: increase of surface density and/or thickness normal to the surface.

(i) *Increase of surface density*

As mentioned already, this is achieved by probes-based capture technology, but with very limited gain in case of small molecules. Herein we introduce an alternative way that makes a target analyte to catalyze some specific reactions to produce a large amount of SPRi-sensitive substance, and the determination is now change to SPRi of the produced substance, rather than the analytes themselves, to largely increase the surface concentration and detection sensitivity. This is particularly applicable to the SPRi determination of miRNA because miRNA can catalyze hybridization of two complementary hairpin DNAs, which is termed hybridization chain reaction (HCR) and demonstrated to work well in fabricating fluorescent and electrochemical sensors. To transfer it to SPRi, the key is how to rule out or lower the spontaneous hybridization of the two DNA chains.

In order to make a miRNA act as a catalyst, it needs to add a miRNA-specific trigger on either one of the two complementary hairpin DNA chains, shorted as H1. The trigger is in fact a section of sequence complementary to the miRNA. To prevent H1 from spontaneous hybridization with its complementary hairpin DNA chain (shorted as H2), the trigger must be added at the stem end of H1 and the stem should be as short as possible. Thus, once the end of H1 with trigger is immobilized on a sensor surface, its loop can cover the short stem and yield high steric hindrance to H2 but low resistance to miRNA. This blocks H1–H2 hybridization but enables H1-miRNA reaction if miRNA appears. The other end of H1 will then be opened and become free due to the formation of H1-miRNA duplex. This free end has not more steric resistance to H2 and H1–H2 hybridization starts to form more stable H2-H1 duplex (Fig. 6.4). As a consequence, miRNA is squeezed out from the H1 chain. The free miRNA will then search other immobilized H1 to initiate next H1-H2 hybridization (the right part in Fig. 6.4). Once starting, this miRNA-catalyzed hybridization reaction will never stop unless miRNA becomes ineffective or H2 and H1 are used up.

In order to maximize the surface density of H1–H2 duplexes, high concentration of H1 is spotted on the sensor and H2 is added excessively, normally at ca. 2–5 folds of H1. In this case, the measured SPRi signal is a function of the quantity of miRNA. This makes the quantification of miRNA possible. Figure 6.5 illustrates a determination of miRNA-15a (a biomarker for multi-tumor diagnosis) via a laboratory-designed H1–H2 duplex as the signal reporter. A linear dynamic range was found between 10 pM and 0.1 μM plotted by the SPRi intensity against the logarithm of the concentration of miRNA. The detection sensitivity was increased for about three orders of magnitude

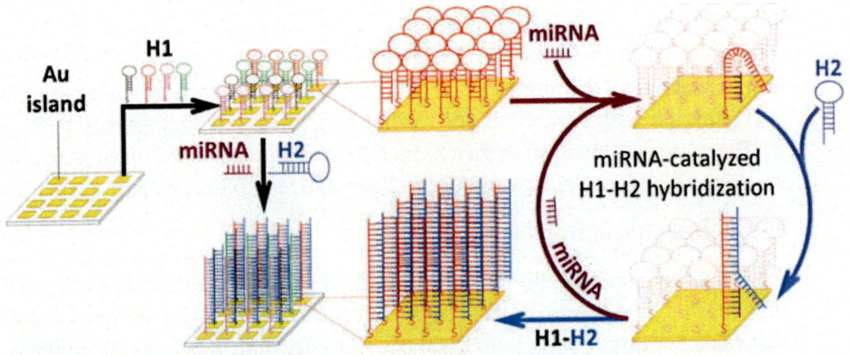

Fig. 6.4 Increase of H1–H2 surface density through a miRNA-catalyzed hybridizationH1. 3′-HS-tagged hairpin DNA; H2. complementary hairpin DNA of H1. Reprinted from Ref. [21] with permission

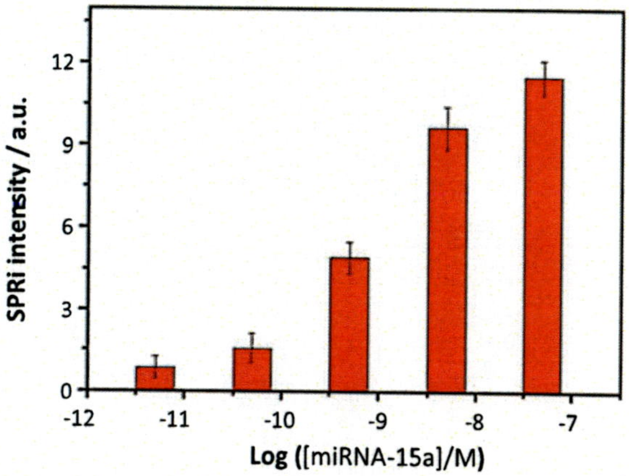

Fig. 6.5 SPRi of miRNA-15a with signals amplified by its catalytic hybridization of two sequence-complementary DNA hairpins H1 and H2. Reprinted from Ref. [21] with permission

compared with the direct concentration of miRNA, with a limit of detection lowered from μM miRNA to at least pM.

(ii) *Increase of thickness*

The surface-density-based amplification can generally increase the detection sensitivity for about 10^3 folds, able to lower the LOD down to pM miRNA. This is however not sufficient in assay the real samples that may contain below fM miRNAs, about three orders of magnitude below the LOD. To further amplify SPRi signals, we need to make use of the extramembranous space of the H1-H2 duplex. There are two probabilities, one is to *in site* lengthen H1–H2 chains, and the other is to add the plasmonic amplifiers at the outer end of H1–H2 duplex. The first strategy looks extremely easy but in practice, the chains of H1 and H2 cannot be prolonged infinitively. In fact, an over-lengthened chain of H1 and H2 will slow down the hybridization and the resultant duplexes easily bend down and fold into smaller-sized structure, and in turn to lower the detection sensitivity. The optimum length of H1–H2 hybrid is dependent not only on the sequence of H1 and H2 but also on the miRNA to be detected. The length of both H1 and H2 must be optimized through experimental test judged by the SPRi signals.

The second technique to further increase the signals can easily be achieved by direct loading the metallic nanoparticles on the upper end of H1–H2 duplex. This needs to insert an extra DNA sequence at the upper end of H2 complementary to the DNA chain immobilized on the nanoparticles. Note, the loading of plasmonic amplifiers may aggravate the bending and folding of H1–H2 duplexes if they are not strong enough or the metallic particles are overweight.

An even smart alternative is to add a rigid molecular rod or network on the top of H1–H2 duplexes to thicken the analyte layer. An atom transfer radical polymerization (ATRP) technique is adoptable to chemically graft a polymer network on the top of H1–H2. By this way, some alkenes like 2-hydroxyethyl methacrylate (HEMA) can be grafted. The polymerization is initiated from a bromoisobutyryl (BiB)-tagged end

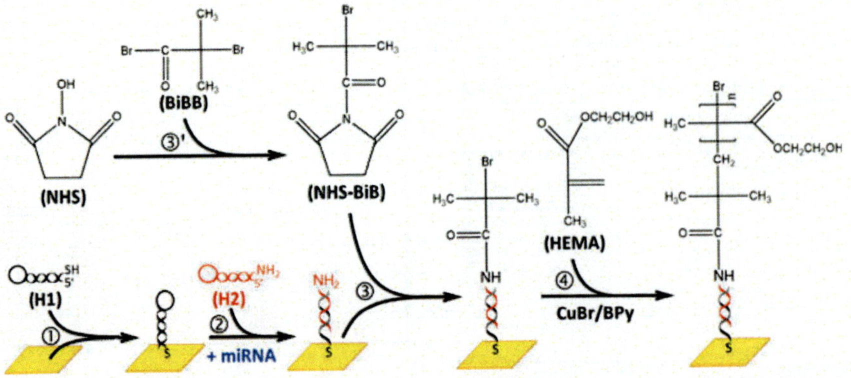

Fig. 6.6 A total reaction route to densify and thicken surficial H1–H2 by hybridization combined with atom transfer radical polymerization H1. 3′-HS-tagged hairpin DNA; H2. H1-complementary hairpin DNA terminated with NH_2 at the 3′-end; BiBB. bromoisobutyryl bromide; BiB. Bromoisobutyryl; HEMA. hydroxyethyl-2-methacrylate; BPy. 2,2′-bipyridine; CuBr/BPy. Cu(I) complex of BPy and/or Br serving as a catalyst.

of DNA (normally at 3′-end) with Cu(I)-2-bipyridyl complex as a catalyst. The total reaction routes are illustrated in Fig. 6.6. There are four key steps:

① assembly of H1 on a gold chip;
② hybridization of H1 with H2 using miRNA as a catalyst;
③ *in situ* introduction of BiB onto the amino terminal with freshly synthesized NHS-BiB (③′); and
④ growth of polymer via ATRP.

Some supposed protocols are as follows for reference:

Approach 6.3 for the preparation of NHS-BiB [39]:

Step 1 cool 0.22 M bromoisobutyryl bromide in diethyl ether in an ice bath for 5 min.
Step 2 dropwise add a mixed solution of 0.43 M N-hydroxysuccinimide and 0.65 M TEA in dioxane in about 30 min.
Step 3 stir the solution at room temperature for 3 h.
Step 4 remove any precipitate by filtration, and wash the clear solution with freshly $NaHCO_3$-saturated solution for three times, and water twice.
Step 5 dry it first with $MgSO_4$, then by evaporation in vacuum to have white solid NHS-BiB (yield 77%).
Step 6 characterize the product by ^1H NMR (in CD_3Cl at 300 MHz): δ 2.858 (s, 4H), 2.077 (s, 6H); ^{13}C NMR (in CD_3Cl at 300 MHz): δ 25.9, 30.9, 51.4, 166.2, 168.9; and by MS: m/z 262.15 and 263.15.

Approach 6.4 for the preparation of BiB-tagged H2 (H2-BiB):

Step 1 mix 3 μL of 100 μmol/L 3′-amino-H2 with 5 μL of 1.0 M $NaHCO_3$/Na_2CO_3 at pH 8.5.

Step 2 add 10 μL of freshly prepared 10 mg/mL NHS-BiB dissolved in DMF.
Step 3 react at room temperature for 60 min to couple the initiator at the 3′ end.
Step 4 remove the excess by gel filtration (e.g., NAP-5 column).

Approach 6.5 or the preparation of hairpin DNA:

Step 1 prepare 2 μM ssDNA in 10 mM Tris-HCl at pH 7.0.
Step 2 heat and maintain the solution at 94 °C for 5 min.
Step 3 naturally cool the solution to room temperature (for ca. 2 h) to form hairpin structure.

Approach 6.6 for the preparation of polymerizing monomer solution:

Step 1 displace the dissolved oxygen in aqueous solutions by bubbling N_2 gas at room temperature for 20 min.
Step 2 dissolve HEMA monomer in the de-O_2 water at a volume ratio of 1:4 and keep bubbling N_2 gas for 30 min.
Step 3 add a solid catalyst mixture of CuBr, 2,2′-bipyridine and ascorbic acid (1:2:1.5, molar ratio) into the monomer solution while bubbling N_2 gas and keep bubbling until use. Note: The solution should be prepared just before use. The reaction cannot be initiated without the addition of a catalyst (the left reaction route in Fig. 6.7), but will proceed stepwise in the presence of freshly prepared Cu(I) complex (the right reaction route in Fig. 6.7), though there exist various equilibrations between each pair of radical and bromo compound.

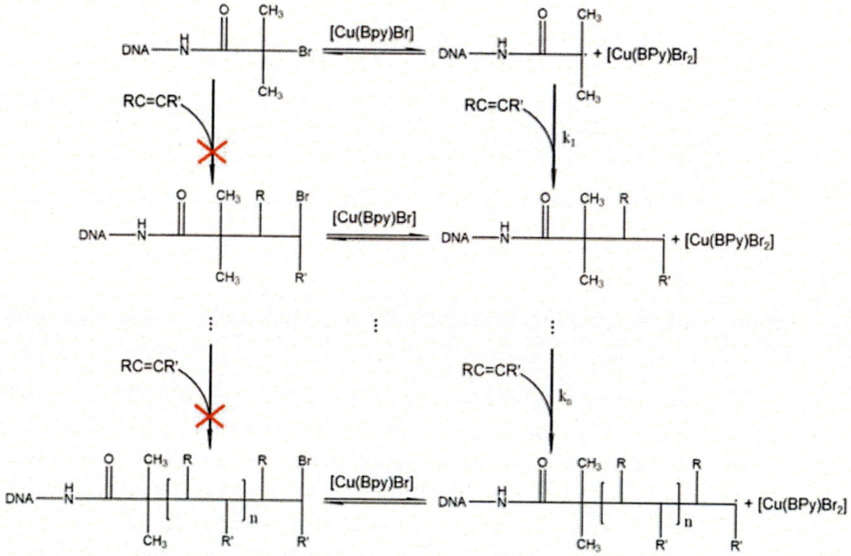

Fig. 6.7 Route for grafting a polymer at the 3′-end of an ssDNA [Cu(BPy)Br]. Cu(I) complex as a catalyst; [Cu(BPy)Br_2]. Oxidized state of [Cu(BPy)Br].

Approach 6.7 for spotting a chip with H1:

Step 1 chemically etch the unwanted parts of a gold-deposited chip, and coat the etched parts with CYTOP, (which can be conducted in batch).
Step 2 ultrasonically clean the chip sequentially in acetone, ethanol and pure water for ca. 20 s each, blow dry with nitrogen gas, and activate the gold surface in air plasma for 3 min.
Step 3 spot the chip with 2 μM 3′-HS-H1 in 10 mM Tris-HCl at pH 7.0, incubate it in a water-vapor-saturated box at room temperature for 4 h to assemble a monolayer of H1 on the gold islands, and wash the chip with Tris-HCl buffer and water, respectively.
Step 4 block the chip in 10 mM aqueous 6-mercapto-1-hexanol (MCH) for 2 h, wash with water and blow dry with nitrogen gas.

Approach 6.8 for direct SPRi of H1-miRNA:

Step 1 spot the H1-immobilized chip with a miRNA sample solution in 10 mM Tris-HCl (pH 8.0) containing 0.2 M NaCl and 10 mM $MgCl_2$, and incubate the chip at 37 °C for 2 h in a vapor-saturated box.
Step 2 wash the chip first with a buffer of 0.1 M PBS at pH 7.0 containing 2.0 mM $MgCl_2$, then with water and blow dry with N_2 gas.
Step 3 install the chip on an SPR imager to record the image(s).

Note, this procedure can be performed in real time. To do so, the H1-spotted chip must first be mounted on the SPR imager, and the miRNA solution is then pumped in until SPRi signal is saturated. Washing and measuring steps are then followed.

Approach 6.9 for indirect SPRi of miRNA through surface densification of H1–H2:

Step 1 spot a H1-immobilized chip with a mixed solution of miRNA and 2 μM (excessive) H2 in 10 mM Tris-HCl at pH 7.0, and incubate it at 37 °C for 2 h in a vapor-saturated box.
Step 2 wash the chip first with 0.1 M PBS at pH 7.0 containing 2.0 mM $MgCl_2$, then with water and blow dry with nitrogen gas.
Step 3 mount the chip onto the SPR imager to record the images.

Approach 6.10 for indirect SPRi of miRNA through surface-densified and thickened H1–H2:

Step 1 perform the hybridization of H1 with H2-BiB instead of H2, and mount the chip on the SPR imager.
Step 2 pump in newly prepared polymerizing monomer solution (diluted to 10% (v/v) HEMA with water), keep reaction at room temperature for 12 min, and pump in 50% (v/v) methanol to wash off the unbound substances.
Step 3 cyclically resume the polymerization by alternatively pumping in the polymerizing monomer solution and 50% methanol until the signals become sufficiently sensitive and stable.
Step 4, record and store the final images after cleaning step.

6.1.2 *Analysis of Nucleic Acid Analogues and Related Substances*

All the approaches are directly transferrable to the analysis of other nucleic acids, such as various RNA and DNA and their fragments. Normally, large RNA and DNA molecules can directly be captured, concentrated and imaged on their complementary probes. In case that the signals are not sufficiently strong, the above-discussed amplification techniques can be utilized with some minute modification to have a section of pairing sequence to load the amplification elements.

This sequence-pairing principle is also extendable to SPRi of other substances with at least one section of pairing sequence, not just nucleic acids. Furthermore, the above-discussed methods are theoretically extendable to the analysis of all chemicals that can be selective captured or labeled with a section of DNA or RNA sequence, which is simply demonstrated in SPRi of progesterone that is a steroid hormone, playing a central role in female reproductive processes such as ovulation and pregnancy, and possibly effects other organs as well. The measurement of its concentration in bodily fluids can assist the diagnosis of early pregnancy and provide insight into other reproductive functions. Although progesterone has not pairing sequence, it can be captured selectively by its aptamers that can be array-spotted on an SPRi sensor chip. The signal can be increased by both techniques of concentration and further amplified. Loaded with streptavidin-coated quantum dots, a linear range between 1.58 ng/mL and 126 μg/mL was obtained, with LOQ down to 1.58 ng/mL (ca. 5 nM) in phosphate buffer.

6.2 Analysis of Saccharides

Saccharides or saccharides are one class of the known essential substances in life systems besides the nucleic acids and proteins. Nevertheless, they are so complicated that they can hardly be fully analyzed by any single analytical tool, no matter how powerful is it. Microarray chip-based SPRi could be a novel alternative that features high throughput, low consumption and significant ability to simultaneously inspect the behaviors of various saccharides under same conditions, which may promote the progress in the research of saccharides. Similar to nucleic acids, saccharides vary in size but differently, this variation normally causes quite a weak change of permittivity and hence weak response of SPRS and SPRi, which is also the reason that we often use saccharides to modify the sensor surface. Further challenge is that the saccharides have in nature no pairable sequences for exploitation of signal amplification methods.

To conquer the challenges, one ready technology is to label the saccharides with a section of DNA sequence or to capture them with DNA probes like aptmer(s). Although this is highly exploitable, we will not further discuss it since the principle has been given already. Herein introduced is an alternative that utilizes lectin-based recognition reactions. Theoretically, all saccharides can be recognized and then captured by their own specific lectins. By use of pre-spotting technology, various saccharides can be imaged.

There are two possible techniques to image saccharides, the first one is to selectively capture and concentrate target saccharides on their specific lectin spots, and the second one is to physically or chemically spot the target saccharides on a sensor chip and then to recognize the spotted saccharides by recognition reaction with their own lectins. The first technique has clear thought and can easily be manipulated; nevertheless, it may result in poor imaging signals because the SPRi signals of lectins are much stronger than saccharides themselves. To have higher sensitivity, the second technique is recommended. However, no matter which technique is adopted, signal amplification must be considered in SPRi of sacchairdes, which is discussed in detail in the following subsection.

6.2.1 Selective Signal Amplification

A strategy that seems ideal in SPRi of saccharides is to selectively capture and amplify the signals. This can remove possible interference of non-target substances. However, this is only applicable to saccharides with available lectins; furthermore, it may complicate the measurement because the selectively captured saccharides need further signal amplification to have sufficient sensitivity. To simplify the manipulation, herein introduced is an easier strategy that captures hydroxyl substances all at once during spotting and then selectively recognizes the target saccharides and amplifies their SPRi signals.

A prerequisite to perform this type of SPRi is to have high-quality sensing chips that can capture saccharides and the capture must not obviously affect the bioactivity of target substances. Although there are physical and chemical approaches, covalently immobilization of saccharides is preferred to perform multiple analysis. At present, there are numerous chemical reactions available for covalent capture of saccharides on a gold surface such as Schiff's base reaction [40, 41] or amidation [42, 43], click reaction [44], Staudinger ligation [45], Diels–Alder reaction [46], Michael addition [47] and streptavidin- and biotin-based immunoreactions [48]. The issue is however that these reactions may affect the bioactivity of the natural saccharides. For example, immobilization of reducing saccharides on surface terminated with hydrazide, aminooxy or aminooxyacetyl [49, 50, 51] may vary their delicate structure: the aminooxy reaction preferentially generates acyclic products [52, 49] while the hydrazide reaction predominantly yields β-anomeric configuration adducts [49, 50].

In order to alleviate the immobilization impact as more as possible, some new chemical reactions have been explored in our laboratory [53]. Among them, cyanuric chloride or CC-related chemistry is the preferred choice. CC is very cheap and has three reactive chlorines that can be replaced stepwise by nucleophilic groups such as hydroxyl and/or amino groups through simply temperature regulation (refer to Sect. 4.6.2.1). It is highly attractive that CC-terminal allows freely rotation of a covalently captured molecule. This endows some extra chances to make the immobilized molecule adjust its orientation for the best recognition. It should however be noted that, to effectively immobilize saccharides, the CC terminals should also be pre-anchored on hydroxyl rather than amino terminals; otherwise, the immobilizing efficiency will be very poor. The reason lies in that the nucleophilicity of amino group is much stronger than hydroxyl.

The chemistry to covalently capture saccharides on a gold sensing surface terminated with CC includes three key steps: modify the sensing surface to have hydroxyl terminal, let the hydroxyl further react with CC, and spot target saccharides on the CC-terminated surface, as illustrated in Fig. 6.8.

To capture hydroxyl compounds on CC-terminated chip, Approach 6.11 is recommended:

Approach 6.11 for immobilization of hydroxyl substances on CC-terminated chip:

Step 1 incubate a cleaned 50-nm gold-coated glass slide in 1 mM MUA in ethanol at room temperature for 8 h, and rinse the chip with ethanol and water sequentially.
Step 2 activate the chip in 75 mM EDC and 15 mM NHS aqueous solution at room temperature for 25 min.
Step 3 block the chip treated with 1 M ethanolamine (EOA, adjusted to pH 8.6 with 5 M HCl) for 1.5 h, wash it with water and blow it dry with nitrogen gas.
Step 4 incubate the chip in 100 mM CC and 100 mM N,N-diisopropylethylamine (DIPEA) in acetone at 4 °C for 6 h, rinse it thoroughly with acetone and blow it dry with nitrogen gas.
Step 5 spot sample solutions (30 mM small hydroxyl analytes or 3 mg/mL polymers like polysaccharide, at pH 9.0 adjusted with 1 M NaOH) on the specified locations, at 15 nL per dot (ca.250 μm round dot), and incubate the spotted chip in a humid chamber at room temperature for 10 h.

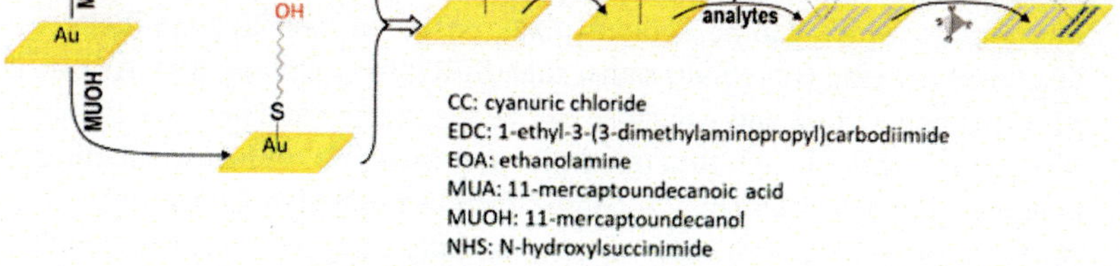

Fig. 6.8 Basic steps to covalently anchor saccharides on CC-terminated gold surface to free their orientation for better recognition with lectins

Step 6 wash the chip by sonication in water for three times and blow it dry with nitrogen gas. Note, the dried chip with saccharide spots can either be imaged directly or be stored at – 20 °C for at least 2 months.

It should be noted that CC chemistry needs basic reaction condition, better at pH above 8.0 otherwise, poor the reaction will offer poor yield and can hardly be regulated by change of temperature. Aqueous solutions are normally alkalized with NaOH while non-aqueous solutions with organic bases. It is also interesting that CC stochastically reacts with all the hydroxyls on saccharide molecules, and all the linking reactions can maintain the anomeric configuration of saccharides [53]. As expected, CC chemistry is not only cost-effective and easy to manipulate but also universally applicable to the capture of substances containing at least one hydroxyl group, which allows to capture both of reducing and non-reducing saccharides. To acquire image, saccharides-spotted chip must be mounted in the flow cell attached onto the bottom of the prism via a thin layer of index-matching oil. To measure strong signals, amplification is required and can be realized using Approach6.12.

Approach 6.12 for amplification of saccharide imaging signals based on concanavalin agglutinin (Con A) recognition reaction:

Step 1 block the saccharides-spotted chip in 1 M EOA at pH 9.0 for 3 h and mount it in SPRi flow cell.
Step 2 pump 25 mM Tris-HCl buffer at pH 7.6, containing 1 mM $CaCl_2$, 1 mM $MnCl_2$, and 0.1% Tween 20, through the flow cell at 60 μL/min until signals become stable.
Step 3 pump 50 μg/mL Con A solution into the flow cell and incubate at 25 °C for 10–20 min.
Step 4 wash away the unbound Con A by pumping in the Tris-HCl buffer to record the amplified images. Note, the chip can be regenerated, for about 10 times, by pumping in 0.1 M phosphoric acid to remove the bounded Con A.

The Approach 6.12 is extendable to other lectins such as peanut agglutinin (PNA). To demonstrate, a chip is spotted with 15 different saccharides and displayed by reaction sequentially with ConA and PNA. As known, ConA can specifically recognize terminal α-D-glucose and α-D-mannose in an order of mannose > glucose, while PNA specifically recognizes terminal β-D-galactose [54]. Figure 6.9 clearly illustrates that a CC-terminated gold chip is able to universally capture all the spotted saccharides, which can be demonstrated, based on their hydrophilicity, by vapor-displayed imaging (Fig. 6.9b) rather than SPRi (for weak signals). As expected, the β1-structured lactose and galactose can be recognized and hence amplified by PNA (Fig. 6.9c), while the α1-structured glucose, mannose, trehalose, maltose, sucrose, malhextose and dextran can be recognized and amplified by ConA (Fig. 6.9d). Differently, all the other saccharides display nothing images without the application of their own lectins. These SPRi data verify that the CC-terminated chips are useful to capture activity-conserved saccharides. Two reasons are responsible for the conservation feature: Firstly, CC is attacked mildly (normally at room temperature) by the far and non-ring hydroxyl groups, which maintains the original cyclic structure and even the anomeric configuration of saccharides. Secondly, CC offers three rotation axes at an angle of 120° that makes the on-CC-captured saccharides revolve easily to adapt their orientation for better "cluster effect" during recognition [55].

Figure 6.9 also reveals a phenomenon that the signals are roughly proportional to the molecular size of saccharides. This can reasonably be attributed to the variation of binding sites that are proportional to hydroxyl groups or monosaccharides. Clearly more binding sites on a saccharide molecule can bind more lectin molecules to produce stronger signals. Another experimentally discovered reason is the differences of recognition dynamics as shown in Fig. 6.10. Mannose has the fastest recognition dynamics than glucose and dextran while the dextran reacts the slowest. This suggests that the large saccharides need a longer reaction time than small ones to reach equilibrium. Nevertheless, difference is within several minutes. The dynamics contribution can be eliminated after reaction for about 15 min. Oppositely, the larger are the saccharides, the faster will they be washed off. This agrees with known results [56]. Thus, SPRi is also a technique for studying the binding affinity of saccharides with lectins.

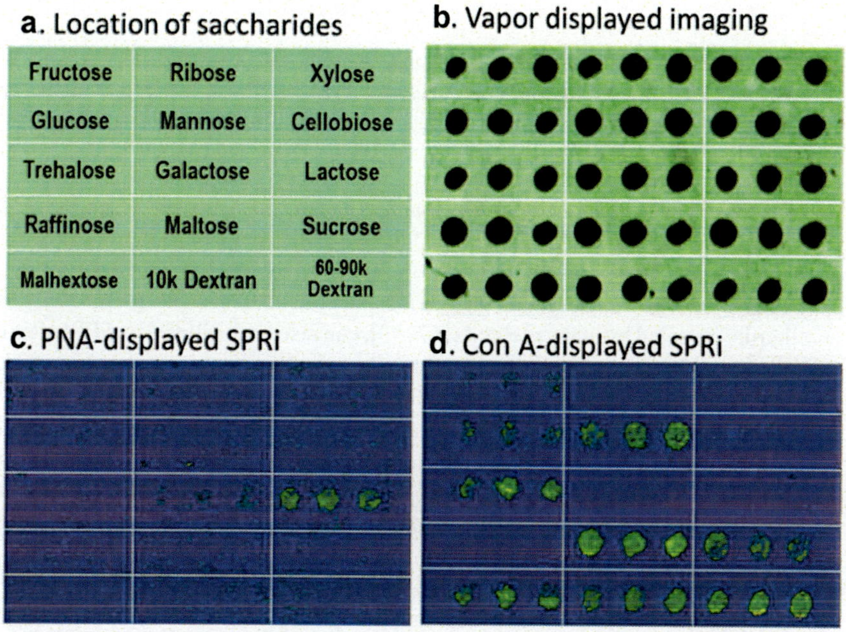

Fig. 6.9 SPRi of spotted saccharides measured on a laboratory-built vapor-displayed imager and SPRi-TX7100. **a** Designed location of saccharides; **b** Vapor-displayed image; **c**, **d** Images amplified with **c** PNA and **d** ConA, respectively

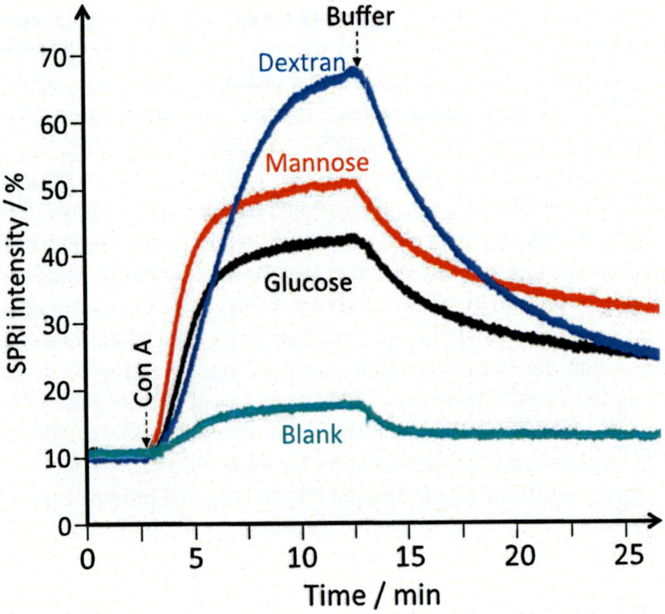

Fig. 6.10 SPRi sensorgrams for reaction dynamic study of saccharides with Con A measured on SPRi-TX7100. The saccharides were pre-immobilized on CC-terminated gold sensing chip, and the signals were recorded in a pumping sequence of Tris–HCl buffer (pH 7.6), 50 μg/mL ConA, and Tris-HCl buffer again

6.2.2 *Cyclic Signal Amplification*

Figure 6.9 shows that small saccharides such as fructose and galactose, which are hardly revealed by many methods [57–59], can also be displayed by SPRi after combination with lectin recognition reaction. Nevertheless, the signals remain weak. This, in addition to imaging loss compared with SPRS [60] may make SPRi inapplicable to the analysis of small saccharides below trace level even if the lectin-display technique is applied [61]. Other measures have thus to be adopted, for example, tagging the saccharides with large molecules [62, 63] and/or nanomaterials [64, 65], e.g., AuNPs or quantum dots [66–68], better in one shot through analyte-induced *in situ* reactions [69, 70]. Dual signal amplification through integral of reactions has also been shown to be effective [35, 71, 72]. Tagging the saccharides with a high refractive index material such as liposomes is also an alternative [73]. However, these measures are not easy to optimize and/or regulate to have a proper range for quantitation. In this section discussed is a sensitivity-adjustable technology that utilizes cyclic recognition reactions to amplify the signals step by step. This stepwise cyclic amplification of signals (SCAOS) technology can facilitate not only the adjustment of amplifying magnification but also optimization of other parameters.

6.2.2.1 Principle of Stepwise Signal Amplification

SCAOS is able to selectively add more mass onto a target analyte through a cyclic recognition reaction that can be operated step by step and stop at any step what you desire. To reduce the disturbance to the analytes as much as possible, a recognition reaction should be utilized. Although there are various pairs of lectin-saccharide recognition reactions, only a few reactions can be controlled stepwise, of which Con A-based recognition is selected because it is directly applicable to the amplification of mannose, glucose, fructose, their oligomers and polymers, and their conjugates. After combination with tagging technology (e.g., labeling analytes with either Con A or its recognizable saccharides such glucose), SCAOS can be universalized to the analysis of other saccharides and non-saccharides. Con A-based SCAOS is also cost-effective since Con A is the cheapest chemical among lectins (2374 ¥ RMB or ca. 400 $ per 500 mg at present).

Although Con A exists as a dimer at pH $\leq$ 5.0, it is normally a tetramer in neutral and alkaline solutions. The tetramer has an approximate tetrahedron conformation with four recognition sites at the four vertexes [74–76]. It can hence recognize more than one α-mannopyranosyl or α-glucopyranosyl residues, which enables stepwise recognition design and manipulation if the residues have also multiple recognition sites. This is easily achieved by use of a cheap, large saccharides such as dextran that is produced in tons and can react with Con A under mild conditions.

To recognize captured saccharides, Con A is applied to form recognition complexes with their targets by its downside two recognition sites, leaving the upper two recognition sites free. Once dextran is applied, it will further react with the Con A to saturate its upper recognition sites. Since a dextran molecule has numerous free recognition sites, it can further recognize with Con A if added again. Clearly, by alternative or cyclic addition of Con A and dextran, the captured saccharide molecules will grow up layer by layer unless either Con A or dextran are not more added. The key steps are illustrated in Fig. 6.11.

6.2.2.2 Practical Imaging Method

SCAOS-combined SPRi or SCAOS-SPRi method can hence be established. Its applicability has been validated in SPRi of dotted small (glucose, mannose and maltose) and large (ovalbumin, OVA) analytes, including the determination of standard glucose and carcinoembryonic antigen (CEA, a broad biomarker for cancer diagnosis). As expected, the amplification can easily be controlled to reach high sensitivity or to stop at any a given step for desiring. It was found that, in practice, the effective steps were usually within 20 steps. The limitation comes first from the restricted propagation depth of SPPs and second from the non-specific adsorption on the sensing chip surface that increases the background signals also with the amplification steps. The maximum amplification of net signals is normally obtained at 5–7 cycles of reactions. The key steps for performing SCAOS-SPRi of saccharides are outlined in Approach 6.13.

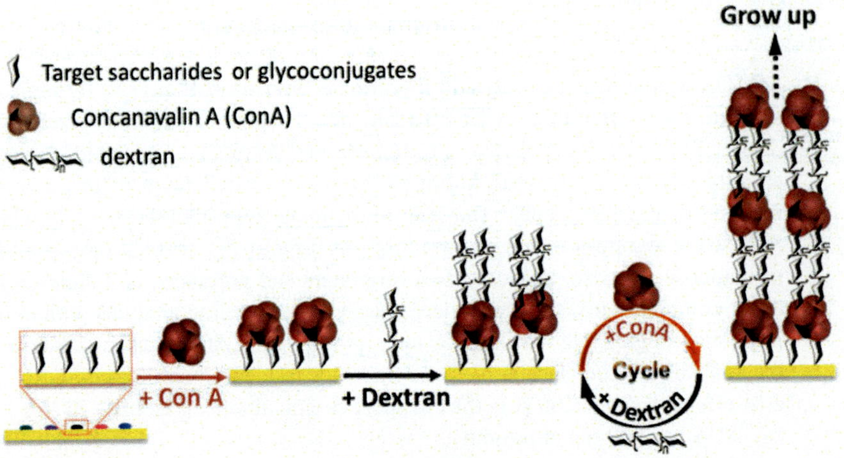

Fig. 6.11 Stepwise or cyclic amplification of signals in SPRi of saccharides measured on SPRi-TX7100. Reprinted from Ref. [77] with permission

Approach 6.13 for SCAOS-SPRi of saccharide samples:

Step 1 spot or load saccharide samples on a CC- or aptamer-terminated gold sensor chip. Note: (i) The aptamer-terminated chips selectively capture their own ligands for example, CAE and its aptamer(s), that can be prepared by deposition HS-aptamer solution (normally 1 μM in a buffer of 10 mM Tris and 1 mM EDTA at pH 7.8) on a cleaned gold chip or by incubation of the chip in an HS-aptamer solution at room temperature for 4 h, and (ii) CC-terminated chips can be prepared by **Aproach 6.11** that universally capture hydroxyl substances.
Step 2 block the chip in 10 mM aqueous MCH for 2 h or overnight for analytes in complex matrices.
Step 3 install the chip in the SPRi flow cell.
Step 4 equilibrate the chip by pumping in a Tris-HCl buffer (25 mM Tris-HCl at pH 7.6 containing 1 mM $CaCl_2$, 1 mM $MnCl_2$, 100 mM NaCl and 0.1% Tween 20, ConA-free) at 30 μL/min until imaging signals become stable.
Step 5 stepwise pump in solutions, at 30 μL/min, in an order of 15 min ConA (usually at 50 μg/mL but variable between 25 and 1000 μg/mL), 10 min Tris-HCl buffer, 10 min dextran (wt. = 40–40000 k, usually at 1 mg/mL but variable between 0.5 and 5 mg/mL) and again 10 min Tris-HCl buffer. Note: the adding order of Con A and dextran must be reversed when the analytes are dextran-recognizable.
Step 6 stop pumping in Con A and/or dextran once the recorded images are clear enough.

6.2.2.3 Key Conditions

To effectively perform SCAOS-SPRi, some key conditions need to be optimized or well-controlled. The first one concerns with the concentration of Con A that can yield much stronger imaging signals than dextran. In fact, it is Con A that serves a signal contributor in SCAOS rather than dextran that acts as a magnification factor. Dextran may be very large, up to at least hundreds of times more than Con A (wt. = 40,000 k), but it produces only a very weak signal. Figure 6.12 shows that a 50-μg/mL Con A solution can yield about 23 folds (or 230 folds per mg/mL) of higher imaging signal than a 0.5-mg/mL dextran solution. It is hence reasonable to first maximize and/or regulate the imaging signals through adjusting the content of Con A. Figure 6.13 shows that Con A increases the imaging intensity like a type of logarithmic function curve, with a turning corner at about 50 μg/mL that looks independent on the mass and concentration of dextran. This sets a concentration threshold for Con A to compromise the operation cost and imaging sensitivity. If the cost is not a problem, more condensed Con A is suggested to be used that can be up to at least 1 mg/mL. Condensed Con A is specially preferred when small dextran is utilized.

However, it should be noted that this does not necessarily mean that an even dilute Con A cannot be used. In contrast, diluted Con A can improve its utilization efficiency in SCAOS. Figure 6.14 shows that the signal increment per unit concentration of Con A decreases obviously with the concentration of Con A and the amplification cycles

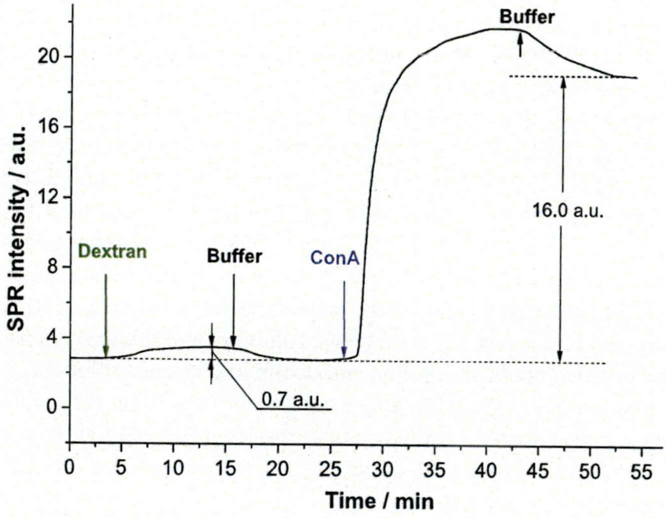

Fig. 6.12 SPRi sensorgram measured from 0.5 mg/mL 40 k dextran, buffer and 50 μg/mL Con A pumped in sequentially. Note, the solution pumping order must not be reversed otherwise the non-specific adsorption of Con A will impact on the measurement. Reprinted from Ref. [77] with permission

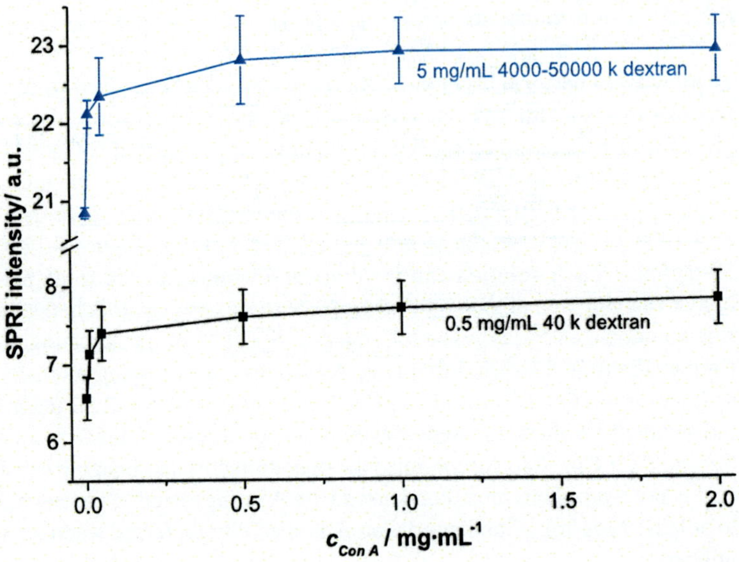

Fig. 6.13 Impacts of dextran concentration and molecular mass on SPRi intensity. Reprinted from Ref. [77] with permission

as well. Although the dilution of Con A can increase the efficiency, it takes more amplification cycles in order to reach a same level of sum signal. For example, to have an imaging intensity of about 40 a.u., about 5, 4 and 3 cycles of SCAOS are required at 25, 50 and 100 μg/mL Con A, respectively. In general, it is not suggested to dilute Con A to below 25 μg/mL.

The second key condition concerns now with the mass and concentration of dextran. As a magnification factor, dextran can multiply the Con A-produced signals by a factor parallel to its binding sites that increase with its molecular mass (Fig. 6.15) and concentration (also refer to Fig. 6.13), or simply proportional to the total number of monosaccharide residues. Thus dextran should be as large as possible to capture more ConA on a same analyte spot, and some concentrated solutions of large dextran would be preferred to reach high magnification of SCAOS-SPRi signals. Unfortunately, this is in practice limited by the solubility and solution viscosity of dextran. In order to speed up its dissolution and obtain low viscous solutions, so as to facilitate the solution pumping and to accelerate on-chip reactions, the dextran mass is better limited to less than 50 million, not surpassing 100 million. In general, a 1.0 mg/mL solution of dextran with molecular mass between 5 and 40 million (or 5000 and 40000 k) can work well.

Although dextran itself yields only weak SPR response, it can largely increase the imaging signals after combined use with Con A, which also gives a similar trend to Con A with a turning corner at about 50 μg/mL. Figure 6.13 reveals that the signal curves gradually progress to a horizontal asymptote as the dextran surpass 0.50–1.0 mg/mL. It is hence suggested to use 1.0 mg/mL dextran or above to perform

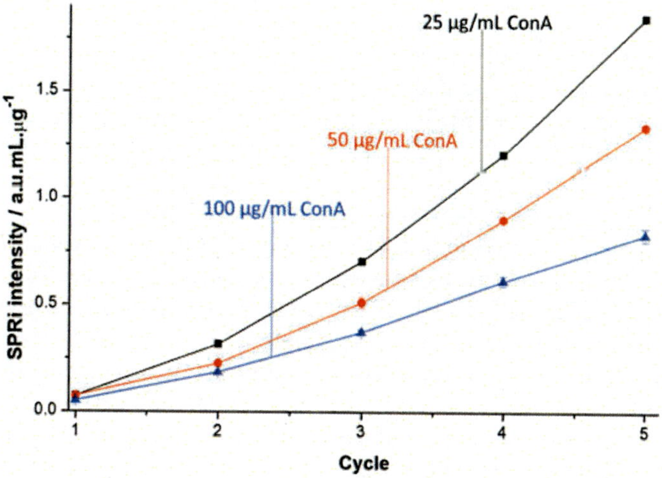

Fig. 6.14 Net contribution of Con A to SCAOS-SPRi intensity increases reversely with the concentration of Con A measured at 1 mg/mL dextran (wt. = 5000–40,000 k)

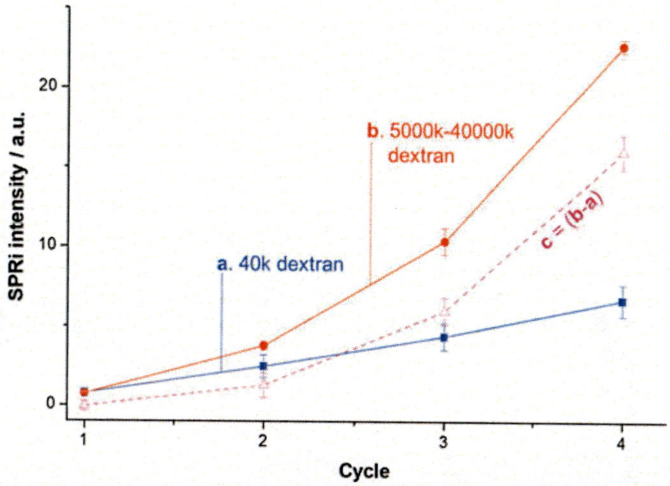

Fig. 6.15 Variation of SPRi intensity with dextran mass and amplification cycles, where the dot line (**c**) is the difference of line (**b**) and line (**a**)

reliable SCAOS-SPRi. Dextran is easily available at a very low cost, and its solution at several mg/mL remains easy to flow through the SPRi measuring system.

The third key condition concerns with the amplification cycle that is the dominating factor to increase the imaging signals in SCAOS-SPRi. Figure 6.16 shows that all the imaging intensity increases with the amplification cycles in a form of flattened S-shape. The curves increase very fast between the cycle 3 and 10, and then urn gradually to plateaus after about 10 amplification cycles. The curves also increase with the spotting concentration and molecular mass of analytes that determine the magnifiable capacity of given dots. Based on these three features, we can decide that, (i) the saturation-associated plateau areas, which are normally stable, can serve for reliable quantitative analysis; (ii) the sensitivity and hence the contrast of images can be regulated by the amplification cycles; and iii) the most sensitive amplification cycles need to be found after deduction of the background that increases in a similar trend to the signal (compare the bottom line with uppers in Fig. 6.16).

To vividly illustrate the most sensitive cycles, Fig. 6.16 is replotted after deduction of the background line measured at 0 μM analytes as shown in Fig. 6.17. By the peak-shaped curves, it is easy to understand that the most sensitive cycles to take the highest contrast images are at the positions of peaks. In the case of spotted glucose, the highest contrast spot images must be taken between the cycle 4 and 7, better at the cycle 6. Nevertheless, five cycles are routinely adoptable to save time and measuring cost since one more cycle will not necessarily increase a large portion of imaging signals. The signals again change gradually to a plateau after about 10 cycles of amplification. Note, in the plateau areas, highly reproducible quantification can be performed but at the cost of lowering some sensitivity.

It should however be noted that the signal-maximal position depends not only on the type of spotted analytes but also on the blocking method used to avoid the non-specific adsorption on the sensor chip surface. The blocking measure fatally determines that whether the experiment can be conducted further. This is the fourth key condition that can never be omitted in performing SPRi experiments.

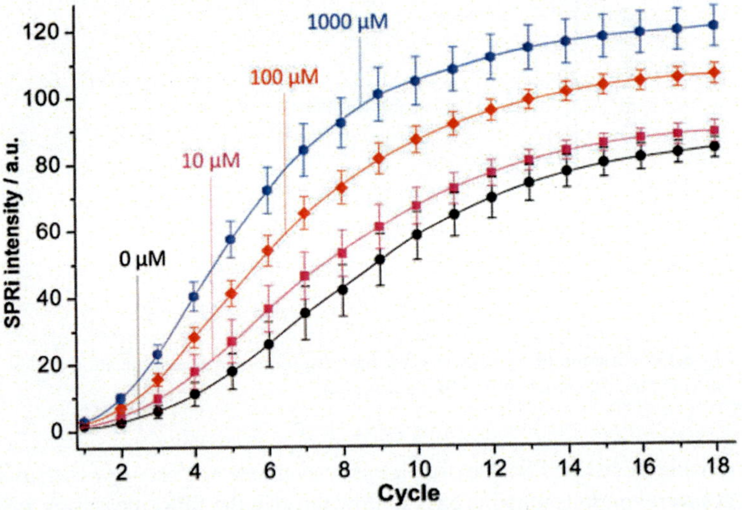

Fig. 6.16 Increase of SPRi intensity against amplification cycle measured after spotting 0, 10, 100 and 1000 μMglucose on a same gold sensor chip that is pre-blocked with ethanolamine. Reprinted from Ref. [77] with permission

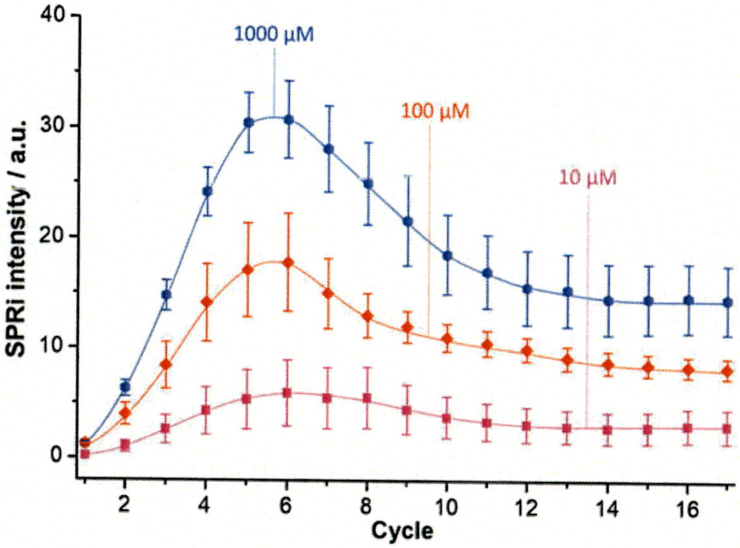

Fig. 6.17 Variation of background-deducted SPRi intensity with amplification cycle calculated from Fig. 6.16 by subtracting the 0 μM glucose signals from the related signals. Reprinted from Ref. [77] with permission

Figure 6.18 shows that the peak signal position varies reversely with the molecular mass of analytes (e.g., OVA ahead of glucose). By comparing Fig. 6.17 with Fig. 6.18, it can be found that a large chemical blocker shifts the peak to more cycles compared with a small blocker, for example, thc large blocker of BSA finds the peak cycle, for the analyte of glucose, at about cycle 9 or 10 that doubles the peak cycle 5 or 6 in use of small EOA as a blocker. Notably, the large blocker of BSA can significantly reduce the peak signals, at least 2 folds compared with the small blocker of EOA in imaging the glucose spots. All these suggest the use of small rather than large chemical blockers to conduct SCAOS-SPRi.

The key reason lies in that SPPs are a nonlinear function of different variables including analytes and non-analytes. The signals of analytes and background (or more exactly the 0-concentration) spots are the sum of SPPs from all composition in the spots, including sensor matter (majorly glass and gold film), non-specifically adsorbed chemicals, and nearby solution (hundreds of nanometers thick). They form different peak-shaped SPRi curves due to the variations of molecular permittivity and mass. They may yield complex combinations, of which the non-specific adsorption is normally a dominator and is also amplified cycle by cycle. Although the non-specific adsorption can greatly be suppressed by use of chemical blockers, they impose their SPPs on that of analytes and in turn shift the SPR value of analytes to a new position after their combination. Unfortunately, BSA is a strong SPR substance that can have a significant portion involved in the measured analyte signals that will vary with the type and concentration of analytes. This explains the variations of the maximal amplification cycles.

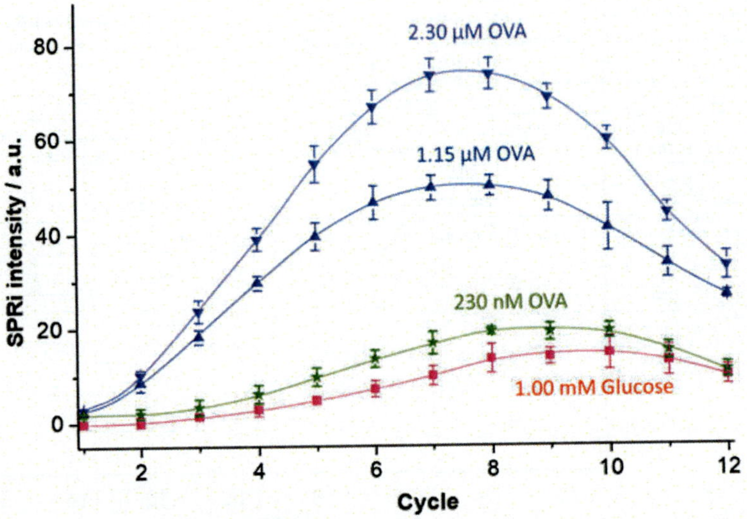

Fig. 6.18 Net increase of SPRi intensity against amplification cycle measured on a chip blocked with 0.4 mg/mL BSA for 3 h

A large chemical blocker can bury the small analytes beneath its adsorption layer and in turn obstructs or even "block" the amplification reaction to occur. This is responsible for the signal loss. To ensure SCAOS performance in the detection of all small analytes, much small chemical blockers need to be used, better at a size similar to or even smaller than that of an analyte. It has been shown that small amines can strongly adsorb on gold surface, and among them, EOA works well in SCAOS-SPRi of small saccharides, and MCH works in SCAOS-SPRi of small DNA fragments like aptamers that may be needed to capture glycans and their conjugates. EOA and MCH are fairly effective at the first two amplification cycles but they may then quickly lose their efficiency. These small blockers may reversibly adsorb on the surface sites and are hence replaceable by other absorptive chemicals such as Con A. This weakens their ability to suppress the non-specific adsorption. This loss can be recouped by combined use of small and large blockers after two cycles of amplification.

In addition to the discussed four key conditions, the normal SPRi running conditions should also be optimized, which has been generally discussed in Chap. 4. They are the selections and/or control of buffer (its composition, concentration, and pH value), flow rate, temperature and sensor chip (e.g., surface modification and cleaning), and development of sample preparation method and spotting technique. After optimization of all the conditions, clear images of spotted and/or spot-captured analytes can be acquired at the maximal amplification cycles as shown in Fig. 6.19.

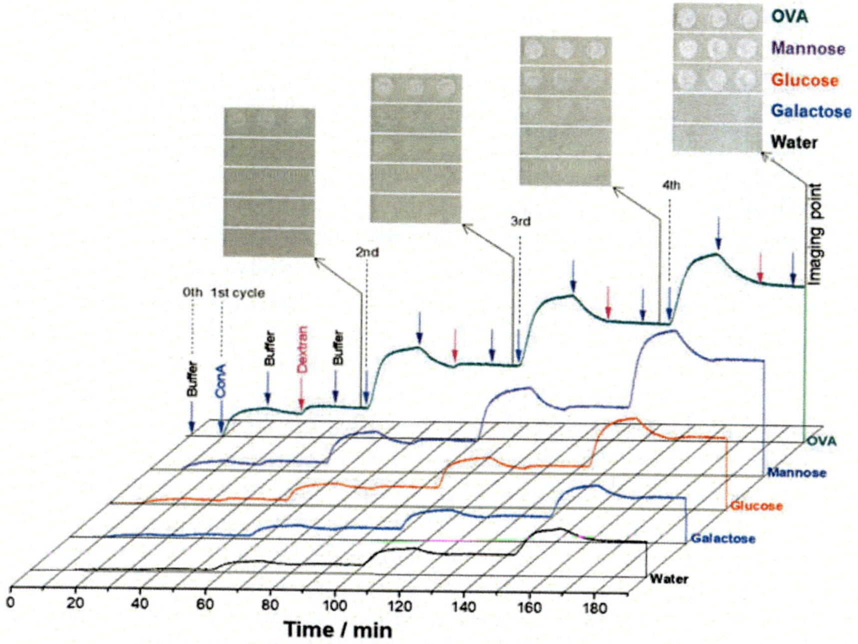

Fig. 6.19 Subtracted dot images (upper) and relative real-time intensity curves (lower) of saccharides spotted at 50 mM Buffer (free of Con A): 1 mM $CaCl_2$, 1 mM $MnCl_2$, 100 mM NaCl, 0.1% Tween 20 and 25 mM Tris–HCl at pH 7.6; Flow rate: 30 μL/min. Reprinted from Ref. [77] with permission.

6.2.2.4 Applications

Based on Con A-dextran recognition pair, SCAOS-SPRi is directly applicable to the display of spot images of saccharides such as mannose and glucose, their oligomers and polymers, and their conjugates [60]. As known, small saccharides can hardly be imaged by common SPRi method but this is not more problem by SCAOS-SPRi. Figure 6.19 shows that, except for OVA (a glycoprotein rich in mannose), there is nearly no image recognized even after reaction with Con A (corresponding to the first or first half cycle of signal amplification). Dim images of mannose and glucose can be obtained after two cycles of signal amplification, and their clear images are acquired at the amplification cycle 4. As expected, the Con A-irrecognizable galactose remains not displayable even at the 4th amplification cycle. Particularly, the large analyte of OVA can also be amplified but seemingly becomes saturated at about cycle 3, which depends in fact on the concentration spotted. Normally condensed saccharide spots need less amplification cycles to image than dilute ones.

Besides image display, SCAOS-SPRi is applicable to the quantitative analysis of both small and large mannose- and glucose-associated analytes that are either immobilized or captured on a gold sensor chip. To perform reliable quantification, the amplification cycles need to be selected or optimized. For demonstration, glucose is exemplified and covalently spotted on a chip with CC as a chemical linker [53]. The glucose spots cannot be imaged directly by SPRi even at a concentration up to 100 mM, while by SCAOS-SPRi, glucose at a spotting concentration of 100 μMglucose can be imaged even after only one cycle of amplification. The spot images are improved at the cycle 2 and more, giving linear working equations between the logarithm of image intensity (log I) and that of concentration (log c) as shown in Table 6.1. It is reasonable that the SCAOS-SPRi-visible spotting concentration decreases reversely with the amplification cycle (referred to LOD in Table 6.1). SCAOS-SPRi has also an upper limit of linearity at about 1000 μM, which is resulted from the limited capture capacity of CC and/or other probes on a chip, from the limited penetration depth of SPPs, and from the upper limit of video camera. This upper limit makes the infinitive amplification of the signals impossible.

The linear range is in general less than two orders of magnitude. In SCAOS-SPRi, a measured image signal is actually a sum of the amplified SPRi response and CCD output. They both have their own linear range. As a consequence, the sum signal changes nonlinearly with the spotting concentration of analytes, usually as a deformed S shape. In this case, data-fitting technique can be adopted to widen the quantification range but at the cost of consuming time since the process is rather tedious. It should be mentioned that the linear correlation coefficient (given by R^2) can surpass 0.99 by either narrowing the concentration range or increase of the amplification cycles, which is also dependent on the size of analytes.

In addition to CC-captured pure saccharides like glucose, SCAOS-SPRi can also determine aptamer-captured analytes. The real applicability can be confirmed by determining CEA that is a broad-spectrum tumor marker [35] and a ConA-recognizable glycoprotein. To determine, CEA is selectively captured by its specific aptamer pre-immobilized on a gold sensing chip and imaged by SCAOS-SPRi. As a protein, CEA is directly detectable by normal SPRi but its concentration should be above 50 μg/mL. Simply after 1 or 2 cycles of signal amplification, SCAOS-SPRi can determine CEA in sample solutions down to about 20 ng/mL as illustrated in the second column in Fig. 6.20. Similar to the immobilized glucose spots, the captured CEA displays clearer and clearer spots images with the amplification cycles, with LOD down to 50 pg/mL in 10% serum solutions.

Table. 6.1 Quantitative features of SCAOS-SPRi tested by glucose spots

Cycle	Regression equation[a]	R^2	Linear range[b]/μM	LOD/μM
1	$y = 0.059x + 0.255$	0.5693	–	~ 100
2	$y = 2.120x + 0.360$	0.9886	50–1000	20
3	$y = 0.1818x + 0.800$	0.9939	10–1000	5
4	$y = 0.1765x + 1.065$	0.9903	5–1000	2.5
5	$y = 0.1470x + 1.290$	0.9986	5–1000	1.5

[a] y denotes log I where I is the imaging intensity with artificial unit (a.u.), and x denotes log c where c is the molar concentration (μM) of glucose

[b] The upper linear limit may vary that is dependent on the amplification cycles

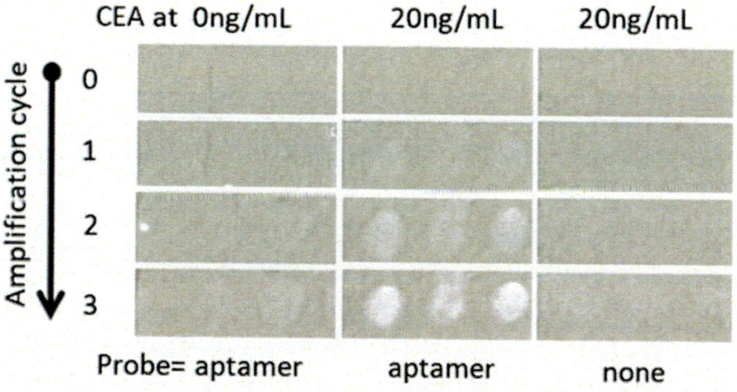

Fig. 6.20 SPRi of aptamer-captured CEA at a solution concentration of 0 (1st column) and 20 ng/mL (middle and the last columns) measured at the amplification cycle 0 (without the addition of Con A), 1, 2 and 3 (the line from top down), using a non-aptamer (last column) as a negative control Aptamer. 5′-SH-$(CH_2)_6$-$(T)_{10}$-ATA CCA GCT TAT TCA ATT-3′; Non-aptamer. 3′-SH-$(CH_2)_6$-$(T)_{10}$-TTTT ACT TTG ATC GTG TCA AAA-5′. Reprinted from Ref. [77] with permission.

The linearity is measurable in each cycle, applicable to the determination CEA in real samples. It is exemplified by detection of CEA in sera from early-stage colon cancer patients, in a linear range from 500 pg/mL to 10 ng/mL (R^2 = 0.9968) at the amplification cycle 5 or from 2.5 to 20 ng/mL (R^2 = 0.9916) at cycle 4. CEA at 7.14 ng/mL is determined form patient serum samples, comparable with 6.82 ng/mL measured by a chemiluminescent method.

In addition to Con A-recognizable substances, SCAOS-SPRi is extendable to the analysis of dextran-recognizable substances and compounds that can be labeled by Con A- or dextran-recognizable substances. Figure 6.21 illustrates some examples, where special attention may be paid to the images of Con A-tagged lysine and dextran-tagged BSA. They are both not recognizable by either dextran or Con A, but they are imaged after they are labeled with either Con A or dextran. Note, this type of labeling strategy is universally combinable with all above-discussed signal amplification methods, not simply limited to SCAOS.

6.3 Analysis of Proteins

Proteins are comparatively easy to be imaged by SPRi to produce either monochromatic (Fig. 6.22a) or color (Fig. 6.22b) images, *though*, the latter images is normally not as clear as the former. The attractive feature for SPRi of proteins lies in its high throughput (Fig. 6.23) that enables the studies of different proteins under same conditions and has since simulated the development of SPRi. Because of the capability of direct readout of label-free proteins [78] and glycoproteins [60] captured or spotted on gold sensor chips [79], SPRi is applicable to the qualitative and/or quantitative (refer to Sect. 4.2.8.2) inspections of membrane protein–ligand recognition,

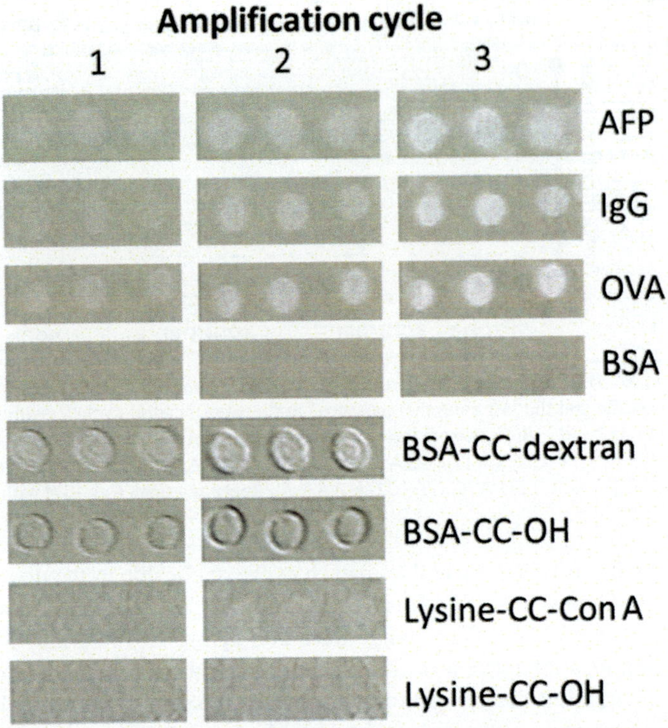

Fig. 6.21 SCAOS-SPRi of native alpha-fetoprotein (AFP), IgG, OVA and BSA, and dextran (10 k)-CC-BSA, HO-CC-BSA, ConA-CC-lysine and HO-CC-lysine spotted at 100 μg/mL on a CC-terminated gold sensor chip. Reprinted with permission from Ref. [77]

immunological reaction, and other reactions such as enzymolysis, hydrolysis and specific or non-specific adsorptions.

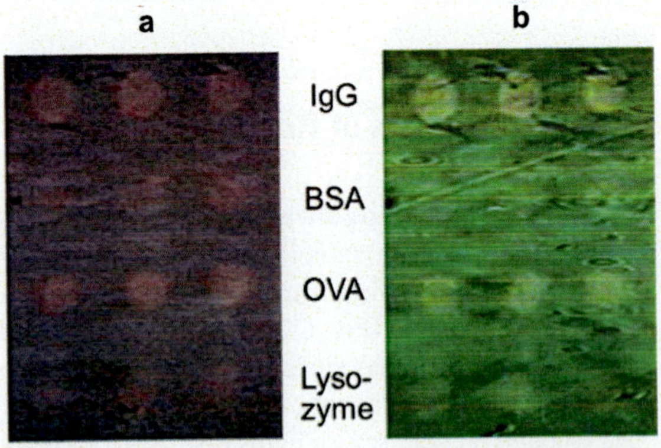

Fig. 6.22 SPRi of proteins spotted a same gold sensor chip measured using a same laboratory-built color SPRi device equipped with a color CCD and white light source. **a** Color CCD-recorded monochromatic image at 632.8 nm; **b** color image with color complementary to red orange

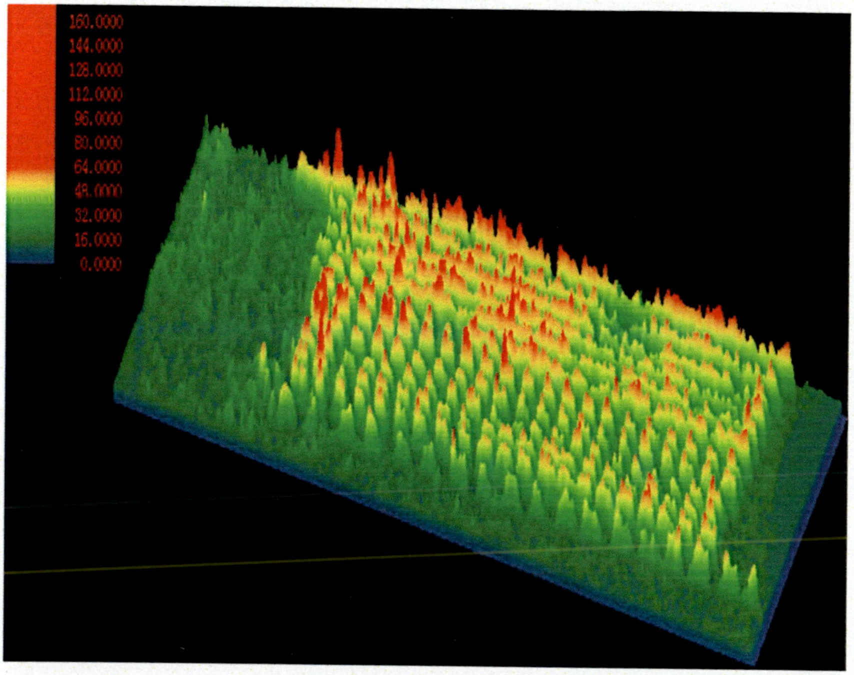

Fig. 6.23 False color image of protein spots recorded by SPRi-PX8100, as a high-throughput readout means

Because SPRi of proteins can have numerous reference and have been or will be discussed in different chapters, this topic is not going to be presented intensively here. The discussion herein will be focused on some issues that may easily be ignored but have serious impact on the performance of SPRi, such as spotting solution, imaging limit, lateral flow and diffusion, and background interference.

6.3.1 Spotting Solution

In addition to real-time monitoring of the behaviors of proteins including their hydrolysis [80] and/or denature (Sect. 8.5), SPRi can sense their subtle changes caused by some rare parameters that may not be of concern usually, for example, protein spotting solutions. While striving to study various signal amplification techniques to improve the sensitivity of SPRi in assaying proteins, we accidently found that protein spotting solution had non-negligible impact on the signal strength of SPRi. The impact was so obvious that we had to conduct in-depth research on it. We can now confirm that it is the pH value of a protein spotting solution that determines the imaging intensity, even more critical than the composition of the spotting solution. Figure 6.24 shows the dependence of imaging intensity on the pH value of spotting solutions, giving an order of imaging intensity as follows:

$$I_{pH2.03} \gg I_{pH11.08} > I_{pH>pI} > I_{pH\approx pI}$$

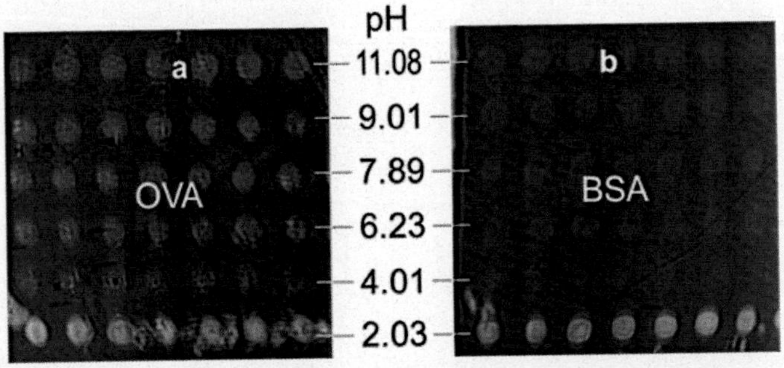

Fig. 6.24 Impact of pH value of spotting solution on SPRi intensity. **a** OVA (pI4.6) dissolved in spotting buffer at 1.0 mg/mL; **b** BSA (pI4.7) dissolved in the same spotting buffer at 1.0 mg/mL; Spotting buffer: 40 mM phosphoric acid, 40 mM acetic acid and 40 mM boric acid adjusted to the indicated pH with NaON; Immobilization chemistry: MUA activated by EDC/NHS; Blocking solution: 1 M ethanolamine

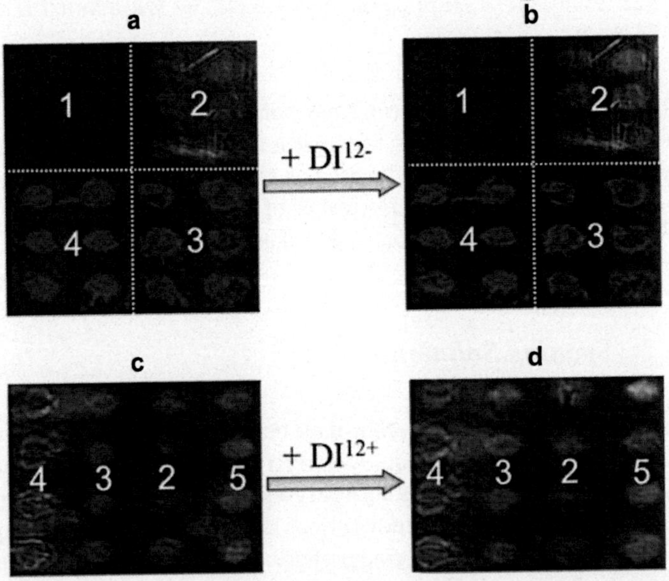

Fig. 6.25 SPRi of proteins spotted at 1.0 mg/mL and measured on SPRi-TX7100. **a** and **c** SPRi of spotted native proteins, **b** SPRi of the protein spots of **a** after electrostatic adsorption of multivalence dendritic anion of DI^{12-} at 2.0 mg/mL, **d** SPRi of the protein spots of **c** after electrostatic adsorption of multivalence dendritic cation of DI^{12+} at 1.5 mg/mL. Spotted analytes: 1, PB buffer at pH7.40; 2, Lysozyme (MW = 14.7 k); 3, Polyglutamine (MW = 14.3 k); 4, Polylysine (MW = 51 k)

This order is more or less universal for spotting proteins based on MUA/EDC/NHS chemistry. It looks reasonable that the most sensitive SPRi is obtained with a pH 2.03 spotting solution but we remain not very clear about why the minimum imaging intensity is at the pH value around pI. It may be attributed to the fast aggregation and sedimentation of proteins at pI. Anyway, for practice, these data suggest the utilization of acidic rather than basic spotting solutions. In both cases, the spotting solution is better adjusted to a pH far away from the pI value. For proteins without any pI information, an acidic spotting solution at around pH2.00 gives the priority to try.

6.3.2 Imaging Limit

The so-called imaging limit here is meant the SPRi detectable size of a protein, which is theoretically dependent on the type of proteins. Based on the fact that SPRS can directly sense proteins with molecular weight > 10 k [81], we can estimate that the SPRi-detectable proteins should be larger than 15 k that is about 1.5 folds of SPRS-detectable proteins. This was validated in our practice as shown in Fig. 6.25a and c. In order to lower the detectable limit of molecular size, down to smaller proteins, peptides and even amino acids in extreme case, signal amplification techniques have to be integrated into SPRi methods.

Accordingly, SCAOS-SPRi can generally be adopted in imaging all types of proteins including the very small proteins, peptides and even amino acids. This can be realized by pre-tagging or *in-situ*-tagging the analytes with mannose or glucose and then cyclic addition of Con A and dextran. For some cases, the plasmonic amplification technology is also adoptable in combination with EDC/NHS chemistry. For peptides and proteins with a size just around the imaging limit, their signals are possible to be amplified by some giant ions simply through electrostatic interaction. This can be exemplified by SPRi of lysozyme with a molecular weight at 14.7 k and polyglutamine with a molecular weight at 14.3 k. They (especially lysozyme) often yields faint images (spot 2 and 3 in Fig. 6.25a and c) compared with polylysine and BSA (spot 4 and 5 in Fig. 6.25a and c). Their signals can be obviously improved by an easy way of strong electrostatic adsorption of large multivalence ions, such as negative or positive dendritic ions as shown in Fig. 6.25b and c, respectively.

6.3.3 Lateral Flow and Diffusion

On an activated or reactive chip surface, it is easy to have a lateral flow and diffusion. We hence need to have effective measures to guarantee that the non-spots area can be free of unwanted reactions. However, this is a serious challenge because it constantly happens during spotting and in the blocking and washing steps. The fluidity of the spotted solution and the mismatched surface tension of the sensor chip will make the dotted composition instable, causing lateral flow and diffusion and in turn resulting in gradually expanding and blurring the periphery of all the sample spots (Fig. 6.24 and Fig. 6.26a, b and c). This effect can universally be suppressed by increasing the viscosity of the spotting solution (e.g., addition of glycerol or cellulose into the solutions) in combination with an appropriate surfactant (e.g., Tween 20) to adjust the surface tension. The effectiveness can be observed by comparison among Fig. 6.26a–d. It should be noted that the type and concentration of tackifier and surfactant have to be re-optimized during SPRi of different proteins (and other analytes as well).

Even with tackifier, the spotted spots will increase their mobility in contact with various solutions. This is just the case during the subsequent washing and blocking steps, where the spots will blur their rear boundary in the direction of insertion (Fig. 6.26a–c). Although this can be suppressed in combination with an appropriate surfactant, the blurring cannot be ideally removed (Fig. 6.26d). A better technique

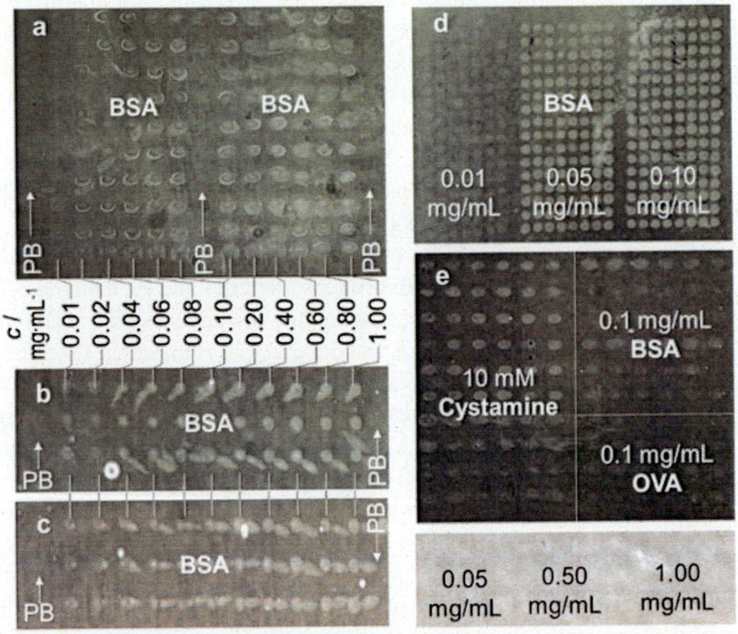

Fig. 6.26 Dependence of image sharpness on the spotting solution and chip-treating process. **a** The chip was spotted with BSA dissolved in PB at pH7.4 containing 30% glycerol, reacted for 12 h, and after the unspotted area was carefully dropped with glycerol, it was washed in stirred water for 30 min and then blocked by sinking in 1 M ethanolamine; **b** The chip spotted with BSA as in **a** was directly blocked by oblique insertion into 1 M ethanolamine; **c** The chip spotted with BSA in PB at pH7.4 containing 10% glycerol was blocked as in **b**; **d** The chip was spotted with BSA in PB at pH7.4 containing 10% glycerol and 0.1% Tween 20 and then treated and blocked as in **a**; **e** The chip was spotted with the indicated analytes dissolved in PB at pH7.4, frozen in liquid nitrogen for 1 min and immediately blocked in 1 M ethanolamine; **f** The chip was spotted with sheep IgG dissolved in 10 mM PB at pH7.0 and treated and blocked as in **e** that is reprinted from Ref. [79] with permission

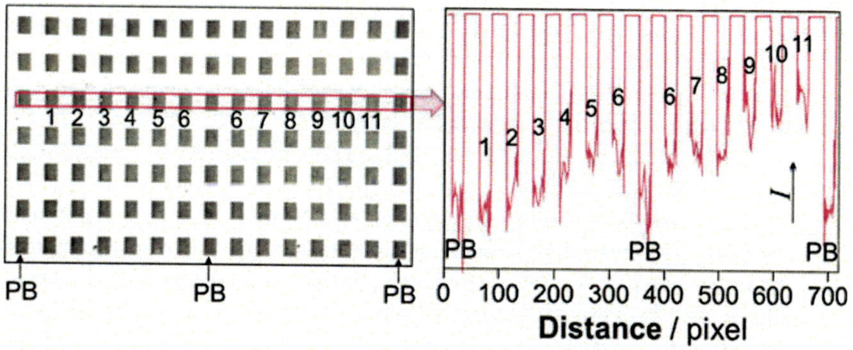

Fig. 6.27 SPRi of BSA spotted on etched microsquare gold islands to produce (left) clear image free of surrounding impact or (right) concentration-dependent intensity curve Spotted sample: 1 = 0.01, 2 = 0.02, 3 = 0.04, 4 = 0.06, 5 = 0.08, 6 = 0.10, 7 = 0.20, 8 = 0.40, 9 = 0.60, 10 = 0.80, and 11 = 1.00 mg/mL BSA dissolved in PB at pH7.4 containing 30% glycerol.

developed and routinely in our laboratory is to freeze the spotted chips in liquid nitrogen just before washing and/or blocking the chip. This frozen technique is independent on the nature of spotting solution and looks universally applicable to both of gray and color SPRi as illustrated in Fig. 6.26e and f.

An alternative technology is to border the spots with either an SPPs-eliminator or a substance with polarity opposite to the spotting solutions. The former can be realized by etching off the gold film along the boundary while the latter can be achieved by coating an oppositely polar or inert material on the boundary. The coating technology is also able to cover or erase the effectiveness of SPPs through shifting the surrounding SPR signals far away from the spots. The coating and etching techniques are both required to couple with lithographic technology and in many case they are also combined with each other. By these bordering techniques, the lateral flow and diffusion effect is not more an issue, and sharp images can easily be acquired with intensity still proportional to the spotted concentration (Fig. 6.27). Clearly, these bordering techniques are also universally applicable. The disadvantage is, however, that the treating process complicates the manipulation and raises technical threshold and cost as well. They are also not possible to eliminate the impacts of the bottom layers of spots, their coverings and the in-spots-penetrated substances.

Another potential alternative technique is to explore solid reactions if available, but this may still take time since we have discovered nothing reported till now, to the best of my knowledge.

6.3.4 Background Interference

Imaging background is a common issue in SPRi but is worsened in imaging proteins since they show stronger adsorptive features than other molecules and easily increase the background or reduce the usable signals of target analytes. This not only causes the reduction of image contrast but also lowers the quantitative ability of SPRi as indicated in Chap. 4. Therefore, the background has to be removed to highlight the net signals of target spots for reliable analysis.

To remove the background, we need to know its sources. By careful observation of a spot on a sensor chip, we can recognize that the background originates from the base beneath a spot, its surroundings, its coverings, and the in-spot penetrating substances. The sources increase not just the intensity of background but also vary the image hue in color SPRi. Obviously, the bordering technology can clear off the surrounding background but cannot remove the remaining three sources (e.g., the base beneath the spots, and the solvent and/or solutions covering and penetrating the spots). Thus background deduction is unavoidable and has become a universal technique in getting rid of the interference of background in SPRi to obtain the real color (Fig. 6.28a, b and a′) or net gray (Fig. 6.28c, d and c′) images of target spots. Background deduction is also a unique technique to flatten the imaging surface and hence to have better quantitative performance (refer also to Sect. 4.2.8).

6.4 Imaging the Fingerprints

As a unique and immutable mark, fingerprints are a "gold standard" for personal identification, originally utilized in forensic investigation and law enforcement, while now expanded to many other occasions such as access control [82, 83]. As known, fingerprints can now be displayed by physical and chemical imaging techniques. In a detective story, the latent fingerprints are often revealed by spraying some powders on a suspicious object. This is a physical imaging technique. It looks so easy but not as sensitive as chemical imaging that needs various chemical principle (e.g., reactions) to make the substances in the fingerprints luminesce [84].

A fingerprint has at least three levels of structure: pattern, minutia points, and pores and ridge contours, which can now be obtained simply physical imaging techniques and analyzed by computer. Chemically imaging the fingerprints focuses then on acquiring the chemical information, including endogenous metabolites and exogenous substances, in addition also to the physical structures.

SPRi is in principle an optical or physical imaging technology but the optical features are induced by related chemicals on or vicinity to the sensor surface, it is hence a comprehensive analytical method able to image the potentially complicated and/or trace chemicals in a fingerprint associating with not only the structure but also health conditions such fingerprint age [85], gender [86], use of illicit drugs [87], contact with explosives [88] or motion state [89].

6.4.1 Basic Considerations

The significant advantages in SPRi of fingerprints still lie in the label-free and high-throughput observation and quantification of fingerprinted chemicals, with also temporal and spatial resolution. Unfortunately, SPRi alone is insufficient to directly image the fingerprint chemicals in our trials because the printed chemicals are usually smaller than 1000 and are of low abundance but very complicated if taking into consideration of only sweat metabolites [90]. Signal amplification can never be avoided in SPRi measurement of fingerprinted chemicals. SCAOS-SPRi is adopted without hesitation.

In order to image the potential metabolites in fingerprints brought out due to sweat, a CC-terminated chip is used that can capture all hydroxyl and amino components, including basically saccharides (majorly glucose), amino acids and peptides. Saccharides involve in many vital physiological processes but remain very "tough" to analyze due to the lack of powerful tools to track their biological functions that vary from cell recognition to metabolic fuel [91, 89]. They are hence suspected to appear in fingerprints and can hopefully be imaged by SPRi.

In practice, glucose is taken as a representative to develop the method for acquisition of high-fidelity images of fingerprints. It is then extended to image other substances that contain ConA- or dextran-recognizable groups or can be labeled with corresponding groups. SCAOS-SPRi can then simultaneously image glucose and amino acids (as representatives) and is applicable to monitoring the exercises-associated metabolic secretions in fingerprints.

6.4.2 Basic Approach

To image the representative chemicals in a fingerprint by SPRi, sensor chip with CC-terminal is prepared and fingerprinted. Before mounted on the measuring chamber to conduct SCAOS-SPRi, the chip may be totally or partially tagged with glucosamine via EDC/NHS chemistry to image the non-saccharides with COOH-terminal by stepwise reactions with con A and dextran. There are in total seven distinct steps as illustrated in Fig. 6.28. The detailed manipulation steps are given in Approach 6.14.

Approach 6.14 for SCAOS-SPRi of chemicals in fingerprints:

Step 1–2 clean and modify a gold sensor chip first with MUOH and then with CC (refer to Sect. 4.5).

Step 3 capture the chemicals from a fingerprint onto the chip by touching the fingertip on the chip under a constant press at 300 g. Note, a special sampler made of a digital tray balance is suggested to avoid the pressure instability and variation caused by subjective pressing. In study of sweating metabolites, the finger should be washed first with medical alcohol, then with pure water and finally sucked dry with a quantitative filter paper or blown dry under nitrogen gas.

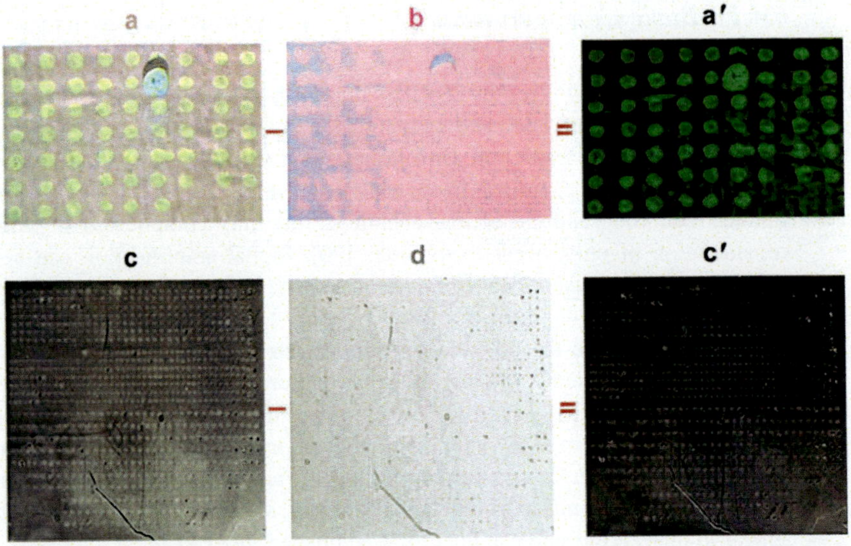

Fig. 6.28 Background detection to have net images of protein spots. **a** Raw color image of IgG spotted at 5 mg/mL; **b** Raw color image of buffer spots; **a′** Background-removed image with a green color different from the raw one; **c** Raw black-and-white image of BSA spotted at 1 mg/mL and recorded under p-light; **d** Raw image of the same BSA spots recorded under s-light; **c′** Image subtracted by **(c–d)**

Step 4 place an isolation wall on the chip to segregate the fingerprint into two parts that will not leak from each other.
Step 5 activate one of the two parts with EDC/NHS and tag the activated part with glucosamine.
Step 6 mount the chip in the flow cell on SPRi device, wash the chip with an imaging buffer and pump in con A and dextran alternatively for 5–7 cycles.
Step 7 record images after the chip is mounted or at any time required.

This approach has been validated to be capable of direct imaging the glucose representatives captured from the fingerprints (Fig. 6.30). The CC-terminal can capture both hydroxyl and amino substances, including both saccharides and non-saccharides, e.g., glucose, mannose (if exists), their oligomers, amino acids, peptides, and even proteins. They are normally at a trace level and usually cannot be directly detected by SPRi alone unless SCAOS technique is integrated.

It is expectable that the trace non-saccharides captured will give nothing images by direct SPRi and SCAOS-SPRi. Nevertheless, they will become imageable if they can be tagged with either ConA- or dextran-recognizable substances. To demonstrate, standard chemicals that may appear in fingerprints are selected and tried, including especially the molecules with hydroxyl and/or amino for CC capture (Fig. 6.31a and b) and carboxyl group for tagging with glucose by esterification or with glucosamine by amidation (Fig. 6.31c). Because the amidation can easily be activated by EDC and NHS under mild conditions, it is a better option in imaging the

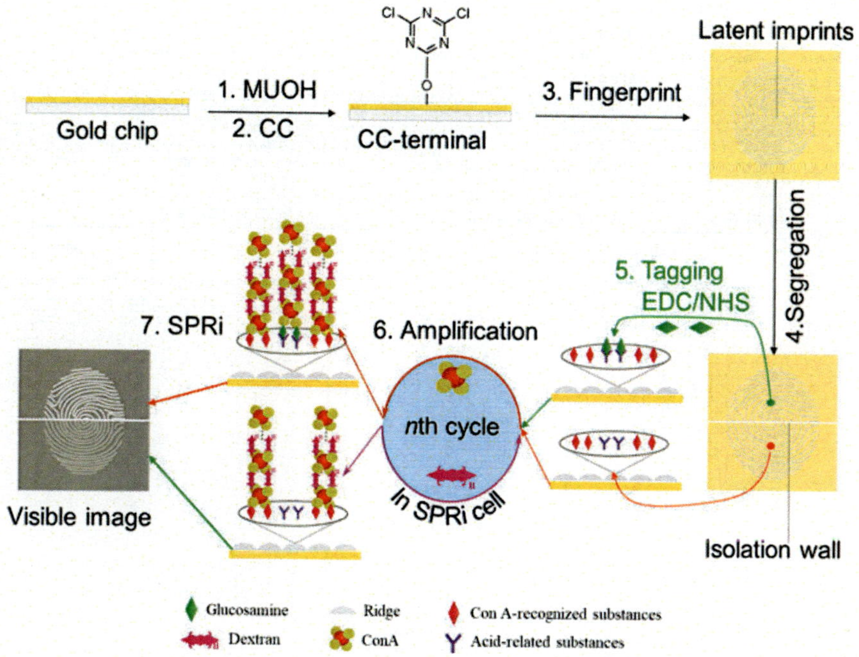

Fig. 6.29 Distinct steps for SCAOS-SPRi of fingerprint. Reorganized from Ref. [89] with permission

fingerprints than the esterification. Therefore, it was adopted in our trials, resulting in expected images (Fig. 6.31d, e and f), where all the spots show nearly invisible images by direct SPRi (Fig. 6.31d), while the con A-recognizable analytes, hIgG, CEA, glucose and maltose, can be imaged by direct SCAOS-SPRi. As expected, all the COOH-containing chemicals are also imaged by SCAOS-SPRi after tagged with glucosamine.

From the practical aspect, this tagging technology is also validated to be capable of imaging the non-saccharides in fingerprints. Figure 6.32a illustrates that the tagged images (the lower halves) are brighter than the tag-free one (the upper halves). Figure 6.32b and c shows the significant variations along the position line L1 and L2 while Fig. 6.32d gives a quantitative comparison using the signals averaged within the square S1 (tag-free) and S2 (tagged), respectively. The increments after amidation are significant, suggesting the real existence of non-saccharides containing carboxyl groups. The existing quantity must be implied in the intensity of S1 and S2 or their difference.

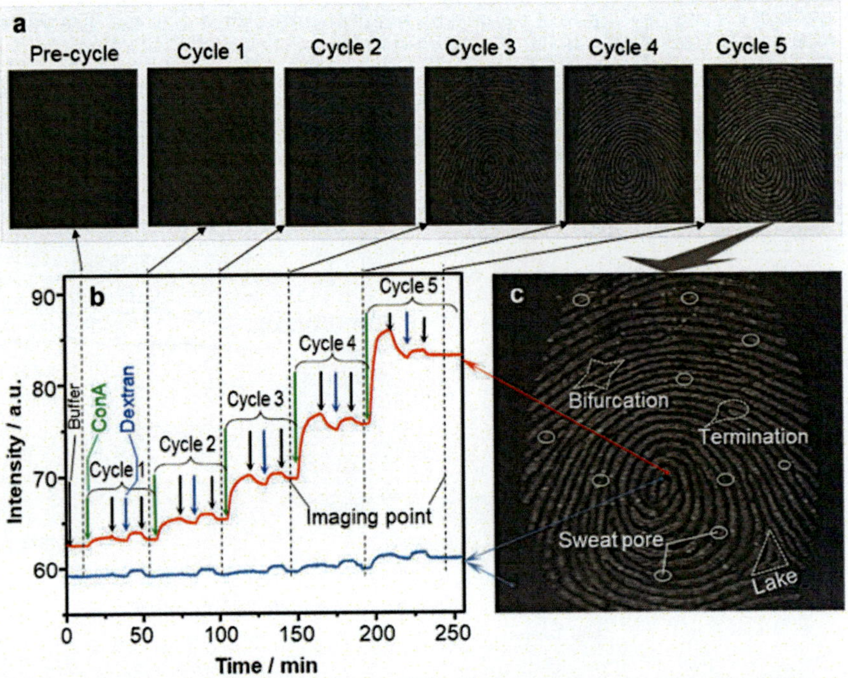

Fig. 6.30 Direct SCAOS-SPRi of saccharides in a fingerprint. **a** Chemical images recorded at the indicated amplification cycles; **b** Intensity variation with the amplification cycles averaged from the ridge (upper curve) and valley (lower curve as a background), respectively; **c** a high-resolution and sharp image recorded at the 5th cycle that shows clearly the first-level detail of whorl, the second-level details of bifurcation, termination and lake, and the third-level details of sweat pores, as indicated, respectively. Reconstructed from Ref. [89] with permission

6.4.3 *Quantitative Analysis*

Quantitative analysis of the fingerprinted biochemicals may be as useful as the analysis of sweat samples in respect disease diagnose, health monitoring, or nutrition tracking. Because the signals of SCAOS-SPRi depends on the amplification cycles, the quantification should be conducted at a given cycle so that the signals can change with the concentration of a representative analyte. To correlate the imaging signals with the quantity of an analyte, averaged signals must be calculated from an interested area by:

$$\overline{S} = \frac{\sum_{S=S_{\mathrm{LOD}}}^{255} n_S(S - \overline{B})}{\sum_{S=0}^{255} n_S} = \frac{1}{N} \sum_{S=S_{\mathrm{LOD}}}^{255} n_S(S - \overline{B}) \tag{6.1}$$

where S is the intensity value of a pixel within the interested image area (normally a rectangle or a square covering more than 6 ridges is selected for easy count of pixels),

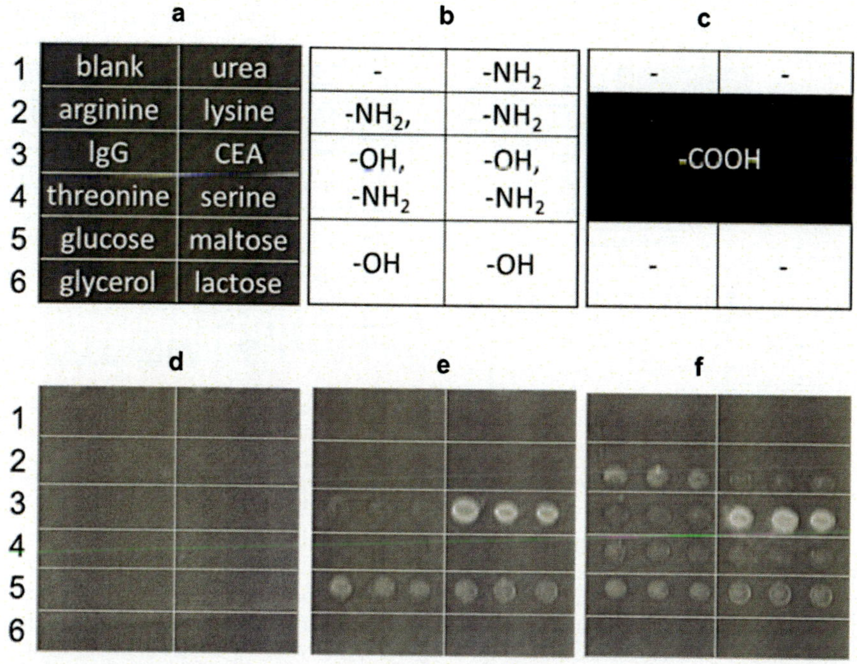

Fig. 6.31 Validating the effectiveness of SPRi and SCAOS-SPRi in detecting the labeled and label-free standard chemicals. **a** Spotting location of 11 standard chemicals potentially included in fingerprints; **b** chemicals with groups for CC capture; **d** chemicals with group for amidation; **d** direct SPRi measurement; **e** SCAOS-SPRi of label-free standard spots; **f** SCAOS-SPRi of glucosamine-tagged standard spots; Blank: A solution of 0.5% NaCl adjusted to pH 6.6 with NaOH; Other spots: Urea, lysine, arginine, human immunoglobulin G (hIgG), carcinoembryonic antigen (CEA), threonine, serine, glucose, maltose, lactate and glycerol solutions at 1800, 7.00, 17.0, 2.00, 0.20, 25.0, 100, 30.0, 30.0, 1000 and 70 μg/mL, respectively. Reorganized from Ref. [89] with permission

the upper limit of 255 indicates that the formula applies to 8-bits digital system; n_s is the number of pixels with a same intensity of S, N is the total pixels enclosed by the rectangle, S_{LOD} denotes the intensity limit of detection with a confidence level above 99%:

$$S_{\mathrm{LOD}} = \overline{B} + 3\sigma_B \tag{6.2}$$

where $\overline{B}$ is the averaged background intensity calculated from a same sized area set close to the images, for reducing bias, four areas are better averaged along the four directions of the fingerprint; σ_B is the standard deviation in measuring the background signals. They can be calculated by

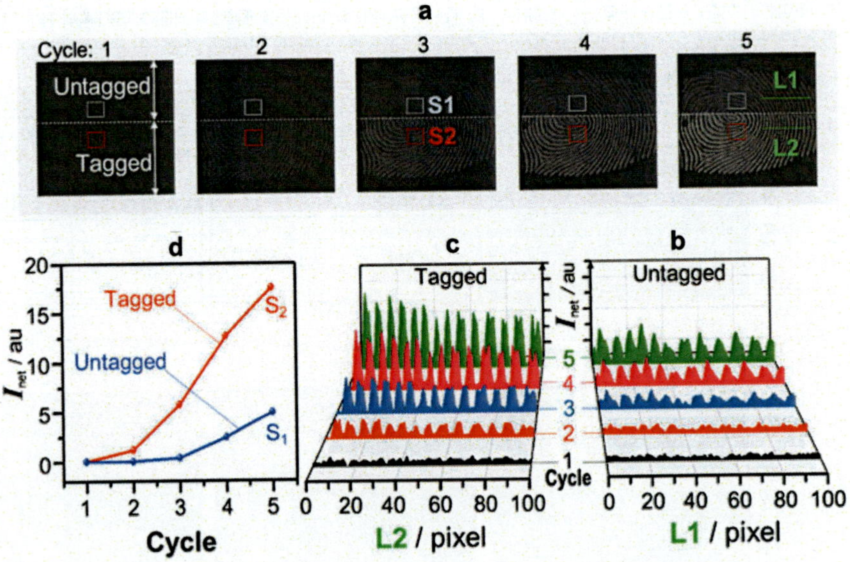

Fig. 6.32 SCAOS-SPRi differentiation of fingerprinted non-saccharides from saccharides by amidation with glucosamine and their image variation with signal amplification cycle. **a** Background-removed chemical imaging, **b** tag-free signal variation along the position line L1 at the indicated amplification cycles; **c** tagged signal variation along L2 at different amplification cycles; **d** net signal increase with amplification cycle plotted by averaging the signals within the square S1 and S2, respectively. Reorganized from Ref. [89] with permission

$$\begin{cases} \overline{B} = \sqrt{\frac{1}{N}\sum_{B=0}^{255} n_B B} \\ \sigma_B = \sqrt{\frac{1}{N}\sum_{B=0}^{255} (B - \overline{B})^2} \end{cases} \quad (6.3)$$

By plotting $\overline{S}$ against lg (c/μM), linear range can be founded at a concentration below 4000 μM glucose or serine as a representative, with linear correlated coefficients above 0.99 at the amplification cycle 5. The lower limit of linear range raises from 0 to 10 μM or even 30 μM as the amplification cycle reduces from 5 to 1.

6.4.4 *Sports Monitoring*

Exercise can accelerate the metabolism of ingested saccharides and is hence beneficial to health. Currently, fitness sports are popular, and running is the easiest and most suitable exercise method for young and old people. We have hence tried to monitor the running-caused chemical changes with SPRi. The basic idea is that exercise will cause sweating that brings metabolites such as saccharides and amino acids out of the sweat pores in the fingerprints, which makes it practicable to assay the fingerprinted chemicals with SPRi. This can be attractive since SPRi or SCAOS-SPRi is a noninvasive and painless method to determine the foregrounded substances. To demonstrate, three levels of running exercise were tried and the fingerprints of volunteers (with informed consent) were assayed. Figure 6.33 shows the concentration curves of saccharides (Fig. 6.33a) and amino acids (Fig. 6.33b) that increase quickly before reaching the peaks at 20 min for the saccharides (quantified with glucose as their representative) or 30 min for the non-saccharides (quantified with serine as the representative). They both declined obviously slowly after the peaks, where the glucose lowers faster than the serine. As a control, their concentrations from no running exercise increase slowly and steadily, without any peak or turning points, the increase is attributed to the continuous evaporation and accumulation of slow sweating. The decrease of the representative glucose could be attributed to the running fatigue and sweat dilution in a prolonged exercise [90, 92]. This is also the reason for the faster decrease of metabolites in the intense running than in the moderate running. Because the catabolism of amino acids (represented by serine) to produce energy mainly takes place during anaerobic phase [90], they will increase when the saccharides are at a low availability [93], and in turn decrease slower than saccharides in the prolonged periods of exercise. The representative glucose has a similar variation tendency to the reported data. Differently, SPRi-measured trend for natural sweating is however different from the iontophoresis-measured one [94]. The precision and reliability of SPRi assay are acceptable according to the intraday and interday relative standard deviations (RSDs), both below 6%.

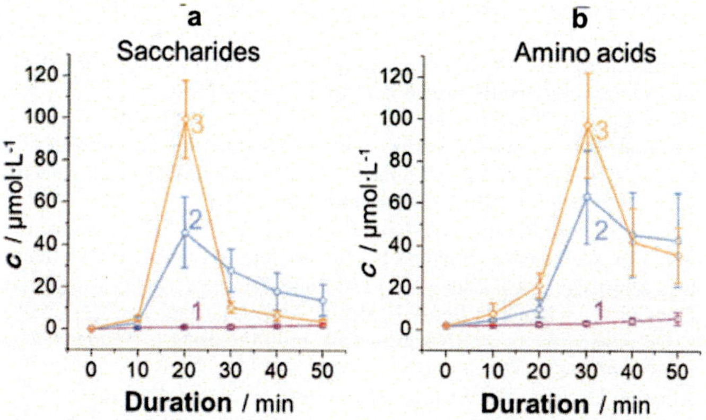

Fig. 6.33 SCAOS-SPRi-measured variations of saccharides and non-saccharides during running exercise. **a** Saccharides measured with glucose as the representative; **b** Amino acids measured with serine as the representative; 1, no running or resting state; 2, moderate running state (at 64–76% of the heart rate maximum that is equal to 220 minus Age [95]; 3, intensive running state (at 77–95% of the heart rate maximum). Reorganized from Ref. [89] with permission

References

1. Lorenzen JM, Kaucsar T, Schauerte C, Schmitt R, Rong S, Hubner A, Scherf K, Fiedler J, Martino F, Kumarswamy R, Kolling M, Sorensen I, Hinz H, Heineke J, van Rooij E, Haller H, Thum T (2014) MicroRNA-24 antagonism prevents renal ischemia reperfusion injury. J Am Soc Nephrol 25:2717–2729. https://doi.org/10.1681/ASN.2013121329
2. Patel V, Williams D, Hajarnis S, Hunter R, Pontoglio M, Somlo S, Igarashi P (2013) miR-17~92 miRNA cluster promotes kidney cyst growth in polycystic kidney disease. Proc Natl Acad Sci USA 110:10765–10770. https://doi.org/10.1073/pnas.1301693110
3. Kato M, Zhang J, Wang M, Lanting L, Yuan H, Rossi JJ, Natarajan R (2007) MicroRNA-192 in diabetic kidney glomeruli and its function in TGF-β-induced collagen expression via inhibition of E-box repressors. Proc Natl Acad Sci USA 104:3432–3437. https://doi.org/10.1073/pnas.0611192104
4. Chau BN, Xin C, Hartner J, Ren S, Castano AP, Linn G, Li J, Tran PT, Kaimal V, Huang X, Chang AN, Li S, Kalra A, Grafals M, Portilla D, MacKenna DA, Orkin SH, Duffield JS (2012) MicroRNA-21 promotes fibrosis of the kidney by silencing metabolic pathways. Sci Transl Med 4:21ra118
5. Bhatt K, Kato M, Natarajan R (2016) Mini-review: emerging roles of microRNAs in the pathophysiology of renal diseases. Am J Physiol-Renal Physiol 310:F109–F118. https://doi.org/10.1152/ajprenal.00387.2015
6. Lu W, Chen Y, Liu Z, Tang W, Feng Q, Sun J, Jiang X (2016) Quantitative detection of microRNA in one step via next generation magnetic relaxation switch sensing. ACS Nano 10:6685–6692. htps://doi.org/https://doi.org/10.1021/acsnano.6b01903
7. de Planell-Saguer M, Rodicio MC, Mourelatos Z (2010) Rapid in situ codetection of noncoding RNAs and proteins in cells and formalin-fixed paraffin-embedded tissue sections without protease treatment. Nat Protoc 5:1061–1073
8. Zhao Y, Chen F, Li Q, Wang L, Fan C (2015) Isothermal amplification of nucleic acids. Chem Rev 115:12491–12545
9. Chen Y, Huang H, Yu X, Qi L (2005) Chiral recognition of dextran sulfate with d- and l-cystine studied by multiwavelength surface plasmon resonance. Carbohyd Res 340:2024–2029
10. Cheng Y, Lei J, Chen Y, Ju H (2014) Highly selective detection of microRNA based on distance-dependent electrochemiluminescence resonance energy transfer between CdTe nanocrystals and Au nanoclusters. Biosens Bioelectron 51:431–436
11. Yan Y, Shen B, Wang H, Sun X, Cheng W, Zhao H, Ju H, Ding S (2015) A novel and versatile nanomachine for ultrasensitive and specific detection of microRNAs based on molecular beacon initiated strand displacement amplification coupled with catalytic hairpin assembly with DNAzyme formation. Analyst 140:5469–5474
12. Causa F, Aliberti A, Cusano AM, Battista E, Netti PA (2015) Supramolecular spectrally encoded microgels with double strand probes for absolute and direct miRNA fluorescence detection at high sensitivity. J Am Chem Soc 137:1758–1761. https://doi.org/10.1021/ja511644b
13. Castoldi M, Schmidt S, Benes V, Hentze MW, Muckenthaler MU (2008) miChip: an array-based method for microRNA expression profiling using locked nucleic acid capture probes. Nat Protoc 3:321–329
14. Alhasan AH, Kim DY, Daniel WL, Watson E, Meeks JJ, Thaxton CS, Mirkin CA (2012) Scanometric microRNA array profiling of prostate cancer markers using spherical nucleic acid–gold nanoparticle conjugates. Anal Chem 84:4153–4160
15. Qavi AJ, Bailey RC (2010) Multiplexed detection and label-free quantitation of microRNAs using arrays of silicon photonic microring resonators. Angew Chem Int Ed 49:4608–4611
16. Qavi AJ, Kindt JT, Gleeson MA, Bailey RC (2011) Anti-DNA:RNA sntibodies and silicon photonic microring resonators: Increased sensitivity for multiplexed microRNA detection. Anal Chem 83:5949–5956
17. Ye LP, Hu J, Liang L, Zhang CY (2014) Surface-enhanced Raman spectroscopy for simultaneous sensitive detection of multiple microRNAs in lung cancer cells. Chem Commun 50:11883–11886

18. Li X, Cheng W, Li D, Wu J, Ding X, Cheng Q, Ding S (2016) A novel surface plasmon resonance biosensor for enzyme-free and highly sensitive detection of microRNA based on multi component nucleic acid enzyme (MNAzyme)-mediated catalyzed hairpin assembly. Biosens Bioelectron 80:98–104
19. Ding X, Cheng W, Li Y, Wu J, Li X, Cheng Q, Ding S (2017) An enzyme-free surface plasmon resonance biosensing strategy for detection of DNA and small molecule based on nonlinear hybridization chain reaction. Biosens Bioelectron 87:345–351
20. Fang S, Lee HJ, Wark AW, Corn RM (2006) Attomole microarray detection of microRNAs by nanoparticle-amplified SPR imaging measurements of surface polyadenylation reactions. J Am Chem Soc 128:14044–14046
21. Hu F, Xu J, Chen Y (2017) Surface plasmon resonance imaging detection of sub-femtomolar microRNA. Anal Chem 89:10071–10077
22. Fasoli JB, Corn RM (2015) Surface enzyme chemistries for ultrasensitive microarray biosensing with SPR imaging. Langmuir 31:9527–9536
23. Wu J, Huang Y, Bian X, Li D, Cheng Q, Ding S (2016) Biosensing of BCR/ABL fusion gene using an intensity-interrogation surface plasmon resonance imaging system. Opt Commun 377:24–32
24. Wang Z, Chen Y (2001) Detection of metal ions using wavelength interrogation surface plasmon resonance spectroscopy with calixarane derivatives as sensing films. Anal Lett 34:2609–2619
25. Wang Z, Chen Y (2001) Analysis of mono- and oligo-saccharides by multi-wavelength surface plasmon resonance (SPR) spectroscopy. Carbohyd Res 332:209–213
26. Chen C, Ridzon DA, Broomer AJ, Zhou Z, Lee DH, Nguyen JT, Barbisin M, Xu NL, Mahuvakar VR, Andersen MR, Lao KQ, Livak KJ, Guegler KJ (2005) Real-time quantification of microRNAs by stem–loop RT–PCR. Nucleic Acids Res 33:e179. https://doi.org/10.1093/nar/gni178
27. Zeng K, Li H, Peng Y (2017) Gold nanoparticle enhanced surface plasmon resonance imaging of microRNA-155 using a functional nucleic acid-based amplification machine. Microchim Acta 184:2637–2644
28. Wang R, Minunni M, Tombelli S, Mascini M (2004) A new approach for the detection of DNA sequences in amplified nucleic acids by a surface plasmon resonance biosensor. Biosens Bioelectron 20:598–605
29. Wei X, Duan X, Zhou X, Wu J, Xu H, Min X, Ding S (2018) A highly sensitive SPRi biosensing strategy for simultaneous detection of multiplex miRNAs based on strand displacement amplification and AuNP signal enhancement. Analyst 143:3134–3140
30. Vaisocherova H, Sipova H, Visova I, Bockova M, Springer T, Ermini ML, Song X, Krejcik Z, Chrastinova L, Pastva O, Pimkova K, Dostalova Merkerova M, Dyr JE, Homola J (2015) Rapid and sensitive detection of multiple microRNAs in cell lysate by low-fouling surface plasmon resonance biosensor. Biosens Bioelectron 70:226–231
31. Liu C, Hu F, Yang W, Xu J, Chen Y (2017) A critical review of advances in surface plasmon resonance imaging sensitivity. Trends Anal Chem 97:354–362
32. Zhou WJ, Chen Y, Corn RM (2011) Ultrasensitive microarray detection of short RNA sequences with enzymatically modified nanoparticles and surface plasmon resonance imaging measurements. Anal Chem 83:3897–3902
33. Foudeh AM, Daoud JT, Faucher SP, Veres T, Tabrizian M (2014) Sub-femtomole detection of 16s rRNA from Legionella pneumophila using surface plasmon resonance imaging. Biosens Bioelectron 52:129–135
34. Wang J, Munir A, Zhu Z, Zhou HS (2010) Magnetic nanoparticle enhanced surface plasmon resonance sensing and its application for the ultrasensitive detection of magnetic nanoparticle-enriched small molecules. Anal Chem 82:6782–6789
35. Hu W, He G, Zhang H, Wu X, Li J, Zhao Z, Qiao Y, Lu Z, Liu Y, Li CM (2014) Polydopamine-functionalization of graphene oxide to enable dual signal amplification for sensitive surface plasmon resonance imaging detection of biomarker. Anal Chem 86:4488–4493
36. Wang Q, Li Q, Yang X, Wang K, Du S, Zhang H, Nie Y (2016) Graphene oxide–gold nanoparticles hybrids-based surface plasmon resonance for sensitive detection of microRNA. Biosens Bioelectron 77:1001–1007

37. Hong X, Hall EA (2012) Contribution of gold nanoparticles to the signal amplification in surface plasmon resonance. Analyst 137:4712–4719
38. Amendola V, Pilot R, Frasconi M, Marago OM, Iati MA (2017) Surface plasmon resonance in gold nanoparticles: a review. J Phys Condens Matter 29:203002. https://doi.org/10.1088/1361-648X/aa60f3
39. Lou X, Lewis MS, Gorman CB, He L (2005) Detection of DNA point mutation by atom transfer radical polymerization. Anal Chem 77:4698–4705
40. Paz JLD, Noti C, Seeberger PH (2006) Microarrays of synthetic heparin oligosaccharides. J Am Chem Soc 128:2766–2767
41. Tully SE, Rawat M, Hsieh-Wilson LC (2006) Discovery of a TNF-α antagonist using chondroitin sulfate microarrays. J Am Chem Soc 128:7740–7741
42. Blixt O, Head S, Mondala T, Scanlan C, Huflejt ME, Alvarez R, Bryan MC, Fazio F, Calarese D, Stevens J, Razi N, Stevens DJ, Skehel JJ, Van Die I, Burton DR, Wilson IA, Cummings RD, Bovin N, Wong CH, Paulson JC (2004) Printed covalent glycan array for ligand profiling of diverse glycan binding proteins. Proc Natl Acad Sci USA 101:17033–17038
43. Paz JLD, Spillmann D, Seeberger PH (2006b) Microarrays of heparin oligosaccharides obtained by nitrous acid depolymerization of isolated heparin. Microarrays of heparin oligosaccharides obtained by nitrous acid depolymerization of isolated heparin. Chem Commun 3116–3118. https://doi.org/10.1039/B605318A
44. Bryan MC, Fazio F, Lee HK, Huang CY, Chang A, Best MD, Calarese DA, Blixt O, Paulson JC, Burton D, Wilson IA, Wong CH (2004) Covalent display of oligosaccharide arrays in microtiter plates. J Am Chem Soc 126:8640–8641
45. Köhn M, Wacker R, Peters C, Schröder H, Soulère L, Breinbauer R, Niemeyer CM, Waldmann H (2003) Staudinger ligation: a new immobilization strategy for the preparation of small-molecule arrays. Angew Chem Int Ed 42:5830–5834
46. Houseman BT, Mrksich M (2002) Carbohydrate arrays for the evaluation of protein binding and enzymatic modification. Chem Biol 9:443–454
47. Park S, Lee MR, Pyo SJ, Shin I (2004) Carbohydrate chips for studying high-throughput carbohydrate-protein interactions. J Am Chem Soc 126:4812–4819
48. Karamanska R, Clarke J, Blixt O, Macrae JI, Zhang JQ, Crocker PR, Laurent N, Wright A, Flitsch SL, Russell DA, Field RA (2008) Surface plasmon resonance imaging for real-time, label-free analysis of protein interactions with carbohydrate microarrays. Glycoconj J 25:69–74
49. Lee MR, Shin I (2005) Facile preparation of carbohydrate microarrays by site-specific, covalent immobilization of unmodified carbohydrates on hydrazide-coated glass slides. Org Lett 7:4269–4272
50. Park S, Lee MR, Shin I (2009) Construction of carbohydrate microarrays by using one-step, direct immobilizations of diverse unmodified glycans on solid surfaces. Bioconjugate Chem 20:155–162
51. Zhou X, Zhou J (2006) Oligosaccharide microarrays fabricated on aminooxyacetyl functionalized glass surface for characterization of carbohydrate–protein interaction. Biosens Bioelectron 21:1451–1458
52. Hatanaka Y, Kempin U, Jong-Jip P (2000) One-step synthesis of biotinyl photoprobes from unprotected carbohydrates. J Org Chem 65:5639–5643
53. Liang K, Chen Y (2012) Elegant chemistry to directly anchor intact saccharides on solid surfaces used for the fabrication of bioactivity-conserved saccharide microarrays. Bioconjug Chem 23:1300–1308
54. Lis H, Sharon N (1998) Lectins: carbohydrate-specific proteins that mediate cellular recognition. Chem Rev 98:637–674
55. Jayaraman N (2009) Multivalent ligand presentation as a central concept to study intricate carbohydrate–protein interactions. Chem Soc Rev 38:3463–3483
56. Wang T, Boer-Duchemin E, Zhang Y, Comet G, Dujardin G (2011) Excitation of propagating surface plasmons with a scanning tunneling microscope. Nanotechnology 22:175201. https://doi.org/10.1088/0957-4484/22/17/175201

57. Bernhard W, Avrameas S (1971) Ultrastructural visualization of cellular carbohydrate components by means of concanavalin A. Exptl Cell Res 64:232–235
58. Villafranca JJ, Viola RE (1974) Proton nuclear magnetic resonance studies of the manganese (II) binding site of concanavalin A: Effect of saccharides on the solvent relaxation rates. Arch Biochem Biophy 165:51–59
59. Moothoo DN, Naismith JH (1999) A general method for co-crystallization of concanavalin A with carbohydrates. Acta Cryst D55:353–355. https://doi.org/10.1107/S0907444998008919
60. Liu W, Chen Y, Yan MD (2008) Surface plasmon resonance imaging of limited glycoprotein samples. Analyst 133:1268–1273
61. Zeng S, Baillargeat D, Ho H-P, Yong K-T (2014) Nanomaterials enhanced surface plasmon resonance for biological and chemical sensing applications. Chem Soc Rev 43:3426–3452
62. Teramura Y, Iwata H (2007) Label-free immunosensing for α-fetoprotein in human plasma using surface plasmon resonance. Anal Biochem 365:201–207
63. Kim S, Lee HJ (2015) Direct detection of α-1 antitrypsin in serum samples using surface plasmon resonance with a new aptamer–antibody sandwich assay. Anal Chem 87:7235–7240
64. Sendroiu IE, Warner ME, Corn RM (2009) Fabrication of silica-coated gold nanorods functionalized with DNA for enhanced surface plasmon resonance imaging biosensing applications. Langmuir 25:11282–11284
65. Spoto G, Minunni M (2012) Surface plasmon resonance imaging: What next? J Phys Chem Lett 3:2682–2691
66. Malic L, Sandros MG, Tabrizian M (2011) Designed biointerface using near-infrared quantum dots for ultrasensitive surface plasmon resonance imaging biosensors. Anal Chem 83:5222–5229
67. Mariani S, Scarano S, Spadavecchia J, Minunni M (2015) A reusable optical biosensor for the ultrasensitive and selective detection of unamplified human genomic DNA with gold nanostars. Biosens Bioelectron 74:981–988
68. Hu WH, Chen HM, Zhang HH, He GL, Li X, Zhang XX, Liu Y, Li CM (2014) Sensitive detection of multiple mycotoxins by SPRi with gold nanoparticles as signal amplification tags. J Colloid Interface Sci 431:71–76
69. Liu Y, Cheng Q (2012) Detection of membrane-binding proteins by surface plasmon resonance with an all-aqueous amplification scheme. Anal Chem 84:3179–3186
70. Hu WH, Chen HM, Shi ZZ, Yu L (2014) Dual signal amplification of surface plasmon resonance imaging for sensitive immunoassay of tumor marker. Anal Biochem 453:16–21
71. Li Y, Lee HJ, Corn RM (2007) Detection of protein biomarkers using RNA aptamer microarrays and enzymatically amplified surface plasmon resonance imaging. Anal Chem 79:1082–1088
72. Yuan PX, Deng SY, Xin P, Ji XB, Shan D, Cosnier S (2015) Mass effect of redox reactions: A novel mode for surface plasmon resonance-based bioanalysis. Biosens Bioelectron 74:183–189
73. Fenzl C, Hirsch T, Baeumner AJ (2015) Liposomes with high refractive index encapsulants as tunable signal amplification tools in surface plasmon resonance spectroscopy. Anal Chem 87:11157–11163
74. Roy B, Das T, Maiti TK, Chakraborty S (2011) Effect of fluidic transport on the reaction kinetics in lectin microarrays. Anal Chim Acta 701:6–14
75. Pallarola D, von Bildering C, Pietrasanta LI, Queralto N, Knoll W, Battaglini F, Azzaroni O (2012) Recognition-driven layer-by-layer construction of multiprotein assemblies on surfaces: a biomolecular toolkit for building up chemoresponsive bioelectrochemical interfaces. Phys Chem Chem Phys 14:11027–11039
76. Huang CF, Yao GH, Liang RP, Qiu JD (2013) Graphene oxide and dextran capped gold nanoparticles based surface plasmon resonance sensor for sensitive detection of concanavalin A. Biosens Bioelectron 50:305–310
77. Liu C, Wang X, Xu J, Chen Y (2016) Chemical strategy to stepwise amplification of signals in surface plasmon resonance imaging detection of saccharides and gycoconjugates. Anal Chem 88:10011–10018
78. Huang H, Chen Y (2006) Label-free reading of microarray-based proteins with high throughput surface plasmon resonance imaging. Biosens Bioelectron 22:644–648

79. Shen G, Han Z, Liu W, Chen Y (2007) Color surface plasmon resonance imaging of protein microdots arrays. Chem Lett 36:926–927
80. Huang H, Chen Y (2006) Surface plasmon resonance imaging studies for proteolytic hydrolysis of proteins. Chem Lett 35:372–373
81. Shankaran DR, Gobi KV, Miura N (2007) Recent advancements in surface plasmon resonance immunosensors for detection of small molecules of biomedical, food and environmental interest. Sens Actuators B 121:158–177
82. Figueroa B, Chen Y, Berry K, Francis A, Fu D (2017) Label-free chemical imaging of latent fingerprints with stimulated Raman scattering microscopy. Anal Chem 89:4468–4473
83. Hai J, Li T, Su J, Liu W, Ju Y, Wang B, Hou Y (2018) Reversible response of luminescent terbium(III)-nanocellulose hydrogels to anions for latent fingerprint detection and encryption. Angew Chem Int Ed 57:6786–6790
84. Wang YL, Li C, Qu HQ, Fan C, Zhao PJ, Tian R, Zhu MQ (2020) Real-time fluorescence in situ visualization of latent fingerprints exceeding level 3 details based on aggregation-induced emission. J Am Chem Soc 142:7497–7505
85. Hinners P, Thomas M, Lee YJ (2020) Determining fingerprint age with mass spectrometry imaging via ozonolysis of triacylglycerols. Anal Chem 92:3125–3132
86. Brunelle E, Huynh C, Alin E, Eldridge M, Le AM, Halamkova L, Halamek J (2018) Fingerprint analysis: moving toward multiattribute determination via individual markers. Anal Chem 90:980–987
87. Brunelle E, Thibodeau B, Shoemaker A, Halamek J (2019) Step toward roadside sensing: noninvasive detection of a THC metabolite from the sweat content of fingerprints. ACS Sens 4:3318–3324
88. Li S, Lu Y, Liu L, Low SS, Su B, Wu J, Zhu L, Li C, Liu Q (2019) Fingerprints mapping and biochemical sensing on smartphone by electrochemiluminescence. Sens Actuator B-Chem 285:34–41
89. Li M, Lee HJ, Condon AE, Corn RM (2002) DNA word design strategy for creating sets of non-interacting oligonucleotides for DNA microarrays. Langmuir 18:805–812
90. Hooton K, Han W, Li L (2016) Comprehensive and quantitative profiling of the human sweat submetabolome using high-performance chemical isotope labeling LC-MS. Anal Chem 88:7378–7386
91. Williams GT, Kedge JL, Fossey JS (2021) Molecular boronic acid-based saccharide sensors. ACS Sens 6:1508–1528
92. Gao W, Emaminejad S, Nyein HYY, Challa S, Chen K, Peck A, Fahad HM, Ota H, Shiraki H, Kiriya D, Lien DH, Brooks GA, Davis RW, Javey A (2016) Fully integrated wearable sensor arrays for multiplexed in situ perspiration analysis. Nature 529:509–514
93. Hargreaves M, Spriet LL (2020) Skeletal muscle energy metabolism during exercise. Nat Metab 2:817–828
94. Zhao J, Lin Y, Wu J, Nyein HYY, Bariya M, Tai LC, Chao M, Ji W, Zhang G, Fan Z, Javey A (2019) A fully integrated and self-powered smartwatch for continuous sweat glucose monitoring. ACS Sens 4:1925–1933
95. McAvoy CR, Moore CC, Aguiar EJ, Ducharme SW, Schuna JM Jr, Barreira TV, Chase CJ, Gould ZR, Amalbert-Birriel MA, Chipkin SR, Staudenmayer J, Tudor-Locke C, Mora-Gonzalez J (2021) Cadence (steps/min) and relative intensity in 21 to 60-year-olds: the CADENCE-adults study. Int J Behav Nutr Phys Act 18:27

Chapter 7
Particle Assays

Many unique advantages make SPRi valuable not only for the studies of biomolecules such as protein, DNA and RNA, carbohydrates and their conjugates, but for the investigations of particles that cover a very wide range of size, starting from macromolecules or nanoparticles up to micro- or even macro-particulates. Some typical examples are bacteria, cells, organelles, virus and various natural or artificial nanoparticles. They form even diverse research objects than molecules, being challenging analytical chemistry. It is in this case that SPRi has been explored, with promising progresses. In this chapter, very basic but unique assays will be discussed.

7.1 Main Challenges

It can be expected that SPRi, as a high-throughput analytical tool, is able to acquire different original information from intact or native particles, such as particle count, morphology, recognition characteristics, degradation or assimilation and excretion of molecules and/or subparticles, under variable desired conditions in real time, especially conducive to the analysis of living cells [1, 2] and cellular secretions [3]. In addition, SPRi can save analytical time and reagents and hence test cost. In short, SPRi can be selected preferentially in biological and medicinal studies.

There are however issues and/or challenges in SPRi of particles. Firstly, SPRi assays are in theory unable to globally display a particle larger than the probing depth of SPPs limited by the evanescent waves, normally within 200 nm in case of gold sensors. In general, only a local (i.e., the contact part) of a particle can be observed unless the particle can change its contact surface in a controlled way to expose all surface. A potential solution is to explore long-range SPRi [4, 5] that is able to reach a depth of ca. 1 μm.

Secondly, the softness of biological particles can increase the contact area that determines the deformation degree. Very soft cells easily fall flat on the sensor

Y. Chen, *Surface Plasmon Resonance Imaging*, Lecture Notes in Chemistry 95,
https://doi.org/10.1007/978-981-99-3118-7_7

surface and produce unreal or even false images. The common principle to overcome this problem is to pack the cells inside some biocompatible gels or to fixate them with chemicals like formaldehyde or glutaraldehyde. These are adoptable in many optical microscopic methods but may cause new issues in SPRi. The gels and fixating reagents will superpose their signals on particles. Although superposed signals can be reduced by mathematic deduction, it may need to combine with some techniques to specifically amplify the target signals (refer to Chap. 6). Fortunately, the deformed images can also be explored for studying the deformation mechanism and/or its impact on cellular functions.

Thirdly, optical interference fringes superimpose on the resulted images when a laser is used as the exciting light source. This also happens in the analysis of discrete particles even without the use of laser. The interference fringes blur images and their boundary, leading to difficulty to recognize the exact position of particles and reducing detection sensitivity as well. A normal solution to this issue is to replace the laser with an incoherent light source like LED. A theoretical solution to this issue is to take away the interference tails through mathematical convolution and deconvolution technology as discussed in Chap. 4.

Fourthly, there is a contradiction between the slow diffusion and fast sinking of particles, which increases with the particle mass. Fast sinking causes transportation difficulty and can block the transporting tubes, while slow diffusion makes the capture of particles difficult. The solution to resist the fast sinking of particles is to increase the viscosity of suspensions but this will further reduce diffusion rate. An alternative solution is to shorten the transportation distance and to widen the pumping tubes together with reduction of values (better value free). Agitation can also speed up the diffusion rate if it is possible, which can also improve the capturing probability.

There are other difficulties in imaging the particles but their impacts are not so typical or relatively small and will not be discussed anymore unless necessary.

7.2 General Strategy to Image Particles

SPRi can observe either crowded or discrete particles. The crowded particles can be treated as densified "molecules" that produce images nearly free of or with negligible interference fringes, while the discrete particles correspond to single particle imaging that produces images with interference fringe pattern(s). Since SPRi of the continuous molecular layers has been discussed already, discrete particles are highlighted to some extent in this chapter.

All particle samples can be analyzed directly or indirectly by SPRi. The direct strategy can observe the contact parts of particles and measure the morphological changes based on RI within a depth of < 200 nm. The acquired morphological information can serve to count or quantify the particles and to analyze the variation of particle shape. The shape-dependent information concerns with particle softness, smoothness, compressibility, division and aggregation. While the count or concentration of particles associates with the determination of their density or mass and capturing efficiency. High contrast and overlap-free images are basic to easily count the particles. This depends very much on the spatial resolution of the imaging method used and the sample applied. Prism-based SPRi can reach a lateral resolution of about 10 μm while microscope lens-based SPRi can increase the resolution to decades of nanometers that is more suitable for the determination of discrete particles. Quantification of particles can also be conducted through SPRi of the particle-related endogenous and/or exogenous molecules. This is a type of indirect determinations that are universally applicable and has been discussed in the previous chapter.

An appropriate sample preparation technology (Phase 1 in Fig. 7.1) may be required to complete either direct or indirect SPRi of particles (Phase 2 in Fig. 7.1). The coupling of sample preparation technology with SPRi opens at least 7 combination routes to image particles as illustrated in Fig. 7.1. The route ① corresponds to the direct capture of target particles on to the sensor surface terminated with specific or non-specific substances (probe 1). Non-specific probes cannot distinguish the identity of the captured particles but can facilitate the count of total particles. In common, specific probes (e.g., antibodies to antigens) are used to selectively capture the target particles on a known position on the sensor surface. The most attractive advantages of this route are its simplicity to manipulate and its minimum loss of particle samples, while its worst drawbacks are its low spatial resolution and low capturing efficiency, due to the interference fringe tails in discrete states, and due to the fast sinking but slow diffusion. Quantitative counting is exacerbated by the unavailability of standard particles for calibration. Usually, this direct assay is not very suitable for quantifying the particles at a very high concentration. To count, the particles must be highly diluted until overlap-free images are recorded. Theoretical reconstruction of the measured images (refer to Chap. 4) is suggested to increase the lateral resolution for revealing more particles.

The route ② concerns with the particle releasing or adsorbing ability. All particles may produce or adsorb molecules through various mechanisms, for example, biological cells can recognize and adsorb various molecules or reversely, secrete and excrete or in general release some specific molecules in response to stimulation or due simply to natural biological reactions. After excretion, the particles remaining on probe 1 are detectable by SPRi together with the excreted molecules that are also captured by their specific probes pre-immobilized nearby. This route can combine with the route ④ that determines the released molecules by common SPRi. The detection can be very sensitive after coupling to a signal amplification measure as discussed in Chap. 6.

The route ③ is actually relative to the purification of dirty particle samples. The trick is to use a specific probe to form complexes with the target particles, better on separable solid surface. After washing, de-complexation and separation (e.g., by centrifugation of filtration), the purified particles are then captured and detected by SPRi as in the route ①. Same to the route ②, this route is better combined with the route ⑦ to obtain reliable data.

The favorite feature of the route ② and ③ is that they can largely improve the sensitivity and reliability of determination due to the combination of molecules-based SPRi. It is known that molecule-based SPRi assays are much easier to perform

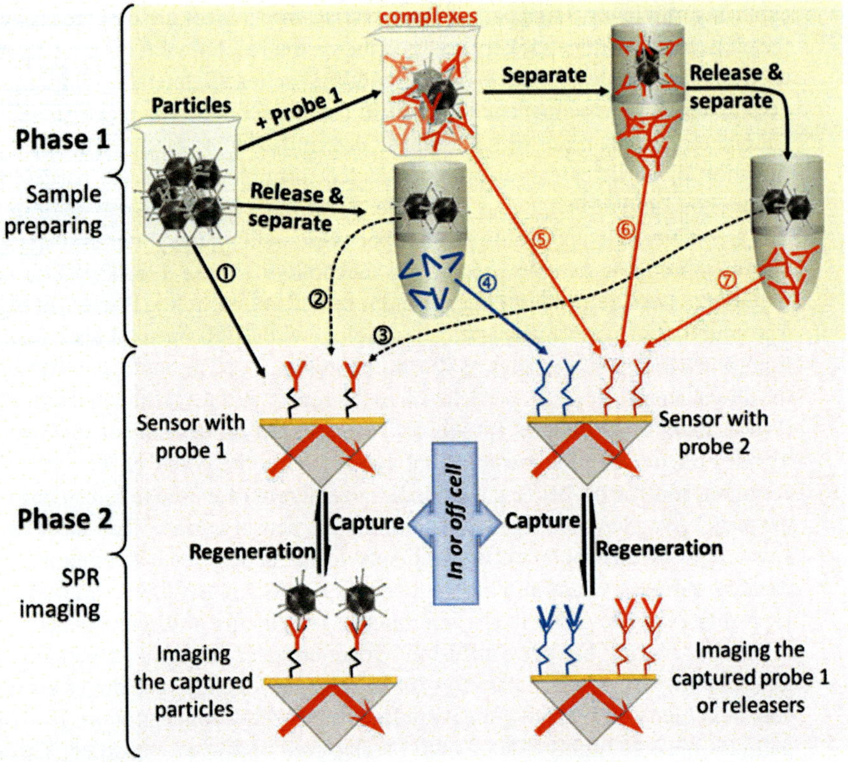

Fig. 7.1 Theoretical routes for SPRi of particles. Probe 1. Specific to the target particles, e.g., antiparticle antibody; Probe 2. Specific to probe 1 and/or the releasers

than the particles-based SPRi, in respect of quantification. In addition, the particles-related reaction processes can also be studied. The obvious shortages are that the analytical procedure is lengthened and the operation cost increases.

The routes ⑤, ⑥ and ⑦ are actually a family of indirect SPRi, where the probe 1 is added excessively to fully form complexes. The reacted products can either be directly injected for SPRi detection as in the route ⑤ or subjected to separation before SPRi analysis as in the routes ⑥ and ⑦. It is suggested to perform SPRi after separation because the direct injection (i.e., the route ⑤) normally yields high background due majorly to the disturbance of the particle complexes. The complexes of the particles with the probe 1 are in theory capable of being co-captured by probe 2. In addition, the free particles may non-specifically adsorb on the sensor surface to further raise the background. At present, these particles can easily be removed by centrifugation or filtration. The liquid phase will then be free of particles, allowing to acquire high quality of data (the route ⑥). The separated particles can be de-complexed to have pure particle samples for direct SPRi (the route ③) and to have released solution applicable to molecules-based SPRi (the route ⑦). The significant advantages of these routes are their flexibility and allowance of free combinations.

Special attention must be paid to the measurement of soft particles such as biological cells whose attaching area on a sensor surface may vary from cell to cell due to either their physicochemical (e.g., absorptivity, softness, total mass or density) or biochemical behaviors, e.g., accumulation and rearrangement of proteins in response to exogenous stimulus [6–9]. These variations are normally SPRi-detectable. For example, SPRi can sense the density change of Chinese hamster ovary cells [10] and the attaching area variation of a rat basophilic leukemia cell line (RBL-2H3) cells, basophils and keratinocytes in response to some exogenous stimulus. This is similar to SPR detection of minute morphological changes during cell apoptosis [11].

As already mentioned, indirect SPRi can largely improve its performance in the study of particles-associated molecules. Normally, the LOD of SPRi in particle detection is around 4×10^{-4} RIU but can be improved by many mechanisms [12–15]. The particle assays will greatly improve their sensitivity by combination with signal amplification technology developed in molecular assays, and/or by coupling to other analytical techniques such as MS, fluorescent spectrometry, Fourier transform infrared (FTIR) [16, 17] and so forth.

7.3 Analysis of Nanoparticles

Nanoparticles or mesoscopic particles are in common the smallest particles and can be produced in either nature or laboratory. The well-known inorganic (or metallic and nonmetallic) NPs are exemplified by AuNPs, AgNPs, carbon dots, quantum dots, silica gels and floating dust as well; while organic NPs include vesicles, viruses, exosomes, organelles and cell fragments. Clearly they are very diverse in size and shape. They are normally observed with high-resolution imaging technology such as electron microscopy, atomic force microscopy, confocal microscopy and so forth [18]. These methods often need special technology to prepare imaging samples in order to increase contrast. For example, the electron microscopy needs to coat the non-conductive particles with conductive substances such metals, the electrical conductivity needs to increase its sensitivity by labeling particles (e.g., viruses) with nanowires [19, 20], the interference pattern detection is enabled in the detection of a single virus by pre-linkage to a waveguide [21], and the microring resonator detection of tiny particles has to be realized by pre-depositing the particles on a microsphere to disturb the evanescent field [22, 23]. Comparatively, SPRi can image and/or detect intact particles, featuring robustness, simplicity and cost-effectiveness besides high throughput and real-time detection. In the following several subsections, SPR imaging of some representative NPs is selectively discussed.

7.3.1 Quantification of Nanoparticles

Having prevalent applications, a large number of NPs have been produced and in turn appeared in environments. The European Community reported that about 1.5 million tons of silica particles were produced per year, with a predicted mean concentration at 5.34 μg/L in fresh surface water [24]. Although the sheer number of total NPs is unclear in fact, it is worldwide believed that NPs would have potential risks to human beings (e.g., causing health problems). An obstacle preventing from insights into the problems is the difficulty to detect and identify the NPs in complex media around us. Although different techniques are available such as MS, surface-enhanced Raman spectroscopy (SERS), electrochemistry and many separation techniques (e.g., chromatography, electrophoresis, centrifugation and filtration), they may introduce new issues while solving the difficulty. After checked, SPRi has been found to have less issues in the analysis of NPs. In terms of size, NPs can also be considered as or compared to some macromolecules like proteins. Therefore, they can be treated as "molecules" when doing SPRi determination.

This idea has been demonstrated to be applicable to the determination of NPs in water samples by use of a 16-bit high-resolution camera and lanthanide-doped yttrium fluoride NPs as testing samples [25]. This type of NPs can easily be synthesized with high uniformity at a controlled size, and dispersed in water for their attractive to negatively charged ligands like BF_4^-. They can also reversely adsorb on the gold surface assembled with a layer dodecanethiol. As a comparison, neither positively nor negatively charged sensing surface can offer this feature. The adsorption was found to be flow-dependent: All the adsorbed NPs were swapped away at 0.4 mg/mL flow rate. This facilitates the regeneration of the chip. At a lower flow rate (e.g., 0.2 mg/mL), the SPR signal is dependent on the concentration of NPs, with a low LOD at ca.30 μg/mL for a commercial SPR instrument and ca. 1.5 μg/mL for a laboratory-build SPR imager.

7.3.2 Counting Nanoparticles

Although very sensitive in detecting the variation of an NP layer down to a level of picometer [26], SPRi was not considered, at the very beginning, as a method for the detection of individual nanoparticles because the measured signal change corresponds to the overall action of many particles on the sensor surface. In theory, the lateral resolution of SPRi, at about 10 μm on gold surfaces [27–29], is also unable to differentiate nanoparticles with each other, but in practice, nanoparticles can be observed as bright spots with a size down to a few tens of nanometers after optimization of measuring conditions in combination with subtraction of background [30, 31]. Thus, SPRi can help to visualize nanoparticles [29–35] for special and/or counting analysis [36, 37].

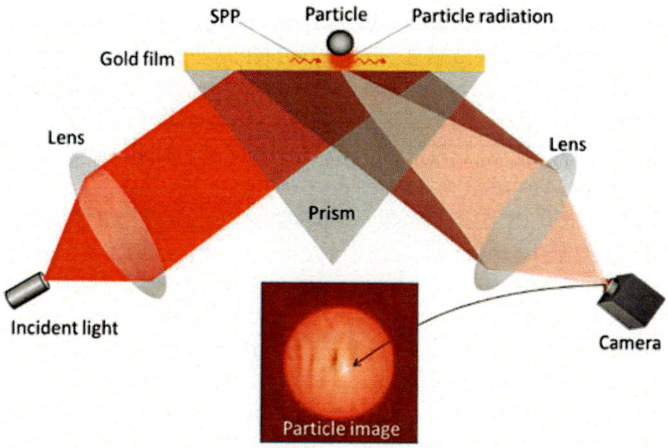

Fig. 7.2 SPRi of a nanoparticle at an incident angle $< \theta_r$ to produce a brighter image with interference fringe pattern. Note, the lateral resolution is adjustable in a certain range by regulating the viewing field (or distance) of the recording camera, for example, a CCD with pixel size of 6.45 × 6.45 μm^2 and 7× magnification lens can observe a 1-μm^2 spot at about 5 μm lateral resolution, or a CMOS with pixel size of 2.2 × 2.2 μm^2 and 3× magnification lens can reach a similar resolution

It should, however, be noted that the reflectivity variation of discrete NPs is actually different from that of a densified film. The discrete NPs can interact with the SPPs of sensor chips and in turn radiate a weak plasmon waves around them. These weak waves are circular at $z = 0$ [38] and observable either by near-field microscopy around cantilever tips positioned near to the sensing layer [39] or by SPRi to produce bright spots as illustrated in Fig. 7.2 [34, 40]. These bright spots can thus be used to determine the NP location and therefore to count the number of the excited NPs. To specifically detect some target NPs, their specific probes have to be anchored on the sensor chip for their capturing via affinity or immunological reactions.

The captured and/or counted NPs on a given field view within a definite time interval must be dependent not only on the binding kinetics and the binding efficiency but also on the NP concentration in a sample analyzed. This allows SPRi to have deeper insight into the single nanoparticles, for example, to intrinsically study the behaviors of single nanoparticle and how these single NPs affect the collective characteristics of NP ensembles or NP materials.

To perform SPRi of NPs, the Kretschmann configuration [41] is commonly adopted. Figure 7.2 illustrates that one particle radiates a weak plasmon wave back into the sensing film after interacts with an SPP. Although weak, the radiated wave may brighten the reflected beam to produce a brighter spot on the image at an angle $< \theta_r$, or reversely, to draw a darker spot at an angle $> \theta_r$. Any type of the spot is however not a real mirror image of the particle but is with interference fringes. Fortunately, the fringes do not affect the counting of the particles after their extreme dilution to avoid their overlap.

For demonstration, sulfate-modified polystyrene (PS) spheres of 200 nm in diameter are negatively charged and deposited on a positively charges gold surface (e.g., modified with aluminum oxychloride). Once the particles pumped in, they will reach the surface due to the Brownian motion and strongly bind on the sensor surface. If the detachment of the absorbed particles is negligible or the sensor surface can be considered as an infinite sink for the particles, the flux Q of the particles toward the surface will be determined by suspension concentration (c) and flow rate (u) according to Refs. [36, 42]:

$$Q = c\left(\frac{uD_{\mathrm{dif}}^2}{\mathrm{d}L}\right)^{1/3} \tag{7.1}$$

where D_{dif} is the diffusion coefficient of the tested particles, d the thickness of the flow cell and L the distance from the inlet to the detection cell. The flux Q is clearly a function of the particle concentration in a sample solution, and in equilibrium state, it is also proportional to the binding rate of the target particles onto the sensor surface for a given system. As indicated in Chap. 4, the SPRi counts should vary exponentially with c or linearly in logarithm expression. Figure 7.3 illustrates a comparison of theoretical [dot line from Eq. (7.2)] with experimental (square dots, with solid fitted line) data for 200 nm PS NPs. The slope of the lines is close to unity over at least three orders of magnitude in respect of the particle concentration with $D_{\mathrm{dif}} = 2.2 \times 10^{-8}$ cm^2/s for PS NPs in water (calculated by Einstein–Stokes diffusion equation). Note that the diffusion coefficient for positive and negative NPs may differ between theoretical and experimental; for example, the experimental D_{dif} was 5.3×10^{-8} cm^2/s and 3.2×10^{-8} cm^2/s, for positive and negative 100 nm PS NPs, respectively [43], while the theoretical value is both equal to 4.4×10^{-8} cm^2/s.

This assay is applicable to the counting analysis of HIV-like particles or viruses through immune-reaction rather than positive–negative statistic interaction (the red circular dot in Fig. 7.3). The measuring duration is about 7 min, much shorter than

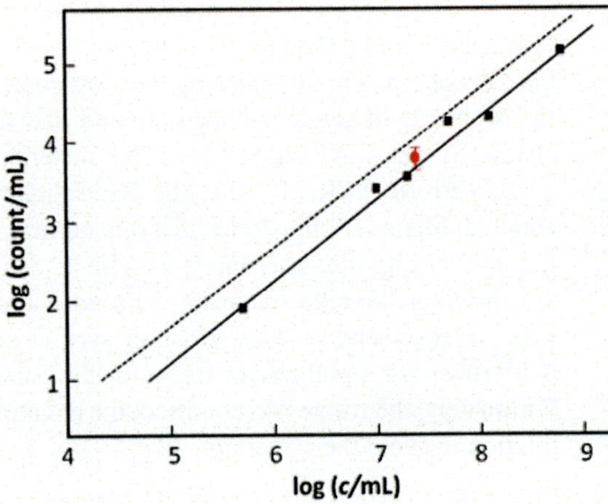

Fig. 7.3 Calibration curve counted from (black square) 200 nm polystyrene particles and (red circular dot) HIV-virus-like particles. Reprinted from Ref. [36] with permission

5-h enzyme-linked immunosorbent assay (ELISA). Figure 7.3 shows that the low LOD measured in 1 h is about 8 × 10^4 viruses /mL (equal to 16 pg/ml of a 24 k protein), where the low LOD is defined as the minimal concentration of particles in a sample solution able to bind 10 particles on the focused sensing area within a given time interval. The measured low LOD value of SPRi is much better than that of sandwich ELISA at ca. 0.1 ng/mL p24 gag protein (or 5 × 10^5 viruses/mL). By the definition, the low LOD of SPRi is also dependent on the focused sensing area. By increase of area from 0.01 to 0.1 cm^2 (e.g., reduction of magnification and/or use of a camera with smaller pixels) and lengthening the measuring time from 1 to 5 h, the low LOD value is able to reach ca.1000 particles/mL.

On the other hand, the upper LOD is determined by the possibility to count the particles, being about the number to assemble a continuous layer within the viewing field. In fact, if two particles overlap on a same area (A_0) of one recognizable spot, double layer of particles forms on the viewing field, which makes the particles indistinguishable. This can be expressed by a formula:

$$Q_{\text{max}} = \frac{\nu}{A_0} \tag{7.2}$$

where ν denotes the frame readout frequency of the video camera. Thus $Q_{\text{max}} \approx 10^9\,\text{mm}^{-2}\,\text{h}^{-1}$ for $\nu = 10$ fps. The maximal concentration allowed is ca.10^{14} particles/mL.

There is another important parameter for SPRi of NPs, the minimum detectable size (MDS). The value of MDS for PS NPs is ca. 40 nm that is determined by the difference of signal amplitude between a particle and its adjacent background. It has been found that the signal amplitude of particles was nearly proportional to the particle diameter between 40 and 1000 nm [36]. To have a better image contrast, the signals are normally acquired at the left part of a SPR curve at a position of ca. $2I_{\text{min}}$. The shot noise of a CCD has serious impact on the imaging contrast. The optimized signal-to-noise (S/N) ratio may be found in the incident angle near to the resonance minimum [44]. It is possible to further reduce the shot noise by increase of the photoelectrons in a pixel. The photon number (n_p) accumulated on the pixels is given by

$$n_p = n_{\text{pe}} n_{\text{pix}} \tau \nu \tag{7.3}$$

where n_{pe} is the number of photoelectrons accumulated in one pixel, n_{pix} is the number of pixels in the observed spot area, and τ is averaged time. Thus, n_p can be increased by having more pixels and increase of averaging time (but loss of time resolution). It should be noted that high n_{pe} will lower the readout rate ν. In general, $n_{\text{pe}}\nu$ maintains at a certain value.

It should be noted that non-specific absorption is a general problem, easily causing false positive data. Nevertheless, the selective SPRi remains advantageous over ELISA since the SPRi-counted signal is more or less size-dependent, which offers a way to minimize the false counting.

7.3.3 Direct Imaging of Exosomes

Exosomes and other vesicles are in nature soft NPs that are different from hard NPs but can also be analyzed with direct SPRi. Herein focused are SPRi of exosomes as a representative. Exosomes, as emerging biomarkers in biopsy, are a group of 30–200 nm endogenous vesicles secreted by most mammalian cells, including tumor and stem cells and tissues [45–51]. They are enclosed inside multivesicular bodies by inward budding of the endodermal membrane to the intraluminal side. After fusing with plasma membrane, the multivesicular bodies release the intraluminal exosomes to the extracellular environment [52, 53]. As a consequence, exosomes carry membrane proteins, e.g., tetraspanin-like CDs (clusters of differentiation antigen molecules) and heat shock proteins (HSPs), cytosol proteins, nucleic acids (e.g., mRNA and miRNA), lipids and glycoconjugates [45, 54, 55]. The included substances are hence dependent on their origin of cells or tissues and physiopathological conditions [56, 57]. They are able to cross many biological barriers like the blood – brain barrier [58], and in turn to transfer some selected substances from their parental to recipient cells [59], which discloses new routes to deliver drugs through the barriers and to study the elusive biochemical processes occurring within many systems, e.g., the central nervous system [60–62]. These show that exosomes may play essential roles in the communication among or between cells [63], tissues and organs through their transportation [64], and concern with some pathologies such as cancer [65, 66] and neurodegenerative [67–69] and inflammatory [70] diseases. They might also play an active role in the propagation or resolution of a specific disease like tissue regeneration [71, 72].

Exosomes have different surface phenotypes or subpopulations dependent on their origin and function. The information on the subtle variation of exosome surface phenotype can predict the metabolic stage of cells and distinguish parent cell types as well, able to assist the diagnosis and therapy of diseases, especially cancers [73] because cancer cells secret high level of exosomes into peripheral blood [74–76]. Another example is the diseases of the central neural system. The proteins and lipids in/on exosomes originated from the central nervous system and traveling in bloodstream may curry important neural information and are hence a new source to discover the biomarkers of neurodegenerative diseases. In fact, the neuronal exosomes have been found to contain pathogenic proteins such as α-synuclein and β-amyloid [77, 78]. It was reported that the concentration of β-amyloid peptides of exosomes in the patient blood of Alzheimer's disease is statistically different from that of health [79] and glial exosomes may serve as mediators of neuroinflammation in multiple sclerosis [14].

However, there is a short of standardized pre-analytical procedures and reproducible methods to study the exosomes [80]. Although there are protocols to determine the exosomes [81, 82] such as SERS immunoassay [83, 84], aptamer/AuNP sensor [85] or CD63-specific aptamer-based fluorescent detection [86], the throughput and sensitivity remain poor (at 1.0×10^5 exosomes/μL in case of fluorescent detection). The heterogeneous experimental settings, the laborious and variable procedures for exosome isolation have yielded diverse knowledge about exosomes [87]. It is hence highly desirable to explore the advantages of SPRi for exosome analysis, such as high throughput, high sensitivity and ability to image native or intact exosomes. In fact, its applicability has been demonstrated in profiling the membrane proteins of exosomes [55, 88] associated with ovarian [89] and breast cancers, and multiple myeloma [90, 91]. To illustrate, herein exemplified is the direct SPRi of neuronal exosomes.

7.3.3.1 Basic Considerations

To sensitively determine the exosomes, especially their phenotypes, SPRi needs to combine with some signal amplification techniques such as extra loading of plasmonic nanoparticles, making use of high refractive index, or deducting the background. There are basically two options to perform direct SPRi of exosomes: (i) utilization of channel pairs for deducting background and (ii) use of a chip spotted with negative and positive controls besides the specific probes to capture target exosomes. The channel-pair-based assays are easy in practice because they do not necessarily need to modify the sensing channels with specific probes. The relative recognition reactions can be conducted prior to sample injection. However, the throughput is low. To increase the analytical throughput, the spots-based assays have to be adopted. Considering that exosomes are very heterogeneous, especially in clinical samples like plasma or urine where different exosomes from various tissues and cells are mixed together, antibodies highly specific to the surface phenotype, e.g., various CDs, epidermal growth factor receptor (EGFR), epithelial cell adhesion molecule (EpCAM) and so forth, should be available and can separately be spotted on the chip with location addressable to recognize the exosomes. The captured exosomes can be differentiated quantitatively by further reaction with other antibodies specific to membrane molecules such as proteins, glycoconjugates and/or lipids. As a consequence, SPRi can offer multiple data for more reliable analysis of the exosomes than other methods that can acquire only one or a bit more data [92].

7.3.3.2 Preparation of Exosomes

Exosomes can be separated and purified from cell culture and body fluids by centrifugation (especially ultracentrifugation) alone [11, 54, 93, 94] or in combination with ultrafiltration and/or density gradients [95, 96], but at the cost of consuming time and losing efficiency [55, 97, 98]. Alternatively, exosomes can be more efficiently isolated with affinity beads [46, 99–101] or size exclusion chromatography (SEC, [102–104]. The following approaches are recommended:

Approach 7.1 For centrifugal isolation of exosomes from the supernatants of cell cultures:

Step 1, centrifugate cell culture supernatants at 10,000 g and 4 °C for 30 min to remove large vesicles, cell debris and intact cells.
Step 2, further centrifugate the cleared supernatants at 100,000 g for 70 min to precipitate the exosomes.
Step 3, wash the precipitated exosomes with a plenty of PBS to remove the adsorbed proteins and other impurities, re-suspend the exosomes in PBS (in common 100 μL per sample) and store them at—80 °C for later use.

Approach 7.2 For centrifugal isolation of exosomes from plasma:

Step 1, dilute a plasma sample with PBS, centrifugate the dilution at 12,000 g and 4 °C for 45 min to remove large particles and various fragments.
Step 2, further centrifugate the supernatant at 110,000 g for 2 h to precipitate exosomes.
Step 3, wash the precipitated exosomes with a plenty of PBS, re-suspend the exosomes in a larger volume of PBS (e.g., 8 mL) and filter the solution through a membrane with 0.22 μm pores to clear off the particulates unable to be removed by centrifugation.
Step 4, centrifugate the filtrate twice at 110,000 g for 70 min to eliminate protein impurity.
Step 5, suspend the cleaned exosomes in 100 μL PBS and stored at – 80 °C before use.

Approach 7.3 For SEC isolation of exosomes from human bloods:

Step 1, centrifugate an anticoagulated human peripheral blood sample first at 1300 g for 10 min to remove large particles including cells, then at 1800 g for 10 min to deplete platelets. Note, the clear serum can be stored at – 80 °C for later use.
Step 2, flush an SEC column with ca. 10 ml PBS (freshly prepared and filtered through a membrane filter with 0.22 μm pores).
Step 3, load 500 μl serum onto the buffer-equilbrated SEC column. Note, frozen samples must be thawed and centrifuged at 10,000 g and 4 °C for 10 min to remove potential aggregates.
Step 4, elute the column with PBS and collect eluent in fractions. Note, the fractions with exosomes have significant ultraviolet absorption at 214 or 280 nm.

It should be noted that, in the isolation of exosomes, primary samples are usually required to be pre-isolated from plasma or biological fluids to prevent the formation of protein complexes and vesicle aggregations. Compared with centrifugation and other separation techniques, SEC does not need PEG or other polymeric additives and can hence reduce the reagent contamination and artifacts [105, 106]. SEC can also reduce the isolation time from hours or even days to about 30 min. Nevertheless, chromatography may co-elute some free proteins in the samples, and the yield remains low.

The prepared exosomes must be characterized in respect of size, morphology and purity. TEM (transmission electron microscope) is often used to have the information on morphology and size distribution. To simplify the characterization, the existence of exosomes is indicated usually by the total amount of membrane proteins or biomarkers analyzed by Western blot. Besides, spectrophotometry, colorimetry and nanoparticle-tracking are also convenient and cost-effective in use. Spectrophotometric characterization measures the absorbance of exosomes at 280 nm, while nanoplasmon-based colorimetry needs to measure the exosomes immediately after incubation with metallic nanoparticles. It has been reported that, after incubation of an exosome suspension with 6 nM spherical gold nanoparticles for 30 min, the level of exosomes was obtained by measuring the LSPR peak ratio at 520 nm over 650 nm [107]. In fact, SPRi itself can also measure the content of exosomes after dilution at a ratio ranging from 2.5 to 130 μL.

7.3.3.3 Preparation of Sensor Chips

Chips with position-addressable array of antibodies (Table 7.1) specific for exosomes and/or their subpopulations can be prepared by many protocols that have been partially discussed in Chap. 4. Herein presented is the preparation of a chip modified with a monolayer of PEG terminated by 80% hydroxyl and 20% carboxyl referring to Approach 7.4.

Approach 7.4 For preparation of SPRi chip for capture and analysis of exosomes:

Step 1, insert a cleaned gold sensor chip in an ethanol solution of 80% HS-C11-PEG2-OH and 20% HS-C11-PEG6-COOH (wt 500) for self-assembly of a hydrophilic layer with mixed terminals of OH and COOH, for 24 h.
Step 2, wash the chip with ethanol, activate the carboxylic terminal in an aqueous solution of 0.2 M EDC and 0.05 M NHS for 15 min.
Step 3, spot, one after another, the required antibodies (Table 7.1) at 50 μg/mL in PBS, with an anti-rat IgG as a negative control. Note, each antibody needs at least 5 spots for statistical assessment.
Step 4, incubated the chip in a chamber with 70–80% relative humidity at room temperature for 4 h or at 4 °C overnight.
Step 5, rinse the chip with PBS followed by water, and block the chip in 1% (w/v) BSA for 2 h then in 1 M ethanolamine at pH 9.0 for 30 min.
Step 6, wash the chip with PBS and water, and blow dry with nitrogen gas. Note, the chip can be stored at 4 °C under protection (e.g., StabilCoat Immunoassay stabilizer to preserve the activity) for later use.
Step 7, mount the chip in the flow cell for SPRi experiments.

To reliably detect exosomes, stable immobilization of active probes on the sensor surface is a prerequisite. It also needs to effectively suppress the non-specific adsorptions of exosomes and unrelated proteins on the chip. Pre-coating the surface with a layer of PEG500 with mixed terminals of 80% OH and 20% COOH is shown to

Table 7.1 Membrane biomarkers of exosomes and related proteins

Biomarker type	Example[a]	Source	Work pH
On membrane			
Generic	CD9	Mammalian cell	5.0
	CD63, CD81		7.4
			7.4
Specific	CD326		
	A33	Colonic epithelial cell	7.4
	MHC-II, CD86	Antigen-presenting cell-derived *T* cell	7.4
	Lactadherin	Immature dendritic cell	7.4
	EGFR, EpCAM	NSCLC tumor cell	7.4
	CD171, ephrin B	Neuron	7.4
	PLP1	Oligodendrocyte	7.4
	GM1	Ganglioside on cell	7.4
Intramembrane			
Heat shock protein	HSP60, HSP70, HSP90, A5	Cell	7.4
Aggrephagy receptor	CCT2	Cell	7.4
Enzyme	GAPDH, enolase 1, adolase 1, PKM2, PGK1	Cell, bacterium	7.4
Ribosomal protein	RPS3	Eukaryotic nucleus	7.4
Signal transducer	ARF1, CDC42	Cell	7.4
Adhesion factor	MFGE8, integrin	Cell	7.4
Skeleton protein	Actin, tubulin	Cell membrane	7.4
Other	Ubiquitin	Eukaryotic cell	7.2
	Rat IgG	B lymphocyte for negative control	7.4
	α-Lactalbumin	Milk for device calibration	7.4

[a]CDs are the clusters of differentiation antigen molecules, named according to the uniformly rule for the leukocyte and other leukocyte differentiation antigens series recommended, in 1986, by the Nomenclature Committee of the World Health Organization; MHC-II, major histocompatibility complex-II; PLP1, sphingomyelin lipid protein 1; GM1, a type of gangliosides; CCT2, chaperonin containing t complex polypeptide 1 subunit 2; GAPDH, Glyceraldehyde-3-phosphate dehydrogenase; PKM2, M2-type pyruvate kinase; PGK1, phosphoglycerate kinase 1; RPS3, ribosomal protein S3; ARF1, ADP-ribosylation factor 1; CDC42, cell division control protein 42; MFGE8, milk fat globule EGF factor 8.

work well assessed with CD171 as a testing sample. EDC/NHS chemistry is shown to be more efficient than streptavidin-based immobilization technique. However, the steps of PEGs-modification, activation and cleaning can be eliminated if a chip with gold islands is utilized.

7.3.3.4 Imaging Methodology

As mentioned, SPRi of exosomes can be conducted in combination with either a channelized flow cell or a full-chip flow cell. The former has at least one channel used for injection of a control sample (e.g., simply a running buffer, a solution of non-target proteins or a suspension at zero concentration of target exosomes) while other channels for samples. This format can simplify the chip preparation but loses throughput in spite that specific probes can be spotted along the channels. The latter format images whole chip pre-spotted with specific probes, featuring extremely high throughput ($> 2 \times 10^4$ spots/cm^2). Its obvious shortages are the need of a spotting technique and potential interference among neighbor spots. To recognize different exosomes on a same chip, highly specific probes must be available, which may not always be achievable.

To measure stable and reliable signals, the chip and the SPRi instrument must be calibrated with standards of known responses, such as a well-characterized anti-α-lactalbumin that can be spotted on the target chip simultaneously. To perform reliable calibration, the chip is normally mounted on the SPR imager in the flow cell and equilibrated, at 25 °C or room temperature, with a flow (at ≤ 10 μL/min) of running buffer until the baseline becomes stable. An exosome sample (usually diluted in the running buffer) is then pumped in until its signal becomes stable, followed by pumping in the running buffer to wash off the unbounded species. The spot images can be taken at a fixed frequency as a video or at some critical points (to save storage space).

It is worth mentioning that the SPRi instrument is better calibrated before every experiment, which can easily be realized by spotting the calibrating chemicals on the measuring chip. In case that the chip is free of calibration spots, the calibration must be performed by flowing (e.g., at 50 μL/min) a standard solution (e.g., 3 mg/mL sucrose) through the SPRi detection cell. The running buffer for exosomes can be of 100 mM HEPES (4-(2-hydroxyethyl)-1-piperazineethanesulfonic acid), 1.5 M NaCl, 30 mM ethylenediaminetetraacetic acid (EDTA) and 0.5% Tween, adjusted to pH 7.4, or of 2 mM KH_2PO_4, 8 mM Na_2HPO_4, 136 mM NaCl, 2.6 mM KCl, 0.5% Tween at pH 7.40 (PBS). The successful capture of exosomes can be achieved by the specific interaction of their membrane markers with their corresponding antibodies. To deduct various interference, the primary images are normally corrected by subtracting the negative control spotted neighboring the samples.

7.3.3.5 Applications

SPRi can easily be utilized to count the number of discrete exosomes or to quantify the abundance of membrane proteins on a type or a subtype of exosome. The former has been discussed already in many parts while the latter is focused herein. There are also two ways to perform selective measurement of the abundance of membrane proteins. Figure 7.4 illustrates an example where the images are recorded from a chip spotted with exosome-specific antibodies and with anti-IgG as a negative control. The image intensity reflects the abundance of the membrane biomarkers specific to CD9, CD171 and CD81, respectively. This opens a way to determine the contents of the membrane markers or the captured exosomes. The exosome-associated total proteins can be detected down to about 1 μg/mL [108]. Note, this data cannot be correlated directly to the real counts of exosomes unless the surface concentration is known for a given exosome. However, even if the measured data of membrane proteins cannot be translated directly to the number of exosomes, the assay can confirm the presence of the biomarkers on/in the measured exosomes. It is interesting that lysed exosomes yield nearly negligible imaging intensity compared with the intact exosomes (Fig. 7.5), which is resulted from their vast mass difference. This implies that free proteins and membrane ghosts in an exosome sample are not necessarily removed to save time and cost. In fact, with a chip spotted with CD9, CD41b (a glycoprotein) and tyrosine kinase receptor, SPRi could directly detect and in turn monitor the abundance change of exosomes in unpurified cell culture supernatant of human hepatoma cell line MHCC97H/L and mouse melanoma cell lines B16-F1/10 [55], giving a positive correlation between the metastatic potential of the cell lines and the level of exosomes.

Direct images of antibody-captured exosomes are shown to be suitable for differentiating the source of exosomes in addition to determining the abundance of the membrane biomarkers as illustrated in Fig. 7.6 [94]. As a consequence, SPRi assay has potential to perform clinical analysis of cancers and/or monitor of the medical treating process. Figure 7.7 shows that non-small cell lung cancer (NSCLC) patients yield significantly higher exosomes than healthy controls, and medical treatment could largely reduce the production of exosomes.

The exosomes captured on a chip through one type of their membrane biomarkers can be used to determine other types or subtypes of biomarker. This can simply be achieved by making the exosomes react with their corresponding free antibodies to further "light up" the images of the spots, which is the second way to determine the membrane biomarkers. This can be exemplified by SPRi of neuronal exosomes captured on a chip spotted with anti-CD171, anti-ephrinB, anti-PLP1 and anti-CD9. To determine CD81 or GM1 on the captured exosomes, anti-CD81 or anti-GM1 is then injected to recognize the membrane proteins. Figure 7.8a shows clearly the CD81-indicated inhomogeneous expression of membrane antigens on the exosomes captured by the four antibodies, giving an abundance order of anti-CD171 > anti-CD9 > anti-PLP1 > anti-ephrinB. Differently, GM1-indicated expression has a different abundance order of anti-PLP1 > anti-CD171 > anti-CD9 > anti-ephrinB (Fig. 7.8b).

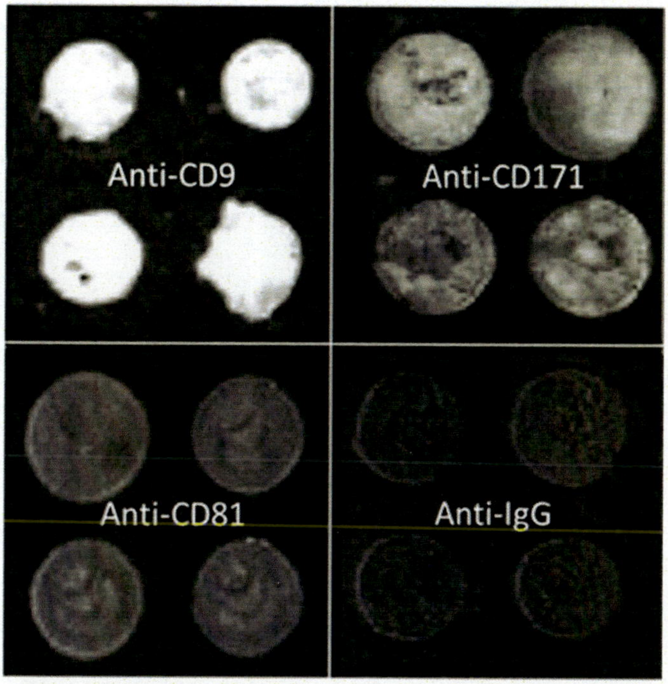

Fig. 7.4 IgG-signal-subtracted plasmonic images acquired after pumping in 500 μL of exosomes (15 μg/mL total proteins) at 10 μL/min across a gold chip spotted with the indicated antibodies as probes. Reconstructed from Ref. [108] with permission

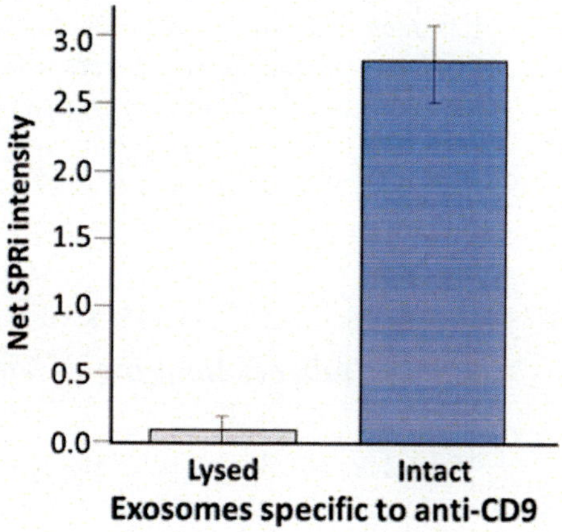

Fig. 7.5 SPRi intensity comparison of lysed exosomes with intact exosomes, captured by anti-CD9 antibody. The error bars indicate the standard deviations averaged over four spots on a same chip. Reconstructed from Ref. [108] with permission

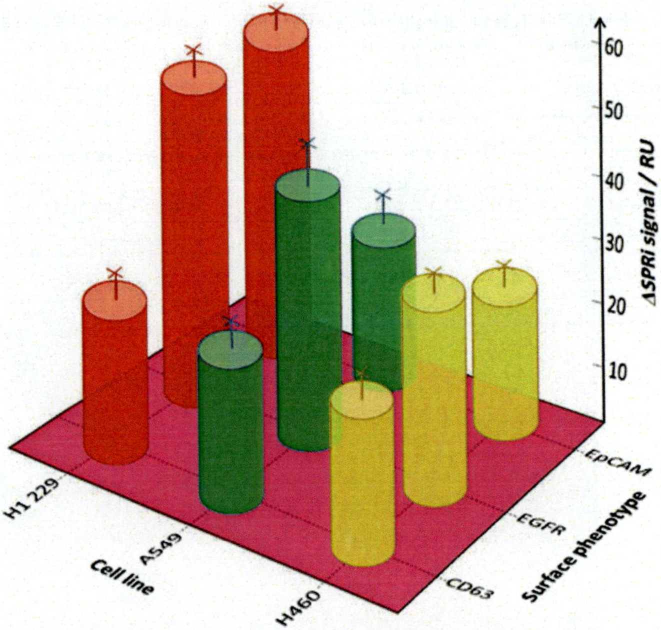

Fig. 7.6 SPRi of NSCLC-related exosomes from cell lines of H1 299, A549 and H460 captured by probes of anti-CD63, anti-EpCAM and anti-EGFR, respectively. The pillar height with standard variation (calculated from three measurements) clearly show different surface phenotypes based on the related membrane biomarkers. Reconstructed from Ref. [94] with permission

It is interesting that the neuronal exosomes are not easily be captured by the immobilized anti-ephrinB, which agrees with some other studies [109]. The different orders have also been validated in analyzing the exosome samples from different volunteers. These exhibit a unique advantage of SPRi assays: able to conduct simultaneous detection, characterization or differentiation of multiple membrane markers and/or subpopulations of exosomes within a same spot or among different spots.

7.3.4 Indirect Imaging of Viruses

Viruses are a striking type of biological NPs. It is these extremely tiny particles in the world that can cause pandemic diseases sun as coronavirus disease 2019 (COVID-19). Rapid detection and identification of viruses in various contaminated samples such as food, drinks, waters and in a patient samples are a prerequisite to efficiently counteract potential epidemics and bioterrorism [110] and critical for human life in many fields, concerning especially health (disease diagnostics and therapeutics) and daily life (monitoring of food and water supplies).

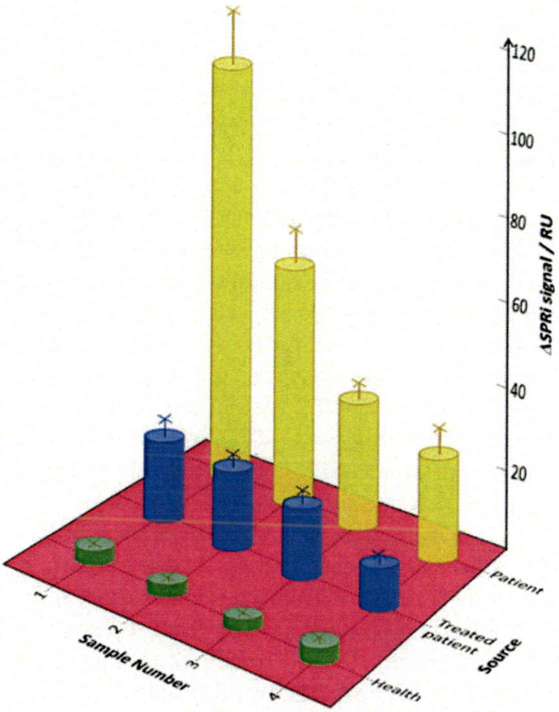

Fig. 7.7 SPRi of NSCLC-related exosomes from NSCLC patients, treated patients and health volunteers. The signal levels averaged over three dots show significant differences among the three types of samples, with $p < 0.5$ (between treated and healthy), $p < 0.05$ (between patient and treated, or patient and healthy) calculated by *t*-test. Reconstructed from Ref. [94] with permission

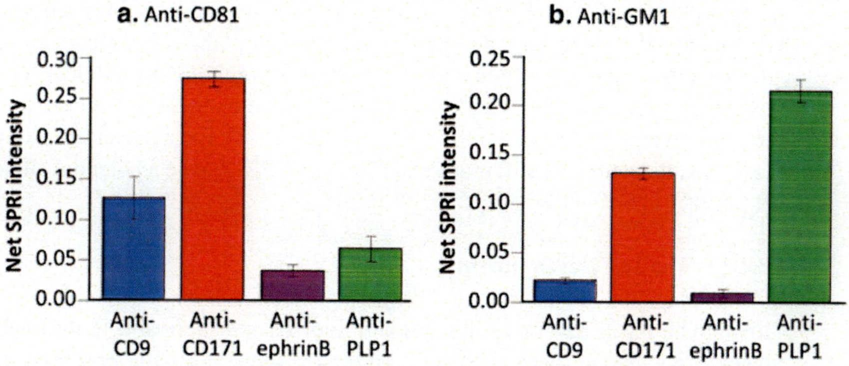

Fig. 7.8 Four-spots-averaged imaging intensity acquired after injection of 500 μL exosomes (20 μg/mL total proteins) at 10 μL/min across an SPRi chip spotted with anti-PLP1, anti-CD171 and anti-ephrinB, anti-CD9 and anti-IgG (negative control). **a** Normalized SPRi intensity measured after further injection of 200 μL anti-CD81 at 25 μL/min. **b** Normalized intensity taken after further injection of anti-GM1. *Note* The normalization is achieved by use of the signal ratio of the injected anti-CD81 or anti-GM1 over the captured exosomes. Replotted based on Ref. [108] with permission

Theoretically, SPRi assays can help to investigate different aspects of viruses, such as viral counting, viral binding, anti-viral drug discovery [111, 112] and so forth, featuring real-time observation of binding events of viruses. Although there are many methods available for the detection, characterization and quantification of viruses, such as polymerase chain reaction (PCR) or reverse transcription-PCR (RT-PCR), and ELISA, they are time-consuming and tedious in sample preparation, and some are short of qualitative and/or quantitative ability. As a promising competitor, SPRi does not exclude the principles used in the indicated methods, on the contrary, it can well combine with them in viral analysis, e.g., coupled with immunological reactions or nucleic acid-based detection principles. The limited lateral resolution [27, 28, 113, 114] has surpassed the border of micrometers [31, 34, 36], making it possible to detect the sensor-captured single viruses. An ideal SPRi assay for viruses is highly specific, rapid, cost-effective and save.

Herein human adenoviruses (hAdVs) are used as an example to discuss their determination by an indirect SPRi approach. hAdVs can infect and replicate in the respiratory tract, gastrointestinal tract, eyes, bladder and liver [115, 116]. They can be prevalent in various waters along the coastal areas or in rivers, swimming pools and drinking water supplies and hence easily cause epidemic, endemic and sporadic infections worldwide. Because their surviving mechanism in the waters remains not yet fully understood [117], hAdVs have been listed as one of the nine microorganisms on the Contamination Candidate List for drinking water by the Environmental Protection Agency of the USA. The World Health Organization (WHO) reported that 1.8 million people died each year from diarrheal diseases, of which 90% were children under the age of five. Over 88% of diarrheal diseases were caused by waterborne or water-related viruses [118].

Adenoviruses are often detected in cell culture through a very complex and lengthy (1–2 weeks) process. Unfortunately, not all groups of viruses can be isolated with regular cell lines [119]. A better option is to use polymerase chain reaction (PCR)-based assays that have been demonstrated to be faster and more sensitive than cell culture in the diagnosis of adenoviral conjunctivitis [120]. Nevertheless, the PCR-based assays remain tedious and time-ineffective (hours to days) because they need complicated extraction and amplification of DNA [121]. Even advanced methods for adenovirus detection are still required [120], and SPRi can be a potential alternative.

7.3.4.1 Method Development

Although all the seven routes illustrated in Fig. 7.1 are applicable to the analysis of adenoviruses, herein focused on is the route ⑥ aiming at reliable and precise determination. Direct SPRi assays are applicable but easily introduce strong background; While indirect SPRi based on the route ⑥ can much reduce the background. Note, the probe 1 will be an anti-viral antibody and the probe 2 is a secondary antibody to the anti-viral antibody. Compared with PCR-based assays [122], this indirect SPRi assay is much simple. It is also sufficiently sensitive, able to reach a limit of detection down to ten plaque forming unit (PFU), applicable to the analysis of aqueous adenoviruses at a content from 13 PFU/mL up to even 7500 PFU/mL [123–125]. The protocol is outlined roughly in Approach 7.5.

Approach 7.5 Critical steps for SPRi of adenovirus:

> **Step 1**, prepare adenovirus samples.
> **Step 2**, add excessive antibody to an adenoviral sample to form complexes.
> **Step 3**, precipitate the formed complexes by centrifugation or filtration.
> **Step 4**, capture the unbound antibody in the free solution onto a chip immobilized with a secondary antibody by either static or dynamic reaction.
> **Step 5**, perform SPRi measurement as usual.
> **Note**, all the experiments concerning with viruses must be performed in a laboratory at a biological safety level 2 or above. Some critical steps will be further discussed in following sections.

7.3.4.2 Preparation of Viral Sample

Viral sample preparation is often unavoidable. The core is to increase the viral copies [126] and purity, which is usually tedious. Viral samples may need dilution and incubation with cells. Following is an approach to prepare viral samples.

Approach 7.6 For incubation and preparation of virus:

> **Step 1**, sequentially dilute a real or standard viral sample to a required concentration.
> **Step 2**, inoculate the viral dilutions onto confluent cells (e.g., A549), and incubate them, for 1.5 h, by gentle shake every 20 min.
> **Step 3**, cover the inoculated cells with 1.25% agarose containing nutrients and antibiotics and incubate them for 5 days.
> **Step 4**, apply a second layer of nutrients onto the cultures, incubate them for another 2 weeks.
> **Step 5**, examine the cultures every day under microscopy during incubation, and count the viral plaques on the day 10.
> **Step 6**, centrifugate the the supernatants normally at 1000 rpm and 4 °C for 5 min, to remove large particles, cells and cell fragments.
> **Step 7**, add an equal volume of anti-viral monoclonal antibody (at the final concentration of 100 ng/mL) to each clear supernatant, allow to react for 30 min.
> **Step 8**, remove the unbound antibodies in each solution by filtration, and use the cleaned solution with antibody-virus complexes for SPRi measurements.

As shown in Fig. 7.1, an incubated viral sample containing antibody-virus complexes and free antibodies can be analyzed directly by SPRi with a sensor chip pre-immobilized with a secondary antibody. This simplifies the manipulation but is not recommended to avoid strong background interference. The viruses and their complexes can also be removed by precipitation and/or centrifugation. The same is to remove the complexes with the secondary antibodies. An adenovirus has a normal size at ca. 90 nm in diameter but is enlarged to above 100 nm after saturation with ca. 240 antibody molecules, which forms a ca. 8-nm shell with (equal to ca. 1 antibody per hexon protein trimer on the viral coat). As a consequence, the resulted complexes can hardly pass through a membrane with 0.1 μm (100 nm) pores, so that they can be separated from the free antibodies in solutions by filtration.

7.3.4.3 Preparation of Measuring Chip

Chip preparation is another critical step. There are various chemical reactions available (refer to Sect. 4.6) for covalent immobilization of probe spots, among them EDS-NHS chemistry is first recommended as shown in Fig. 7.9. Proteins like antibodies are normally spotted at 0.1–1.0 mg/mL and 0.1–0.5 μL per spot depending on the sensitivity and resolution of imager. Immediately after spotting, the chip is usually incubated at room temperature for 2–4 h in a humidified chamber. Critically, the spotted chips have to be blocked normally with 1 mg/ml BSA to suppress the nonspecific adsorption sites [127]. The blocking can be enhanced with small molecules like ethanol amine in some cases.

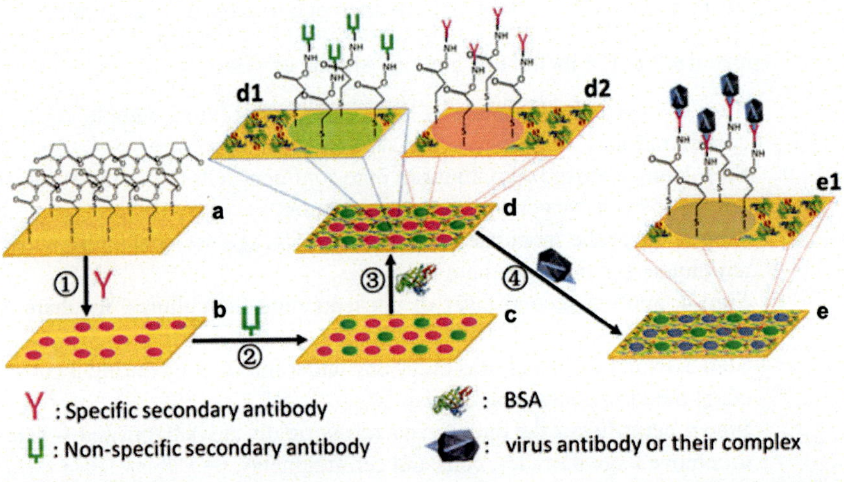

Fig. 7.9 Schematic steps for preparing a virus-measuring sensor chip via NHS-EDS chemistry. ① Spot specific secondary antibodies on an activated chip. ② Further spot non-specific antibodies for negative control. ③ Block chip with BSA. ④ Capture anti-virial antibody or virus-antibody complex on the chip. **a** Chip with NHS terminal. **b** Chip spotted with specific antibody. **c** Chip with specific and non-specific antibody. **d** Chip blocked with BSA. **e** Chip with captured viral complex or secondary antibody; Insertion d1 and d2. Binding structure of non-specific and specific antibodies; Insertion e1. Captured target viruses on the specific probes

7.3.4.4 Determination of Virus

SPRi determination can be realized statically or dynamically with chip exposure to buffer or air. Static determination under air can produce less deformed images with high sensitivity, but this method greatly deviates from physiological conditions. It is not suggested unless necessary in some special cases. In practice, flow tests are the most commonly used technique as in the analysis of molecules. Stop-flow or intermittent flow measurements are also frequently used. By either method, virus-free running buffer should first flow through the system often at a constant flow rate for more than 10 min (e.g., a rate at 15 μL/min is able to deliver 195 μL buffer in 13 min) until baseline becomes stable. A standard or real viral solution can thus be injected to initiation of capturing reactions. Once the reactions reach an equilibrium or steady state, the unbounded antibodies are washed out to taken the images of captured viruses. The flow rate is better optimized to compromise the capturing reaction, measuring speed and imaging quality. If the capturing reaction is not sufficiently fast after taking measures, stop-flow incubation has to be conducted during the process of detection. The signals are usually acquired at a given frequency to plot dynamic sensorgram and to produce video. Images may be taken intermittently at the time points of interest to save space of storage.

It should be noted that the light incident angle has to be adjusted to increase the image contrast and to make the visual field fully cover all the analytical surface. If the incident angle is too large to take a sufficiently wide image, the buffer can be evacuated to image against air at a smaller angle, on the condition that the analytical object can survive in the air for a while enough to measure the signals. Air bubbles can easily appear in the flow path and detection cell, especially at corners and hydrophobic sites. They not only impacts on buffer flow but may also add false signals to the images. To suppress the bubbles, all the solutions are better degassed inline or just prior to injection.

To ensure accurate determination, highly uniform chips of same quality should be used but to save cost, we usually choose to reuse the same chip. This is realized through regeneration of the used chip. Immune chips can easily be regenerated by flash the chip with about 200 μL of 5 mM NaOH at a flow rate of 5–20 μL/min. In common, antibody-immobilized chips can be regenerated for about 20 times and have a shelf life of about half month if stored at 4 °C. To facilitate immunochemical reaction, the running buffer at pH 7.0–7.6 is often selected, e.g., 20–50 mM PBS or HEPES. Besides the chip quality, the dosage and incubation time to capture the antibody on the adenoviral hexon have to be optimized [127]. The probe (i.e., the secondary antibody) concentration for immobilization needs to be optimized, usually at a level of μg/mL (e.g., 0.2 μg/mL tested with 100 ng/mL anti-viral antibody). Higher concentration makes the chip overcrowded while lower concentration reduces the density and binding amount of immobilized probes, thus both higher and lower concentrations will lose the detection sensitivity.

7.3.4.5 Main Features

The main features of indirect SPRi include (i) the decrease of the measured signals for the excessive antibody with the increase of adenoviral counts; (ii) an appropriate linear working range (e.g., between 88 and 27% antibody or 10 PFU/mL and 5000 PFU/mL virus, tested with 200 μg/mL of 100 ng/mL adenoviral antibody after 30 min incubation), where the LOQ at 10 PFU/mL is the same as that of two-steps PCR assay but ten folds higher than that of fluorescence assays; (iii) short analytical time in half an hour (compared with several hours for PCR assays or 3 days for fluorescence assays); and (iv) the extendibility to the measurement of other viruses like rotavirus and lentivirus by simply replacement of the specific probes.

7.3.5 Imaging of Liposomes

Different form exosomes, liposomes can be constructed artificially from amphiphilic lipids to form bilayered spherical surround by self-assembly in an aqueous phase, while similar to the membrane of exosomes, the lipid surrounding membrane serves as a barrier to hold an internal cavity and separate its internal and external aqueous phases. This makes liposomes able to deliver drugs after encapsulation. By co-assembly of stimuli-sensitive lipids in the bilayer, the liposomes can imitate the release of cells to respond to endogenous and/or exogenous stimuli. It has been demonstrated that, in addition to prism-based SPRi, high RI lens-based SPRi can also serve as a label-free tool to study the soft liposomes in aqueous phase. On an 11-Mercaptoundecylamine- and streptavidin-functionalized gold surface, the diffraction patterns of a single liposome can be visualized and analyzed with a phenomenological theory through the field enhancement effect emerging from the surface binding of the particles. After building up an empirical cumulative distribution function (ECDF), the number of liposomes containing encapsulated gold nanoparticles could be determined. The detectable concentration for the nano-sized particles is typically between 50 fM and 60 pM because the nanoscale imaging resolution is only possible in very dilute suspensions.

7.3.5.1 Basic Challenges and Solutions

As predicted by theory, SPRi can measure very diluted or even single nanoparticles but due to the scattered field interference fringe tails, the images normally have low lateral resolution. The images have also low contrast because particle-scattered field is much weaker than SPP field. In addition, the shot and mechanical noises will further blur the image contrast. Presently, in order to increase the image contrast and in turn to increase the accuracy in recognizing the particle, some new techniques have been explored and used, including signal amplification, noise filtration (e.g., by FFT) and theoretical reconstruction of the raw image with deconvolved particle permittivity or with differential signals in combination with Fourier transform technology (refer to Chap. 4). The latter two techniques are somewhat tedious but effective in practice. Hence, it remains waiting for faster technology to improve the quality of plasmonic images in respect of discrete particles.

7.3.5.2 Preparation of Liposomes

Generally, liposomes can be formed, under sonication at 60 °C, by hydration of a newly dried film with an aqueous solution such as PBS buffer or a buffer together with substances to be encapsulated. The film can be formed by evaporation of a hydrophobic solution composed of pure or mixed lipids dissolved in chloroform. The solvent is normally evaporated under a nitrogen gas stream, by gradually warming for a certain time. There are nested bilayers in the liposomes. To increase the encapsulation capacity, the nested bilayers must be reduced as much as possible. This can be achieved by several approaches, among them extruding through some porous membranes are an easy way to manipulate. The extruding may need to repeat for decades of times. The unified liposome suspensions can be stored at 4 °C for about 2 weeks. Before use, the stock suspensions must be eluted through a gel filter column (e.g., Sephadex G-50) to remove excess nanoparticles and impurities. The liposome fractions can be detected by UV light scattering. Approach 7.7 gives an example to prepare liposomes.

Approach 7.7 For preparation of AuNPs-encapsulated liposomes [128, 129]:

Step 1, prepare 1.5 μM lipids in chloroform at a ratio of 90:3:7:0.5 of 1,2-dipalmitoyl-*sn*-glycero-3-phosphocholine, 1,2-distearoyl-snglycero-3-phosphoethanolamine-*N*-[amino(polyethylene glycol)-2000], 1,2-dipalmitoyl-sn-glycero-3-phosphoethano-lamine-*N*-(biotinyl), and *N*-(fluorescein-5-thiocarbamoyl)-1,2-dihexadecanoyl-sn-glycero-3-phosphoethanolamine.
Step 2, gradually warm up the solution to 60 °C and keep for 30 min to evaporate the solution.
Step 3, hydrate the dried film with 750 μL AuNPs in PBS (13.7 mM NaCl, 2.7 mM KCl and 11.9 mM phosphate at pH 7.4) to form ca. 2.1 mM liposomes (total lipids).
Step 4, repeatedly extrude the resulting liposomes through double-stacked polycarbonate membranes with 400 nm pores at 60 °C for 11 times, and through membrane with 200 nm pores for 11 times.

7.3.5.3 Measuring Method

SPRi is usually performed on an inverted microscope equipped with a 100 × high numerical aperture oil immersion objective (e.g., NA = 1.49). The p-polarized exciting red light (e.g., 633 nm, 750 nm or 814 nm with a power at > 1 mW) is normally used. The reflected beam is collected with a CCD or CMOS camera to acquire the raw images at 0.01–0.1 s exposure times. To have clear images, the liposomes are better diluted to about 20 pM.

To improve the image contrast, the raw images at (*x*, *y*) may be treated according to Eq. (4.51), pixel by pixel, at an interval of $(t_{j+1} - t_j) = m$ s ($m \geq 1$) depending on the appearing duration of a liposome or the images are reconstructed according to Sect. 4.3.2.3 to have more precise bright spots (Fig. 4.11g) for confident counting. The noises are usually filtered with 2D Fourier transformation (2D FT) and reversed 2D FT. Briefly, a spatial difference image at (*x*, *y*) is Fourier-transformed into a frequency image at (ω_x, ω_y). This frequency image is a ring at $z = 0$, with a squared radius of $k^2 = {\omega_x}^2 + {\omega_y}^2$, for a particle adsorbed on the sensor surface. The noises can be filtered from the image by masking the inside and outside areas of the ring. The sharpened image is obtained by reverse transform. Viitala et al. have explored an SPRi protocol for the detection of diffraction patterns of liposomes in aqueous phase without any additives, and for the analysis of the relationship of the diffraction pattern with the size distribution of liposomes, which was extended to determination of the encapsulated number of AuNPs to withdraw the information of encapsulation efficiency [129].

7.4 Analysis of Cells

Similar to NPs especially exosomes and liposomes, biological cells can be imaged by SPRi in either discrete or continuously crowded state. Cell analysis may concern with many different aspects, herein confined will be the analysis of cell-related events, typically including the count of cells alone, study of cell binding (through reaction, interaction, recognition in nature or in response to stimulation) and quantification of cell-produced free or fixed molecules (by cell expression, excretion or secretion naturally or in response to stimulation). The latter two aspects can be merged into one type as measurement of cells together with molecules.

7.4.1 Challenges and Related Considerations

There are however challenges in SPRi of cell samples. Cells cannot be imaged completely due to the limited penetration depth of SPPs. Most of the cells are at a size ranging from 1 to 15 μm, about the same order of the lateral propagation length of SPPs (ca. 10 μm) but at least tenfold larger than their propagation depth (< 200 nm for gold film). Living cells are not static, but constantly change under biological and/or non-biological conditions. For example, mammalian cells change their protein expression when exposed to a liquid flow, usually a special flowing system with CO_2 supply may be required [130, 131]. Cell suspensions are also prone to clog the liquid delivery system. This can contaminate the flowing system, easy to produce "memory effect".

In spite of the challenges, SPRi is able to image the cells via their membrane along the chip-contacted part(s). Its high throughput enables the simultaneous observation of dozens of crowded or secrete intact cells. SPRi can hence provide disturbance-free visual information on the native states of cellular events [132]. Because antibodies can be used in natural state and/or free of conjugation, SPRi can avoid the conjugation effect in cell-involved immunological reactions, allowing more accurate depiction of immunological interactions under the *in vivo*-like conditions. Furthermore, real-time SPRi is able to dynamically follow the cells in response to stimulation and related steps. Therefore, stop-flow techniques or time-lapse sampling is not necessarily required or can be omitted [133].

In short, SPRi is able to address the 3D localization of cells and to measure the cell–sensor gap [134], cell excretion or secretion and cell expression), and it is applicable to: (i) counting the cells, (ii) investigating the intercellular interaction, e.g., cell interactions [135–137] and cell bindings [138–140] and (iii) observing the intracellular organelle movement [141], intracellular signal transduction [136, 142] and intracellular activity [2]. They measurements are simply based on the refractive index changes [143, 144]. In the analysis of cells, SPRi can also act as a normal SPR but this will not be highlighted in this chapter.

7.4.2 Analysis of Cells Only

Usually numerous crowded cells are analyzed to obtain statistically significant data in such as clinical assay, drug screening, bioactivity study and so forth. Very crowded cells can be measured by SPRi in a way the same as in the measurement of molecules. Imaging signals vary as cells approach the sensing surface through the coupling of cellular refractive indexes with SPPs. Different from molecules and discrete NPs, cells may saturate SPRi signals especially after the cells fully cover the sensing chip. Cells constantly coexist with molecules but in this subsection, only cells themselves are focused.

7.4.2.1 Monitoring the Capture and Release of Cells

Both deposition and detachment of cells on the sensor surface can cause significant variation of SPRi signals. This enables SPRi to follow cell adhering on and detaching from the sensor. There are different chemistry and biochemistry that allow reversal capture and release of cells. With a biochip immobilized with DNA probes in combination with specific hybridization reaction, SPRi becomes able to monitor the specific capture and orthogonal release of primary spleen cells [145]. DNA and RNA probes can easily be immobilized on gold surface by sulfhydryl chemistry or electropolymerization and can be paired through A-T and G-C hydrogen bonds. When the free complementary chain is tagged on a cell-specific antibody, capture of cells on the chip is achieved through pairing and immunological reactions. This is accompanied with an increase of SPRi signals. By denaturing the paired dsDNA stems or the immune complexes, the captured cells will be released from the probes on the chip. This will be followed by decrease of SPRi intensity if the released cells are washed off by flowing buffer. The DNA strands can also be cut with a unique restriction enzyme to release the cells. By addressable probe arrays, the increase and decrease spots possess the ability to identify the captured or released type of cells, allowing selectively study of the heterogeneity of cells.

7.4.2.2 Monitoring the Division or Apoptosis of Cells

As in the SPRi of cell adherence or detachment, cell division or apoptosis lead to increase or decrease of cells on the sensor surface, which in turn causes reflectivity or imaging intensity increase or decrease, respectively. SPRi can thus vividly and dynamically study the dividing or apoptotic processes of intact living cells. As known, cell division and apoptosis are both critical to cell fate. The observable study of these fatal events is undoubtedly significant in life science and clinical medicine, especially in cancer treatments.

Cell division is accompanied by morphological variation together with size expansion, which is observable by SPRi at a level of single cells. In addition to the image change during cell division, the intensity curve of single cell will increase with time within a given field of vision. This has been validated by use of VU1D9 (an IgG1)-producing hybridoma cells deposited on a recombinant human epithelial cell adhesion molecule (rhEpCAM)-spotted sensor [146]. As indicated in the circle area in Fig. 7.10, a single hybridoma cell at around 266 min starts to divide, leading to a fast increase of the intensity curve with time. The curve starts to slow down somewhat at about 350 min until it drops suddenly at about 700 min when the divided cells rapidly detach from the chip until completion at about 740 min. From the images, it can be observed that the dividing cell roughly doubles its size at about 383 min and becomes two juxtaposed cells at 450 min. Both the sensorgram and images can clearly reveal the dividing process in real time.

Reversely, the apoptosis is usually accompanied by at least loss of cells and/or degradation and laceration of cell membrane. SPRi is hence able to assess the cell apoptotic process, which is particularly meaningful in therapeutic studies of cancers. As known, numerous compounds are excreted during apoptosis, and a larger number of potential therapeutic chemicals have been studied as well. Unfortunately, few drugs entered into the therapeutic practice, due majorly to the inefficacy and toxicities. The lack of powerful tools to track the process is another obstacle. In spite that many methods have been reported and are available commercially, it is still challenging to simultaneously measure the apoptotic variations of cell and nuclear morphology, DNA content, cell membrane permeability, mitochondrial membrane potential changes and cytochrome C localization and release. Presently apoptosis is studied by the end point measurements of some molecules, normally free of cells. SPRi appeared in the very time and has since been tried because it can in theory screen

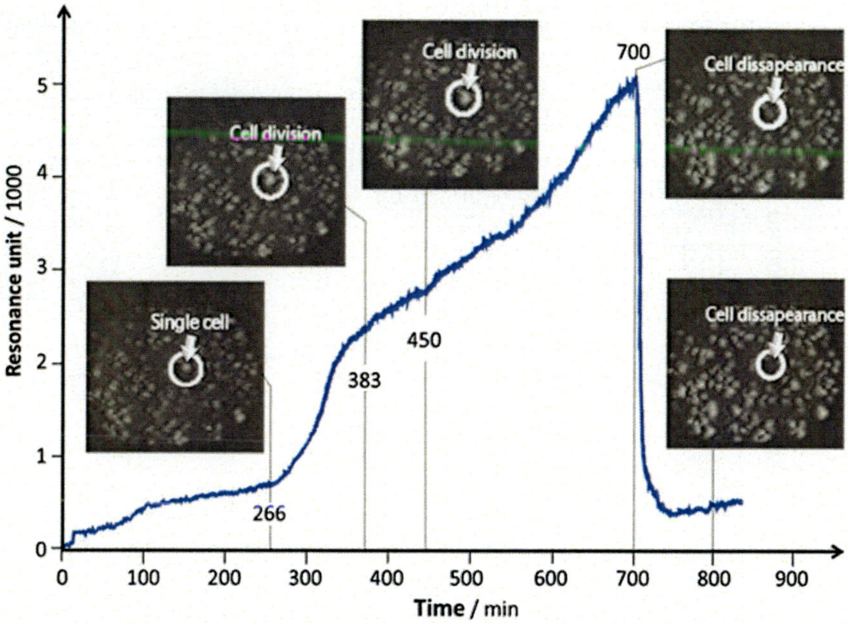

Fig. 7.10 SPRi of cell division and subsequent cell detachment tested with VU1D9-producing hybridoma cells deposited on rhEpCAM-spotted sensor. Reconstructed from Ref. [146] with permission

various potential therapeutic drugs under the appearance of intact cells, allowing simultaneous acquisition of multiple parameters.

As expected, SPRi of apoptotic breast cancer cells has been accessed by capture of the living cells (cell line MCF7) on a chip spotted with monoclonal antibodies of anti-cytochrome C, anti-EpCAM and anti-CD49e [147]. Clear image of the captured cells on the anti-EpCAM spot can be observed (Fig. 7.11a) while the response of the captured cells to the presence of paclitaxel can be measured from the mixed spot of anti-EpCAM and cytochrome C (the top curve in Fig. 7.11) and from the image variation between A and B or simply from C in Fig. 7.11. In a bit detail, signals keep nearly unchanged that are observed on the spots of anti-CD49e, anti-cytochrome C or coupling buffer free of proteins. This implies none of MCF7 cells captured on these spots and hence no cytochrome C released. Differently, the anti-EpCAM spot responds vary fast at the beginning phase of paclitaxel injection and soon turns to a saturated state over a long time, which reflects the stable binding of MCF7 cells on anti-EpCAM. The highest responding signal curve in Fig. 7.11 measured from the mixed spot of anti-EpCAM and anti-cytochrome C increase also dramatically in the first several minutes, showing an immediate release of cytochrome C from the bound cells once exposed to paclitaxel. After ca.18 h of exposure to paclitaxel, the curve starts to jumps, and after reaching a peak at ca. 21 h, it decreases slowly until nearly close to that of anti-EpCAM spot.

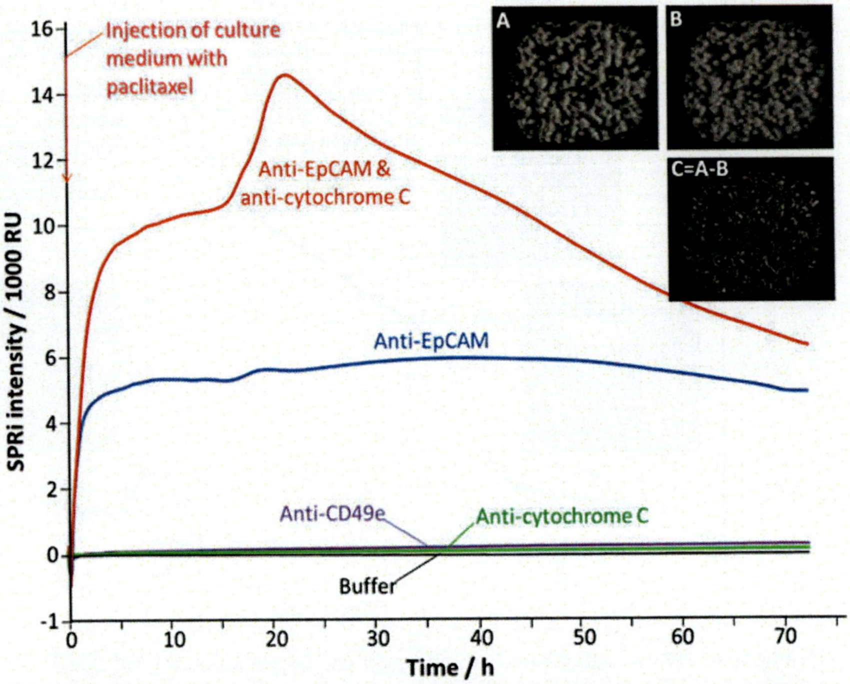

Fig. 7.11 SPRi of 72-h apoptosis of MCF7 cells induced by cytostatic paclitaxel. The cells were captured, respectively, on the spots of protein-free buffer, anti-cytochrome C, anti-CD49e, anti-EpCAM, and mixed anti-EpCAM and anti-cytochrome C, where the concentration of each protein was at 10 μg/mL. **a** Image of the captured MCF7 cells taken at the beginning of injection paclitaxel. **b** Image of the captured MCF7 cells taken at the end of experiment. **c** Difference of the images **a** and **b** to show the apoptosis-caused decrease of cells. Reconstructed from Ref. [147] with permission

The measuring principle is as follows: Cytochrome C is a hemoprotein normally binding to cardiolipin in the inner mitochondrial membrane. It initiates an apoptosis after oxidization by mitochondrion-produced oxygen species and detachment from the cardiolipin into the cytoplasm, and further binding apoptotic protease activating factor 1 and activating caspase 9. During a cascade of apoptotic events, the cytochrome C leaks from cells, becoming measureable by SPRi together with the cells. Detection of early apoptosis enables rapid therapy screening using live cells.

7.4.2.3 Discrimination and Determination of Cells

SPRi is suitable for discriminative and quantitative detection of discrete and crowded cells captured on antibodies-anchored sensor chips, somehow the same as the case of exosomes. For crowded cell samples, the quantitation can be achieved by correlation the imaging intensity with cell concentration, normally using half logarithm mathematics. The discriminative determination of cells by SPRi can be demonstrated using mixed cells of J774 (a murine macrophage cell line) and HL60 (a human promyelocytic cell line as a control). With isotypic controls and macrophage-specific antibodies as probes, an electrochemical copolymerization-combined SPRi assay could distinguish down to 5% J774 cells from the mixed HL60 cells [148]. The sensitivity was improved for 10 folds using a polyethylene oxide-modified chip to immobilize the antibodies. The assay has also been accessed with complex blood cellular samples. The subpopulations of peripheral blood mononuclear cells captured on a chip grafted with four different specific antibodies were measured to have different relative abundance, able to quantitatively distinguishing the cell types.

It is worth of mentioning that very crowded cells normally do not produce observable interference fringes (compare Fig. 7.12a with b) but yield resonance-absorbing minimum that is the same as molecules. Horii et al. measured that the resonance angle could shift more than 2° for a sensor loaded with crowned cells [149]. The shifted angle ($\Delta\theta_r$) is a function of cell counts, also the same as molecular concentration. A quantitative working curve can hence be built by plotting $\Delta\theta_r$ against cell count (better against the logarithm of cell concentration to have linear curves). It is interesting that direct SPRi of intact living cells can be realized by incubation of cells on the chip *in situ* or offline.

Discrete cells can directly be counted from the acquired images (Fig. 7.12b). To have countable images, the cell suspension needs sufficient dilution in a way the same as SPRi of nanoparticles. The interference fringe (Figs. 5.12b and 7.12b) pattern complicates the identification of cell location. To suppress the fringes, the images have to be reconstructed by removal of the fringe pattern through Fourier transform technology. For large and soft cells, the membrane may have different distance away the chip surface, which will further complicate the counting because it is hard to accurately recognize the boundary of a cell. This issue is however easy to solve by use of a common optical microscopy instead of SPRi.

7.4.2.4 Imaging a Single Cell

Single cell imaging is an extreme case of discrete cells. With modest clarity and resolution, SPRi is a vivid means to monitor the behavior of an individual cell and can be used for cell structural assessments. Many optical microscopic tools are available at present, and some are of super-resolution. Nevertheless, we remain lack of visualization tools able to observe the structure at a level down to molecules, or to monitor and track the differentiation, growth and death or apoptosis of intact living cell in real time. Presently used sensitive optical methods usually need to label the target cells with fluorophores, optical absorbing groups and/or nanoparticles. Although label-free optical microscopic techniques are available such as noninvasive Raman spectroscopic assays, they are commonly not sensitive enough; as a consequence, they need to further treat the cell samples with silver or gold nanoparticles to enhance the sensitivity.

Differently, single cell SPRi can focus observation on an intact living cell. Because completely free of labeling, SPRi possesses an ability to acquire artifacts-free images

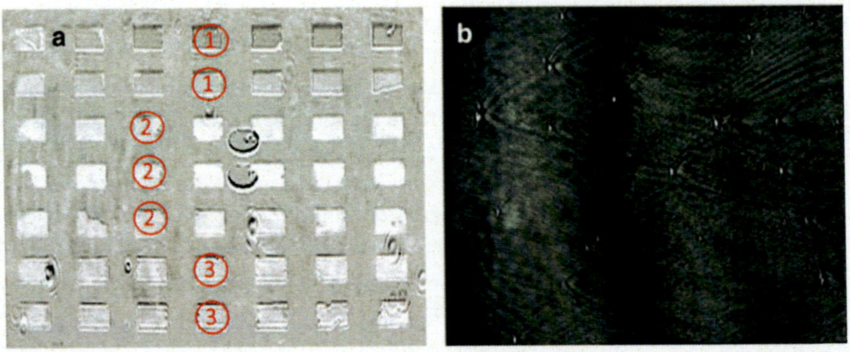

Fig. 7.12 SPRi of **a** crowded (> 200 cells/μL) and **b** discrete (< 5 cells/μL) Jurkat cells (TIB-152, human acute T cell leukemia). **a** Measured on model SPRi-TX7100 with a chip pre-spotted with ① no probe as a control, ② specific aptamer, ③ aptamer-ssDNA hybrid. **b** Measured on the device shown in Fig. 3.13 with a chip physically deposited with diluted cells, where the vertical dark and wide stripes are caused by the thin layer interference of laser while the horizontal parabolic interference fringes are scatted SPPs caused by cells. Jurkat cell preparation: Cultured cells (1.2 mL) are centrifuged at 800 rpm for 4 min, washed 3 times with pH7.0 PBS and suspended in the same PBS buffet

from a living cell based on its membrane and intracellular refractive index distribution; thus, it is unique to study the intracellular compositional changes associated with the fluctuation of local refractive index. This has been validated in imaging the single human mesenchymal stem cell [150] that can attach to a wide variety of surfaces, not mentioning to gold sensor. Figure 7.13a illustrates an intensity-based fibroblastic morphology of one human mesenchymal stem cell fixed with formaldehyde to retain its fine structure. The image intensity depends on the coupling strength among the SPPs, cell reflected wave vectors and the refractive index. In common, the refractive index changes with substances in the membrane and inside the cell, in the normal direction of the membrane. The false color image (Fig. 7.13b) shows more clearly the SPP coupling strength, from weak in green to strong in red. In particular, the interference fringes around the cell (Fig. 7.13a, b and especially c) may offer additional information, for example, the decay length of SPPs that might concern with some abnormal changes of a cell. With finely nuanced resolution of refractive index, single cell SPRi can reach a level of subcellular resolution.

7.4.2.5 Determination of Spacing Distance Between Cell and Chip

A large soft cell may lie one side of its membrane on the sensing chip surface, at a different spacing distance (d_s), which produces different reflection intensity and uneven images. This is an issue in SPRi of single cells but it can also be explored as a method to extract the value of d_s that is useful for better understanding of cellular adhesion dynamics and cell-to-surface interactions. Although some *in vitro*

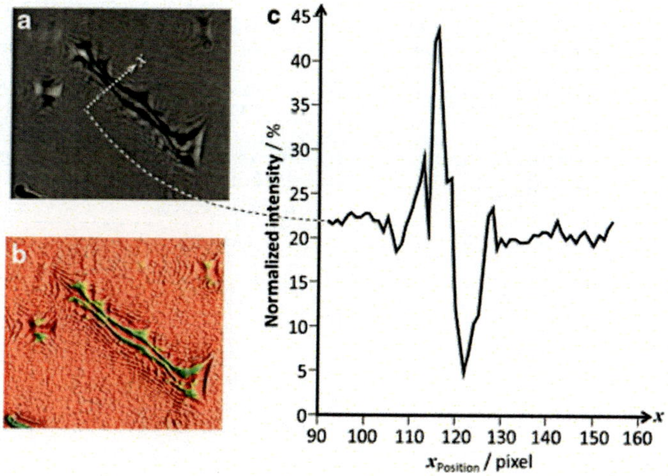

Fig. 7.13 SPRi of a formaldehyde-fixed human mesenchymal stem cell after *in situ* culture and fixation. **a** Black-and-white image taken in air at the optimized incident angle of 31° (at the least standard deviation of relative intensity ± 2.3). **b** False color image to facilitate recognition (weak in green while strong in red). **c** Normalized intensity variation along *x* line or pixel position. Reconstructed from Ref. [150] with permission

measuring tools have been explored such as fluorescence interferometry [151], interference reflection microscopy [152, 153], and total internal reflection fluorescence microscopy [154], label-free SPRi has unique advantages to image a single cell with improved depth resolution [155]. Considering that the precision depends on the validity of a resonance model, lens-based SPRi is a better choice than the prism-based one in extracting d_s.

Theoretically, if the cell membrane variation dominantly affects the resonance characteristics, its reflectivity or the imaging intensity calculable from Fresnel coefficient equations contains the spacing distance information. As a consequence, d_s can be reckoned with some assumptions. Considering a cell with cytosol ($n = 1.36$) surrounded by a membrane with a known thickness and refractive index (e.g., 7.5 nm thick, $n_m = 1.50 \pm 0.04$), once it lies on a buffer-covered 50-nm gold film deposited on a glass (e.g., BK7 with $n = 1.515$), a layer structure can be figured out as illustrated in Fig. 7.14 where the separation layer is simply the running buffer.

By the illustrated model, it can be predicted that the resonant angle minimum (θ_r) may differ from part to part within a single cell depending on the compositional variation of the membrane and cytosol. This change was validated in SPRi of a primary human aortic endothelial cell: θ_r is 74° at the cell center but shifts to 71° at the cell boundary [155]. Furthermore, an experimental θ_r is usually wider than the theoretical one, after interpixel averaging. This broadening includes both external (e.g., optical aberration and non-planar beam propagation) and internal (e.g., intracellular distribution variations of molecules and membrane) factors. Suppose that the external factor can be fixed or omitted (e.g., neglecting the roughness of the sensor surface),

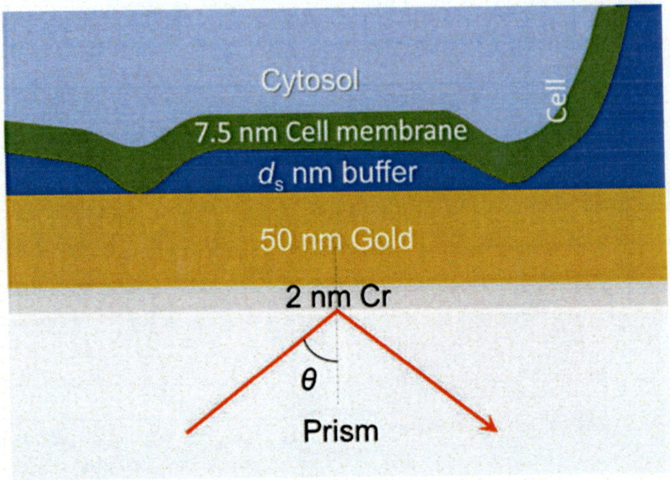

Fig. 7.14 Schematic layer structure for a soft cell deposited on a gold sensor chip with a buffer-spaced distance averaged at d_s. The cells can be deposited by physical sedimentation or by culturing in an appropriate medium (often supplied with 5% CO_2) on the gold film for > 20 min at 37 °C

and the resonant variation can thus be ascribed to only the internal variations of cell membrane, cellular organelles and molecules. The relation of d_s with θ_r can thus be obtained by fitting the measured data with a rational function as shown in Fig. 7.15 [156]. By this plot, a topological plasmonic image (Fig. 7.16a) in false color can be translated to spacing distance-based 2D (Fig. 7.16b), 3D (Fig. 7.16c) pictures or 1D digital curve along a given lateral direction (Fig. 7.16d). Clearly d_s does not change linearly with θ_r. The nearly flat bottom in Fig. 7.16d indicates the closest contact of the cell membrane to the sensor surface at d_s = 40–60 nm, which agrees with literature values [157–160]. Indeed, it is not zero since there is still a buffer layer.

The precision of d_s extracted can be tested by varying the membrane thickness (d_m) between 4.5 and 10.5 nm and its refractive index (n_m) between 1.46 and 1.54. In general, a thicker and/or denser membrane yields a larger d_s. The value of d_s increases from 39.7 to 67.8 nm as d_m is thickened from 4.5 to 10.5 nm measured at $\theta_r = 74°$, giving averaged $\overline{d}_s = 53 \pm 15$ nm with ca. 28% error referred to the center d_s at $d_m = 7.5$ nm and $n_m = 1.50$. Thus, the precision is on the order of 15 nm. The precision is affected by many factors including CCD, model used, motor repeatability, the thickness of the gold and chrome films and RIs of all the materials, of which the system-related parameters are more predictable than the target cells (e.g., membrane thickness and index). The system influences on the precision can further be improved by use of a longer exciting optical beam to have deeper penetrating depth; e.g., 785 nm wave can penetrate deeper than 633 nm, by 1.24 folds, which is given by

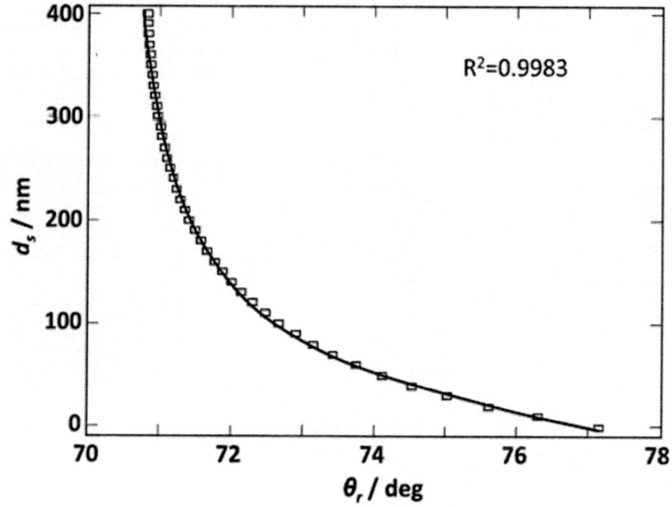

Fig. 7.15 Dependence of cell-to-sensor spacing distance on the measured resonance angle based on Nelder model tested with a primary human aortic endothelial cell. Reprinted from Ref. [155] with permission

$$\delta_{I,1z} = \frac{\lambda}{4\pi\sqrt{n_0^2 \sin^2\theta_r - n_1^2}} \tag{7.4}$$

Although not ideal, SPRi is an easy and quick tool to quantify the cell-to-sensor spacing distance. The problems that a cell has not uniform spatial resonance is however waiting for deeper insights.

7.4.3 Analysis of Cells Together with Molecules

Undoubtedly, the measured SPRi signals come from not only cells but also running buffer coupled with the sensing system. In cells only, the composition remains extremely complex and changes spatiotemporally, from cell environment to membrane and cytosol, where the surrounding biomolecules include not only the buffer but more importantly the cell secretion or excretion in response to stimulations. They overlap together unless they are purposely separated prior to imaging. These have led to the establishment of a unique class of SPRi assays to analyze the cells and molecules all together.

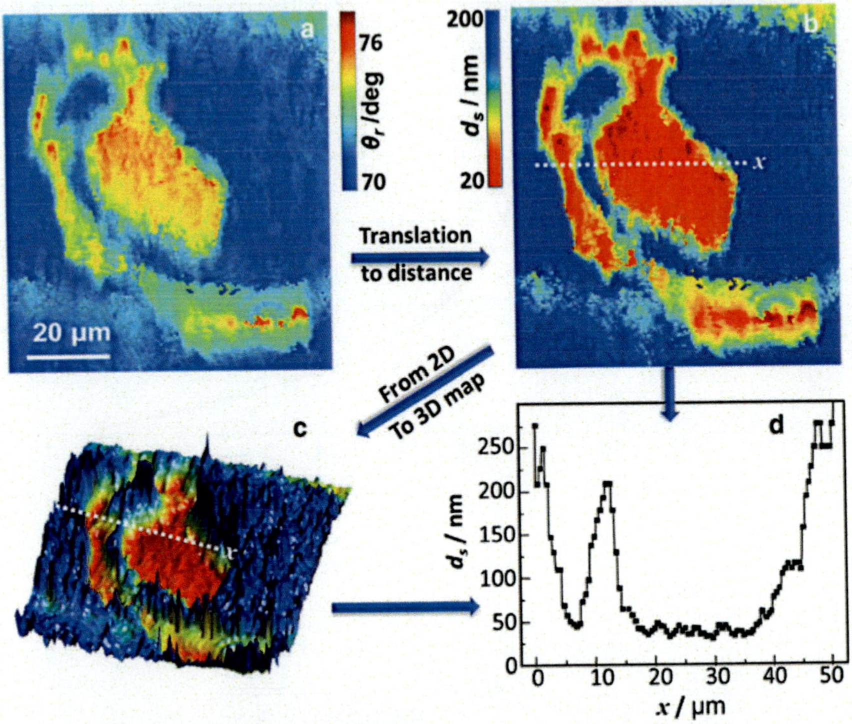

Fig. 7.16 SPRi of a primary human aortic endothelial cell at an imaging area of 78 × 91 μm^2. **a** Resonance angle image. **b** 2D image pictured by cell membrane-sensor spacing distance. **c** 3D image of (**b**). **d** Curve of (**b** or **c**) along the dashed line (x), where the lowest flat line is the contacting region of the cell on the sensor surface. Reconstructed from Ref. [155] with permission

7.4.3.1 Observation of Allergic Reaction at Cell Level

There are numerous stimuli that can make cell respond and in turn express, secrete and/or excrete molecules, for example, allergic reaction. With the ability to simultaneously observe cells and molecules, SPRi suits to monitor, track or dynamically study the related cell events and changing processes. This is in principle the same as SPR but features incomparably high throughput and increased information concerning with compositional and morphological changes.

The feasibility of SPRi to the analysis of allergic reactions can be demonstrated with a type of mast cells, RBL-2H3, in response to allergic stimulation. The intact RBL-2H3 cells have a resonance angle at 52.2° in a crowded state before stimulation but shift the angle to 52.4° within 1 min after stimulated with 100 ng/mL antigen [149]. This is due to the superposition of three factors: degranulation of cells, release of histamine and change of Ca^{2+} content. The angle variation could cause ca. 26% increase of the reflection intensity. It can be inferred that this SPRi assay would be superior to the typical assays based on high-performance liquid chromatography (HPLC) that needs to combine with fine sample pre-treatments to measure the released histamine or related molecular events. SPRi is also advantageous over some fluorescence assays that need fluorescent dyes to observe the Ca^{2+} variation or to measure fluorescence-free substances. It is interesting that SPRi study of the allergic reactions at a cellular level is more sensitive than at a molecular level that needs removal of cells. More significant is that SPRi is able to track the discrete cells in response to stimulation. This was validated with blood basophils in response to anti-IgE [133, 143, 144, 161]. This assay could differentiate the allergic type of patients in combination with different antigens. In addition, many more cells such as mouse keratinocytes and human epidermal carcinoma (A431) cells have been studied by SPRi in response to either specific or non-specific stimuli, e.g., antigen, phorbol ester or epidermal growth factor [133, 143, 144, 149, 161].

7.4.3.2 Analysis of Cell-Excreted Free Molecules

Cells can excrete different types of molecules, concerning with many fundamental cell events such as intercellular communication. This is of primary importance in living systems full of cells. Monitoring and quantifying the cell-excreted molecules are invaluable for the study of fundamental biology and medical applications.

SPRi offers two strategies to measure the cell-related molecules. The first strategy the easiest to think of is direct SPRi of the molecules after removal of the cells. While the second is more informative that dynamically analyze the target molecules while the cells are secreting and/or excreting. Since the first strategy has been discussed, herein discussed will be the second one. The core is to image the cells together with the appearing molecules.

The quantity of the excreted molecules can be extracted from the signal difference $\Delta\overline{I}$ that is calculated by deducting the signal before cell excretion. This method has been shown to be applicable to the determination of proteins (e.g., antibodies) excreted from either crowded or discrete cells. Tested with murine hybridoma B-cell line F10, dynamic observation of the cell excretion process can be achieved by SPRi without the removal of the coexisted cells. This allows the determination of *in situ* produced antibody in real time. This is advantageous over the standard ELISA assays that need to start a measurement at 30–60 min after incubation.

In another study, single hybridoma cell-based excretion is analyzed for a long period of time at 37 °C, with recombinant human epidermal growth factor receptor-2 (HER2) as a negative control [146]. The hybridoma cell line is so selected that can produce a type of monoclonal antibody, VU1D9 (IgG1 subtype), specific to EpCAM. With a chip blocked with 1% BSA buffered at pH 4.5 followed by 100 mM 2-aminoethanol buffered at pH8.0, the images are acquired every 60 s at a same angle and the net signals $\Delta\overline{I}s$ are calculated by deducting the very beginning imaging signals (only cells on chip) from the later signals. By sedimentation of a dilute cell suspension on a chip spotted with 10 μg/mL rhEpCAM, the discretely bound cells (after wash off the unbound cells) can be tracked in real time as shown in

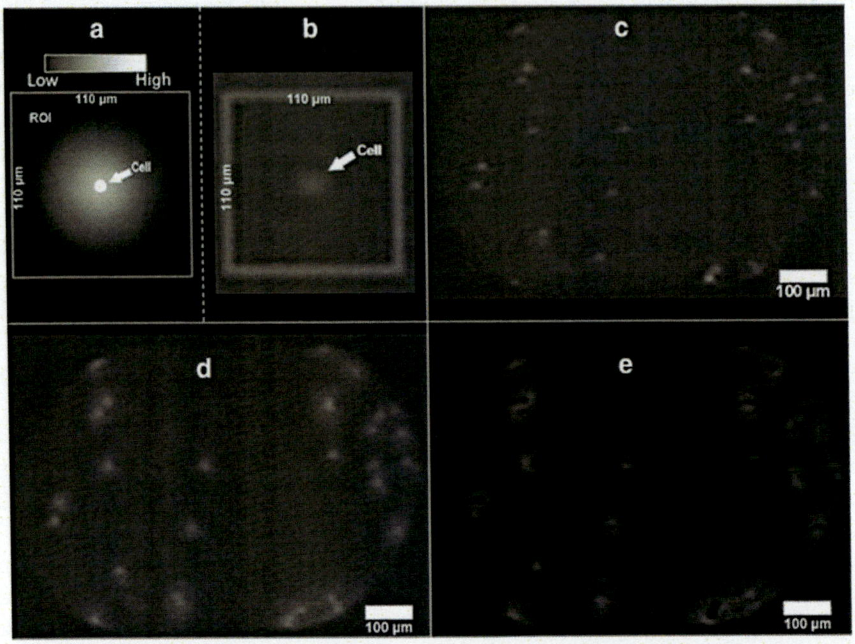

Fig. 7.17 SPRi of VU1D9-producing hybridoma cells deposited on rhEpCAM-spotted sensor. **a** A designed square vision (110 × 110 µm^2 or 20 × 20 pixels) centered at a cell of interest (the whitest dot). **b** Real region with a cell taken before cell excretion. **c** Image taken before the deposited cells excrete the target molecules. **d** Image taken after cell excretion for 60-h, where the hazy "fog" is the excreted antibodies. **e** Deducted image of (**c**) and (**d**), giving the cell position in black spots. Reconstructed from Ref. [146] with permission

Fig. 7.17, where the cell excretion-caused imaging variation can be observed clearly (C and D) and the net changes can be extracted by deduction (*E*). The quantitative variation can be calculated within a given vision field around a cell, for example, 110 × 110 µm^2 (covering 20×20 pixels). The results (Fig. 7.18) reveal that all the captured cells can produce a measurable amount of the target antibody, and their production increases with excretion time for over 60 h without any flow of solutions (mass transfer by diffusion and Brownian motion). As can be anticipated that there is individual difference in respect of the cellular secretion yield (Fig. 7.18). Tested in 10 h, Stojanović et al. calculated that the production averaged over 53 cells was 0.3 pg cell^{-1} h^{-1}.

7.4.3.3 Analysis of Molecules on Cell Membrane

The basic idea to analyze the molecules on cell membrane is based on the fact that SPRi signal will increase with the deposition of cells on the sensor chip. Although the measured signal consists of many portions, the portion corresponding to the

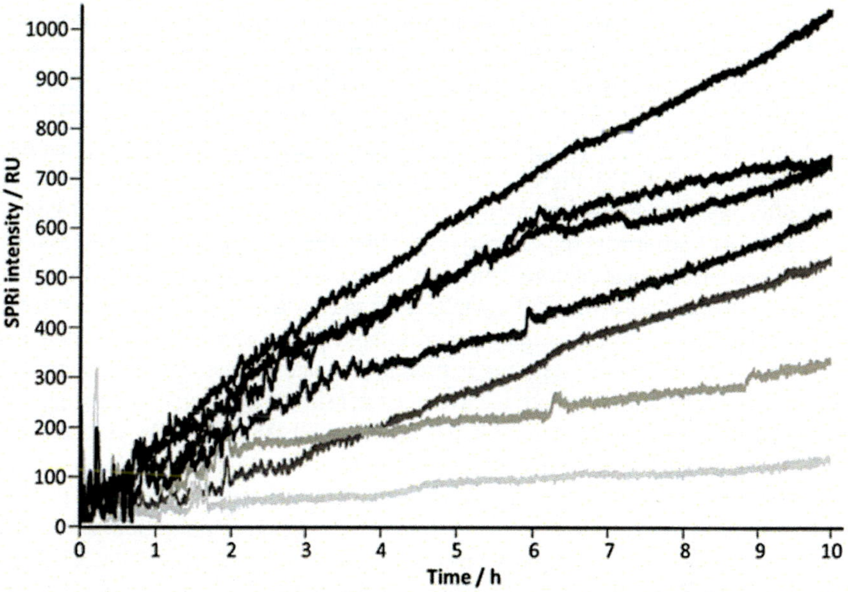

Fig. 7.18 Sensorgrams measured from seven individual VU1D9-producing cells within a vision field of 110 × 110 μm^2. Replotted from Ref. [146] with permission

saturated specific adsorption will keep constant after removal of the free and the non-specific adsorbed cells (e.g., by blocking techniques and mathematic deduction). As a consequence, the measured signal will be proportional to the sum of all specific sites (N_P) within the penetrating depth of SPPs. If the adsorbed membrane area takes a ratio of α over the total membrane surface, N_P can be calculated by

$$\because \log(\overline{I}_P - \overline{I}_0) = \log(\Delta\overline{I}_P) = A' \log N_p + B'$$
$$\therefore \log N_p = A \log(\Delta\overline{I}_p) + B \qquad (7.5)$$

where $\overline{I}_p$ is the averaged imaging intensity over a focused area on the sensor surface immobilized with a target probe, while $\overline{I}_0$ is the averaged intensity over a same area around the location of negative control or the location free of probe, and A, A', B and B' are constants to be calibrated, which can be determined with known sites of standard cells. In case that the calibration cannot be realized because of the unavailability of standard cells, normalization of the imaging data among different types of cells can serve as an alternative way to perform quantitative comparison. In both cases, the SPRi data can be used to study the structure of a cell membrane.

The so-called specific sites can be the different types of recognition molecules on/in the cell membrane such as antigen or biochemical containing carboxyl, hydroxyl or amino group. The specificity depends very much on what type of probes is utilized.

Quantification of molecules on the cell membrane can be exemplified by the determination of on-membrane antigen (e.g., EpCAM, human breast adenocarcinoma cell line MCF-7 and human breast adenocarcinoma cells SKBR3). In order to use an SPRi cell analysis protocol [162], the anti-EpCAM needs to be immobilized at different concentration [163] on a sensor chip that is in turn used to capture the cells containing membrane EpCAM. Figure 7.19 illustrates an image (insertion) taken at about 30 min after injection and gravitational sedimentation of cells (breast cancer cell lines of HS578T), where the sensorgram reveals that the adsorption of flowing cells does not stop after 30 min of capture, especially at the locations immobilized with high concentration of antibody (e.g., 13 μg/mL). This means that 30 min capture is not sufficient or multilayer adsorption may occur. In spite of the slow adsorption dynamics, the condition of 30 min capturing is tried to evaluate the membrane antigen density as shown in Fig. 7.20, giving the averaged density of the membrane antigen EpCAM at 100%, 26% and 7% for MCF7, SKBR3 and HS578T cells, respectively. They roughly agree with the data of 100%, 43.9% and 0.4% measured by flow cytometer. We believe that the difference between these two sets of data would be resulted from that the SPRi assay is performed before the adsorption equilibration, which can easily be observed in either Fig. 7.19 or Fig. 7.20 in the case of MCF7 and SKBR3 at high concentration of anti-EpCAM immobilized. More accurate measurement should be performed by calibration methods rather than the normalized one.

7.4.3.4 Monitoring of Intracellular Signal Transduction

Cells in a living system are surrounded by both of molecules and cells. The surrounding substances are highly complex and constantly changing. To work harmoniously, the cells need to perfectly communicate with them and make timely responses, which concerns with cell communication and cell signal transduction. Once a cell recognizes, interacts and/or reacts with the surrounding cells or molecules excreted from near or distant cells, it turns the signals inside the cell, causing intracellular changes of molecular functions, so as to change some intracellular metabolic processes. As a result, the cell growth rate is affected or cell death is even induced. In other words, the combination of extracellular factors with membrane or nuclear receptors can trigger a series of intracellular biochemical reactions and protein-protein interactions, and initiation of some genes to express. Clearly, the cascade of intracellular signal transduction must be initiated after the extracellular signals transmit across the membrane to the interior of the cells. It should be noted that hydrophobic molecules can easily pass through the cell membrane, directly bind to cytoplasmic or nuclear receptors and then change the gene transcriptional activity. This process is termed cell signal transduction. Up to now, there are different intracellular signal transduction modes and pathways, forming a complex network system with multiple levels of cross-regulation. Therefore, to clarify the mechanism of cell signal transduction means to recognize the expression and regulation mode of cell proliferation, differentiation, metabolism and death in the whole life process.

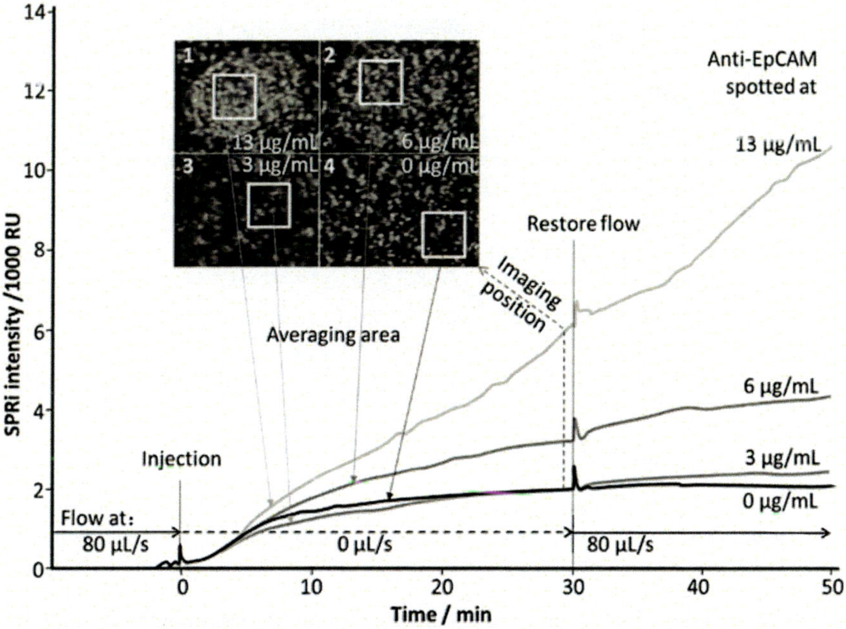

Fig. 7.19 SPRi of a breast cancer cell line of MCF7 captured by anti-MCF7 immobilized at the indicated concentrations. A chip coated with 100 nm hydrogel-like gel (G-type) was spotted with anti-EpCAM in sodium acetate buffer (pH4.5) at 0, 3, 6 and 13 mg/mL, and blocked with 1% BSA and/or 100 mM 2-aminoethanol for 15 min each. The cells (600 mL, suspended in running PBS at ca. 2×10^6 mL^{-1} and agitated) were then injected into the SPRi flow chamber at 80 mL/s and captured onto the chip by stopping the flop for 30 min. 1, 2, 3, 4. Image taken just before restoring the buffer flow; □. Area with equal pixels for averaging the imaging intensity to draw the sensorgrams; Running buffer. PBS containing 0.0003% Tween 20 and 0.25% EDTA; Chip regeneration. Flow of glycine HCl at pH2.0 for 1 min. Reconstructed from Ref. [163] with permission

Clearly, it is a prerequisite to observe, in real time, the extra- and intracellular process of signal transduction, which is necessary in not only basic researches of cell biology but also biomedical and pharmaceutical applications. The conventional methods usually have tedious labeling and washing process to read out a signal. This may be alternated by label-free SPRi assays.

As above-illustrated, SPRi can measure molecular and cellular recognition events in real time and can determine the cell-produce free or on-membrane molecules; it is thus suitable for insight into the cell signal transduction. This can be exemplified by SPRi observation of the translocation of protein kinase C (PKC) in PC12 cells. PKC is a family of Ser/Thr kinases that play an essential role in many cellular processes, including control of fundamental autonomous activities such as cell survival, proliferation and differentiation [164]. In nervous system, PKC associates with the regulation of axonal and dendritic growth, neural cell adhesion molecule-mediated neurite outgrowth, neurotransmission and axonal regeneration[165–167]. It is hence critical to analyze the translocation of PKC activity from cytosolic to membrane fractions of

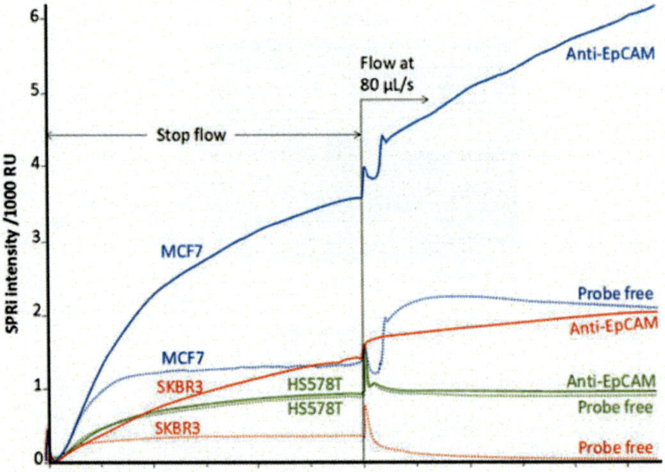

Fig. 7.20 Overlaid SPRi sensorgram of MCF7, HS578T and SKBR3 cells captured on chips spotted with and without 20 μg/mL anti-EpCAM. Reconstructed from Ref. [163] with permission

intact cells, which is presently realized with tools of immunoblotting and fractionations of cytosol and membrane [168, 169]. In the specific case of blood cells, the intercellular signaling cascades are mainly mediated by the extracellular secretion of proteins, which shall act as firing events for downstream cells [170]. The secreting activity can be monitored at either a molecular or a cellular level. The former is realized by first labeling the soluble proteins then reading out with a colorimetric (e.g., ELISA) assay, while the latter is achieved by culturing the cells on surface-sensitive plates, capturing and labeling the secreted molecules after the cells are washed off. Labeling individual cells in solution via their secreted products makes it possible to analysis of cell hybridomas with flow cytometry [171]. Although genetic fusion-based labeling of a target molecule with green-fluorescent protein can be realized in living cells, these labeling-washing-based approaches refer to endpoint analysis, thus hardly access to the kinetic insights of the secretion processes. In addition, these labeling and washing steps complicate the manipulation and optimization process, and in turn impact on time, signal, throughput and applicability. Hence, it is desired to have an accurate, noninvasive and real-time alternative tool and SPRi is a novel one to conduct the analysis of cell signal transduction.

SPRi can spatiotemporally investigate signal molecular recruiting to and dissociating from cell membrane dynamically triggered by external stimuli. This was confirmed by use of a model nerve precursor PC12 cells that can be depolarized by stimulation with a high concentration of K^+ and phorbol-12-myristate-13-acetate (PMA, a PKC activator) and inhibited by staurosporine. The PC12 cells are depolarized by chemical stimulation with KCl, accompanied by a sudden increase of discrete cell signal (Fig. 7.21a), where Hanks' solution (0.137 M NaCl, 5.37 mM KCl, 1.26 mM $CaCl_2$, 4.17 mM $NaHCO_3$, 0.44 mM KH_2PO_4 and 1.89 mM glucose) is used as negative control (with no detectable signal change) and for the deduction of background. The stimulation-caused variation can be vividly observed from the images taken before (Fig. 7.21b) and after (Fig. 7.21c) stimulation. As illustrated in Fig. 7.21a, all the single cells (line 1–5) almost synchronously respond to the stimulation while the cell-free location (line 6) does not. The response patterns of the discrete cells are much similar but response strength is heterogeneous or has individual difference. This signal pattern is similar to that of PKC translocation in a cell line from Rattus norvegicus adrenal pheochromocytoma, PC12 cells [172]. The excess cell-to-cell variation may be attributed to the differences in physical condition of the particular cells.

In addition to the individual cell difference, the SPRi intensity of each cell also synchronously varies with the concentrations of KCl as shown in Fig. 7.22. The signal

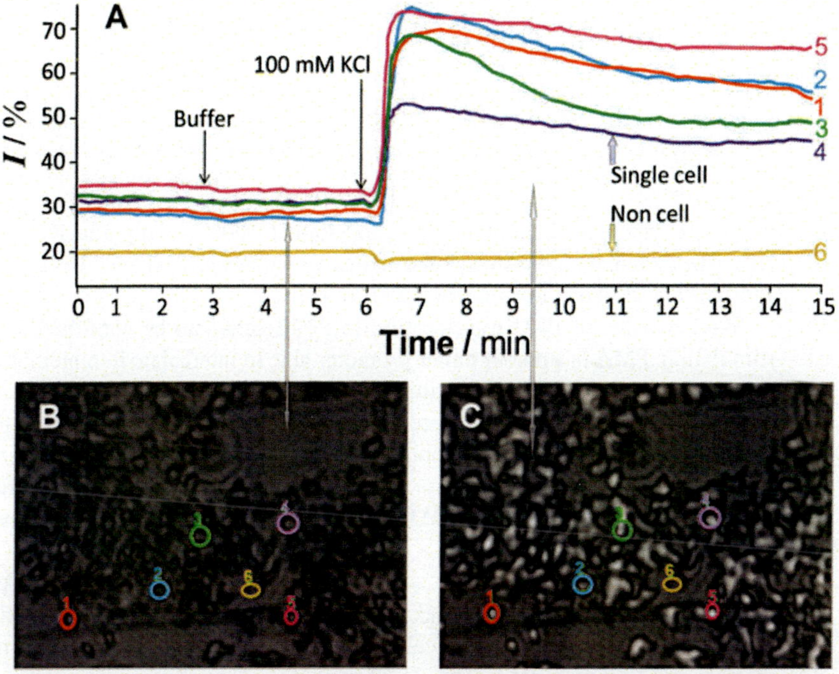

Fig. 7.21 SPRi of PC12 cells stimulated with 100 mM KCl. Experiment was performed by filling a suspension of newly grown cells at 4×10^5 cells/mL onto a gold chip sealed in a square well, incubating the cells at 5% CO_2 and 37 °C for 24 h, and monitoring the imaging baseline on an SPR imager until it became stable. The chip was then washed twice with Hanks' solution at 37 °C to remove the culture medium, then mounted on the imager and covered with 270 μL Hanks' solution. To stimulate the cells, an aliquot of 30 μL KCl or PMA (phorbol-12-myristate-13-acetate) solution was gently added and SPR imaging signals were recorded by cooled CCD camera with a 7× magnification lens. Note, PKC can be inhibited by pretreating (before PMA stimulation) the cells with 100 nM staurosporine (a PKC inhibitor) for 30 min. **a** Dynamic curves measured at different imaging locations. **b** Image taken after the injection of Hanks' solution but before the injection of KCl. **c** Image taken after the injection of 100 mM KCl. 1–5 one cell; (6) Control without a cell. Reconstructed from Ref. [142] with permission

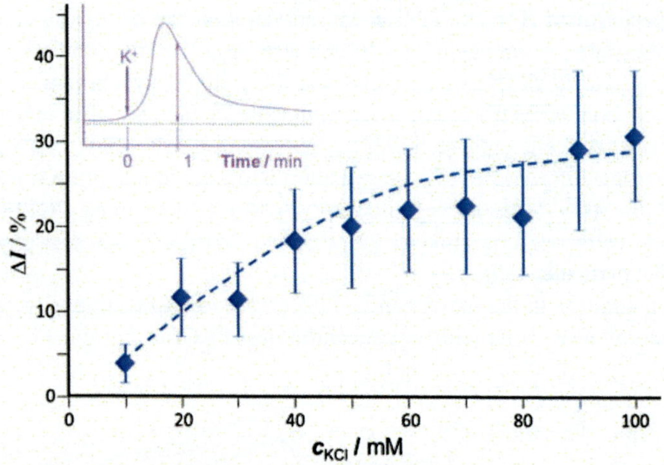

Fig. 7.22 Dependence of relative imaging intensity (ΔI) of discrete PC12 cells on the stimulating concentration of KCl (c_{KCl}). The signal is read at 1 min after injection of KCl (above insertion), while other conditions are the same as in Fig. 7.16. Reconstructed from Ref. [142] with permission

increase is fast at the beginning phase and starts to slow down gradually at 60 mM KCl. The increment looks to become zero at very concentrated KCl. In general, KCl at 40 mM is sufficient to simulate the PC12 cells.

SPRi analysis of PKC translocation in PC12 cells can be confirmed by PMA stimulation. PMA is a potent tumor promoter able to intercalate into membrane and activate intracellular PKC by recruiting it to cell membranes. Similar signal variation pattern of SPR images of PC12 cells taken before and after addition of PMA is obtained but at a slower rate, without a sudden increase along the sensorgram since PMA needs to reach the cell interior. The signal also varies with the concentration of PMA between 0 and 1000 nM. PMA at 500 nM is shown enough to stimulate PC12 cells.

The inhibition of PKC can be assessed by a similar SPRi assay. By taking the inhibitor of staurosporine as an example, its interfering with the stimulation can be observed through treating the cells first with staurosporine (e.g., at 100 nM) for 30 min, and then with PMA. The signal reduction of the staurosporine-treated cells can be measured, until to a basal level.

It should be highlighted that SPRi allows to dynamically monitor the translocation of label-free PKC induced by the stimulation of intact Cells, which facilitates the acquisition of reliable data under conditions the same as or close to physiological environments.

7.4.3.5 Screening of Pharmaceutical Cell Line

Reliable and rapid screen of productive cell lines is crucial in biopharmaceutical production of antibodies (e.g., recombinant monoclonal antibodies). This remains a challenge due not only to the highly heterogeneous yield of different cell lines but also to the existence of numerous cell libraries. The presently used methods like limiting dilution, which is widely used for the simple and cost-effective cloning of single cells, are laborious and time-consuming and need a long term of cell culture, leading to low probability of monoclonality and high dependence of efficiency on the total number of cells. This surpasses the present screening ability. It is hence desirable to have a method able to simultaneously isolate single cells and sensitively detect their product(s) in real time. This is now achievable with SPRi that has been demonstrated to be feasible for label-free detection of biomolecules in real time excreted from discrete or crowded intact cells, with a high capturing rate, for example, at 99.1% of the excreted product captured while only 0.9% diffusion into the bulk solution [173].

In order to retrieve the screened cells of interest, the cells must be free or movable (and hence not really imaged by SPRi) rather than captured firmly on the sensor surface. This can be achieved by combination of self-sorting microwell technology [174] with SPRi [175]. Briefly, a tray with an array of cylindrical microwells (Fig. 7.23a) is fabricated, and a microhole is drilled through the bottom of each well, with a size somewhat smaller than a cell to be hosted (Fig. 7.23b). After each well is filled with one cell at most, the microwell tray is placed on a sensor chip that is pre-immobilized with probes, filled with a buffer and sealed in a frame (Fig. 7.23c). The cells in the microwells will then excrete molecules that diffuse down onto the sensor surface. The molecules of interest will be captured on the probes and detected by SPRi. The strongly imaged spots (Fig. 7.23d) indicate the cells with high production, and the corresponding microwells are punched down to harvest the selected cells after disassembly of the tray (Fig. 7.23e). This method enables fast isolation of numerous productive cells. For a chip with effective area of 8 $\times$8 mm^2, it can host about 6400 microwells with 70 μm diameter and 360 μm depth. This implies that the screening throughput can reach 6400 cells per run. Furthermore, this SPRi-based selection shortens the screening time from weeks to only hours [175]. It is due to the unique ability to fast screen intact living cells with high throughput that SPRi can become a very promising cell screening tool.

7.5 Analysis of Bacteria

Bacteria (or germs in English) in a broad sense are prokaryotes, including two groups of eubacteria and archaea, which refers to a large class of primitive unicellular organisms with only pseudonucleus that is a nuclear region where the DNA is not tightly wrapped by membrane. While in a narrow sense, the bacteria are a kind of prokaryotic microorganisms mostly reproduced in the form of dichotomy with thin and short shapes, and simple structures. Bacteria were first discovered by Antony van

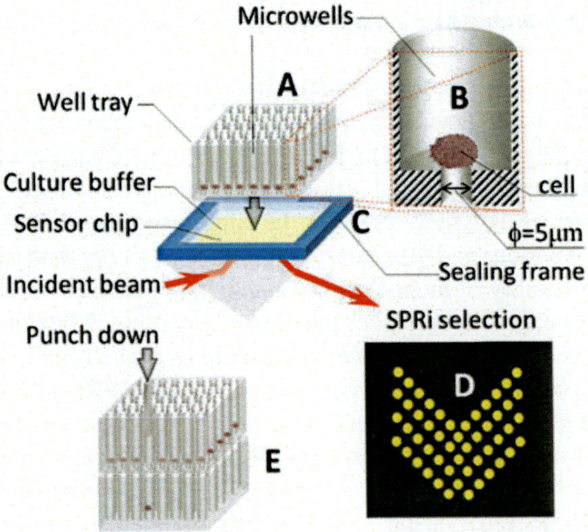

Fig. 7.23 Key steps to screen target cells in combination of cell sorting with SPRi. **a** Microwells filled with one cell each well at most. **b** One cell held in the well bottom with an even smaller through hole. **c** The well tray assembled down to the sensor chip surrounded with a sealing frame and filled with an appropriate volume of culture buffer. **d** Image taken just beneath each microwell to record the cell-excreted molecules that diffuse down through the hole and captured by their probes immobilized on the chip. **e** Collection of each productive cell by punching it down through the bottom hole for further culture to increase their copies

Leeuwemhoek in 1683, and confirmed, by Louis Pausteur, that they could be reproduced from their parents existing in the air. The term of bacteria was coined by Christian Gottfried Ehrenberg in 1828 from a Greek word of βακτηριον that means "little stick".

Bacteria have a size between 0.2 and 5 μM. They are mainly composed of cell membrane, cytoplasm, ribosomes and other parts, and some of them also have special structures such as capsule, flagella, pili and so on. Bacteria can be classified as cocci, bacilli and spirillum based on their shape, as autotrophic and heterotrophic bacteria according to lifestyle, as aerobic and anaerobic bacteria by the demand of oxygen, or as psychrophilic, thermophilic and normothermic bacteria via their survival temperature.

Bacteria are the largest number of individuals in nature and the main participants in the material cycle of nature. They widely distribute in soil and water (including hot springs and even radioactive waste) or live in symbiosis with other organisms. Our human body also carries a considerable number of bacteria, about ten times the total number of human cells in the body and on the epidermis. However, there are so many kinds of bacteria that only a small part of them have been studied and named by scientists. One of the challenges is the lack of advanced tool to analyze them. Any new tool is invaluable if available or exploitable. SPRi is hence tried with new insights into bacteria, evolving till now three basic categories: detection, identification and assess of bacteria and/or bacteria-related processes.

7.5.1 *Imaging of Crowded Bacteria*

Similar to NPs and cells, crowded bacteria can be analyzed directly or indirectly by SPRi. Except that the sample preparation, the analytical procedure is in fact much the same as discussed previously. Since bacteria distribute highly widely, herein discussed will be focused on two types of them, symbiosis with plants and with animals. The main mechanism for specific detection is always based on the immunochemistry, with either antibody or antigen as a probe.

7.5.1.1 Direct Detection

The basic strategy to directly and selectively detect bacteria by SPRi is based on immunochemistry. This can be realized using a sensor chip immobilized with specific monoclonal antibodies as probes to capture the target cells and with polyclonal antibodies as selective amplifier to further increase the detection sensitivity, where the latter technique is also termed sandwich detection. Both channelized and spotting styles of SPRi are usable. Cultured or extracted bacteria are normally subjected to SPRi measurements after they are captured on a spotted or channelized chip by flowing technology. Many types of bacteria have been tested and found to bc quantitative. The LOD of the unamplified assays can reach 1–10 million cells/mL and can be improved for about ten folds after combined with the sandwich technique.

To demonstrate, *Acidovorax avenae* subsp. *citrulli* (*Aac*) is tested with SPRi, by use of an unrelated bacterium of *Xanthomonas camprestris pv. Vesicatoria* as a negative control. *Aac* can transmit devastating crop disease by seed infection and can cause bacterial fruit blotch in watermelons and other cucurbits. The traditional tool to determine *Aac* is ELISA, having an LOD down to 5×10^4 colony forming units (CFU)/mL. As an alternative, SPRi (with 7 channels) was shown to be capable of simplifying the determination. Both intact whole and broken cells of *Aac* are detectable at a concentration down to 10^6 CFU/mL captured on its monoclonal antibody, 11E5, optimized at 10 μg/mL. By further amplification with a polyclonal antibody binding to the membrane protein complex of *Aac*, the LOD can be lowered to 5×10^5 CFU/mL as expectation. Although this LOD remains tenfold higher than that of ELISA, it makes SPRi able to directly detect the intact *Aac* in naturally infected plants. The real applicability has been tested by SPRi of sap samples from the watermelon leaves of naturally infected plant. The sap can be prepared by grinding 0.1 g plant leaves in 4 mL buffer of 137 mM NaCl, 3 mM KCl, 20 mM Na2HPO$_4$, 1 mM KH_2PO_4 and 0.05% Tween20 (pH 7.4) and centrifuged at 2000 rpm for 5 min. The undiluted and diluted (1:400–800 in the buffer) sap samples can be directly injected for SPRi tests. An obvious advantage of SPRi over ELISA is that the chip can be regenerated for at least five cycles. The regeneration is simply achieved by flowing wash away the *Aac* with 10 mM glycine at pH 2.0, at a flow rate of 30 L/min [176].

The selectivity can be validated by detection of *Candida albicans* in a mixture of *Escherichia coli*, *Streptococcus mutans*, *Staphylococcus aureus*, *β-streptococci* and *Lactobacillus casei*. By capture of the target bacterial cells on a channelized chip with specific antibodies [177], the LOD can reach about 10^7 cells/mL detected just after the capture of the target bacteria and reached about 10^6 cells/mL after amplification with polyclonal antibodies.

Clearly, this type of SPRi can simultaneously detect more than one bacteria in complex matrices. This can be accessed by SPRi of food pathogens, *Cronobacter* and *Salmonella* [178]. These food pathogens can especially affect infants and neonates by ingesting contaminated powdered infant formula. Therefore, regulations require the removal of pathogens from these food products. This is need of fast and simultaneous methods to overcome the shortage of the present reference and alternative methods that may take several days to finish analysis. There is also no validated method to do simultaneous detection. Fortunately, SPRi is just able to do such analysis as demonstrated in the detection of four Cronobacter species and eight non-specific bacterial species with one sensor chip spotted with antibody arrays. Although different durations were observed in the detection of different bacteria, such as 24 and 8 min for *C. sakazakii* and *S. Typhimurium*, respectively, SPRi in deed enables the fast and simultaneous detection of multiplex pathogens at a concentration below 30 CFU cells in 25 g powders, within one day.

Prior to SPRi measurements, bacterial samples are often offline cultured, isolated and purified. The bacterial samples can also be cultured directly on the sensor chip while performing SPRi as in the analysis of cells. It is usual that the culture of bacteria takes a long time while in SPRi, this procedure can be simplified by coupling the culture process with SPRi testing in real time, which is also termed capture-culture-measure (CCM) approach [179]. This coupling allows specific determining the low levels of pathogenic bacteria. On an antibody-modified sensor chip, *E. coli* in spring water, and *Salmonella enterica* in raw cow milk and ground meat can be detected by SPRi in real time. Very trace bacteria can also be assayed based on the SPRi starting response time *vs.* the initial concentration of bacteria in samples. The assay can detect bacteria at a concentration down to 2.8 CFU/mL, applicable to the specific and rapid detection of *E. coli O157:H7* in various food processing stress conditions and complex food matrices [180].

SPRi can offer additional advantages in integration with other optical imaging techniques such as Raman microscopy and fluorescence imaging, which provides multiplexed, comparable, even reliable or sensitive results under nearly identical conditions. In fact, SPRi hybridized with fluorescence imaging allows to detect single pathogen cells of *E. coli O157*: *H7* [181]. The protocol is nearly the same as SPRi of single cells by use of an antibody-immobilized sensor chip, the only difference lies in that the bacteria for multiplexed image must be labeled fluorescently. The superiority of the fluorescence images is that it can distinguish the living and dead pathogens multiply tagged with green-fluorescent SYTO®9 and red-fluorescent propidium iodide.

7.5.1.2 Indirect Detection

In case that characteristic molecules related to excreted by target bacterial cells are available, the direct SPRi determination of bacteria can be alternated by indirect format that determines the molecules rather than bacteria to speed up the analytical process. An example is given in the determination of *Legionella pneumophila* able to produce high content of 16S rRNA [182], allowing specific and sensitive detection more than 90% of legionellosis. With a sensor chip immobilized with RNA-paired DNA probes, *L. pneumophila* in complex environment samples, particularly those containing amoebae, can be determined with SPRi after the capture of the extracted 16S rRNA, with an initial concentration of the bacteria lowered to about 2.5×10^5 CFU/mL [183, 184]. The direct detection limit of 16S rRNA is at a sub-femtomole level extracted from *L. pneumophila*, and the total analytical time can be reduced to about 3 h. In short, this indirect SPRi assay is versatile and extendable to the detection of other types of bacteria.

7.5.1.3 Identification of Bacteria

Both direct and indirect SPRi assays are exploitable for bacterial identification by use of a sensor chip spotted with a set of specific probes. Indirect assays need to previously know the identifiable molecules closely associated with bacteria, and its reliability depends on the availability of specific probe set for the molecules. Direct assays are more desirable because they directly identify the target bacteria to save time and increase accuracy. Reliable and fast detection and identification of bacteria are the prerequisite for the treatment of pathogenic diseases. As known, bacteria appear everywhere, and foodborne diseases happen after eating pathogenic contaminated foods. Normal methods for food control need long-term bacterial culture to increase the copy number of bacteria to enable detection. Alternatively, direct SPRi is able to reduce the analytical procedure while increase the throughput. For example, *E. coli* strains have differential recognition patterns after interaction with specific lectins, antibodies, bacteriophages and even carbohydrates, among which carbohydrate probes are desirable, if work, for their low cost and easy availability. Carbohydrate-based SPRi identification of five closely related *E. coli* strains has been validated to be possible [185]. With a sensor chip spotted with seven carbohydrates of glucose, galactose, mannose, fucose, maltose, N-acetylglucosamine and sialic acid that are conjugated based on pyrrole chemistry, *E. coli* strains, including the enterohemorrhagic *E. coli* O157:H7, can be differentiated by the recognition pattern or fingerprinting because each type of the bacteria interacts in different ways with the carbohydrate probes. The detection and discrimination could be finished in less than 10 h from an initial bacterial concentration of 10^2 CFU/mL.

7.5.1.4 Track the Nanomotion of Discrete Bacterial Cells

As mentioned already, interaction of bacteria with some specific molecules may cause undesirable events such as infection and foodborne illness. Most of the currently adopted methods need to culture bacteria prior to imaging, which takes several days to weeks due to the slow growth of bacteria. A faster method is hence highly desirable. In clinical treatments, quick differentiation of bacteria and fast search of narrow spectrum antibiotics for the earliest possible treatments can reduce morbidity and mortality rates. The extreme case is to image a single bacterial cell that is completely free of cell culture. Since SPRi can image discrete particles, it must be able to track the interaction process of single bacterium with other particles including molecules.

In theory, it can be predicted that living bacteria must move in suspensions more freely and longer distance than the deaths. The motion of a particle in suspensions consists of autonomous and passive components. The passive motion comes majorly from particle and molecular collisions due to Brownian motion that must be the same for same-sized bacteria while the autonomous motion includes diffusion for all particles and walking only for living bacteria. It is due to autonomous walking that a living bacterium can move a longer distance than a dead one.

This expectation has been demonstrated in SPRi observation of the vertical (*z*-) motion of individual *E. coli* O157:H7 cells tethered on the sensor chip surface, with precision at a nanometer level [186]. Figure 7.24 shows two typical motions for (A) living and (B) dead bactrium against time, where the death is induced by the addition of polymyxin B, an antibiotic able to penetrate the outer membrane of Gram-negative bacilli to cause death. The motion was quantified by translation of the SPR image contrasts into vertical distances along *z*-axis above the chip surface. It is clear that, for the same bacterium, its motion range is significantly reduced from living to dead states. The nanomotion amplitude of a living bacterium seems to be affected somewhat by the suspension conditions (Fig. 7.24c), which is reasonable since the bacteria can adsorb or desorb substances from or to its surroundings. However, this type of variation is much smaller than the death-caused change. This measurement can be extended to the study of the nanomotion of other bacteria, for example, uropathogenic *E. coli* CFT073 strain killed by either polymyxin B or streptomycin.

Besides the vertical motion, the lateral motion may also be visible in SPRi of discrete bacteria. Due to the lateral resolution limitation (at about 10 μm), the bacteria must not be tethered on the chip so that the bacterium can freely move and walk to a sufficiently long or spatially distinguishable distance. This method remains waiting for exploitation. In general, a free bacterium will walk three-dimensionally, unless specifically marked, the walking bacterium can hardly be followed. However, this method is worthy of exploitation because they can seemingly possess the advantages of easy manipulation and intuitive extraction of distance.

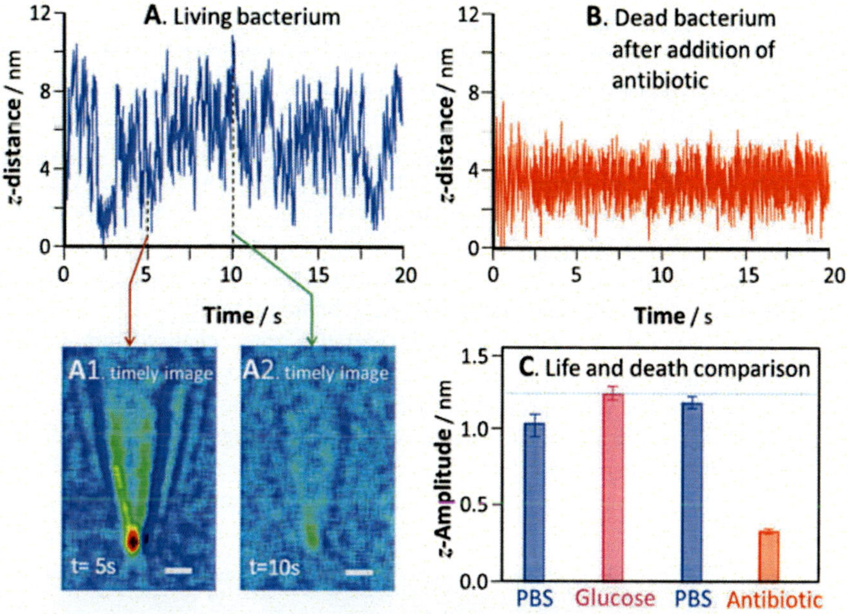

Fig. 7.24 SPRi determination of the motion of tethered pathogenic *E. coli* O157:H7 with 680 nm superluminescent diode as an excitation source. The bacteria were tethered on antibody that was pre-immobilized on a 47-nm thick gold film self-assembled with a monolayer of PEG/PEG-COOH via EDS-NHS chemistry. **a** Vertical (*z*-axis) motion curve of a living bacterium; (A1) Images taken at 5 s (close to sensor surface); (A2) Images taken at 10 s (far away from the sensor surface). **b** Motion curve of the bacterium killed by addition of polymyxin B that can penetrate the outer membrane of Gram-negative bacilli. **c** Motion amplitude of the imaged bacterium primarily in PBS buffer, then in a glucose solution, back to PBS, and finally killed in antibiotic. Reconstructed from Ref. [186] with permission

7.5.1.5 Monitoring the Bacterium-Protein Interaction

Interaction of bacteria with molecules is in some extend more important than the interaction between molecules. This can be considered to be the first gate to kill or secure bacteria. Bacterial walking or motion is potentially a mechanism to seek the protecting and/or to avoid the killing molecules. It is thus significant to track the interaction between bacteria and molecules to understand the pathogenesis, antibiotic resistance, antimicrobial action and immune evasion. The interaction information can be acquired either with grouped bacteria or with single bacterium. The former information represents statistical average behavior that can easily be analyzed with many normal methods while the latter information reflects the inherent heterogeneity from bacterium to bacterium and is hard to acquire till now, which needs a unique or sensitive tool to track and/or quantify the interactions of bacteria with molecules at the single bacterium level. As a label-free and real-time tool, SPRi possesses an inherent advantage to study the binding kinetics of single bacterium interacting with molecules or more specifically ligands.

This has been successfully assessed with *E. coli* O157:H7 strain [187], where the discrete bacteria are captured on an antibody-modified sensor chip and allowed to further interact with free antibodies to monitor the binding and dissociating process. The antibody used is the goat anti-*E. coli* O157:H7 IgG polyclonal antibody (Ab157) while that of negative control is goat anti-*E. coli* O145 IgG polyclonal antibody (Ab145). The full binding and dissociating process is recorded in real time as a video. From the distinct plasmonic images or video of discrete bacteria as illustrated in Fig. 7.25, the binding kinetic can be vividly followed and the related constants, k_{on}, k_{off} and K_D (=k_{off}/k_{on}), can be determined, at a single bacterium level, based on the first-order kinetic equation by fitting technology.

As expected, the binding and dissociating rates change from bacterium to bacterium as illustrated in Fig. 7.26, where A, B and C show the distributions of k_{on}, k_{off} and K_D values, respectively [187]. Among the observed 33 bacterial cells, k_{on} and k_{off} have a variability about two orders of magnitude but seemingly do not have a statistic center, possibly due to the number of the observed cells is not enough. Astonishing, the distribution of K_D has a center at about 3.9 nM that is close to the

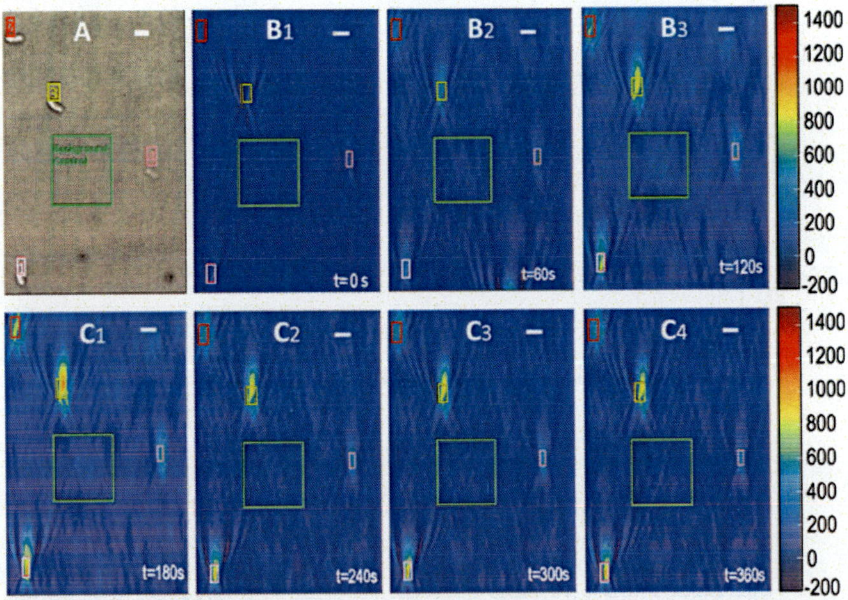

Fig. 7.25 Time-difference SPR images of captured discrete *E. coli* O157:H7 during binding and dissociating Ab157 antibody. The antibody was immobilized on a chip coated with PEG/PEG-COOH via EDS-NHS chemistry. The capture of bacteria was initiated by a 330-μl/min flow of 10 μg/mL Ab157 in PBS for 3 min while the dissociation was started at the binding end by flowing in PBS for > 4 min □. Area to average background signal; □Single cell captured; Color bar. Imaging intensity; Scale bar. 2 μm. **a** Image taken by transmission bright-field microscope to indicate the position of the captured cells; (B1–B3) Images taken during the binding phase at 0, 60 and 120 s, respectively; (C1–C4) Images taken during the dissociation phase at 180, 240, 300 and 360 s, respectively. Reconstructed from Ref. [187] with permission

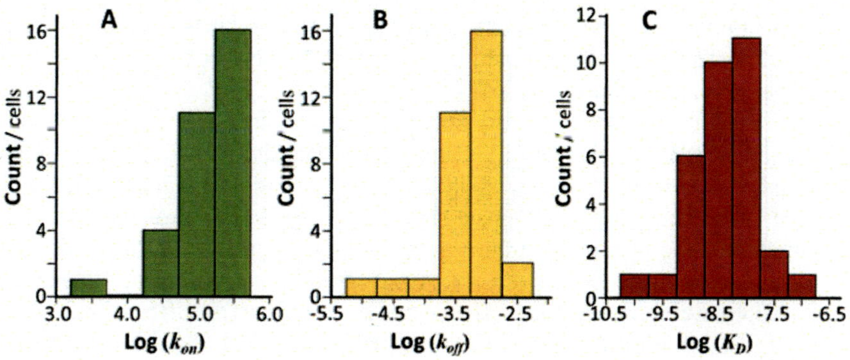

Fig. 7.26 Distributions of **a** binding rate, k_{on}, **b** dissociating rate, k_{off}, and **c** dissociation constant, K_D measured from 33 individual *E. coli* O157:H7 cells in reaction with Ab157 antibody. Reconstructed from Ref. [187] with permission

values reported in literature [188]. The very wide distribution of K_D, near four orders of magnitude, suggests the great heterogeneity of individual bacteria in the recognition of a same type of antibody. This heterogeneity starting from outside the cells may elicit differentiated intracellular events.

7.5.1.6 Study of Bacterial Biofilms

As a tool originally invented from metallic films, SPRi is undoubtedly applicable to the study of bacterial films formed by highly crowded bacteria. However, this advantage has not yet attracted enough attention even after it has been demonstrated to be unique in monitoring the formation of biofilms on a gold sensor surface with some model bacteria such as *E. coli* and *Pseudomonas aeruginosa* [2, 189]. It is known that the most species of living bacteria tend to form slimy biofilms on some surfaces such as plastics and metals, which is responsible for about 60% of hospital-acquired infections [190]. Once adhering to some surfaces, bacteria form biofilms through excretion or discretion of exopolysaccharides, peptides and metabolites [191, 192]. The bacteria in the biofilms will significantly increase their resistance to antibacterial chemicals, by a factor up to 1000 times, compared with them in suspensions [193]. It can never be overemphasized to study the conditions responsible for forming and/or destroying the biofilms *in situ* and in real time. This is what SPRi is good at and must be easier and more cost-effective than fluorescent and scanning electron microscope (SEM) techniques.

With *E. coli* and *Pseudomonas aeruginosa* as testing samples, SPRi has been validated to be capable of observing the process to form biofilms on a gold sensor surface [2, 189]. Similar to molecules, SPRi of highly crowded bacterial films is free of the interference fringes. From this type of real-time images, the biofouling can be observed to increase with time and saturated at about 10 h. Because *P. aeruginosa* and *S. aureus* are hydrophobic, they are expected to easily attach and accumulate on hydrophobic surface like aging bare gold. It must be known that a freshly cleaned gold surface is usually hydrophilic, but the surface soon becomes hydrophobic during the experimental manipulations.

Reasonably, SPRi can also be used to observe the removal of the biofilms, for example, the detachment of a mature *E. coli* biofilm from the sensor chip surface under a detergent flow (e.g., SDS). A more interesting idea must be the utilization of surface coatings to eliminate the harmful biofilms from the source, that is, to inhibit the formation of biofilms from the very beginning. Abadian et al. found that bare gold surface, protein (e.g., BSA and casein) and antibiotic (e.g., penicillin with streptomycin) coatings had different inhibition ability [189]. The casein-coated gold surface exhibited the stronger anti-bacterial ability than BSA while the bare gold is the weakest. It is unexpected that the penicillin/streptomycin cocktail showed only short-term effect to inhibit the formation of biofilms, shorter than 24 h referred to the bare gold. It seems that there is a linkage between the inhibition effect of a coating and the bacteria to be inhibited. SPRi images (taken every 3 min) revealed that, compared with bare gold, a lubricin coating had over 90% clearance after 24 h on-chip incubation while the coating of bovine submaxillary mucin only led to approximately 7% reduction [194].

As all the above-discussed, SPRi can perform from single to fully crowded particles, ranging from nanometers up to micrometers, offering extra- and intraparticle information in respect of composition, interaction, signal transduction, subparticle movement and so forth. The unique advantages include the ability to *in situ* and in real-time inspect intact and/or objects (e.g., cells, bacteria and related organelles). It is applicable to the static or dynamic tracking or monitoring of various reactions, with various possibilities to amplify the detection sensitivity.

References

1. Hide M, Tsutsui T, Sato H, Nishimura T, Morimoto K, Yamamoto S, Yoshizato K (2002) Real-time analysis of ligand-induced cell surface and intracellular reactions of living mast cells using a surface plasmon resonance-based biosensor. Anal Biochem 302:28–37
2. Abadian PN, Kelley CP, Goluch ED (2014) Cellular analysis and detection using surface plasmon resonance techniques. Anal Chem 86:2799–2812
3. Milgram S, Bombera R, Livache T, Roupioz Y (2012) Antibody microarrays for label-free cell-based applications. Methods 56:326–333
4. Vala M, Etheridge S, Roach JA, Homola J (2009) Long-range surface plasmons for sensitive detection of bacterial analytes. Sens Actuators B Chem 139:59–63
5. Vala M, Robelek R, Bocková M, Wegener J, Homola J (2013) Real-time label-free monitoring of the cellular response to osmotic stress using conventional and long-range surface plasmons. Biosens Bioelectron 40:417–421
6. Kosaihira A, Ona T (2008) Rapid and quantitative method for evaluating the personal therapeutic potential of cancer drugs. Anal Bioanal Chem 391:1889–1897
7. Chabot V, Cuerrier CM, Escher E, Aimez V, Grandbois M, Charette PG (2009) Biosensing based on surface plasmon resonance and living cells. Biosens Bioelectron 24:1667–1673
8. Lee SH, Ko HJ, Park TH (2009) Real-time monitoring of odorant-induced cellular reactions using surface plasmon resonance. Biosens Bioelectron 25:55–60

9. Nishijima H, Kosaihira A, Shibata J, Ona T (2010) Development of signaling echo method for cell-based quantitative efficacy evaluation of anti-cancer drugs in apoptosis without drug presence using high-precision surface plasmon resonance sensing. Anal Sci 26:529–534
10. Chen H, Huang J, Lee J, Hwang S, Koh K (2010) Surface plasmon resonance spectroscopic characterization of antibody orientation and activity on the calixarene monolayer. Sens Actuators B 147:548–553
11. Maltais JS, Denault JB, Gendron L, Grandbois M (2012) Label-free monitoring of apoptosis by surface plasmon resonance detection of morphological changes. Apoptosis 17:916–925
12. Lahav A, Auslender M, Abdulhalim I (2008) Sensitivity enhancement of guided wave surface plasmon resonance sensors. Opt Lett 33:2539–2541
13. Krasnykov O, Karabchevsky A, Shalabney A, Auslender M, Abdulhalim I (2011) Sensor with increased sensitivity based on enhance doptical transmission in the infrared. Opt Commun 284:1435–1438
14. Shalabney A, Abdulhalim I (2011) Sensitivity enhancement methods for surface plasmon sensors. Lasers Photon Rev 5:571–606
15. Shalabney A, Abdulhalim I (2012) Figure of merit enhancement of surface plasmon resonance sensors in the spectral interrogation. Opt Lett 37:1175–1177
16. Ziblat R, Lirtsman V, Davidov D, Aroeti B (2006) Infrared surface plasmon resonance: a novel tool for real time sensing of variations in living cells. Biophys J 90:2592–2599
17. Yashunsky V, Shimron S, Lirtsman V, Weiss AM, Melamed-Book N, Golosovsky M, Davidov D, Aroeti B (2009) Real-time monitoring of transferrin-induced endocytic vesicle formation by mid-infrared surface plasmon resonance. Biophys J 97:1003–1012
18. Filipe V, Hawe A, Jiskoot W (2010) Critical evaluation of nanoparticle tracking analysis (NTA) by nanosight for the measurement of nanoparticles and protein aggregates. Pharm Res 27:796–810
19. Patolsky F, Zheng G, Hayden O, Lakadamyali M, Zhuang X, Lieber CM (2004) Electrical detection of single viruses. Proc Natl Acad Sci 101:14017–14022
20. Wanekaya AK, Chen W, Myung NV, Mulchandani A (2006) Nanowire-based electrochemical biosensors. Electroanalysis 18:533–550
21. Ymeti A, Greve J, Lambeck PV, Wink T, van Hövell SWFM, Beumer TAM, Wijn RR, Heideman RG, Subramaniam V, Kanger JS (2007) Fast, ultrasensitive virus detection using a young interferometer sensor. Nano Lett 7:394–397. https://doi.org/10.1021/nl062595n
22. Ramachandran A, Wang S, Clarke J, Ja SJ, Goad D, Wald L, Flood EM, Knobbe E, Hryniewicz JV, Chu ST, Gill D, Chen W, King O, Little BE (2008) A universal biosensing platform based on optical micro-ring resonators. Biosens Bioelectron 23:939–944
23. Vollmer F, Arnold S, Keng D (2008) Single virus detection from the reactive shift of a whispering-gallery mode. Proc Natl Acad Sci 105:20701–20704
24. Giese B, Klaessig F, Park B, Kaegi R, Steinfeldt M, Wigger H, von Gleich A, Gottschalk F (2018) Risks, release and voncentrations of engineered nanomaterial in the environment. Sci Rep 8:1565. https://doi.org/10.1038/s41598-018-19275-4
25. Wunderlich L, Hausler P, Märkl S, Bierl R, Hirsch T (2021) Nanoparticle determination in water by LED-excited surface plasmon resonance imaging. Chemosensors. 9:175–184. https://doi.org/10.3390/chemosensors9070175
26. Jung LS, Campbell CT, Chinowsky TM, Mar M, Yee SS (1998) Quantitative interpretation of the response of surface plasmon resonance sensors to adsorbed films. Langmuir 14:5636–5648
27. Berger CEH, Kooyman RPH, Greve J (1994) Resolution in surface plasmon microscopy. Rev Sci Instrum 65:2829–2836
28. Berger CEH, Kooyman RPH, Greve J (1999) Surface plasmon propagation near an index step. Opt Comm 167:183–189
29. Rothenhäusler B, Knoll W (1988) Surface plasmon microscopy. Nature 332:615–617
30. Weichert F, Gaspar M, Timm C, Zybin A, Gurevich EL, Engel M, Müller H, Marwedel P (2010) Signal analysis and classification for plasmon assisted microscopy of nanoobjects. Sens Actuators B 151:281–290

31. Zybin A, Kuritsyn YA, Gurevich EL, Temchura VV, Überla K, Niemax K (2010) Real-time detection of single immobilized nanoparticles by surface plasmon resonance imaging. Plasmonics 5:31–35. https://doi.org/10.1007/s11468-009-9111-5
32. Brockman JM, Nelson BP, Corn RM (2000) Surface plasmon resonance imaging measurements of ultrathin organic films. Annu Rev Phys Chem 51:41–63
33. Homola J (2008) Surface plasmon resonance sensors for detection of chemical and biological species. Chem Rev 108:462–493
34. Wang J, Munir A, Zhu Z, Zhou HS (2010) Magnetic nanoparticle enhanced surface plasmon resonance sensing and its application for the ultrasensitive detection of magnetic nanoparticle-enriched small molecules. Anal Chem 82:6782–6789
35. Piliarik M, Homola J (2008) Self-referencing SPR imaging for most demanding high throughput screening applications. Sens Actuators B 134:353–355
36. Gurevich EL, Temchura VV, Überla K, Zybin A (2011) Analytical features of particle counting sensor based on plasmon assisted microscopy of nano objects. Sens Actuators 160:1210–1215
37. Victoria S (2012) Application of surface plasmon resonance (SPR) for the detection of single viruses and single biological nano-objects. J Bacteriol Parasitol 3:7. https://doi.org/10.4172/2155-9597.1000e110
38. Valle PJ, Ortiz EM, Saiz JM (1997) Near field by subwavelength particles on metallic substrates with cylindrical surface plasmon excitation. Opt Commun 137:334–342
39. Hecht B, Bielefeldt H, Novotny L, Inouye Y, Pohl DW (1996) Local excitation, scattering, and interference of surface plasmons. Phys Rev Lett 77:1889–1892
40. Konopsky VN, Kouyanov KE, Novikova NN (2001) Investigations of the interference of surface plasmons on rough silver surface by scanning plasmon near-field microscope. Ultramicroscopy 88:127–138
41. Kretschmann E (1971) Die bestimmung optischer konstanten von metallen durch anregung von oberfliichenplasmaschwingungen. Z Phys 241:313–324
42. Sjölander S, Urbaniczky C (1991) Integrated fluid handling system for biomolecular interaction analysis. Anal Chem 63:2338–2345
43. Ruenraroengsak P, Florence AT (2005) The diffusion of latex nanospheres and the effective (microscopic) viscosity of HPMC gels. Int J Pharm 298:361–366
44. Zybin A, Boecker D, Mirsky VM, Niemax K (2007) Enhancement of the detection power of surface plasmon resonance measurements by optimization of the reflection angle. Anal Chem 79:4233–4236
45. Théry C, Zitvogel L, Amigorena S (2002) Exosomes: composition, biogenesis and function. Nat Rev Immunol 2:569–579
46. Koga K, Matsumoto K, Akiyoshi T, Kubo M, Yamanaka N, Tasaki A, Nakashima H, Nakamura M, Kuroki S, Tanaka M, Katano M (2005) Purification, characterization and biological significance of tumor-derived exosomes. Anticancer Res 25:3703–3707
47. Ho MM, Ng AV, Lam S, Hung JY (2007) Side population in human lung cancer cell lines and tumors is enriched with stem-like cancer cells. Cancer Res 67:4827–4833
48. El Andaloussi S, Mäger I, Breakefield XO, Wood MJA (2013) Extracellular vesicles: biology and emerging therapeutic opportunities. Nat Rev Drug Discov 12:347–357
49. Kourembanas S (2014) Exosomes: vehicles of intercellular signaling, biomarkers, and vectors of cell therapy. Annu Rev Physiol 77:13–27
50. Sáenz-Cuesta M, Osorio-Querejeta I, Otaegui D (2014) Extracellular vesicles in multiple sclerosis: what are they telling us? Front Cell Neurosci 8:1–9. Article 100. https://doi.org/10.3389/fncel.2014.00100
51. Yáñez-Mó M, Siljander PR-M, Andreu Z, Zavec AB, Borràs FE, Buzas EI, Buzas K, Casal E, Cappello F, Carvalho J, Colás E, da Silva AC, Fais S, Falcon-Perez JM, Ghobrial IM, Giebel B, Gimona M, Graner M, Gursel I, Gursel M, Heegaard NHH, Hendrix A, Kierulf P, Kokubun K, Kosanovic M, Kralj-Iglic V, Krämer-Albers E-M, Laitinen S, Lässer C, Lener T, Ligeti E, Linē A, Lipps G, Llorente A, Lötvall J, Manček-Keber M, Marcilla A, Mittelbrunn M, Nazarenko I, Nolte-'t Hoen ENM, Nyman TA, O'Driscoll L, Olivan M, Oliveira C, Pállinger É, del Portillo HA, Reventós J, Rigau M, Rohde E, Sammar M, SánchezMadrid F, Santarém

N, Schallmoser K, Ostenfeld MS, Stoorvogel W, Stukelj R, Van der Grein SG, Vasconcelos MH, Wauben MHM, De Wever O (2015) Biological properties of extracellular vesicles and their physiological functions. J Extracell Vesicles 4:27066–27126

52. Harding C, Heuser J, Stahl P (1983) Receptor-mediated endocytosis of transferrin and recycling of the transferrin receptor in rat reticulocytes. J Cell Biol 97:329–339
53. Pan BT, Teng K, Wu C, Adam M, Johnstone RM (1985) Electron microscopic evidence for externalization of the transferrin receptor in vesicular form in sheep reticulocytes. J Cell Biol 101:942–948
54. Raposo G, Stoorvogel W (2013) Extracellular vesicles: exosomes, microvesicles, and friends. J Cell Biol 200:373–383
55. Zhu L, Wang K, Cui J, Liu H, Bu X, Ma H, Wang W, Gong H, Lausted C, Hood L, Yang G, Hu Z (2014) Label-free quantitative detection of tumor-derived exosomes through surface plasmon resonance imaging. Anal Chem 86:8857–8864
56. Willms E, Johansson HJ, Mäger I, Lee Y, Blomberg KEM, Sadik M, Alaarg A, Smith CIE, Lehtiö J, El Andaloussi S, Wood MJ, Vader P (2016) Cells release subpopulations of exosomes with distinct molecular and biological properties. Sci Rep 6:22519. https://doi.org/10.1038/srep22519
57. Gualerzi A, Niada S, Giannasi C, Picciolini S, Morasso C, Vanna R, Rossella V, Masserini M, Bedoni M, Ciceri F, Bernardo ME, Brini AT, Gramatica F (2017) Raman spectroscopy uncovers biochemical tissue-related features of extracellular vesicles from mesenchymal stromal cells. Sci Rep 7:9820. https://doi.org/10.1038/s41598-017-10448-1
58. Ha D, Yang N, Nadithe V (2016) Exosomes as therapeutic drug carriers and delivery vehicles across biological membranes: current perspectives and future challenges. Acta Pharm Sin B 6:287–296
59. Boriachek K, Islam M N, Möller A, Salomon C, Nguyen N-T, Hossain MSA, Yamauchi Y, Shiddiky MJA (2018) Small 14:1702153. https://doi.org/10.1002/smll.201702153
60. Aryani A, Denecke B (2016) Exosomes as a nanodelivery system: a key to the future of neuromedicine? Mol Neurobiol 53:818–834
61. Jarmalavičiūtė A, Pivoriūnas A (2016) Exosomes as a potential novel therapeutic tools against neurodegenerative diseases. Pharmacol Res 113:816–822
62. García-Romero N, Carrión-Navarro J, Esteban-Rubio S, Lázaro-Ibáñez E, Peris-Celda M, Alonso MM, Guzmán-De-Villoria J, Fernández-Carballal C, de Mendivil AO, García-Duque S, Escobedo-Lucea C, Prat-Acín R, Belda-Iniesta C, Ayuso-Sacido A (2017) DNA sequences within glioma-derived extracellular vesicles can cross the intact blood-brain barrier and be detected in peripheral blood of patients. Oncotarget 8:1416–1428
63. Camussi G, Deregibus MC, Bruno S, Cantaluppi V, Biancone L (2010) Exosomes/microvesicles as a mechanism of cell-to-cell communication. Kidney Int 78:838–848
64. Tkach M, Théry C (2016) Communication by extracellular vesicles: where we are and where we need to go. Cell 164:1226–1232
65. Yang C, Robbins PD (2011) The roles of tumor-derived exosomes in cancer pathogenesis. Clin Dev Immunol 2011:842–849. https://doi.org/10.1155/2011/842849
66. Tickner JA, Urquhart AJ, Stephenson SA, Richard DJ, O'Byrne KJ (2014) Functions and therapeutic roles of exosomes in cancer. Front Oncol 4:127. https://doi.org/10.3389/fonc.2014.00127
67. Lim Y-J, Lee S-J (2017) Are exosomes the vehicle for protein aggregate propagation in neurodegenerative diseases? Acta Neuropathol Commun 5:64. https://doi.org/10.1186/s40478-017-0467-z
68. Quek C, Hill AF (2017) The role of extracellular vesicles in neurodegenerative diseases. Biochem Biophys Res Commun 483:1178–1186
69. Soria FN, Pampliega O, Bourdenx M, Meissner WG, Bezard E, Dehay B (2017) When cooperation was efficient or inefficient. Functional near-infrared spectroscopy evidence. Front Neurosci 11:26. https://doi.org/10.3389/fnsys.2017.00026
70. György B, Szabó TG, Pásztói M, Pál Z, Misják P, Aradi B, László V, Pállinger E, Pap E, Kittel A, Nagy G, Falus A, Buzás EI (2011) Membrane vesicles, current state-of-the-art: emerging

role of extracellular vesicles. Cell Mol Life Sci 68:2667–2688. https://doi.org/10.1007/s00018-011-0689-3
71. Cobelli NJ, Leong DJ, Sun HB (2017) Exosomes: biology, therapeutic potential, and emergingrole in musculoskeletal repair and regeneration. Ann N Y Acad Sci 1410:57–67
72. Zhang S, Chuah SJ, Lai RC, Hui JHP, Lim SK, Toh WS (2018) MSC exosomes mediate cartilage repair by enhancing proliferation, attenuating apoptosis and modulating immune reactivity. Biomaterials 156:16–27. https://doi.org/10.1016/j.biomaterials.2017.11.028
73. Wu X, Chen H, Wang X (2012) Can lung cancer stem cells be targeted for therapies? Canc Treat Rev 38:580–588
74. Lobb RJ, van Amerongen R, Wiegmans A, Ham S, Larsen JE, Möller A (2017) Exosomes derived from mesenchymal non-small cell lung cancer cells promote chemoresistance. Int J Canc 141:614–620
75. Van Niel G, D'Angelo G, Raposo G (2018) Shedding light on the cell biology of extracellular vesicles. Nat Rev Mol Cell Biol 19:213–228
76. Wang N, Song X, Liu L, Niu L, Wang X, Song X, Xie L (2018) Circulating exosomes contain protein biomarkers of metastatic non-small-cell lung cancer. Canc Sci 109:1701–1709. https://doi.org/10.1111/cas.13581
77. Grey M, Dunning CJ, Gaspar R, Grey C, Brundin P, Sparr E, Linse S (2015) Acceleration of α-synuclein aggregation by exosomes. J Biol Chem 290:2969–2982
78. Rajendran L, Honsho M, Zahn TR, Keller P, Geiger KD, Verkade P, Simons K (2006) Alzheimer's disease β-amyloid peptides are released in association with exosomes. Proc Natl Acad Sci U S A 103:11172–11177
79. Fiandaca MS, Kapogiannis D, Mapstone M, Boxer A, Eitan E, Schwartz JB, Abner EL, Petersen RC, Federoff HJ, Miller BL, Goetzl E (2015) Identification of preclinical Alzheimer's disease by a profile of pathogenic proteins in neurally derived blood exosomes: a case-control study. J Alzheimer's Dement 11:600−607.e1. https://doi.org/10.1016/j.jalz.2014.06.008
80. Witwer KW, Buzás EI, Bemis LT, Bora A, Lässer C, Lötvall J, Nolte-'t Hoen EN, Piper MG, Sivaraman S, Skog J, Théry C, Wauben MH, Hochberg F (2013) Standardization of sample collection, isolation and analysis methods in extracellular vesicle research. J Extracell Vesicles 2:20360. https://doi.org/10.3402/jev.v2i0.20360
81. Ferhan AR, Jackman JA, Park JH, Cho NJ, Kim DH (2018) Nanoplasmonic sensors for detecting circulating cancer biomarkers. Adv Drug Deliv Rev 125:48–77
82. Shao H, Im H, Castro CM, Breakefield X, Weissleder R, Lee H (2018) New technologies for analysis of extracellular vesicles. Chem Rev 118:1917–1950
83. Kwizera EA, O'Connor R, Vinduska V, Williams M, Butch ER, Snyder SE, Chen X, Huang X (2018) Molecular detection and analysis of exosomes using surface-enhanced Raman scattering gold nanorods and a miniaturized device. Theranostics 8:2722–2738
84. Li T-D, Zhang R, Chen H, Huang Z-P, Ye X, Wang H, Deng A-M, Kong J-L (2018) An ultrasensitive polydopamine bi-functionalized SERS immunoassay for exosome-based diagnosis and classification of pancreatic cancer. Chem Sci 9:5372–5382
85. Jiang Y, Shi M, Liu Y, Wan S, Cui C, Zhang L, Tan W (2017) Aptamer/AuNP biosensor for colorimetric profiling of exosomal proteins. Angew Chem Int Ed 56:11916–11920
86. Yu X, He L, Pentok M, Yang H, Yang Y, Li Z, He N, Deng Y, Li S, Liu T (2019) An aptamer-based new method for competitive fluorescence detection of exosomes. Nanoscale 11:15589–15595. https://doi.org/10.1039/C9NR04050A
87. Tang Y-T, Huang Y-Y, Zheng L, Qin S-H, Xu X-P, An T-X, Xu Y, Wu Y-S, Hu X-M, Ping B-H, Wang Q (2017) Comparison of isolation methods of exosomes and exosomal RNA from cell culture medium and serum. Int J Mol Med 40:834−844
88. Sina AAI, Vaidyanathan R, Dey S, Carrascosa LG, Shiddiky MJA, Trau M (2016) Real time and label free profiling of clinically relevant exosomes. Sci Rep 6:30460. https://doi.org/10.1038/srep30460
89. Im H, Shao H, Park YI, Peterson VM, Castro CM, Weissleder R, Lee H (2014) Label-free detection and molecular profiling of exosomes with a nano-plasmonic sensor. Nat Biotechnol 32:490–495

90. Grasso L, Wyss R, Weidenauer L, Thampi A, Demurtas D, Prudent M, Lion N, Vogel H (2015) Molecular screening of cancer-derived exosomes by surface plasmon resonance spectroscopy. Anal Bioanal Chem 407:5425–5432
91. Di Noto G, Bugatti A, Zendrini A, Mazzoldi EL, Montanelli A, Caimi L, Rusnati M, Ricotta D, Bergese P (2016) Merging colloidal nanoplasmonics and surface plasmon resonance spectroscopy for enhanced profiling of multiple myeloma-derived exosomes. Biosens Bioelectron 77:518–524
92. Van der Pol E, Coumans F, Varga Z, Krumrey M, Nieuwland R (2013) Innovation in detection of micro particles and exosomes. J Thromb Haemostasis 11:36–45. https://doi.org/10.1111/jth.12254
93. Raposo G, Nijman HW, Stoorvogel W, Liejendekker R, Harding CV, Melief CJ, Geuze HJ (1996) B lymphocytes secrete antigen-presenting vesicles. J Exp Med 183:1161–1172
94. Fan Y, Duan X, Zhao M, Wei X, Wu J, Chen W, Liu P, Cheng W, Cheng Q, Ding S (2020) High-sensitive and multiplex biosensing assay of NSCLC-derived exosomes *via* different recognition sites based on SPRi array. Biosens Bioelectron 154:112066. https://doi.org/10.1016/j.bios.2020.112066
95. Théry C, Boussac M, Veron P, Ricciardi-Castagnoli P, Raposo G, Garin J, Amigorena S (2001) Proteomic analysis of dendritic cell-derived exosomes: a secreted subcellular compartment distinct from apoptotic vesicles. J Immunol 166:7309–7318
96. Ji H, Erfani N, Tauro BJ, Kapp EA, Zhu HJ, Moritz RL, Lim JW, Simpson RJ (2008) Difference gel electrophoresis analysis of Ras-transformed fibroblast cell-derived exosomes. Electrophoresis 29:2660–2671
97. Tauro BJ, Greening DW, Mathias RA, Ji H, Mathivanan S, Scott AM, Simpson RJ (2012) Comparison of ultracentrifugation, density gradient separation, and immunoaffinity capture methods for isolating human colon cancer cell line LIM1863-derived exosomes. Methods 56:293–304
98. Rood IM, Deegens JK, Merchant ML, Tamboer WP, Wilkey DW, Wetzels JF, Klein JB (2010) Comparison of three methods for isolation of urinary microvesicles to identify biomarkers of nephrotic syndrome. Kidney Int 78:810–816
99. Clayton A, Court J, Navabi H, Adams M, Mason MD, Hobot JA, Newman GR, Jasani B (2001) Analysis of antigen presenting cell derived exosomes, based on immuno-magnetic isolation and flow cytometry. J Immunol Methods 247:163–174
100. Valadi H, Ekstrom K, Bossios A, Sjostrand M, Lee JJ, Lotvall JO (2007) Exosome-mediated transfer of mRNAs and microRNAs is a novel mechanism of genetic exchange between cells. Nat Cell Biol 9:654–659
101. Ostrowski M, Carmo NB, Krumeich S, Fanget I, Raposo G, Savina A, Moita CF, Schauer K, Hume AN, Freitas RP, Goud B, Benaroch P, Hacohen N, Fukuda M, Desnos C, Seabra MC, Darchen F, Amigorena S, Moita LF, Thery C (2010) Rab27a and Rab27b control different steps of the exosome secretion pathway. Nat Cell Biol 12:19–30
102. Baranyai T, Herczeg K, Onódi Z, Voszka I, Módos K, Marton N, Nagy G, Mäger I, Wood MJ, El Andaloussi S, Pálinkás Z, Kumar V, Nagy P, Buzás EI, Ferdinandy P, Giricz Z, Kittel Á (2015) Isolation of exosomes from blood plasma: qualitative and quantitative comparison of ultracentrifugation and size exclusion chromatography methods. PLoS ONE 10(12):e0145686. https://doi.org/10.1371/journal.pone.0145686
103. Welton JL, Webber JP, Botos L-A, Jones M, Clayton A (2015) Ready-made chromatography columns for extracellular vesicle isolation from plasma. J Extracell Vesicles 4:27269. https://doi.org/10.3402/jev.v4.27269
104. Mol EA, Goumans M-J, Doevendans PA, Sluijter JPG, Vader P (2017) Higher functionality of extracellular vesicles isolated using size-exclusion chromatography compared to ultracentrifugation. Nanomedicine 13:2061–2065
105. Gámez-Valero A, Monguió-Tortajada M, Carreras-Planella L, Franquesa M, Beyer K, Borràs FE (2016) Size-exclusion chromatography-based isolation minimally alters extracellular vesicles' characteristics compared to precipitating agents. Sci Rep 6:33641. https://doi.org/10.1038/srep33641

106. Nakai W, Yoshida T, Diez D, Miyatake Y, Nishibu T, Imawaka N, Naruse K, Sadamura Y, Hanayama R (2016) A novel affinity-based method for the isolation of highly purified extracellular vesicles. Sci Rep 6:33935. https://doi.org/10.1038/srep33935
107. Maiolo D, Paolini L, Di Noto G, Zendrini A, Berti D, Bergese P, Ricotta D (2015) Colorimetric nanoplasmonic assay to determine purity and titrate extracellular vesicles. Anal Chem 87:4168–4176
108. Picciolini S, Gualerzi A, Vanna R, Sguassero A, Gramatica F, Bedoni M, Masserini M, Morasso C (2018) Detection and characterization of different brain-derived subpopulations of plasma exosomes by surface plasmon resonance imaging. Anal Chem 90:8873–8880
109. Mustapic M, Eitan E, Werner JK, Berkowitz ST, Lazaropoulos MP, Tran J, Goetzl EJ, Kapogiannis D (2017) Plasma extracellular vesicles enriched for neuronal origin: a potential window into brain pathologic processes. Front Neurosci 11:278. https://doi.org/10.3389/fnins.2017.00278
110. Bally M, Gunnarsson A, Svensson L, Larson G, Zhdanov VP, Höök F (2011) Phys Rev Lett 107:188103-1–5. https://doi.org/10.1103/PhysRevLett.107.188103
111. Rich RL, Myszka DG (2003) Spying on HIV with SPR. Trends Microbiol 11:124–133
112. Caygill RL, Blair GE, Millner PA (2010) A review on viral biosensors to detect human pathogens. Analyt Chim Acta. 681:8–15. https://doi.org/10.1038/ki.2010.278
113. Rothenhäusler B, Knoll W (1987) Plasmon surface polariton fields versus TIR evanescent waves for scattering experiments at surfaces. Opt Commun 63:301–304
114. Rothenhäusler B, Knoll W (1987) Total internal diffraction of plasmon surface polaritons. Appl Phys Lett 51:783–785
115. Uhnoo I, Svensson L, Wadell G (1990) Enteric adenoviruses. *Baillière's*. Clin Gastroenterol 4:627–642
116. Desselberger U, Gray J (2003) Perspectives in medical virology, vol 9. Elsevier, Amsterdam
117. Kundu A, McBride G, Wuertz S (2013) Adenovirus-associated health risks for recreational activities in a multi-use coastal watershed based on site-specific quantitative microbial risk assessment. Water Res 47:6309–6325
118. Pond K (2005) Water recreation and disease. In: Plausibility of associated infections: acute effects, sequelae and mortality, 1st edn. World Health Organization, IWA Publishing, London
119. Li D, He M, Jiang SC (2010) Detection of infectious adenoviruses in environmental waters by fluorescence-activated cell sorting assay. Appl Environ Microbiol 76:1442–1448
120. Percivalle E, Sarasini A, Torsellini M, Bruschi L, Antoniazzi E, Grazia Revello M, Gerna G (2003) A comparison of methods for detecting adenovirus type 8 keratoconjunctivitis during a nosocomial outbreak in a Neonatal Intensive Care Unit. J Clin Virol 28:257–264
121. Puig M, Jofre J, Lucena F, Allard A, Wadell G, Girones R (1994) Detection of adenoviruses and enteroviruses in polluted waters by nested PCR amplification. Appl Environ Microbiol 60:2963–2970
122. Amano Y, Cheng Q (2005) Detection of influenza virus: traditional approaches and development of biosensors. Anal Bioanal Chem 381:156–164
123. Jiang S, Noble R, Chu W (2001) Human adenoviruses and coliphages in urban runoff-impacted coastal waters of southern California. Appl Environ Microbiol 67:179–184
124. Prescott AB, Barkely TU (2008) Trends in water resources research. Nova Science Publishers, New York
125. Wong K, Onan BM, Xagoraraki I (2010) Quantification of enteric viruses, pathogen indicators, and *Salmonella* bacteria in class B anaerobically digested biosolids by culture and molecular methods. Appl Environ Microbiol 76:6441–6448. https://doi.org/10.1128/AEM.02685-09
126. Jiang H, Patel PH, Kohlmaier A, Grenley MO, McEwen DG, Edgar BA (2009) Cytokine/jak/stat signaling mediates regeneration and homeostasis in the *Drosophila* midgut. Cell 137:1343–1355
127. Yildirim N, Li D, Long F, Gu AZ (2013) IEEE sensors conference. IEEE, Baltimore, pp 1–5
128. Novotny L, Hecht B, Pohl DW (1997) Interference of locally excited surface plasmons. J Appl Phys 81:1798–1804

129. Viitala L, Maley AM, Fung HWM, Corn RM, Viitala T, Murtomäki L (2016) Surface plasmon resonance imaging microscopy of liposomes and liposome-encapsulated gold nanoparticles. J Phys Chem C 120:25958–25966
130. Chachisvilis M, Zhang Y-L, Frangos JA (2006) G protein-coupled receptors sense fluid shearstress in endothelial cells. Proc Natl Acad Sci USA 103:15463–15468
131. Chen K, Obinata H, Izumi T (2010) Detection of G protein-coupled receptor-mediated cellular response involved in cytoskeletal rearrangement using surface plasmon resonance. Biosens Bioelectron 25:1675–1680
132. Li BB, Clements WR, Yu XC, Shi K, Gong Q, Xiao YF (2014) Single nanoparticle detection using split-mode microcavity Raman lasers. Proc Natl Acad Sci 111:14657–14662
133. Yanase Y, Hiragun T, Yanase T, Kawaguchi T, Ishii K, Hide M (2013) Application of SPR imaging sensor for detection of individual living cell reactions and clinical diagnosis of type I allergy. Allergol Int 62:163–169
134. Toma K, Kano H, Offenhäusser A (2014) Label-free measurement of cell–electrode cleft gap distance with high spatial resolution surface plasmon microscopy. ACS Nano 8:12612–12619
135. Peterson AW, Halter M, Tona A, Bhadriraju K, Plant AL (2009) Surface plasmon resonance imaging of cells and surface-associated fibronectin. BMC Cell Biol 10:16. https://doi.org/10.1186/1471-2121-10-16
136. Wang W, Wang S, Liu Q, Wu J, Tao N (2012) Mapping single-cell–substrate interactions by surface plasmon resonance microscopy. Characterization of micropatterned lipid membranes on a gold surface by surface plasmon resonance imaging and electrochemical signaling of a pore-forming protein. Langmuir 28:13373–13379
137. Wang W, Yang Y, Wang S, Nagaraj VJ, Liu Q, Wu J, Tao N (2012) Label-free measuring and mapping of binding kinetics of membrane proteins in single living cells. Nat Chem 4:846–853
138. Giebel K-F, Bechinger C, Herminghaus S, Riedel M, Leiderer P, Weiland U, Bastmeyer M (1999) Imaging of cell/substrate contacts of living cells with surface plasmon resonance microscopy. Biophys J 76:509–516
139. Soon CF, Khaghani SA, Youseffi M, Nayan N, Saim H, Britland S, Denyer MCT (2013) Interfacial study of cell adhesion to liquid crystals using widefield surface plasmon resonance microscopy. Colloids Surf B 110:156–162
140. Peterson AW, Halter M, Tona A, Plant AL (2014) High resolution surface plasmon resonance imaging for single cells. BMC Cell Biol 15:35
141. Yang Y, Yu H, Shan X, Wang W, Liu X, Wang S, Tao N (2015) Label-free tracking of single organelle transportation in cells with nanometer precision using a plasmonic imaging technique. Small 11:2878–2884
142. Shinohara H, Sakai Y, Mir TA (2013) Real-time monitoring of intracellular signal transduction in PC12 cells by two-dimensional surface plasmon resonance imager. Anal Biochem 441:185–189
143. Yanase Y, Araki A, Suzuki H, Tsutsui T, Kimura T, Okamoto K, Nakatani T, Hiragun T, Hide M (2010) Development of an optical fiber SPR sensor for living cell activation. Biosens Bioelectron 25:1244–1247
144. Yanase Y, Hiragun T, Kaneko S, Gould HJ, Greaves MW, Hide M (2010) Detection of refractive index changes in individual living cells by means of surface plasmon resonance imaging. Biosens Bioelectron 26:674–681
145. Bombera R, Leroy L, Livache T, Roupioz Y (2012) DNA-directed capture of primary cells from a complex mixture and controlled orthogonal release monitored by SPR imaging. Biosens Bioelectron 33:10–16
146. Stojanović I, van der Velden TJG, Mulder HW, Schasfoort RBM, Terstappen LWMM (2015) Quantification of antibody production by surface plasmon resonance imaging. Anal Biochem 485:112–118
147. Stojanović I, van Hal Y, van der Velden TJG, Schasfoort RBM, Terstappen LWMM (2016) Detection of apoptosis in cancer cell lines using surface plasmon resonance imaging. Sens Biosens Res. 7:48–54

148. Cortès S, Villiers CL, Colpo P, Couderc R, Brakha C, Rossi F, Marche PN, Villiers MB (2011) Biosensor for direct cell detection, quantification and analysis. Biosens Bioelectron 26:4162–4168
149. Horii M, Shinohara H, Iribe Y, Suzuki M (2011) Living cell-based allergen sensing using a high resolution two-dimensional surface plasmon resonance imager. Analyst 136:2706–2711
150. Wilkop T, Manivannan N, Balachandran W, Ray AK (2020) Surface plasmon resonance for human bone marrow cells imaging. IEEE Sensors J 20:11625–11631
151. Braun D, Fromherz P (1998) Fluorescence interferometry of neuronal cell adhesion on microstructured silicon. Phys Rev Lett 81:5241–5244
152. Holt MR, Calle Y, Sutton DH, Critchley DR, Jones GE, Dunn GA (2008) Quantifying cell–matrix adhesion dynamics in living cells using interference reflection microscopy. J Microsc 232:73–81
153. Matsuzaki T, Ito K, Masuda K, Kakinuma E, Sakamoto R, Iketaki K, Yamamoto H, Suganuma M, Kobayashi N, Nakabayashi S, Tanii T, Yoshikawa HY (2016) Quantitative evaluation of cancer cell adhesion to self-assembled monolayer-patterned substrates by reflection interference contrast microscopy. J Phys Chem B 120:1221–1227
154. Dos Santos MC, Déturche R, Vézy C, Jaffiol R (2014) Axial nanoscale localization by normalized total internal reflection fluorescence microscopy. Opt Lett 39:869–872
155. Son T, Seo J, Choi I-H, Kim D (2018) Label-free quantification of cell-to-substrate separation by surface plasmon resonance microscopy. Opt Commun 422:64–68
156. Nelder JA (1966) Inverse polynomials a useful group of multi-factor response functions. Biometrics 22:128–141
157. Izzard CS, Lochner LR (1976) Cell-to-substrate contacts in living fibroblasts: an interference reflexion study with an evaluation of the technique. J Cell Sci 21:129–159
158. Truskey GA, Burmeister JS, Grapa E, Reichert WM (1992) Total internal reflection fluorescence microscopy (TIRFM). II. Topographical mapping of relative cell/substratum separation distances. J Cell Sci 1992 103:491–499
159. Davies PF, Robotewskyj A, Griem ML (1993) Endothelial cell adhesion in real time: measurements in vitro by tandem scanning confocal image. J Clin Invest 91:2640–2652
160. Davies PF, Robotewskyj A, Griem ML (1994) Quantitative studies of endothelial cell adhesion: directional remodeling of focal adhesion sites in response to flow forces. J Clin Invest 93:2031–2038
161. Yanase Y, Hiragun T, Yanase T, Kawaguchi T, Ishii K, Hide M (2012) Evaluation of peripheral blood basophil activation by means of surface plasmon resonance imaging. Biosens Bioelectron 32:62–68
162. Schasfoort RBM, Bentlage AEH, Stojanovic I, van der Kooi A, van der Schoot E, Terstappen LWMM, Vidarsson G (2013) Label-free cell profiling. Anal Biochem 439:4–6. https://doi.org/10.1016/j.ab.2013.04.001
163. Stojanović I, Schasfoort RBM, Terstappen LWMM (2014) Analysis of cell surface antigens by surface plasmon resonance imaging. Biosensors Bioelectronics. 52:36–43
164. Larsson C (2006) Protein kinase C and the regulation of the actin cytoskeleton. Cell Signal 18:276–284
165. Kolkova K, Stensman H, Berezin V, Bock E, Larsson C (2005) Distinct roles of PKC isoforms in NCAM-mediated neurite outgrowth. J Neurochem 92:886–894
166. Amadio M, Battaini F, Pascale A (2006) The different facets of protein kinases C: old and new players in neuronal signal transduction pathways. Pharmacol Res 54:317–325
167. Teng F, Tang B (2006) Axonal regeneration in adult CNS neurons: signaling molecules and pathways. J Neurochem 96:1501–1508
168. Min DS, Ahn BH, Rhie DJ, Yoon SH, Hahn SJ, Kim MS, Jo YH (2001) Expression and regulation of phospholipase D during neuronal differentiation of PC12 cells. Neuropharmacology 41:384–391
169. Sena CM, Santos RM, Standen NB, Boarder MR, Rosario LM (2001) Isoformspecific inhibition of voltage-sensitive Ca^{2+} channels by protein kinase C in adrenal chromaffin cells. FEBS Lett 492:146–150

170. Hernandez-Fuentes MP, Warrens AN, Lecher RI (2003) Immunologic monitoring. Immunological Rev 196:247–264
171. Manz R, Assenmacher M, Pflüger E, Miltenyi S, Radbruch A (1995) Analysis and sorting of live cells according to secreted molecules, relocated to a cell-surface affinity matrix. Proc Natl Acad Sci U S A 92:1921–1925
172. Fontainhas AM, Obukhov AG, Nowycky MC (2005) Protein kinase Ca modulates depolarization-evoked changes of intracellular Ca^{2+} concentration in a rat pheochromocytoma cell line. Neuroscience 133:393–403
173. Stojanović I, Baumgartner W, van der Velden TJG, Terstappen LWMM, Schasfoort RBM (2016) Modeling single cell antibody excretion on a biosensor. Anal Biochem 504:1–3
174. Swennenhuis JF, Tibbe AG, Stevens M, Katika MR, van Dalum J, Tong HD, van Rijn CJ, Terstappen LW (2015) Self-seeding microwell chip for the isolation and characterization of single cells. Lab Chip 15:3039–3046
175. Abali F, Stevens M, Tibbe AGJ, Terstappen LWMM, van der Velde PN, Schasfoort RBM (2017) Isolation of single cells for protein therapeutics using microwell selection and surface plasmon resonance imaging. Anal Biochem 531:45–47
176. Puttharugsa C, Wangkam T, Huangkamhang N, Gajanandana O, Himananto O, Sutapun B, Amarit R, Somboonkaew A, Srikhirin T (2011) Development of surface plasmon resonance imaging for detection of Acidovorax avenae subsp citrulli (Aac) using specific monoclonal antibody. Biosens Bioelectron 26:2341–2346
177. Yodmongkol S, Thaweboon S, Thaweboon B, Puttharugsa C, Sutapun B, Amarit R, Somboonkaew A, Srikhirin T (2016) Application of surface plasmon resonance biosensor for the detection of Candida albicans. Jpn J Appl Phys 55:02BE03
178. Morlay A, Piat F, Mercey T, Roupioz Y (2016) Immunological detection of Cronobacter and Salmonella in powdered infant formula by plasmonic label-free assay. Lett Appl Microbiol 62:459–465
179. Bouguelia S, Roupioz Y, Slimani S, Mondani L, Casabona MG, Durmort C, Vernet T, Calemczuk R, Livache T (2013) On-chip microbial culture for the specific detection of very low levels of bacteria. Lab Chip 13:4024–4032
180. Mondani L, Roupioz Y, Delannoy S, Fach P, Livache T (2014) Simultaneous enrichment and optical detection of low levels of stressed Escherichia coli O157:H7 in food matrices. J Appl Microbiol 117:537–546
181. Zordan M D, Grafton M M G, Acharya G, Reece LM, Cooper CL, Aronson AI, Park K, Leary JF (2009) Detection of pathogenic *E. coli* O157:H7 by a hybrid microfluidic SPR and molecular imaging cytometry device. Cytometry Part A 75A:155–162. https://doi.org/10.1002/cyto.a.20692
182. McKillip JL, Jaykus LA, Drake M (1998) rRNA stability in heat-killed and UV-irradiated enterotoxigenic *Staphylococcus aureus* and *Escherichia coli* O157:H7. Appl Environ Microbiol 64:4264–4268
183. Foudeh AM, Daoud JT, Faucher SP, Veres T, Tabrizian M (2014) Sub-femtomole detection of 16s rRNA from *Legionella pneumophila* using surface plasmon resonance imaging. Biosens Bioelectron 52:129–135
184. Foudeh AM, Trigui H, Mendis N, Faucher SP, Veres T, Tabrizian M (2015) Rapid and specific SPRi detection of *L. pneumophila* in complex environmental water samples. Anal Bioanal Chem 407:5541–5545
185. Bulard E, Bouchet-Spinelli A, Chaud P, Roget A, Calemczuk R, Fort S, Livache T (2015) Carbohydrates as new probes for the identification of closely related Escherichia coli strains using surface plasmon resonance imaging. Anal Chem 87:1804–1811
186. Syal K, Iriya R, Yang YZ, Yu H, Wang SP, Haydel SE, Chen HY, Tao NJ (2016) Antimicrobial susceptibility test with plasmonic imaging and tracking of single bacterial motions on nanometer scale. ACS Nano 10:845–852
187. Syal K, Wang W, Shan XN, Wang SP, Chen HY, Tao NJ (2015) Plasmonic imaging of protein interactions with single bacterial cells. Biosens Bioelectron 63:131–137

188. Medina MB, van Houten L, Cooke PH, Tu SI (1997) Real-time analysis of antibody binding interactions with immobilized E. coli O157:H7 cells using the BIAcore. Biotechnol Tech 11:173–176
189. Abadian PN, Goluch ED (2015) Surface plasmon resonance imaging (SPRi) for multiplexed evaluation of bacterial adhesion onto surface coatings. Anal Methods 7:115–122
190. Chicurel M (2000) Slimebusters. Nature 408:284–286. https://doi.org/10.1038/35042737
191. Zmantar T, Chaieb K, Miladi H, Mahdouani K, Bakhrouf A (2006) Detection of the intercellular adhesion loci (ICA) in clinical *Staphylococcus aureus* strains responsible for hospital acquired auricular infection. Ann Microbiol 56:349. https://doi.org/10.1007/BF03175030
192. Fang Y, Shen Z, Li L, Cao Y, Gu LY, Gu Q, Zhong XQ, Yu CH, Li YM (2010) A study of the efficacy of bacterial biofilm cleanout for gastrointestinal endoscopes. World J Gastroenterol 16:1019–1024. https://doi.org/10.3748/wjg.v16.i8.1019
193. Gilbert P, McBain AJ (2001) Biofilms: Their impact on health and their recalcitrance toward biocides. Am J Infect Control 29:252–255
194. Aninwene GE, Abadian PN, Ravi V, Taylor EN, Hall DM, Mei A, Jay GD, Goluch ED, Webster TJ (2015) Lubricin: a novel means to decrease bacterial adhesion and proliferation. J Biomed Mater Res 103:451–462

Chapter 8
Process and Bioprocess Analysis

8.1 Introduction to Process Analysis

8.1.1 Process, Instruments and Methods

Generally speaking, there are two typical types of processes before us, the intermittent and the continuous. The intermittent processes can be exemplified by fermentation or reaction in tanks, steelmaking in furnaces and so forth, while the continuous processes can be found everywhere, for example, atmosphere system, water flow, life growth, chemical changes and industrial and agricultural productions. All the processes contain fast and/or slowly changing parameters that may need to be measured in real time or with a delay, depending on our aim (i.e., study or control of the processes). Usually, physical and chemical (including biochemical) parameters are of interest, which can now be measured with either contact or contactless sensors and/or analyzers. The contactless measurements are generally realized by optical, electromagnetic, electronic and mechanical principles, while the contact measurements are often achieved through separation techniques. Physical parameters are somehow easer to measure with contactless sensors than chemicals and biochemical. The non-physical parameters are often buried in complex environments; therefore, advanced sample preparation and separation techniques are required to "reveal" the parameters by removal of the background impact. Usually, single and simple parameters are often easy to measure in real time, while multiple and complex parameters may be measured with various delays.

From the point of view of instrumentation, there are also two types of equipment corresponding to intermittent and continuous measurements. Most sensors can acquire and send out continuous signals for timely control, while all separation-based instruments can measure parameters only intermittently except for ultrafast separation that may cause only a negligible delay.

Y. Chen, *Surface Plasmon Resonance Imaging*, Lecture Notes in Chemistry 95,
https://doi.org/10.1007/978-981-99-3118-7_8

After combination of the instruments with the processes, four categories of process analytical methods can be deducted:

(i) Intermittent and delayed measurement of parameters;
(ii) Intermittent and timely measurement of parameters;
(iii) Continuous and delayed measurement of parameters;
(iv) Continuous and timely measurement of parameters.

From the point of view of instrument location, another four combinations are available that appear more often in industrial and related research fields:

(i) Offline analysis, which measures parameters in center laboratories after sampling;
(ii) At-line analysis, which measures parameters near a process line soon after sampling;
(iii) Online analysis, which measures parameters on a bypass pipeline for parallel sampling;
(iv) Inline analysis, which measures parameters inside a process line or reactor.

Note, the terms of online and inline analysis are used controversially at present, for example, an assay with a sensor head inserting in the process line or directly inside the reactor may be called an online analysis (http://www.cvg.ca/Presentations/2007/PAT%20Solution.pdf). In this book, we recognize this as an inline analysis because it can be considered as a further development of online analysis [1]. In other word, the online analysis is used more generally than the inline analysis, and it can become inline only when the measuring point is in the main process line or reactor. The measuring point can be accessed by either physical insertion of a probing head or a contactless sensor through a window across the process main stream.

Clearly, the offline and at-line analysis can only be used for the intermittent and delayed measurement of parameters. Oppositely, the online or inline analysis may be suitable for both intermittent and continuous measurement of parameters in real time, depending on the features of the analytical instrument used. In common, ream-time or shortly delayed data can be used to adjust process variables by feedback or feedforward control, and the long-term delayed data are used only for forward control.

8.1.2 Application of SPRi to Process Analysis

Strictly speaking, industrial process analytical technology (PAT) is used in production sites that need analyzers to acquire signals for timely or delayed control. The instruments used in PAT have to be normalized to facilitate data transfer and/or comparison, which is required among different production units, factories or companies. A mature PAT must consist of all equipment and media necessary for realizing measurements, including at least sample treatment unit (e.g., for sampling and sample conveying or transferring, conditioning and/or recirculating), analyzer, controller and infrastructural equipment [2, 3]. An ideal PAT can automatically conduct intermittent or continuous sampling, sample pretreatment, chemical and signal analysis and output of control signals. The whole system must run stably for a long term and can be calibrated and maintained inline, online or at least at line.

Therefore, SPRi needs to be standardized and normalized in respect of instrumentation and data processing and conveying. Although SPRi can conduct continuous and real-time analysis, featuring high throughput and fast measurement, it remains only a potential type of analyzer for industrial PAT, to the best of our knowledge. Nevertheless, SPRi has been successfully applied to different non-strict process analyses. The typical examples are the kinetic analysis of chemical reaction and/or recognition processes, drug screening process, (biological) membrane formation or detachment processes, cell excretion or secretion process and clinical diagnosis, most of which have been discussed in previous chapters.

8.1.2.1 Offline, At-Line, *In Vitro* or *In Situ* Process Analysis with SPRi

Offline or at-line process analysis with SPRi is the easiest format in practice. It needs to conduct SPRi analysis simply after sampling at some required time points, regularly or irregularly. In more details, offline SPRi analysis is meant to perform SPRi of samples in a centralized laboratory, much the same as ordinary SPRi. The advantages include the utilization of sophisticated instrumentation, sharing use of various kinds of facility to save operation cost, availability of mature analytical methods and protocols, having expertise consultation from specialists. In a central laboratory, it is also easy to modify method in time or to maintain instrument whenever it is required. Oppositely, the disadvantages include the time delay from sampling to sample analysis and data reporting, additional administrative cost and inapplicability to the fast processes or the processes needing timely control.

Alternatively, at-line SPRi will largely reduce the delay, but a dedicated imager should be available and installed in close proximity to the process line, which can also be called near line SPRi analysis, enabling near timely control. At-line assays can avoid the long distance sample delivery and in turn reduces potential sample deterioration and loss. Nevertheless, the dedicated instrumentation needs time and cost to fabricate and validate. In addition, the measurable parameters must be reduced to more generic ones.

In life science, these two types of process analysis may be referred to *in vitro* analysis that has been toughed in the analysis of molecules and particles, ordinarily used in clinic analysis. Although at-line and offline SPRi sacrifice the features of continuous sampling and real-time analysis, they allow to conduct *in vitro* analysis by copying or simulating biological or biochemical reaction conditions in the detection cell, which is suitable for kinetic researches and measurements as have been discussed in previous chapters.

8.1.2.2 Online and Inline Process Analysis with SPRi

Online SPRi needs to couple it with automatic sampling and sample conditioning techniques, in two categories:

(i) Intermittent SPRi that sampling and injection of only a portion of the process stream into SPRi imager for analysis and control of the process at the required time point;
(ii) Continuous SPRi that permits the sample to continuously flow through the detection cell, ideally at an equal linear speed with the material flow of the process, to acquire real-time dada for timely control of the process.

Such an online analysis, with proper interfaces, is also suitable for the study of living things. In this case, the sampling should not impact seriously on the living systems, better to a negligible level, and the best to a disturbance-free or non-destructive level. It is just due to the ability to perform continuous assays that makes online SPRi advantageous over the offline and at-line formats. Nevertheless, the online analysis requires the establishment of a separate or bypass analytical pipeline to enable sampling from the process main stream at a matching linear flow speed, temperature and pressure. This is also the best solution for an analyzer needing frequent cleaning, calibration or validation.

Inline SPRi aims at performing the direct analysis inside the producing pipeline or container or *in vivo* analysis for a living system. SPRi instruments must be further modified to enable sensing the substances deep inside the production line or container. It has thus following features:

(i) The sensor or sampling head is inserted in the process stream or reactor through direct or indirect contact (e.g., a transparent window for optical propagation);
(ii) It allows *in vivo* continuous and real-time sensing for timely control without shutdown process;
(iii) It can selectively sense chemical variation in the main stream;
(iv) It is easy and reliable to regenerate the sensor and related analytical parts.

Inline analysis can thus be superior to the other three analytical formats. Online and inline SPRi assays can both continuously acquire data for continuous control of process, while offline and at-line analyses are limited to intermittent or discontinuous measurements and evaluations, not applicable to the timely control (except for forward control) of fast processes.

Because few practical examples have been reported, we will discuss some potential examples that seemingly close the above-discussed concepts and/or characters, in spite that may be involved in previous chapters.

8.2 At Line or Inline Screening of Aptamers

Aptamers are a type of nucleic acids such as a section of single-strand DNA [4, 5] or RNA [6, 7]. They have different usages, for example, as antibody alternatives [8], to build autonomous nanorobots [9], to detect biomarkers [10], to image [11] or capture [12] cells, to deliver drugs [13] and to treat cancers [5]. They are often obtained through a process of systematic evolution of ligands by exponential enrichment (SELEX) [14, 15] of some chemically synthesized pools that have different oligonucleotides, with a typical length of 70–100 bases [16, 17]. The early time SELEX used offline screening methods such as capillary electrophoresis (CE) [18], magnetic beads [19], liquid chromatography (LC) [20] and microfluidics [21, 22] to search some pre-aptamers for further selection and amplification.

Nucleic acids can be amplified by the well-known PCR technique. To perform PCR, each chain in the primarily selected nucleic acids, which has a random core region at a fixed length (30–60 nt), should be flanked with a PCR primer at ~ 20 nt [23]. The issue is that the primer flanks may hybridize with the random region to form unwanted secondary structures and in turn to disturb the selection process [24, 25]. Thus, "minimal primer" SELEX [26] and even "primer-free" SELEX [27, 28] have been tried, but they remain in needs of additional enzyme ligations and fixed-sequence hydrolysis steps, which takes a longer time than the typical SELEX process. In addition, those steps may bring in other byproducts of enzymatic reactions. Enzyme-free method has also been tried to minimize the interference [29]. Nevertheless, the offline screening method is unable to assess the binding interaction between pool and target.

A better alternation is to explore inline screening technique. SPRi-based SELEX methods (SPRi-SELEX) [30, 31] have since been explored after SPR-SELEX [32] to perform inline, real-time and high throughput analysis of critical nucleic acids. SPRi-SELEX tremendously increases the throughput because simultaneous positive and negative screenings of nucleic acids are enabled. The negative screening is unavoidable to isolate the potential aptamers from the very complicated matrixes coexisted in the pools [33]. Similar to SPR, the native sensitivity of SPRi does not allow SPRi-SELEX to detect the selected aptamers in the primary or first round of screening [34, 35]. The practical SPRi-SELEX needs to increase it sensitivity with new technique, e.g., LSPR [36, 37] or by signal amplification, e.g., use of silver decahedra NPs [30, 31].

SPRi-based real-time assessment of binding events between targets and nucleic acids in a pool SPRi-SELEX is schematically shown in Fig. 8.1 that can be scaled up to a production level, serving for inline and real-time SPRi. To facilitate operation, the nucleic acids may hybridize on nanoprobes [38] so that the non-hybridized or free nucleic acid can easily be separated by washing the measuring cell. The captured nucleic acid chains are then desorbed and subjected to PCR to amplify the copies. The amplified copies are released by denature in 50 mM NaOH for next cycle of SELEX.

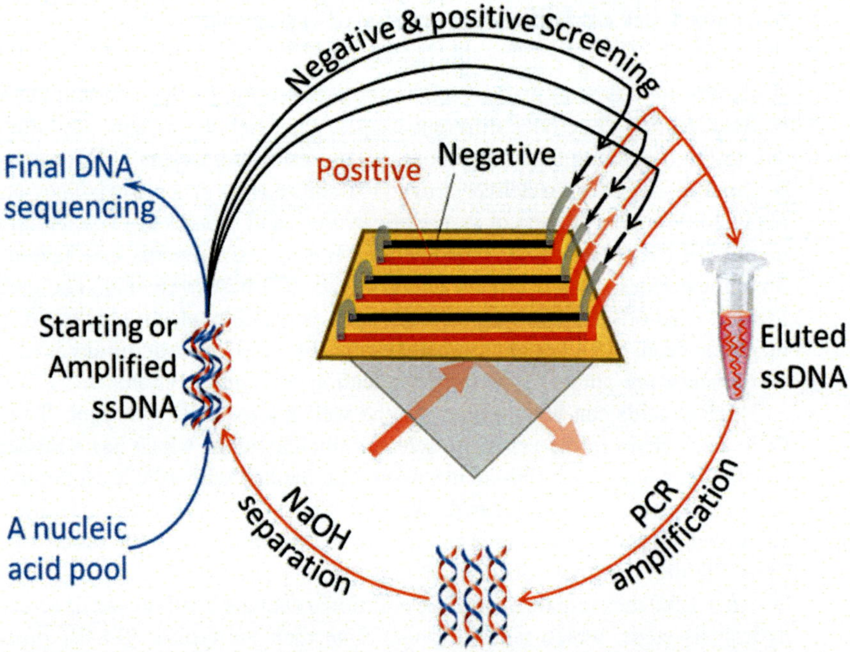

Fig. 8.1 Simplified process for inline SPRi-SELEX of aptamer(s). The PCR amplification may be conducted on some magnetic beats to simplify ssDNA separation. The templates are the complementary ssDNA to the target aptamers in the eluent and they are prepared by a short-time PCR of the eluent. The amplified ssDNA may be further incubated with some amplifiers like nanoparticles to improve SPRi sensitivity

To perform inline SPRi-SELEX, the sensor chip must be divided into several channel pairs, with either odd or even channels for negative screening and even or odd channels for positive screening. The target probes are separately immobilized in the positive screening channels normally by EDC/NHS-based amidation chemistry, while counter proteins for negative screening are immobilized inside the negative channels by the same chemistry. For example, in the screening of aptamer against lactoferrin, the even channels are immobilized with 50 μg/mL lactoferrin and the odd channels are immobilized with a mixture of 100 μg/mL BSA, 5 μg/mL casein, 50 μg/mL α-lactalbumin and 50 μg/m β-lactalbumin in 10 mM acetate buffer at pH 5.0.

Each pair of the odd and even channels can be connected in series, with inlet of the odd channel connected to the outlet of the even one. During SELEX, the nucleic acids from a pool are pumped first into the negative channel and then into the positive channel, where the negative channels are used to capture those potentially unwanted nucleic acids and other protein-absorptive substances, while the positive channels are used to capture the targeting chains in the solution from the negative channel.

The captured chains in the positive screening channels are further washed with an appropriate buffer to remove the weakly adsorbed nucleic acids. The strongly binding chains are considered to be the target molecules and are then collected and sent to amplification by PCR to produce sufficient copies. The amplified chain sample is then sequenced, and once the sequence is confirmed, the sample is guided to SPRi system for next cycle of selection. Following reference approach outlines the key steps for SELEX of aptamers.

Approach 8.1 The major steps for SELEX of aptamers, where the cleaning and drying steps have been eliminated:

Step 1, immobilize related probes in the positive and negative SPRi channels.
Step 2, block the remaining active sites first with BSA for 10–30 min, then with 1 M ethanolamine hydrochloride at pH 8.5 for about 10 min.
Step 3, inject a nucleic acid library into the inlet of the negative channels and flow it out from the positive channels.
Step 4, wash out the weakly absorbed substances.
Step 6, release the strongly binding nucleic acids and amplify them by PCR.
Step 7, denature PCR product in NaOH to separate single-strand DNA.
Step 8, inject the amplified single-strand DNA for the second cycle of SELEX.
Step 9, repeat the step 1 to step 8 until sufficient target aptamer molecules are obtained.

Prior to SPRi, the system is normally primed with a running buffer until the baseline becomes stable. The system is then washed with a regeneration solution (e.g., 2 mol/L NaCl) and primed again with the running buffer. These washing steps are performed alternatively for three times.

The library to be screened should contain above nmol/L (better at μmol/L) nucleic acids to ensure SPRi detection. DNA is normally denatured in buffer at ca. 95 °C for 4–5 min and re-folding at room temperature for another 4–5 min. To enhance SPRi detection, the folded nucleic acid library can be fixed on some nanoparticles (e.g., AuNPs, silver decahedra nanoparticles or Ag_{10} [30, 31] by hybridization with their complementary strands previously immobilized on the NPs. This NPs-nucleic acid library can then be injected into SPRi system for screening.

There are various protocols to perform PCR. Herein suggested is a simple and easy approach:

Approach 8.2 For PCR of aptamer candidates:

Step 1, pre-denature a DNA sample with forward and reverse primers, dNTP, ddNTP and enzymes at ca. 94 °C for 5 min. Note, a aliquot of 50 μL solution is amplified twice by the PCR. The first PCR is conducted by mixing 23 μL SPRi eluent with 25 μL of 2×Taq Master Mix, 1 μL of 20 μmol/L forward primer and 1 μL of 20 μmol/L biotin-labeled reverse primer. The second PCR is performed with the first PCR product added with 5 μL template, 25 μL 2 × SYBR® Premix ExTaqTM II, 1 μL tetramethyl rhodamine (TAMRA)-labeled forward primer (20 μM), 1 μL biotin-labeled reverse primer (20 μM) and 18 μL *diethyl pyrocarbonate* (DEPC)-treated water (free of DNase, RNase and proteinase).

Step 2, circulate a temperature program of ca. 94 °C → 60.5 °C → 72 °C for 30 s each and 72 °C for another 5 min. Note, the total amplification takes about 10 cycles that are better optimized for a specific reaction. The optimization can be achieved by monitoring the concentration of the target nucleic acid(s).

The selected aptamers need to be confirmed by DNA sequencing, which can now be performed by automated DNA sequencer or by some commercial service centers. By SPRi-SELEX, aptamers with $K_a = 1 \times 10^6$ mol^{-1} L can easily be selected.

In this inline process analysis, SPRi majorly acts as a tool to isolate the correct sequence of aptamers through selective capture. Other sensorable parameters such as concentration, temperature, reaction kinetic constant and so forth remain not measured and used for control. They are waiting for exploration.

8.3 Pharmaceutical Process Analysis

The development and commercialization of a new drug may take over 10 years and require sizeable investments (at millions of US dollars). There may be 80–85% of products that fail during developmental processes, even during the expensive clinical trials [39]. One of the critical measures is to evaluate the related products, which needs efficient evaluation methodology. Thus, any methodological advance will tremendously impact on the progresses of pharmaceutical science and industry.

SPRi can be utilized in many pharmaceutical processes because it can measure the concentration of relative substances and screen various binding events. The unique superiority of SPRi has emerged in drug binding studies, especially in searching and assessing the interactions of small drug molecules with proteins [40–45]. SPRi can simulate physiological or designed conditions at will, which facilitates the measurement of useful parameters that can match with living organisms or with the practical production systems.

However, the intrinsic sensitivity of SPRi needs to be improved before it can be utilized in a practical process for the study or production of small drug molecules that yield usually low plasmonic resonance response. The common strategy to improve the detectability is to increase the binding sites per unit area via surface modification techniques. The detectability of small drugs can also be improved by increase of its pseudo molecular weight through specific loading technology, e.g., polymerization, cyclic reactions or addition of particles such as NPs. These techniques are more often termed signal amplification that has been discussed in previous chapters. Herein discussed will be only the technology to increase the binding sites to save space.

To capture small drug molecules on a sensor surface as more as possible, the gold sensing surface needs to be pre-modified with proteins-inert substance(s) having multiple sites to enable the immobilization of multiple probes. The most often used "inert" substances are hydrophilic macromolecules such as dextran, chitosan, PVA, PEG and so forth. There are ready-for-use commercial chips that are coated with dextran with or without charged terminals. Many negatively charged proteins can be electrostatically attracted on carboxyl terminals to further enable immunocapture [46] of small drug molecules. Such a charged chip is easily regenerated by simply flowing in a solution buffered at a pH able to neutralize the charges on either proteins or the terminals, for example, with 10 mM glycine–HCl at pH 2.0 to neutralize the carboxyl groups.

The real applicability of SPRi in the study of small drugs has been assessed with a drug FK506 (tacrolimus). According to Langmuir model of 1:1 adsorption, the SPRi response is governed by [47]:

$$I_{\mathrm{SPRi}} = I_{\mathrm{Probe}} \frac{M_{\mathrm{Drug}}}{M_{\mathrm{Probe}}} \tag{8.1}$$

where I is the maximum SPRi intensity, and M denotes molecular mass. Equation (8.1) shows that the drug signal increases with its captured mass and also with drug molecular number on its target protein but decreases with the probe or protein mass. Within a given area of interest, proper increase of the active number of probes will certainly raise the capture rate of small drugs. This can be demonstrated by measuring biotin on a dextran-coated chip, with streptavidin as its probe. Although the streptavidin has extremely high affinity toward the biotin, with $K_d = 4 \times 10^{-14}$ M [48], the signal of the captured biotin is buried in the strong background if measured on a normal chip. While on a dextran-modified chip to increase the capturing sites, the biotin could gain 4.4×10^2 RIU signal tested with 4 nM sample solution, in spite that the streptavidin itself yielded a strong signal up to 3.3×10^4 RIU. The demonstration was performed by sequential pumping in biotin sample solutions dissolved in PBS at pH 7.4 (containing 1.0% dimethylsulfoxide (DMSO)) at different concentrations. The flow rate was set at 2.0 μL/s and kept flow for 300 s to complete either association or dissociation.

Note, when absorptive substances such as proteins are involved in SPRi measurements, the hydrophobic gold surface must be inertized by modification with hydrophilic substances such as MUA or dextran. In the detection of FK506, dextran can both deactivate the chip surface and sterically increase the sites for immobilizing the probes. Figure 8.2 shows the images (a, b and c) and histogram (d) variations of FK506 captured by its binding protein 12 (FKBP12) immobilized on the dextran-coated (a), bare (b) and MUA-coated (c) chips, respectively. It is clear that the dextran-coated chip offers much higher surface density of FKBP12 to produce stronger signals than the bare and MUA-coated chips (d and e). As a consequence, injection of 20 nM FK506 can yield 96 RIU response on the dextran chip but no signals on bare or MUA-coated gold chip. The binding of FK506 on the dextran-coated chip (e) becomes strong enough to calculate the binding constant, giving K_d = 0.24 nM that is close to the literature value of 0.6 ± 0.2 nM [49].

Dextran can be immobilized on either bare or modified gold chips. The modifier can be a mixture of MUA and MUOH or PEG-COOH and PEG-OH [49]. The ratio of COOH to OH must be adjusted to have a proper surface concentration of COOH. In order to enable electrostatic action at a higher pH, these weak COOH groups are better changed to stronger ones through esterification with succinic anhydrate. Such

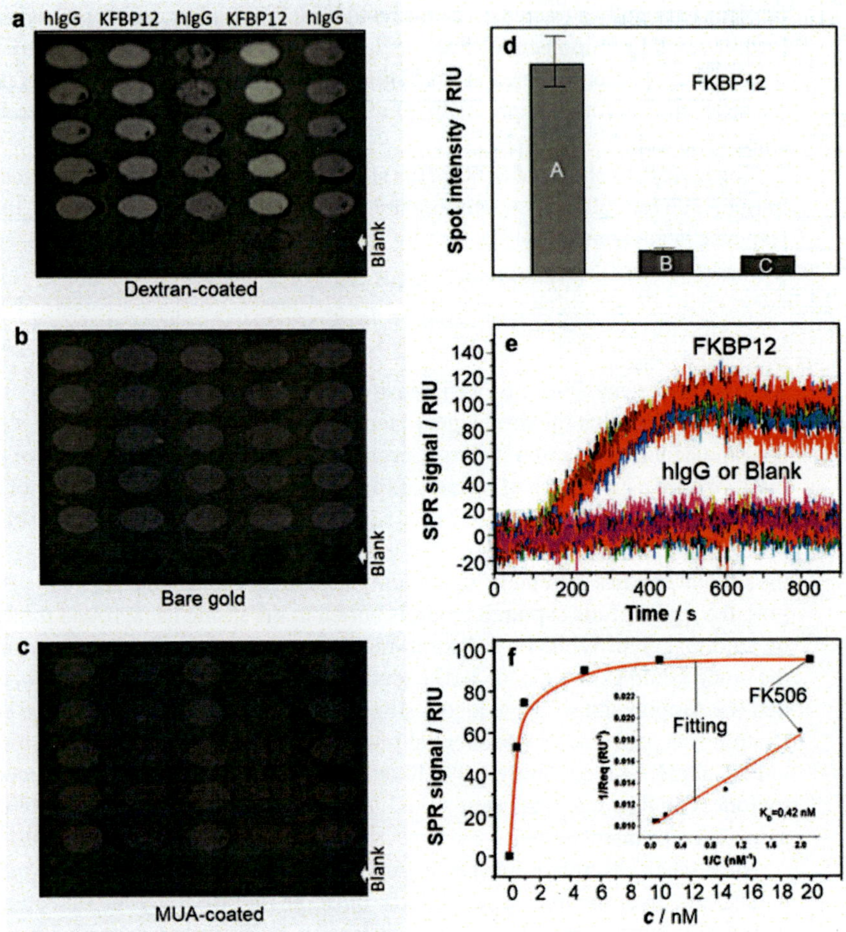

Fig. 8.2 SPRi of FK506 captured on its target protein of FKBP12 immobilized on **a** 0.01% dextran-coated, **b** bare and **c** MUA-coated gold chips that give **d** different immobilization intensities of FKBP12 and **e** selective SPR response **f** dependent on the concentration of FK506. Reconstructed from Ref. [46] with permission

chips are ready to be spotted with the target proteins at pH4.5 and then incubated at 4 °C for 2 h in a chamber with above 60% humidity.

The dextran-modified sensor surface is undoubtedly a way to increase the binding sites and in turn to improve the detection sensitivity of small drugs, but its surface concentration should be controlled within a certain range that should be optimized dependent on the molecular size of dextran and the subsequent binding molecules. The regulation of the surface concentration can be achieved by changing the content of either its precursor or dextran itself. It should, however, be noted that the target proteins may be deactivated, partially or event totally. The main reason lies in the mutual masking and steric shielding effects of the immobilized proteins and the captured drug molecules. It can be attributed partially to the non-specific adsorption of unwashed proteins, which happens occasionally, but this issue can be eliminated by increase of washing strength and/or changing the washing process. Inappropriate washing may also lead to partial denaturation of and/or damage to the protein probes, which cannot be avoided completely because biological molecules are quite delicate and perishable.

The above example reveals that SPRi can serve for inline or online process analysis in the development and production of small drugs based on its ability to directly measure the concentration of target analytes in real time under the producing conditions without any bias. It will be more suitable to act as a process analytical tool in the case of large protein drugs that have strong SPRi signal. Nevertheless, the author has not yet found or self-tried a practical application, to be honest. Although this may take time, SPRi is promising and worth of exploitation.

8.4 Clinical (Process) Analysis

Clinical analysis is a special process analysis, where the target analytes are involved in the sick processes. The relative analytical techniques impact not only on life or health quality but more obviously on the diagnostic level of diseases. This is a common concern of all our mankind. Human beings have never stopped developing and adopting advanced methodology and tools. As an advanced tool, SPRi has naturally been considered in this field, with the advantage of measuring data under the physiological conditions. Theoretically, SPRi can be performed in offline and online formats but not inline or *in vivo* manner at present.

Modern diagnosis and monitor of diseases rely very much on the information of functional biomolecules. It is necessary to analyze the level changes of the related biomolecules in patients in response to treating drugs or quantitatively, to the drug dosage to ensure safety and efficacy [50]. This is normally achieved through the analysis of biofluids with standardized protocols, e.g., ELISA. The routine methods adopted in clinical laboratories include also separation and identification techniques such as chromatography, electrophoresis, optical or mass spectrometry. They are broadly applied to the analysis of different classes of biomolecules, with high throughput to manage a large number of samples, but also at the costs of finance and time. These limit the applications of the techniques to average and local hospitals, especially the hospitals in rural and remote areas. Although their samples might be sent to some advanced central laboratories for reliable analysis, the time delay is fatal in emergency or rapid medical actions. In many medicinal cases, it needs to frequently and/or continuously monitor and/or measure the applied drugs, their metabolites or some endogenous molecules [51]; thus, fast clinical assays should be available. This has led to the bloom of exploitations of various novel analytical tools and methods, for examples, many sensors and point-of-care diagnostic devices [52–55].

Among them, SPRi is superior to others for its applicability to the analysis of differently sized intact molecules and particles of clinical interest [56] such as amino acids, peptides and proteins (especially, antigens, antibodies and enzymes), nucleic acids (e.g., microRNA, oligonucleic acids, single or double-strand DNA), carbohydrates, viruses and cells. It has been reported that SPRi can detect small drugs, hormones, cancer markers, antibodies and other biomolecules [57]. The early clinical studies with SPRi enrolled only a few patients (ca. 10). This is, however, not sufficient. A greater number of patients are necessary to be studied to effectively validate SPRi for real clinical applications. Another issue is that SPRi cannot detect drugs below ng/mL or nmol/mL, but this issue has been removed through combination of SPRi with signal magnification techniques.

SPRi assays are normally used to analyze solution or biofluidic samples like whole blood, plasma, sera, urine, synovial fluid, cerebrospinal fluids or secretions, saliva and ascites fluid. Furthermore, the samples can be extended from liquid phase to solid or semi-solid phase such as tissue sections, white or red blood cells and even stools. These latter samples can be imaged either by direct contact of the sample surface with the sensor chips or after extraction to become solution state. It is worth of notification that some fluidic samples such bloods or sera are normally diluted with a running buffer, to a ratio of about 10% to minimize the non-specific adsorption effects. Dilution of a sample may lose some original information and increases the complexity of point-of-care assays. Thus, minimal or even free of sample handling is desirable, which puts forward challenges to SPRi.

8.4.1 Key Challenges

The key challenges of SPRi in clinical applications include (1) how to assay whole blood, (2) how to stably and sensitively determine trace targets, (3) how to eliminate background interference and (4) how to have advanced SPRi instruments for easy use. Ideally, direct detection of analytes in whole blood would be achieved without sample preparation to significantly simplify the manipulation. Since biological molecules are in most case fragile or easily denatured during sample preparation, storage and even analysis. They hence need to be kept in low temperature to save their activity and to stabilize SPRi measurement as well. It is better to sample and to handle the samples just before use to ensure the acquisition of information from natural substances. This is not easy to realize in clinical applications. The instrumentation associates with the development of not only the experimental prototypes but also commercialized devices, which has been more or less discussed in Chap. 3. The background is often not stable but varies from place to place and from time to time, which has to be considered seriously.

Because most biomarkers circulate in bloodstream, the corresponding sera are analyzed the most often in clinical analysis and appeared at the highest frequency in publications. However, serum samples remain extremely complex, where the abundant proteins can impede the proper function of SPRi through their numerical superiority to compete the occupancy of the sensing surface. This issue is ubiquitous to surfaces in contact with biofluids. Surface chemistry should be developed to minimize the non-specific adsorption of proteins and applied prior to real analysis [58–60]. The most often used technique is polyhydroxy surface chemistry (refer also to Sect. 4.6 and 8.3). For example, dextran-modified surface can excellently minimize the non-specific adsorption of many types of proteins in saline solutions and is hence used in commercial SPRi kits to stabilize the baselines. Other usable techniques include membrane cloaking [61], pretreatment of the sensing surface with blank serum [62], addition of dextran to samples [63], depletion of background proteins [64] and extraction of target analytes [65, 66]. Their drawback is that they complicate the sample handling and/or analytical process. Surface chemistry does not necessarily simple, but it can be pre-programmed or commercialized to have ready-for-use chips. In addition to dextran, PEG can also reduce the non-specific adsorption of proteins by a factor up to 99% [67–71]. At this level of non-specific adsorption, direct analysis of crude biofluids becomes realizable [72–86].

Low fouling surface chemistries sometimes limit the capture of biomolecular receptors or show decreased performance of the surface-bonded molecular receptors, which is somewhat similar to non-specific adsorption. In addition, the analyte may also bind to the bulk proteins in biofluids and to the free antigen as in the case of prostate-specific antigen [87]. The bulk protein binding reduces the effective concentration of target analytes through the block, cloaking or even occupation of the binding site(s) and in turn impedes the capture of target analytes, which lowers the SPRi detection sensitivity. For example, in the analysis of CXCL12 (e.g., stromal cell-derived factor-1 or SDF-1) in urine [88], CXCL12 can bind to glycosaminoglycans in either the urine or the blood of patients and in turn interferes with antibody recognition and quantification as illustrated in Fig. 8.3a, leading to significant decrease of SPR signal in serum than in PBS. The similar binding does not necessarily affect the capture of target analytes; in contrast, it increases the molecular weight of the receptor complex, which in turn increases the refractive index shift, leading to a slightly increase of SPR response in the adsorptive serum than in PBS (Fig. 8.3b). Thus, the use of pooled biofluids should be careful because the background response is highly variable with individuals or with the sources of pooled biofluids [89].

The level of non-specific adsorption is different from patient to patient [90], from serum to serum and even from infants to adults [91]. Hence, the calibrations in pooled human biofluids to assess the general performances of SPRi at the development stage may not readily be applicable to the actual clinical samples. In order to solve this problem, enzyme-based secondary detection can be used, but the cost is to increase the complexity of SPRi measurement and to sacrifice the feature of "real-time" measurement. Alternatively, the use of a reference surface immobilized with a different antibody is often adopted, but it can only partially compensate the change that can become significant due to a slight difference in the non-specific adsorption between surfaces [92]. The most ideal way to compensate for the background variation is to control the analysis with an analytes-free serum from the same individual.

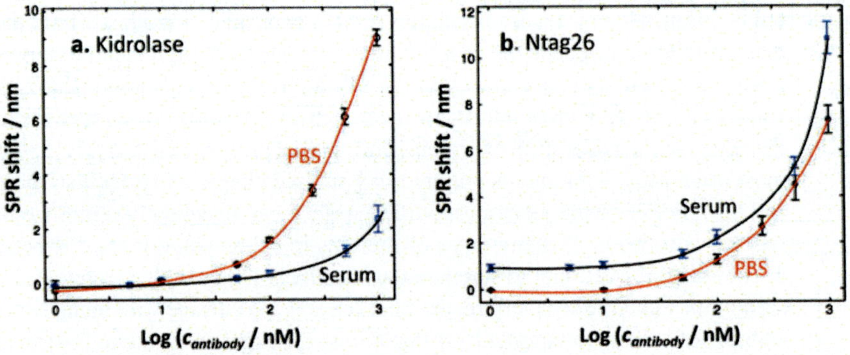

Fig. 8.3 Calibration curves for anti-L-asparaginase recognizing the asparaginase-immobilized on sensor surface. The calibrations were conducted in either PBS or undiluted human serum with two different types of asparaginase of **a** Kidrolase and **b** Ntag26, where the response is lower in serum with Kidrolase, but slightly larger with Ntag26. Adapted from [90] with permission

This needs to deplete the analytes, for example, by excessive addition of the same antibody as that immobilized on the capture surface. In this case, the non-specific adsorption and bulk contributions of serum become identical in the reference and measuring spots, able to isolate specific signal from the interference. This method may remain questionable even if the analytes can be depleted completely because all the recognition reactions are not unidirectional but of equilibration.

There are so many trials and examples in clinical diagnosis and related analysis; herein, we cannot discuss all of them but only limited to a few representatives such as allergy.

8.4.2 Analysis of Type I Allergy

Allergy is a special immune reaction process in response to maybe not harmful invasion. Hence, it is also called hypersensitivity reaction. Presently, we adopt offline or online strategy to study and to diagnose allergy. SPRi can be used in both cases.

Type I allergy is one of many types of allergies that have serious impact on human daily life such as repeated atopic dermatitis and food allergy. It is an IgE-associated disease caused by abnormal release of allergic substances from basophils (in blood) and mast cells (in connective and mucosal tissues). The mechanism is that once the allergen (i.e., antigen) binds to IgE in blood, it will be recognized and captured by the high-affinity IgE receptor (FcεRI, the Fc segment receptor of immunoglobulin) located on the surfaces of basophils and mast cells. This leads to cross-linking of the FcεRI and release of preformed and newly synthesized inflammatory mediators, e.g., histamine, arachidonic acid metabolites and cytokines, causing allergic symptom [93–95]. The allergen binding is clearly the initial step to provoke mast cell and basophils-induced allergic reactions. In order to avoid anaphylactic shock and aggravations of allergic diseases, it is critical to identify the antigens and their binding-induced allergic reactions. The antigens can now be detected by various immunological methods, e.g., ELISA. However, it is even crucial to learn if the binding complexes of antigen and IgE will cause the activation of mast cells and basophils because the potential allergical reactions are more decisive than the IgE–antigen binding itself in the diagnosis of type I allergy [96–98]. Both of these cases can easily be studied or monitored with offline or at-line SPRi under physiological conditions.

First, there is no problem for SPRi to characterize the binding candidates of allergens by *in situ* capture of them on an IgE-immobilized sensor by many methods that have been discussed in previous chapters. Furthermore, SPRi can follow sequential events of intracellular signaling rather than just the binding of ligands on cell surface and thus has the potential to do clinical diagnosis [94, 99–110]. A research example is conducted with an RBL-48 cell line as testing sample [111]. The RBL-48 cell line can be activated by an allergen called anti-human IgE antibody (anti-IgE). This enables SPRi to detect IgE in $< 1\ \mu$L peripheral blood in response to various antigens, without the use of fresh human basophils to facilitate its applications.

With the setup illustrated in Fig. 8.4a, the response of RBL-48 cells, after treated with human serum overnight, to a multivalent anti-IgE can be monitored in real time. The cross-linkage of IgE antibody-FcεRI complex on the cell surface and the release of histamine can be measured, giving ca. 35% net reflectivity calculated by time-differential images shown in the lower line of Fig. 8.4b. As a comparison, the cells without the treatment of human serum gave no response to anti-IgE antibody (the upper images in Fig. 8.4b). The RI in RBL-48 cell areas keeps unchanged when stimulated by anti-IgE if the cells are cultured without human serum; oppositely, the signal increases in a few minutes after injection of anit-IgE for the cells cultured with human serum.

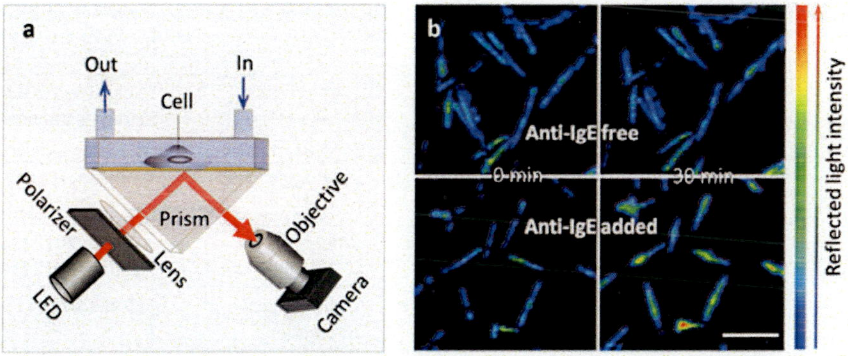

Fig. 8.4 **a** SPRi setup for **b** imaging the cells that are *in situ* cultured in human serum and then stimulated with (lower images) or without (upper images) anti-IgE inside the flow chamber. The LED emits 640 nm light, the sensor chip is a glass slide deposited with 1 nm Cr and 49 nm gold, the prism has a refractive index of 1.72 and the objective lens (×0.6) is fixed on CMOS camera. Reconstructed from Ref. [111] with permission

The data revealed that only the donor 3 was found, as expectation, to have positive response among the four donors (donor 1 with pollen allergy, donor 2 without allergic disease, donor 3 with atopic dermatitis (AD) and donor 4 with mild asthma). The RBL-48 cells increased their RI in response to cedar pollen antigen after culturing with the serum of donor 1, but remained unchanged after culturing with the serum of donor 2 or without any human serum. RBL-48 cells cultured with the serum of donor 3 with AD response to sweet antigen, but those cultured with the serum of donor 4 without AD showed no response.

For its ability to simultaneous detect multiple spots, SPRi can recognize the allergic antigens and to detect the stimulated release of biochemicals, which is favorable for clinic diagnosis (Table 8.1). More information on SPRi of cell excretion can be founded in Chap. 7.

In short, cell-based SPRi assays can be referred to Chap. 7, while the allergic reaction can be performed according to Approach 8.3.

Approach 8.3 For *in situ* measurement of allergic reaction with SPRi:

Step 1, culture the target cells (e.g., RBL-48) on a modified SPRi chip covered with or without human peripheral blood serum.
Step 2, mount the chip with the freshly cultured cells in SPRi flow chamber (Fig. 8.4a) kept at 37 °C, and start to record the imaging signals.
Step 3, wash the chip with 1,4-piperazinediethanesulfonic acid or PIPES buffer.

Table 8.1 Collection of SPRi assays for some cell-associated processes

Cell	Analysis of	First author, year
LS102.9, mouse B-type lymphocytes; 13G7, mouse *T*-type lymphocytes	Cell surface antigen	Suraniti 2007 [112]
J774, murine macrophage cell line; HL-60, human promyelocytic leukemia cell line; human PBMC, peritoneal blood mononucleated cell		Cortès 2011 [113]
Human red blood cells		Schasfoort 2013 [114] Houngkamhang 2013 [115]
HS578T, SKBR3, MCF7, human cancer cell lines	(EpCAM)	Stojanović 2014 [116]
RBL-2H3, rat basophilic leukemia cell line PAM212, mouse keratinocyte cell line	Allergy	Yanase 2010 [117] Yanase 2012 [107]
RBL-2H3, rat basophilic leukemia cell line	Cell–antigen interaction	Horii 2011 [118]
vSMC, rat aortic vascular smooth muscle cell line	Cell–matrix interaction	Peterson, 2009, 2010 [119, 120]
PC12, rat adrenal pheochromocytoma	Stimulating cells	Shinohara 2013 [121]
A549, human type II alveolar epithelial cell line	Cell death	Zhang 2013 [122]

Step 4, stimulate the cells with IgE antibody to release histamine and other molecules.
Step 5, collect the image signal at a frequency of 10–24 frame/s.
Step 6, dispose or regenerate the SPRi sensor chip after each assay.
Step 7, plot the sensorgram with or illustrate and analyze the net images after subtraction of the background (cell-free) signal.

8.4.3 Diagnosis of Other Diseases

Human diseases (e.g., cancers, rheumatoid arthritis, hepatitis, etc.) are all related to some abnormal life processes. Similar to allergy, these processes can be studied, tracked or monitored with offline or at-line or more exactly *in situ* SPRi assays. Some factors shown up in these processes can cause unique changes of SPRi signals and hence can be detected with SPRi. For example, the activation of epidermal growth factor receptor or EGFR on epidermal cells like keratinocytes in some cancers can cause SPRi signal variation that is different from the changes of other receptors, e.g., FcεRI on mast cells and basophils. An impaired pattern of SPRi signal on EGFR may reveal disordered intracellular signal transductions of abnormal cells, such as cancer cells. It was found that 5/6 carcinoma cell lines showed mono or biphasic change of signals [102]. Based on this type of response, SPRi can be applied to the diagnosis of many diseases. Nevertheless, we still need to accelerate the transition from proof-of-concept stage to practice because there are only quite limited examples available (Table 8.2). This also requires the active participation of medicinal doctors and patients, which is out of the scope of this book and we hence stop here.

Table 8.2 Brief collection of clinical analysis of human samples with SPRi

Disease	Analyte	First author, year
Preterm birth	Fibronectin in cervicovaginal secretions	Chen 2012 [123]
Prostate cancer	Prostate-specific antigen in serum	Erturk 2008 [124]
Bladder cancer	Podoplanin in serum or urine	Gorodkiewicz 2012 [125] Sankiewicz 2016 [126]
Leukemia	Cathepsin G in white blood cells	Gorodkiewicz 2012 [127]
Rheumatoid arthritis	Anti-citrullinated protein antibodies in serum	Lokate 2007 [128]
Rheumatoid arthritis or lung disease	Cathepsin G in saliva or in endometrial tissue	Gorodkiewicz 2012 [127] Grzywa 2014 [66]
β-thalassemia	Genomic DNA in blood	D'Agata 2011 [129]

8.5 Reaction Process Analysis

Chemical reactions are ubiquitous and indispensable in our body and our environment (up to the whole universe). The reaction scale covers a large range, e.g., in an organelle a living body, a reaction bottle in a laboratory or an industrial reaction tank. No matter how large is the reaction scale, monitor and control of the reaction process are essential, and SPRi is a novel tool able to observe and study the process [130] with high throughput and in real time. Since there are numerous reactions with variable reaction processes, it is impossible to enumerate all of them, but we have better to confine the discussion within three typical processes of recognition (without chemical conversion), reaction (with new chemicals produced) and denaturation (or renaturation). Because the topics of recognition and chemical reaction have been discussed in Chaps. 5, 6 and 7, only the denaturing process will be further discussed below, with macromolecules as examples.

Macromolecules, especially biomacromolecules, have highly ordered structure (up to at least fourth level) that will undergo deconstruction or allosteric phenomena due to the changes of their existing conditions. Such structural changes may affect their recognition or chemical reaction ability and are termed denaturation that can easily be encountered in the study of DNA, protein and their conjugates.

To demonstrate, the denaturation of various proteins was monitored with SPRi in combination with spotting technology. The whole process can be conducted and monitored *in situ* with SPRi and recorded in real time to produce a video for post-experimental studies. Figure 8.5 illustrates an example where an RNase, a kindey-shaped dimer protein (Fig. 8.5a), is spotted on a chip by chemical immobilization through a linking arm of 8–10 carbon chain to free the rotation of RNase and imaged in a buffer of 0.1 M Tris–HCl at pH 7.4 by SPRi before denaturation as shown in Fig. 8.5b. This protein can be denatured in 6 M guanidine hydrochloride but still maintains its molecular integrity (without the breakage of S–S bond) in the presence of dithiothreitol (DTT). The denaturation leads to a significant decrease of the image intensity for all the related spots (Fig. 8.5c) due to the structural collapse that makes the molecules soft and prone to fall flat on the sensor surface. Interestingly, the RNase molecules can be renatured when the denaturing solution is replaced back to the 0.1 M Tris–HCl buffer (Fig. 8.5d). The renaturation is accompanied by the recovery of the image intensity. This can be confirmed either by visual comparison of the image (d) with (b) or by *t*-test ($p < 0.05$) that was accomplished by dividing the total 35 spots into five groups based on their spotting concentration.

It is clear that this SPRi assay allows offline simulation of the whole process of denaturation and renaturation, which can help to set up and/or optimize the process. Nevertheless, the only parameter of image intensity is not sufficient for process control. More parameters are better measured by modification of the assay. Considering that all reaction-involving events are critically dependent on temperature, reactant concentration and/or material feeding speed, we have tried to combine these parameters into one SPRi assay. This was validated to be realizable in combination with the spotting technique. Figure 5.4 shows a reaction catalyzed by an enzyme

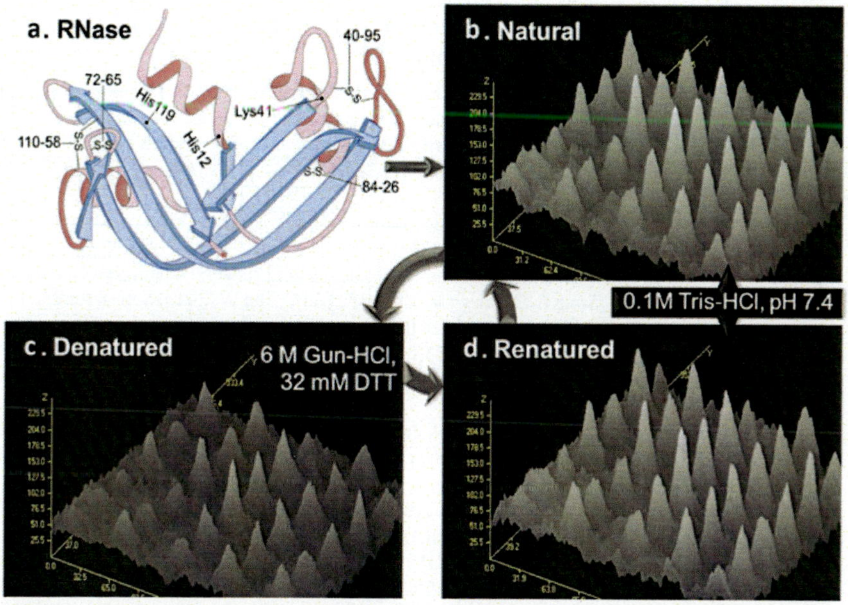

Fig. 8.5 SPRi of reversible denaturation and renaturation process of protein RNase measured on SPRi-TX7100. The denaturation and renaturation were monitored and judged from the imaging intensity variation, which agrees with the known fact for RNase. **a** Crystal structure of RNase that is a kidney-shaped dimer (2 × 124, mw 13,700) protein with four S–S bonds as indicated in the picture; **b**, **c**, **d** Images of intact, denatured and renatured protein spots, respectively; Gun: Guanidine; DTT: Dithiothreitol; Tris: tris(hydroxylmethyl)aminomethane

from eight different sources. The variation is significant. With the same SPRi instrument fabricated in our laboratory equipped with a special control software edited in this laboratory, it is able to simultaneously observe up to 2000 reactions. If such SPRi assays can be really utilized in industrial processes, the feedback control of the processes can be speeded up largely, with a huge amount of equipment costs saved. This looks very promising and is highly attractive, but further exploration and validation are still required before taking action.

If the throughput is not an essential issue, channelized sensor chips are suggested for timely monitor of the reactions in order to avoid the cross-interference between or among the different reactions.

In conclusion, SPRi has a great potential to involve in PAT, with various unique advantages that are not available from the analytical tools presently used. It is worth of effort to make SPRi a practical tool in PAT, which cannot be overemphasized.

References

1. Bonastre A, Ors R, Capella JV, Fabra MJ, Peris M (2005) In-line chemical analysis of wastewater: present and future trends. Trends Anal Chem 24:128–137
2. Rathore AS, Bhambure R, Ghare V (2010) Process analytical technology (PAT) for biopharmaceutical products. Anal Bioanal Chem 398:137–154. https://doi.org/10.1007/s00216-010-3781-x
3. IEC TR 63176, Process analysis technology systems as part of safety instrume. Edition 1.0, 2019-01
4. Liu JW, Cao ZH, Lu Y (2009) Functional nucleic acid sensors. Chem Rev 109:1948–1998
5. Zhu G, Zheng J, Song E, Donovan M, Zhang K, Liu C, Tan W (2013) Self-assembled, aptamer-tethered DNA nanotrains for targeted transport of molecular drugs in cancer theranostics. Proc Natl Acad Sci USA 110:7998–8003
6. Burnett JC, Rossi JJ (2012) RNA-based therapeutics: current progress and future prospects. Chem Biol 19:60–71
7. Filonov GS, Moon JD, Svensen N, Jaffrey SR (2014) Broccoli: rapid selection of an RNA mimic of green fluorescent protein by fluorescence-based selection and directed evolution. J Am Chem Soc 136:16299–16308
8. Mayer G, Ahmed M-SL, Dolf A, Endl E, Knolle PA, Famulok M (2010) Fluorescence-activated cell sorting for aptamer SELEX with cell mixtures. Nat Protoc 5:1993–2004
9. Douglas SM, Bachelet I, Church GM (2012) A logic-gated nanorobot for targeted transport of molecular payloads. Science 335:831–834
10. Xu L, Yan W, Ma W, Kuang H, Wu X, Liu L, Zhao Y, Wang L, Xu C (2015) SERS encoded silver pyramids for attomolar detection of multiplexed disease biomarkers. Adv Mater 27:1706–1711
11. Paige JS, Nguyen-Duc T, Song W, Jaffrey SR (2012) Fluorescence Imaging of cellular metabolites with RNA. Science 335:1194–1194
12. Shen Q, Xu L, Zhao L, Wu D, Fan Y, Zhou Y, OuYang WH, Xu X, Zhang Z, Song M (2013) Specific capture and release of circulating tumor cells using aptamer-modified nanosubstrates. Adv Mater 25:2368–2373
13. Mo R, Jiang T, DiSanto R, Tai W, Gu Z (2014) ATP-triggered anticancer drug delivery. Nat Commun 5:3364. https://doi.org/10.1038/ncomms4364
14. Willner I, Zayats M (2007) Electronic aptamer-based sensors. Angew Chem Int Ed 46:6408–6418. https://doi.org/10.1002/anie.200604524
15. Kimoto M, Yamashige R, Matsunaga K, Yokoyama S, Hirao I (2013) Generation of high-affinity DNA aptamers using an expanded genetic alphabet. Nat Biotechnol 31:453–457
16. Sefah K, Shangguan D, Xiong X, O'Donoghue MB, Tan W (2010) Development of DNA aptamers using Cell-SELEX. Nat Protoc 5:1169–1185
17. Duan N, Gong WH, Wu SJ, Wang ZP (2017) An ssDNA library immobilized SELEX technique for selection of an aptamer against ractopamine. Anal Chim Acta 961:100–105
18. Luo Z, Zhou H, Jiang H, Ou H, Li X, Zhang L (2015) Development of a fraction collection approach in capillary electrophoresis SELEX for aptamer selection. Analyst 140:2664–2670
19. Hünniger T, Wessels H, Fischer C, Paschke-Kratzin A, Fischer M (2014) *Just in time*-selection: a rapid semiautomated SELEX of DNA aptamers using magnetic separation and BEAMing. Anal Chem 86:10940–10947
20. Müller J, El-Maarri O, Oldenburg J, Pötzsch B, Mayer G (2008) Monitoring the progression of the in vitro selection of nucleic acid aptamers by denaturing high-performance liquid chromatography. Anal Bioanal Chem 390:1033–1037
21. Wang Q, Liu W, Xing Y, Yang X, Wang K, Jiang R, Wang P, Zhao Q (2014) Screening of DNA aptamers against myoglobin using a positive and negative selection units integrated microfluidic chip and its biosensing application. Anal Chem 86:6572–6579
22. Spiga FM, Maietta P, Guiducci C (2015) More DNA–aptamers for small drugs: a capture–SELEX coupled with surface plasmon resonance and high-throughput sequencing. ACS Comb Sci 17:326–333

23. Bawazer LA, Newman AM, Gu Q, Ibish A, Arcila M, Cooper JB, Meldrum FC, Morse DE (2014) Efficient selection of biomineralizing DNA aptamers using deep sequencing and population clustering. ACS Nano 8:387–395
24. Wen JD, Gray DM (2004) Selection of genomic sequences that bind tightly to Ff gene 5 protein: primer-free genomic SELEX. Nucleic Acids Res 32:e182. https://doi.org/10.1093/nar/gnh179
25. Lai YT, DeStefano JJ (2011) A primer-free method that selects high-affinity singlestranded DNA aptamers using thermostable RNA ligase. A primer-free method that selects high-affinity single-stranded DNA aptamers using thermostable RNA ligase. Anal Biochem 414:246–253
26. Pan W, Xin P, Clawson GA (2008) Minimal primer and primer-free SELEX protocols for selection of aptamers from random DNA libraries. Biotechniques 44:351–360
27. Jarosch F, Buchner K, Klussmann S (2006) In vitro selection using a dual RNA library that allows primerless selection. Nucleic Acids Res 34:e86. https://doi.org/10.1093/nar/gkl463
28. Tsao S-M, Lai J-C, Horng H-E, Liu T-C, Hong C-Y (2017) Generation of aptamers from A primer-free randomized ssDNA library using magnetic-assisted rapid aptamer selection. Sci Rep 7:45478. https://doi.org/10.1038/srep45478
29. Ouellet E, Lagally ET, Cheung KC, Haynes CA (2014) A simple method for eliminating fixed-region interference of aptamer binding during SELEX. Biotechnol Bioeng 111:2265–2279
30. Jia W, Li H, Wilkop T, Liu X, Yu X, Cheng Q, Xu D, Chen H-Y (2018) Silver decahedral nanoparticles empowered SPR imaging-SELEX for high throughput screening of aptamers with real-time assessment. Biosens Bioelectron 109:206–213
31. Jia W, Lu Z, Yang H, Li H, Xu D (2018) Elimination terminal fixed region screening and high-throughput kinetic determination of aptamer for lipocalin-1 by surface plasmon resonance imaging. Anal Chim Acta 1043:158–166
32. Ngubane NA, Gresh L, Pym A, Rubin EJ, Khati M (2014) Selection of RNA aptamers against the M. tuberculosis EsxG protein using surface plasmon resonance-based SELEX. Biochem Biophys Res Commun 449:114–119
33. Hong SL, Wan YT, Tang M, Pang DW, Zhang ZL (2017) Multifunctional screening platform for the highly efficient discovery of aptamers with high affinity and specificity. Anal Chem 89:6535–6542
34. Misono TS, Kumar PKR (2005) Selection of RNA aptamers against human influenza virus hemagglutinin using surface plasmon resonance. Anal Biochem 342:312–317
35. Dausse E, Barre A, Aime A, Groppi A, Rico A, Ainali C, Salgado G, Palau W, Daguerre E, Nikolski M, Toulme JJ, Di Primo C (2016) Aptamer selection by direct microfluidic recovery and surface plasmon resonance evaluation. Biosens Bioelectron 80:418–425
36. Pelossof G, Tel-Vered R, Willner I (2012) Amplified surface plasmon resonance and electrochemical detection of Pb^{2+} ions using the Pb^{2+}-dependent DNAzyme and hemin/G-quadruplex as a label. Anal Chem 84:3703–3709
37. Zeng S, Baillargeat D, Ho H-P, Yong K-T (2014) Nanomaterials enhanced surface plasmon resonance for biological and chemical sensing applications. Chem Soc Rev 43:3426–3452
38. Li MH, Choi SK, Leroueil PR, Baker JR Jr (2014) Evaluating binding avidities of populations of heterogeneous multivalent ligand-functionalized nanoparticles. ACS Nano 8:5600–5609
39. Polastro ET (1996) Managing primary process development. In: Barnacal PA (ed) Pharmaceutical manufacturing international. Sterling Publications Ltd., London, pp 67–70
40. Karlsson R, Kullman-Magnusson M, Hämäläinen MD, Remaeus A, Andersson K, Borg P, Gyzander E, Deinum J (2000) Biosensor analysis of drug–target interactions: direct and competitive binding assays for investigation of interactions between thrombin and thrombin inhibitors. Anal Biochem 278:1–13
41. Myszka DG, Rich RL (2000) Implementing surface plasmon resonance biosensors in drug discovery. Pharm Sci Technol Today 3:310–317
42. Banaszynski LA, Liu CW, Wandless TJ (2005) Characterization of the FKBP rapamycin FRB ternary complex. J Am Chem Soc 127:4715–4721

43. Kanoh N, Kyo M, Inamori K, Ando A, Asami A, Nakao A, Osada H (2006) SPR imaging of photo-cross-linked small-molecule arrays on gold. Anal Chem 78:2226–2230
44. Campbell CT, Kim G (2007) SPR microscopy and its applications to high-throughput analyses of biomolecular binding events and their kinetics. Biomaterials 28:2380–2392
45. Rich RL, Myszka DG (2007) Higher-throughput, label-free, real-time molecular interaction analysis. Anal Biochem 361:1–6
46. Li S, Yang M, Zhou W, Johnson TG, Wang R, Zhu J (2015) Dextran hydrogel coated surface plasmon resonance imaging (SPRi) sensor for sensitive and label-free detection of small molecule drugs. Appl Surf Sci 355:570–576
47. Di Primo C, Lebars I (2007) Determination of refractive index increment ratios for protein–nucleic acid complexes by surface plasmon resonance. Anal Biochem 368:148–155
48. Holmberg A, Blomstergren A, Nord O, Lukacs M, Lundeberg J, Uhlén M (2005) The biotin–streptavidin interaction can be reversibly broken using water at elevated temperatures. Electrophoresis 26:501–510
49. DeCenzo MT, Park ST, Jarrett BP, Aldape RA, Futer O, Murcko MA, Livingston DJ (1996) FK506-binding protein mutational analysis: defining the active-site residue contributions to catalysis and the stability of ligand complexes. Protein Eng 9:173–180
50. McKeating KS, Aube A, Masson J-F (2016) Biosensors and nanobiosensors for therapeutic drug and response monitoring. Analyst 141:429–449
51. Rogers ML, Boutelle MG, Cooks RG, Pemberton JE (2013) Real-time clinical monitoring of biomolecules. Annu Rev Anal Chem 6:427–453
52. Von Lode P (2005) Point-of-care immunotesting: approaching the analytical performance of central laboratory methods. Clin Biochem 38:591–606
53. Justino CIL, Rocha-Santos TA, Duarte AC (2010) Review of analytical figures of merit of sensors and biosensors in clinical applications. Trends Anal Chem 29:1172–1183
54. Rusling JF, Kumar CV, Gutkind JS, Patel V (2010) Measurement of biomarker proteins for point-of-care early detection and monitoring of cancer. Analyst 135:2496–2511
55. Zhang W, Guo S, Pereira Carvalho WS, Jiang Y, Serpe MJ (2016) Portable point-of-care diagnostic devices. Anal Methods 8:7847–7867
56. Mariani S, Minunni M (2014) Surface plasmon resonance applications in clinical analysis. Anal Bioanal Chem 406:2303–2323
57. Homola J (2008) Surface plasmon resonance sensors for detection of chemical and biological species. Chem Rev 108:462–493
58. Blaszykowski C, Sheikh S, Thompson M (2012) Surface chemistry to minimize fouling from blood-based fluids. Chem Soc Rev 41:5599–5612
59. Vaisocherova H, Brynda E, Homola J (2015) Functionalizable lowfouling coatings for label-free biosensing in complex biological media: advances and applications. Anal Bioanal Chem 407:3927–3953
60. Vaisocherova H, Sipova H, Visova I, Bockova M, Springer T, Ermini ML, Song X, Krejcik Z, Chrastinova L, Pastva O, Pimkova K, Dostalova Merkerova M, Dyr JE, Homola J (2015) Rapid and sensitive detection of multiple microRNAs in cell lysate by low-fouling surface plasmon resonance biosensor. Biosens Bioelectron 70:226–231
61. Phillips KS, Han JH, Cheng Q (2007) Development of a "membrane cloaking" method for amperometric enzyme immunoassay and surface plasmon resonance analysis of proteins in serum samples. Anal Chem 79:899–907
62. Masson J-F, Battaglia TM, Khairallah P, Beaudoin S, Booksh KS (2007) Quantitative measurement of cardiac markers in undiluted serum. Anal Chem 79:612–619
63. Trabucchi A, Guerra LL, Faccinetti NI, Iacono RF, Poskus E, Valdez SN (2012) Surface plasmon resonance reveals a different pattern of proinsulin autoantibodies concentration and affinity in diabetic patients. PLoS ONE 7:e33574
64. Lewis KB, Hughes RJ, Epstein MS, Josephson NC, Kempton CL, Kessler CM, Key NS, Howard TE, Kruse-Jarres R, Lusher JM, Walsh CE, Watts RG, Ettinger RA, Pratt KP (2013) Phenotypes of allo- and autoimmune antibody responses to FVIII characterized by surface plasmon resonance. PLoS ONE 8:e61120

65. Socher I, Andrei-Selmer C, Bein G, Kroll H, Santoso S (2009) Low-avidity HPA-1a alloantibodies in severe neonatal alloimmune thrombocytopenia are detectable with surface plasmon resonance technology. Transfusion (Malden, MA, U S) 49:943–952
66. Grzywa R, Gorodkiewicz E, Burchacka E, Lesner A, Laudanski P, Lukaszewski Z, Sienczyk M (2014) Determination of cathepsin G in endometrial tissue using a surface plasmon resonance imaging biosensor with tailored phosphonic inhibitor. Eur J Obstet Gynecol Reprod Biol 182:38–42
67. Vaisocherova H, Yang W, Zhang Z, Cao Z, Cheng G, Piliarik M, Homola J, Jiang S (2008) Ultralow fouling and functionalizable surface chemistry based on a zwitterionic polymer enabling sensitive and specific protein detection in undiluted blood plasma. Anal Chem 80:7894–7901
68. Vaisocherova H, Zhang Z, Yang W, Cao Z, Cheng G, Taylor AD, Piliarik M, Homola J, Jiang S (2009) Functionalizable surface platform with reduced nonspecific protein adsorption from full blood plasma-Material selection and protein immobilization optimization. Biosens Bioelectron 24:1924–1930
69. Bolduc OR, Clouthier CM, Pelletier JN, Masson J-F (2009) Peptide self-assembled monolayers for label-free and unamplified surface plasmon resonance biosensing in crude cell lysate. Anal Chem 81:6779–6788
70. Bolduc OR, Lambert-Lanteigne P, Colin DY, Zhao SS, Proulx C, Boeglin D, Lubell WD, Pelletier JN, Fethiere J, Ong H, Masson J-F (2011) Modified peptide monolayer binding His-tagged biomolecules for small ligand screening with SPR biosensors. Analyst 136:3142–3148
71. Rodriguez Emmenegger C, Brynda E, Riedel T, Sedlakova Z, Houska M, Alles AB (2009) Interaction of blood plasma with antifouling surfaces. Langmuir 25:6328–6333
72. Haimovich J, Czerwinski D, Wong CP, Levy R (1998) Determination of anti-idiotype antibodies by surface plasmon resonance. J Immunol Methods 214:113–119
73. Thaler M, Metzger J, Schreiegg A, Denk B, Gleixner A, Hauptmann H, Luppa PB (2005) Immunoassay for sex hormone-binding globulin in undiluted serum is influenced by high-molecular-mass aggregates. Clin Chem 51:401–407
74. Lee C-Y, Gamble LJ, Grainger DW, Castner DG (2006) Mixed DNA/oligo (ethylene glycol) functionalized gold surfaces improve DNA hybridization in complex media. Biointerphases 1:82–92
75. Dutra RF, Mendes RK, Lins da Silva V, Kubota LT (2007) Surface plasmon resonance immunosensor for human cardiac troponin T based on self-assembled monolayer. J Pharm Biomed Anal 43:1744–1750
76. Carlsson J, Gullstrand C, Westermark GT, Ludvigsson J, Enander K, Liedberg B (2008) An indirect competitive immunoassay for insulin autoantibodies based on surface plasmon resonance. Biosens Bioelectron 24:876–881
77. Bolduc OR, Masson J-F (2008) Monolayers of 3-mercaptopropylamino acid to reduce the nonspecific adsorption of serum proteins on the surface of biosensors. Langmuir 24:12085–12091
78. Bolduc OR, Pelletier JN, Masson J-FSPR (2010) biosensing in crude serum using ultralow fouling binary patterned peptide SAM. Anal Chem 82:3699–3706
79. Garay F, Kisiel G, Fang A, Lindner E (2010) Surface plasmon resonance aided electrochemical immunosensor for CK-MB determination in undiluted serum samples. Anal Bioanal Chem 397:1873–1881
80. Wang R, Lajevardi-Khosh A, Choi S, Chae J (2011) Regenerative surface plasmon resonance (SPR) biosensor: Real-time measurement of fibrinogen in undiluted human serum using the competitive adsorption of proteins. Biosens Bioelectron 28:304–307
81. Brault ND, Gao C, Xue H, Piliarik M, Homola J, Jiang S, Yu Q (2010) Ultra-low fouling and functionalizable zwitterionic coatings grafted onto SiO_2 via a biomimetic adhesive group for sensing and detection in complex media. Biosens Bioelectron 25:2276–2282
82. Brault ND, White AD, Taylor AD, Yu Q, Jiang S (2013) Directly functionalizable surface platform for protein arrays in undiluted human blood plasma. Anal Chem 85:1447–1453

83. Arvinte T, Palais C, Green-Trexler E, Gregory S, Mach H, Narasimhan C, Shameem M (2013) Aggregation of biopharmaceuticals in human plasma and human serum Implications for drug research and development. mAbs 5:491–500
84. Jang HR, Wark AW, Baek SH, Chung BH, Lee HJ (2014) Ultrasensitive and ultrawide range detection of a cardiac biomarker on a surface plasmon resonance platform. Anal Chem 86:814–819
85. Cappi G, Spiga FM, Moncada Y, Ferretti A, Beyeler M, Bianchessi M, Decosterd L, Buclin T, Guiducci C (2015) Label-free detection of tobramycin in serum by transmission-localized surface plasmon resonance. Anal Chem 87:5278–5285
86. Tokarzewicz A, Romanowicz L, Sveklo I, Gorodkiewicz E (2016) The development of a matrix metalloproteinase-1 biosensor based on the surface plasmon resonance imaging technique. Anal Methods 8:6428–6435
87. Jiang ZX, Qin Y, Peng Z, Chen SH, Chen S, Deng CY, Xiang J (2014) The simultaneous detection of free and total prostate antigen in serum samples with high sensitivity and specificity by using the dual-channel surface plasmon resonance. Biosens Bioelectron 62:268–273
88. Vega B, Calle A, Sanchez A, Lechuga LM, Ortiz AM, Armelles G, Rodriguez-Frade JM, Mellado M (2013) Real-time detection of the chemokine CXCL12 in urine samples by surface plasmon resonance. Talanta 109:209–215
89. Pereira AD, Rodriguez-Emmenegger C, Surman F, Riedel T, Alles AB, Brynda E (2014) Use of pooled blood plasmas in the assessment of fouling resistance. RSC Adv 4:2318–2321
90. Aubé A, Charbonneau DM, Pelletier JN, Masson J-F (2016) Response monitoring of acute lymphoblastic leukemia patients undergoing l-asparaginase therapy: Successes and challenges associated with clinical sample analysis in plasmonic sensing. ACS Sens 1:1358–1365
91. Cornelius RM, Archambault JG, Berry L, Chan AKC, Brash JL (2002) Adsorption of proteins from infant and adult plasma to biomaterial surfaces. J Biomed Mater Res 60:622–632
92. Springer T, Bockova M, Homola J (2013) Label-free biosensing in complex media: a referencing approach. Anal Chem 85:5637–5640
93. Hide M, Yanase Y, Greaves MW (2007) Cutaneous mast cell receptors. Dermatol Clin 25:563–575
94. Yanase Y, Hide I, Mihara S, Shirai Y, Saito N, Nakata Y, Hide M, Sakai N (2011) A critical role of conventional protein kinase C in morphological changes of rodent mast cells. Immunol Cell Biol 89:149–159
95. Siraganian RP (2003) Mast cell signal transduction from the high-affinity IgE receptor. Curr Opin Immunol 15:639–646
96. Griese M, Kusenbach G, Reinhardt D (1990) Histamine release test in comparison to standard tests in diagnosis of childhood allergic asthma. Ann Allergy 65:46–51
97. Valent P, Hauswirth AW, Natter S, Sperr WR, Bühring HJ, Valenta R (2004) Assays for measuring in vitro basophil activation induced by recombinant allergens. Methods 32:265–270
98. Sturm EM, Kranzelbinder B, Heinemann A, Groselj-Strele A, Aberer W, Sturm GJ (2010) CD203c-based basophil activation test in allergy diagnosis: characteristics and differences to CD63 upregulation. Cytometry B Clin Cytom 78:308–318
99. Hide M, Tsutsui T, Sato H, Nishimura T, Morimoto K, Yamamoto S, Yoshizato K (2002) Real-time analysis of ligand-induced cell surface and intracellular reactions of living mast cells using a surface plasmon resonance-based biosensor. Anal Biochem 302:28–37
100. Suzuki H, Yanase Y, Tsutsui T, Ishii K, Hiragun T, Hide M (2008) Applying surface plasmon resonance to monitor the IgE-mediated activation of human basophils. Allergol Int 57:347–358
101. Tanaka A, Tanaka T, Suzuki H, Ishii K, Kameyoshi Y, Hide M (2006) Semipurification of the immunoglobulin E-sweat antigen acting on mast cells and basophils in atopic dermatitis. Exp Dermatol 15:283–290
102. Tanaka M, Hiragun T, Tsutsui T, Yanase Y, Suzuki H, Hide M (2008) Surface plasmon resonance biosensor detects the downstream events of active PKCβ in antigen-stimulated mast cells. Biosens Bioelectron 23:1652–1658

103. Nishijima H, Kosaihira A, Shibata J, Ona T (2010) Development of signaling echo method for cell-based quantitative efficacy evaluation of anti-cancer drugs in apoptosis without drug presence using high-precision surface plasmon resonance sensing. Anal Sci 26:529–534
104. Yanase Y, Suzuki H, Tsutsui T, Hiragun T, Kameyoshi Y, Hide M (2007) The SPR signal in living cells reflects changes other than the area of adhesion and the formation of cell constructions. Biosens Bioelectron 22:1081–1086
105. Yanase Y, Suzuki H, Tsutsui T, Uechi I, Hiragun T, Mihara S, Hide M (2007) Living cell positioning on the surface of gold film for SPR analysis. Biosens Bioelectron 23:562–567
106. Yanase Y, Araki A, Suzuki H, Tsutsui T, Kimura T, Okamoto K, Nakatani T, Hiragun T, Hide M (2010) Development of an optical fiber SPR sensor for living cell activation. Biosens Bioelectron 25:1244–1247
107. Yanase Y, Hiragun T, Yanase T, Kawaguchi T, Ishii K, Hide M (2012) Evaluation of peripheral blood basophil activation by means of surface plasmon resonance imaging. Biosens Bioelectron 32:62–68
108. Yanase Y, Hiragun T, Yanase T, Kawaguchi T, Ishii K, Hide M (2013) Application of SPR imaging sensor for detection of individual living cell reactions and clinical diagnosis of type I allergy. Allergol Int 62:163–169
109. Yanase Y, Hiragun T, Ishii K, Kawaguchi T, Yanase T, Kawai M, Sakamoto K, Hide M (2014) Suface plasmon resonance for cell-based clinical diagnosis. Sensors 14:4948–4959
110. Hiragun T, Yanase Y, Kose K, Kawaguchi T, Uchida K, Tanaka S, Hide M (2012) Surface plasmon resonance-biosensor detects the diversity of responses against epidermal growth factor in various carcinoma cell lines. Biosens Bioelectron 32:202–207
111. Yanase Y, Hiragun T, Yanase T, Kawaguchi T, Ishii K, Kumazaki N, Obara T, Hide M (2014) Clinical diagnosis of type I allergy by means of SPR imaging with less than a microliter of peripheral blood. Sens BioSens Res 2:43–48
112. Suraniti E, Sollier E, Calemczuk R, Livache T, Marche PN, Villiers MB, Roupioz Y (2007) Real-time detection of lymphocytes binding on an antibody chip using SPR imaging. Lab Chip 7:1206–1208
113. Cortès S, Villiers CL, Colpo P, Couderc R, Brakha C, Rossi F, Marche PN, Villiers MB (2011) Biosensor for direct cell detection, quantification and analysis. Biosens Bioelectron 26:4162–4168
114. Schasfoort RBM, Bentlage AEH, Stojanovic I, van der Kooi A, van der Schoot E, Terstappen LWMM, Vidarsson G (2013) Label-free cell profiling. Anal Biochem 439:4–6. https://doi.org/10.1016/j.ab.2013.04.001
115. Houngkamhang N, Vongsakulyanon A, Peungthum P, Sudprasert K, Kitpoka P, Kunakorn M, Sutapun B, Amarit R, Somboonkaew A, Srikhirin T (2013) ABO blood-typing using an antibody array technique based on surface plasmon resonance imaging. Sensors 13:11913–11922
116. Stojanović I, Schasfoort RBM, Terstappen LWMM (2014) Analysis of cell surface antigens by surface Plasmon resonance imaging. Biosensors Bioelectronics 52:36–43
117. Yanase Y, Hiragun T, Kaneko S, Gould HJ, Greaves MW, Hide M (2010) Detection of refractive index changes in individual living cells by means of surface plasmon resonance imaging. Biosens Bioelectron 26:674–681
118. Horii M, Shinohara H, Iribe Y, Suzuki M (2011) Living cell-based allergen sensing using a high resolution two-dimensional surface plasmon resonance imager. Analyst 136:2706–2711
119. Peterson AW, Halter M, Tona A, Bhadriraju K, Plant AL (2009) Surface plasmon resonance imaging of cells and surface-associated fibronectin. BMC Cell Biol 10:16. https://doi.org/10.1186/1471-2121-10-16
120. Peterson AW, Halter M, Tona A, Bhadriraju K, Plant AL (2010) Using surface plasmon resonance imaging to probe dynamic interactions between cells and extracellular matrix. Cytometry A 77:895–903
121. Shinohara H, Sakai Y, Mir TA (2013) Real-time monitoring of intracellular signal transduction in PC12 cells by two-dimensional surface plasmon resonance imager. Anal Biochem 441:185–189

122. Zhang LL, Chen X, Wei HT, Li H, Sun JH, Cai HY, Chen JL, Cui DF (2013) An electrochemical surface plasmon resonance imaging system targeting cell analysis. Rev Sci Instrum 84:085005. https://doi.org/10.1063/1.4819027
123. Chen CY, Chang CC, Yu C, Lin CW (2012) Clinical application of surface plasmon resonance-based biosensors for fetal fibronectin detection. Sensors 12:3879–3890
124. Erturk G, Ozen H, Tumer MA, Mattiasson B, Denizli A (2016) Microcontact imprinting based surface plasmon resonance (SPR) biosensor for real-time and ultrasensitive detection of prostate specific antigen (PSA) from clinical samples. Sens Actuators B 224:823–832
125. Gorodkiewicz E, Charkiewicz R, Rakowska A, Bajko P, Chyczewski L, Niklinski J (2012) SPR imaging biosensor for podoplanin: Sensor development and application to biological materials. Microchim Acta 176:337–343
126. Sankiewicz A, Guszcz T, Mena-Hortelano R, Zukowski K, Gorodkiewicz E (2016) Podoplanin serum and urine concentration in transitional bladder cancer. Cancer Biomark 16:343–350
127. Gorodkiewicz E, Sieńczyk M, Regulska E, Grzywa R, Pietrusewicz E, Lesner A, Łukaszewski Z (2012) Surface plasmon resonance imaging biosensor for cathepsin G based on a potent inhibitor: development and applications. Anal Biochem 423:218–223
128. Lokate AMC, Beusink JB, Besselink GAJ, Pruijn GJM, Schasfoort RBM (2007) Biomolecular interaction monitoring of autoantibodies by scanning surface plasmon resonance microarray imaging. J Am Chem Soc 129:14013–14018
129. D'Agata R, Breveglieri G, Zanoli LM, Borgatti M, Spoto G, Gambari R (2011) Direct detection of point mutations in nonamplified human genomic DNA. Anal Chem 83:8711–8717
130. Huang H, Chen Y (2006) Surface plasmon resonance imaging studies for proteolytic hydrolysis of proteins. Chem Lett 35:372–373

Chapter 9
Challenges and Prospects

SPRi has been demonstrated to have many advantages over other analytical tools as exemplified in Chaps. 5–8 that have made SPRi unreplaceable in many fields, especially in molecular recognition-based analysis like selective screening, affinity studies, immune reactions, single or discrete particle imaging, and so forth. Keeping these in mind, SPRi is also facing some critical issues and challenges after a fairly long term of exploration and development. It remains quite vague about the future of SPRi and its frontiers of researches and applications. We are trying to figure out possible routes and chance in future SPRi developments, basically through the analysis of several essential challenges in combination with prospection based on our limited and superficial judgment.

9.1 Basic Challenges

It is hard to judge the research frontiers of SPRi in future without superb wisdom, experiences and skills, but we feel obligated and responsible to contribute our efforts to its development. We are herein trying to speculate on the frontiers through pairing its features with shortcomings. Like a double-edged sword, a factor that can provide amazing advantages will always be accompanied by its opposite issues or even challenges. Based on the theory of photoexcited SPPs and our investigations as well, some essential advantages can easily be deduced, and so deduced are the issues or challenges. They are collected in Table 9.1.

Among the listed issues/challenges, the essential challenges come from identification, sensitivity, lateral resolution, image distortion, sensing depth, mechanisms integration and film quality. Most of them have more or less found solutions that have been discussed in previous chapters such as the issues concerning with sensitivity, image distortion and film fabrication. While the remaining issues like lateral resolution, identification, comprehensive utilization of the resonant mechanisms need further study to find effective solutions. In following discussions, we will summarize

Y. Chen, *Surface Plasmon Resonance Imaging*, Lecture Notes in Chemistry 95,
https://doi.org/10.1007/978-981-99-3118-7_9

Table 9.1 Basic features and challenges of SPRi based on photoexcited SPPs

Key parameter/format	Advantages	Issues/challenges
Metal film With negative ε Appropriate thickness Flat and smooth surface **Excitation** Visible light Near-field configuration **Sample** With positive ε or n **Propagation** Depth < 200 nm Length ~ 10 μm	**Film or chip optional** Modified, spotted, channelized **Configuration optional** Prism, microscope object Grating, fiber, particles, etc. **Unique measuring feature** Label-free or non-destructive Transparency independent Imaging conditions at will Universally applicable High throughput Nanometer vertical resolution > 10 μm lateral resolution	Few optional metals Limited chip life Near- and far-field isolation, monotonous mechanism Insufficient sensitivity (limited by n and film quality) Weak in identification Image distortion Limited sensing depth Low lateral resolution

the discussed measures to conquer the issues or challenges. Some prospections will also be given on future development that may hopefully become new frontiers of SPRi.

9.2 Issue on Sensitivity

SPRi is sensitive but remains not sufficient in the direct detection of various low abundance molecules. This is essential to extend its application range and to improve its analytical reliability. Although SPRi sensitivity has been continuously increased [1] during improvement of methodology and removal of application barriers, it is still desired to further raise the sensitivity to greatly widen its applicable range. Since a lot of effort has been paid on the enhancement of SPRi sensitivity, it will become more and more difficult to further increase the sensitivity in future. To have a prospect, we briefly summarize some potential enhancing ways.

The extremely effective way to increase the imaging sensitivity what we have discussed is signal amplification. In principle, SPRi sensitivity increases with the difference of refractive index (Δn) between an analyte and a reference and in turn with the molecular mass of the target analytes. Different reactions have since been explored to add mass onto the target analyte as much as possible. The very effective ways include the selective load of particles and/or macromolecules physically or chemically (e.g., through cyclic or polymeric reactions). Unfortunately, the mass cannot be added infinitively, which is limited by the size of the added substances and the sensing depth. The signals will be saturated if the added substances are larger than the sensing depth. The further development of signal amplification technology is becoming more and more difficult. At present, the valuable research topics may be to explore easier, more efficient and cost-effective loading technology and loading materials with greater gain (e.g., with high refractive index).

The basic way to improve the sensitivity is background subtraction. The original images are not the net signals of analytes but superposed upon a high background. Therefore, the net information must be uncovered through deducting the background from the real-time-measured images. To deduct the background as clean as possible that usually varies with running conditions and time, the subtrahend must be an analytes-free image measured either at the beginning ($t = 0$) of measurement or simultaneously with analytes from their vicinity. The subtraction results in either temporal or spatial difference images (for further analysis and publication). In practical, the spatial difference images are more often adopted than the temporal, to suppress also the influence of time-dependent factors. Although this image-difference technology may improve the imaging contrast and sensitivity, it has little room for further exploration.

The third way to improve SPRi sensitivity is noise elimination or denoise. Signal amplification and background deduction do not necessarily suppress the noises, but oppositely may bring in additional noise(s). Therefore, signal denoise is in general necessary. There are presently mathematical (rather than physical or hardware) measures such as arithmetic or geometric averaging operation and transformation between time and frequency domains. The averaging operation is simple and easy, but the useful signals may also be averaged or reduced; differently, the transformation has technical barrier but can keep the required signals through setting a frequency threshold to cut off the noises and through combining with convolution operation. Fortunately, mathematic transformation can now be realized with matured computation programs. Furthermore, many mathematical methods are available such as Fourier transform, wavelet transform, Laplace transform and so forth. In present SPRi analysis, the most often used technique is FFT. The issue is however that we are still short of real-time denoise technology. This may be a topic for future research. It looks also an attractive topic to decompose and analyze all the imaging frequencies, which may reveal some not-yet-recognized signals or allows to realize multi-level analysis of samples from limited experimental data, to save time and cost.

The forth way to improve the sensitivity may be pointed to hardware, including the change of excitation configuration and the replacement of the angle or wavelength interrogation by phase interrogation. SPRi configuration is quite fixed and is dependent on the excitation means and interrogation factor. For example, phase interrogating SPRi can improve the sensitivity for about 10^3 folds. By use of interferometric technology, irrelevant noises can be filtered out through phase-locked amplification technique, so that SPRi reached an LOD down to 10^{-8} RIU. The disadvantages are also obvious such as redesign and re-fabrication of SPRi instrument, which has technical barriers, takes time and may cost high.

The fifth way to improve the sensitivity lies in metal film quality. SPRi sensitivity depends greatly on the quality of gold sensing film, especially its thickness and surface smoothness. The presently adopted sensor chips usually consist of a very thin gold film that is hardly be coated on some glass slides with currently available technology, always with a deviation of several nanometers in thickness from the optimal resonant thickness (ca. 50 nm). The film is further degraded by the need of an intermediate adhesive layer (e.g., ca. 2 nm Cr or Ti) that can evidently reduce the sensitivity and by the rough or unsmooth surface resulted from the deposition-caused aggregation and/or crystallization of the gold atoms into nanoparticles at a size up to about 5 nm. The thickness deviation, together with adhesive layer, can directly reduce and/or changes the sensitivity from film to film; While the surface roughness easily causes local scattering, emission and reflection of SPP waves and the scattered SPP waves may further couple with the propagating SPPs. Both mechanisms can increase the signal noise, blur the recorded images and hence reduce the sensitivity. Clearly, these factors can hardly be removed completely. The current remedy includes surface modification and use of more sensitive metals such as Ag. To resist oxidation, Ag films have to be protected by further depositing a layer of noble metal such as Au or by coating with a layer of inertial polymer(s). This refers to the exploration and utilization of multi-layered sensing films, which increases not only the sensing sensitivity but also sensing depth. By coupling the plasmonic metal film with waveguide structure, the sensing depth may be lengthened to μm level, thus allowing the loading of thicker amplification substances onto target analytes. Comprehensive study of these film fabrication techniques may open a novel gate to improve the detection sensitivity. Total replacement of metal sensing films is another field worthy of deep investigation. Bloch layers have been shown to be adoptable and can further be explored, which have still numerous chances. One advantage of Bloch layers lies in the utilization of insulating materials that will facilitate the coupling of SPRi to electromethodology such as electrophoresis. In short, sensing films offer a lot of chances to further explore sensitive SPRi assays and even novel sensing methods.

9.3 Removal of Image Distortion and Related Issues

Image distortion and light spot center shift along *x* direction typically appeared in use of glass prism coupling configuration as has been discussed in Chap. 3. Some practically solving measures have also been explored. The most often used technology is to re-shape the image by computer imaging software, but it needs to neglect the spot shifting issue. The second measure what we suggested is to change the video orientation rather than to replace the prism-based configuration, that is, to arrange the camera head parallel to the prism bottom instead of vertically or at an angle. This measure can remove the issue of image distortion but not that of the width shrinking. The third measure is to replace the glass prism with a grating. Different from the second measure, this measure can remove the light spot shifting issue but not the image distortion. Grating coupling is also accompanied by significant loss of sensitivity. The fourth measure is to use fiber arrays as the coupling media to confine the distortion and light center shift within the cross-sections of the fibers. Fiber array coupling is flexible but seriously loses the imaging sensitivity.

The fifth measure what we have proposed to fix the light center spot and to improve the sensitivity is to replace the glass or hard prism with a soft one [2]. In Chap. 3, we have shown that a liquid prism is able to remove the non 90° light (Fig. 3.5). Herein, we show that it can produce much clearer plasmonic images (Fig. 9.1a) than a glass prism (Fig. 9.1b) even if the liquid has a much lower refractive index (1.5511) than the glass (1.717). The liquid prism could also improve the imaging sensitivity without the regulation of incident angle, which can simply be realized through adjusting the liquid composition to change the refractive index. However, the liquid prism needs a leak-proof container with soft arms. A theoretically ideal solution to this new issue is to explore highly transparent soft "glasses". The possibly promising materials may include liquid glass and soft but isotropic polymers that is waiting for exploration. Similar to glass prisms, liquid prisms remain unable to remove the imaging distortion issue, but it can reduce the distortion degree at an incident angle as small as possible and by use of liquid with refractive index as high as possible.

A comprehensive solution that looks attractive is to explore microhemispherical lens array (Fig. 9.2a) that may combine the advantages of fiber array, grating and glass prim. In theory, a hemispherical lens can always allow a light ray to vertically enter and exit the spherical surface when it propagates along the radial direction (Fig. 9.2b), and the image distortion will either be removed by bottom-parallel orientation of the

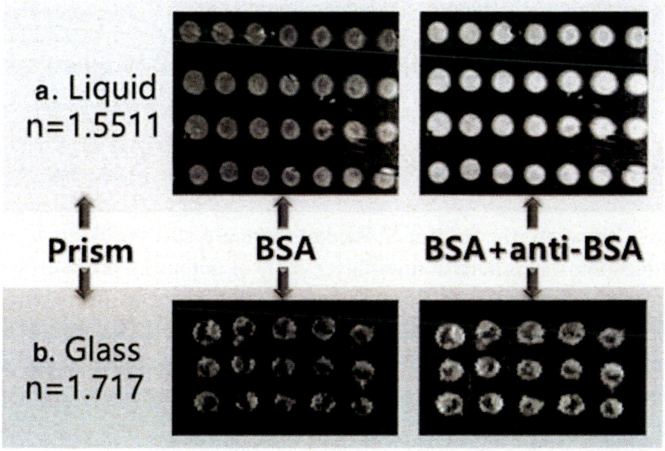

Fig. 9.1 Improvement of imaging contrast by use of a liquid prism rather than a glass measured on SPRi-PX8100. **a** Imaged with a liquid prim with a refractive index at $n = 1.5511$. **b** Imaged with a glass prism with $n = 1.717$

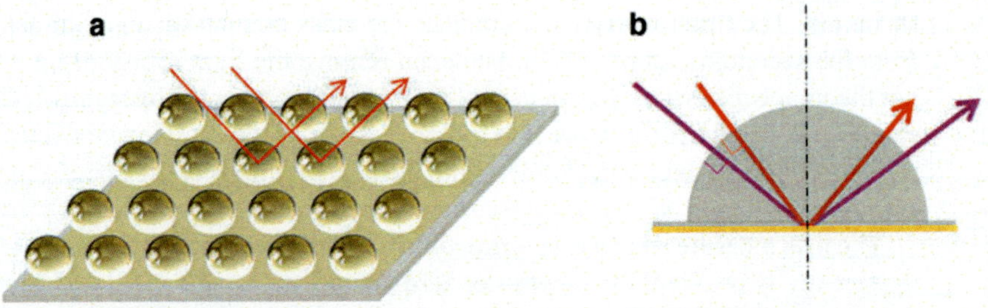

Fig. 9.2 Schematic configuration for SPRi via microhemispherical lens array **a** to avoid the light spot center shifting **b** as changing the incident angle for a better shape of images

camera (Fig. 3.11P3) or be confined within the hemisphere by common arrangement of the camera toward the light beams. It looks practical because there are various commercial microarray lenses available. The expected problem is the light interference from adjacent lenses (e.g., non-collimated light, excessive facula and/or diffuse reflection). The potential solutions include: (1) the use of arrayed excitation light source (e.g., LED array, slit or diaphragm or single longitudinal mode fiber arrays); (2) collimation of the light beams as much as possible and (3) sufficient separation of the micro lenses; and (3) coupling with the lens-matched microarrayed gold films.

A completely different idea that has been briefly introduced in Sect. 4.1.1 is to perform plasmonic imaging in a way much the same as ordinary optical microscopy, that is, to light up various imaging elementals as in the case of LSPR sensing [3, 4]. Similar to plasmonic amplification, analytes on a surface or in a transparent body can be lighted separately with their probes tagged with AuNPs, AgNPs or other metal NPs.

9.4 Sensor Films-Associated Issues

9.4.1 Preparation Challenge

Sensor chips are the root of SPRi, but there are still problems unsolved. The most troublesome one is hard to smoothly, evenly or thickness-precisely (ideally at an error < 0.5 nm), compactly and firmly deposit ca. 50 nm Au films on glass slides. This depends not only on the quality of glass substrate (e.g., with smooth, flat and defect-free surface) and the purity of gold or relative materials but more radically on the regulation precision of the deposition technique used. Currently, high-quality optical glass prisms and slides and ultrapure gold are commercially available. The issue lies in the short of convenient deposition technology. Although there are currently quite many coating techniques such as vacuum evaporation deposition (VED), vacuum magnetic sputtering (VMS), chemical vapor deposition (CVD), sol–gel coating, spraying, electroplating and so forth, few have been adopted in the practical preparation of SPRS and SPRi sensor chips, except for VED, VMS and possibly VCD.

VED is at present the easy-to-use technology for preparing the SPRi sensor chips at comparatively low cost and fairly high precision (with deviation of several nanometers), with easy manipulation and enough flexibility (e.g., *in situ* change of deposition conditions for multiple coatings, with optional evaporation sources including resistance heating, induction evaporation, electron beam gun and vacuum arc). The major disadvantages include that (i) the film thickness is location-dependent (even under rotation), (ii) the films need reformation by heating or other techniques to improve their firmness and smoothness, (iii) the deposition repeatability is insufficient or needs a great improvement and (iv) an adhesive layer of ca. 2 nm Cr or Ti is unavoidable, which increases imperfection of the films and reduces resonance strength or imaging sensitivity. As a consequence, all the VED-fabricated chips have to be checked by optical transmission or SPR technique to screen out qualified films. Furthermore, VED is in fact not very suitable for the coating of non-metal materials.

VMS is presently the most advanced and the most flexible coating technology, with optional operation modes (e.g., direct or alternative current sputtering and unbalanced sputtering), able to combine with various types of chemical reactions. VMS allows fairly precise deposition of different materials (with an error of several nanometers) at a designed thickness to obtain fairly compact and uniform films with quite strong adherence on glass substrates. Nevertheless, it has high technique and cost barriers that prevent it from becoming a common technology.

CVD can deposit noble metals, alloys, ceramics and inorganic compounds onto wrinkled or curved surfaces under vacuum or ambient pressure. The related reactions (normally decomposition) usually occur at a medium or high temperature but can be reduced by plasma- and/or laser-assisted techniques. Unfortunately, VCD is rarely used for the preparation of SPRi sensor chips because it needs volatile materials. In addition, its control factors are not fully revealed. Therefore, VCD is currently used to coat micrometer films on large substrates rather than nanometer films on centimeter surfaces. Nevertheless, VCD is expected to have a broad space for further exploration.

AS mentioned, a defect, unsmooth sensor chip will raise the noise level due to the random reflection and scattering of SPPs, while a film with uneven and suboptimal thickness will lead to unexpected shallowing and/or broadening of the resonant absorption dip along the lateral direction of the films. As a consequence, SPRi loses its imaging contrast and sensitivity owing to the shallowed and broadened resonance peak and the superposition of the raised noise and background as well. The present solutions to these issues include (i) reforming the films right after deposition and (ii) modifying the film surface, normally with soft spackling matter (e.g., polysaccharides) to smoothen the surface. In fact, the reformation is not universally applicable to non-metallic substances; therefore, the surface modification technology is routinely utilized in practice because it is comparatively easy to develop, with many options, and still under developing. Nevertheless, we still desire to explore new materials and methods for the fabrication of ideal sensor chips.

9.4.2 *Limited Sensing Depth and Stability*

Another troublesome problem in the preparation of sensor chips is the restricted option of metals. Presently, Au is the most often used metal for it can resist oxidation and is inert in numerous solutions (except for aqua regia and ionic liquids). Even though, the Au-based chips are not stable enough in practice, with fairly limited number of regeneration (ca. 10 times in most cases). Furthermore, Au films can only offer a sensing depth of about 180 nm, not surpassing 200 nm, which makes SPRi unable to fully image > 200 nm particles. Therefore, SPRi is a weak tool to image microparticles such as cells that have a height of several to tens of microns. Most of the cell volume or their inner content will be outside the sensing range.

The sensing depth (and hence the imaging ability) can be improved to above 230 nm by replacement of Au with Ag. Unfortunately, the increment is quite limited. Furthermore, Ag is susceptible to oxidation in air and solutions. Hence, it must be utilized under protection by covering under a thin layer of inert substance such as ca. 5 nm Au or organic polymers. A more exciting measure to increase the sensing depth to a level of micrometer is to explore long-range SPRi (LRSPRi) methodology. LRSPRi can be realized via a symmetric substrate configuration, e.g., dielectric/metal/dielectric or metal/dielectric/metal, to reduce the confinement of surface plasmon mode and in turn to increase the penetrating depth of z mode [5, 6]. In theory, LRSPRi can be expected to have better images of cells with a greater sensitivity due to the sharper resonance dip [7] and the contribution of the cell bulk refractive index variations [8], similar to SPRS of bacteria [9, 10]. As expected, very high sensitivity and image contrast can be obtained at a smaller resonant angle after mediation with antimonene [11]. LRSPRi has larger space to explore than the common SPRi; however, it increases the difficulty to prepare high-quality sensing films. As known, it has technical barrier to deposit a qualified waveguide layer on a flat substrate at a thickness of hundreds of nanometer to micrometers, especially in use of hydrophobic materials such as Teflon or CYTOP that is not easy to adhere evenly, smoothly and firmly on glass slides. It is waiting for the emergence of new technology to lower the barrier.

9.4.3 *Janus-Like Metal Conductance*

The electric conductance of metal films is Janus-like property, with both advantages and disadvantages for SPRi. On the one hand, it facilitates the hyphenation of SPRi with electrochemistry to empower its ability, and on the other hand, it causes electrolytic damage to the films and/or electrolytic bubbles in the cell; therefore, it prevents the direct coupling of SPRi with electroseparation techniques like electrophoresis. The challenge is that the metal film will take most of the current in contact with a solution, making only limited current through the solution. This may lead to poor or even no electric separation of analytes and hence prevents *in situ* coupling of

SPRi with chip-based electrophoresis. To couple, it is necessary to sufficiently reduce the metal-conducted current by some insulating measures. The basic measure is to disintegrate the whole metal film into islands surrounded by and covered under insulating material(s) such as Cytop or Teflon. This can presently be realized without much issue [12], but the quality depends very much on the quality of the coated insulating film that in turn depends on surface modification technology. A measure to completely avoid the metal current is to replace the SPPS with Bloch surface waves [13–16]. Bloch surface waves can work similarly to SPPs, but this means to completely abandon SPRi. In addition, Bloch surface waves also need advanced technology to precisely prepare layered films.

Metal films are also accompanied by heating effect under the illusion of light, which impacts on the behavior of molecules, especially the thermosensitive molecules, and complicates the elucidation of the measured signals. It is hence worth of mentioning that the flow cell of SPRi is better with thermostatic control with temperature adjustable.

In short, the ideal solution to the preparation of sensor chips is to explore easier, cheaper and more effective nanofilm manufacturing technology with higher accuracy. It is hard to predict what kind of new technology will emerge, but we are expecting that some new techniques can integrate the principle of molecular self-assembly so as to improve the regulation accuracy of film preparation to angstrom level or a monomolecular layer.

9.5 Comprehensive Utilization of SPPs

In principle, surface plasmons and their excited SPPs can cause at least optical absorption, scattering and dark excitation of fluorescence and Raman scattering. They are currently explored separately rather than comprehensively. SPRi is an example that uses majorly the resonance absorption phenomenon, which only partially combines with the near-filed scattering mechanism (i.e., in imaging the discrete particles). It looks highly difficult to integrate all the mechanisms so as to invent new and powerful SPPs-based methods. This is a greatly attractive research topic, worthy of deep exploration because it means the possibility to perform multi-parameter and multi-level identification and determination of very limited samples.

Identification or qualitative analysis is one of the root tasks in analytical chemistry. Unfortunately, this is just the weakness of SPRi as indicated in Table 1. At present, SPRi needs probes to recognize the target molecules. Its identification reliability depends very much on the stability and specificity of the used probes that may change their characteristics in variable and complex environments. If dark background Raman and even infrared spectroscopies can be integrated into SPRi, it will possess the inherent identification ability. The imaging parameters will also be increased accordingly. By this integration technology, SPRi may also add options to increase its imaging sensitivity by dark excitation of fluorescence, to perform scattering imaging via bright or dark field.

The above-discussed is intrinsic integration that is highly interesting and has got some absorbing studies [17], but the advancing pace remains little due majorly to the need of invention of disruptive coupling configuration. To accelerate the progress and to widen the applicability of SPRi, non-intrinsic integration measures have also appeared as comparatively easy alternatives, that is, the coupling of SPRi to other analytical tools, for example, coupling to mass spectrometer or other qualitative analytical methods to acquire identification ability, coupling to electrochemistry to sensitize detection and/or add more detection parameters and coupling with capillary or chip-based electrophoresis to act as a universal but multichannel detector. This type of coupling or integration techniques can also become highly interesting and has been studied for quite a long term. They are being discussed in following subsections with some prospects.

9.5.1 In Situ *Coupling of SPRi with Electrochemistry*

Coupling of SPRi with electrochemistry (EC) [18–22] started quite early for it can easily be realized by merging the electrochemical working electrode with SPRi sensing film as illustrated in Fig. 9.3, parallel to SPR detection of electrochemically induced chemical changes [23–25]. As a consequence, the coupled system can simultaneously collect the dynamic or real-time changes of electron transfer and permittivity happening on the sensor surface or in its close vicinity. This also allows optical observation or continuous monitoring of the changes of electrochemical redox reaction(s) with lateral resolution. Electrochemistry such as cyclic voltammetry can follow redox reactions and can be used to evaluate the quality of self-assembled monolayers. Nevertheless, it cannot offer lateral resolved information unless scanning electrochemical microscopy is used, and it is difficult to simultaneously measure a sample at different locations except for the use of working electrode microarray that corresponds to the use of an arrayed gold sensing film. This coupling can be expected to have space limitation and needs to sacrifice lateral resolution. In addition to SPRi monitoring of electrochemical reactions or electrochemically stimulated responses (e.g., for living cells), the electrochemical reaction can act reversely, that is used to spot substances [19], to assist capture of targets [26] and/or modifying the sensor surface.

Although electrochemical-SPRi has developed for quite a long time, it remains worth of further exploration. Two important fields may need to pay a great attention: (i) its application to the study of living cells in either crowded or discrete states, including stimulated responses, where the key issue has to be further removed is to break through the limitation of SPRi sensing depth; and (ii) exploration and addition of more analytical functions to it to empower its ability and to expand its applicability, where other electrochemical methods other than cyclic voltammetry are waiting for exploitation, the similar is the exploration of various SPPs-related excitation and

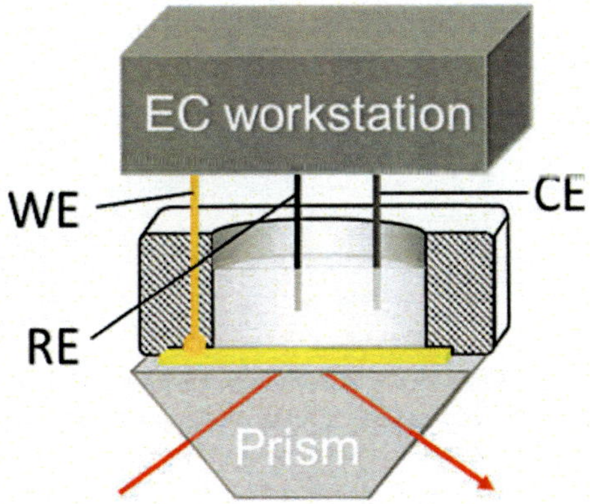

Fig. 9.3 *In situ* coupling of SPRi with electrochemistry by use of the sensor film as a working electrode (WE) based on three-electrode system having counter electrode (CE) and reference electrode (RE)

operation modes. In addition, the coupling strategy of separating the working electrode from the sensing film remains waiting for exploration, which is more flexible, with promising opportunity.

9.5.2 Coupling of SPRi with Other Optical Imaging Methods

As an optical method, SPRi has inherent technical convenience to couple with other optical technologies, such as Raman spectrometry/microscopy, infrared spectrometry, bright/dark field optical transmission/scattering microscopy, fluorescent microscopy, laser confocal microscopy and so forth. These candidates will largely enrich SPRi coupling family and produce mutually comparable imaging signals for deep inspections. Unfortunately, few researches have been reported till now and no commercialized instrument is available yet. The related devices must be established in laboratories and can easily be realized because the couplings have no insurmountable technical barrier except for space congestion.

9.5.3 Coupling of SPRi with MS, MSi and Other Identification Techniques

To greatly increase the identification ability, SPRi has been coupled with MS and/or MSi for decades of years [27–32]. There are in practice two coupling strategies for SPRi-MS/MSi: The first one is to conduct MS analysis of SPRi-eluted solutions, and the second one is to perform on-chip MS of the captured substances. The former is the easiest to realize because it does no need to change the hardware of SPRi and MS devices and their analytical methods; therefore, both their optimal or well-established approaches can be utilized directly, including the use of variable ion sources, for example, MALDI [33, 34], surface-enhanced laser desorption/ionization or SELDI

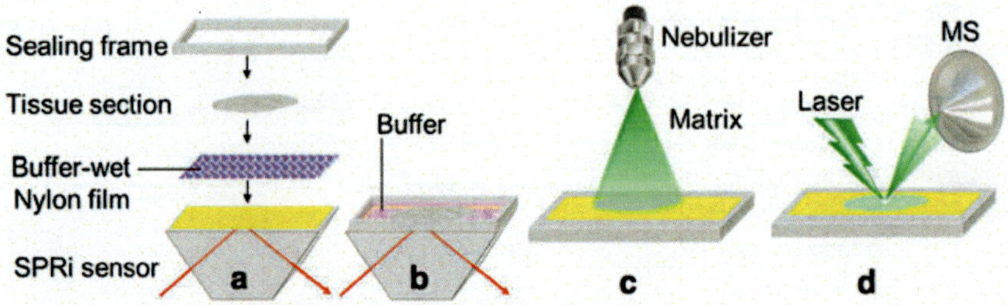

Fig. 9.4 Critical steps to perform SPRi-MS or MSi tissue sections. **a** A way to assemble a tissue section on an SPRi sensor chip with a buffer-saturated nylon diaphragm. **b** Immediate SPRi of the transferred molecules (note, additional buffer may be supplemented if the nylon diaphragm is dried too fast). **c** Spraying a matrix onto the sensor film for MALDI MS (better after its detachment from the SPRi instrument). **d** MS or MSi of the chemicals on the chip

[35] and electrospray ionization or ESI [35, 36]. This strategy will, however, reduce lateral and temporal resolutions due to the need of eluting the analytes in separate channels rather than on spots [27, 28]. This type of coupling may also need sample collection and spatial transfer, easily causing sample contamination and material loss, which challenges the quantitative analysis and the applications of the coupled methods to assaying the very limited samples. In addition, it will be hard to elute the high-affinity or strongly captured components for MS analysis.

Differently, the latter needs free of or much less sample treatments, able to reduce sample waste. On-chip MS or MSi is better performed by demounting the sensor chip from the SPRi flow cell (Fig. 9.4) to facilitate measurements and to protect SPRi device as well. The obvious disadvantage is that the on-chip MS or MSi will destroy the sensor chip and limit the selection of ionization sources except for the most often adopted MALDI [27, 28, 36–41]. MALDI also facilitates MSi measurements [41, 42].

SPRi-MSi is a mutually beneficial technique because SPRi enables MSi to quantify difficulty, while MSi helps SPRi to identify unknown molecules. Currently, SPRi-MSi is majorly applied to imaging and quantifying the microarrays based on affinity interactions. In practice, SPRi-MSi can also be used to image tissue sections but needs to combine with high-fidelity transfer technique [43] to enable both quantification and characterization [41]. The critical steps to conduct SPRi-MSi of a tissue section are illustrated in Fig. 9.4.

Clearly, current SPRi-MSi does not directly image the tissue sections. In fact, it images the section prints, that is, the molecules transferred from the tissue sections. Therefore, the imaging fidelity depends very much on lateral mass transfer effect that is mainly caused by molecular diffusion and liquid mobility. The lateral mass transfer will cause adjacent interference and in turn yields unreal or false images. Thus, high-fidelity section printing technology must be invented to perform reliable SPRi and

SPRi-MSi. In theory, there are three possible measures to transfer molecules from a tissue section onto a sensor chip or simply to print the tissue section:

(i) Contact transfer. It is performed by pressing a sample slice directly on the sensor surface and then conducting SPRi and MSi in sequence. It can fast transfer or print the molecules from the microprotrusions on a surface onto the sensor chip. This can be exemplified in SPRi of fingerprints [44], but MSi of the fingerprints failed because the transferred molecules are too trace to be measured by MS and MSi in our laboratory. However, this does not imply that the contact transfer has no value for further exploration. In contrast, it deserves deeper exploration to facilitate fast SPRi-MSi. It is highly expected to find or invent ways to speed up molecular transfer with high fidelity;

(ii) Free solution film-spaced transfer. The contact transfer is not really suitable for smooth flat tissue sections; otherwise, super strong background will be recorded and submerges the target signals. Therefore, the transfer of molecules from the smooth surface must thus be separated by a certain distance. This can in theory be realized by insertion of a 200–1000 μm thick free solution film between the two surfaces. In order to have a uniform liquid diaphragm, a frame spacer is better utilized. The frame spacer can easily be enclosed by a fused silica capillary or a glass fiber with outer diameter of ca. 400 μm or surrounded with blades (at about 200 μm thickness per slide). In theory, the thinner is the liquid film, the faster will be the transfer. However, too thin a liquid diaphragm may cause contact issue for a soft tissue section that may collapse downward in the middle part. A liquid film at 400 μm thickness or above is suggested for soft tissue sections. To prevent the middle collapse, the section is better adsorbed on a hard support, with its free surface contacting the liquid film. A serious issue encountered is that the liquid diaphragm cannot resist the lateral mass transfer. In order to prevent the liquid from flowing, the transfer should be performed on a vibration-proof and tilting-free platform. Despite all these, the molecular diffusion is unavoidable, so that the liquid film must be applied to the transfer of slow diffusing substances (e.g., macrobiomolecules like proteins) rather than fast diffusing small molecules. Even in the case of proteins, the diffusion rate can be faster than 5.47 μm/s [41] that easily results in unreal images;

(iii) Permeable immobile membrane-spaced transfer. To reduce the liquid mobility and molecular diffusivity, the liquid film must be replaced by an immobile but liquid-permeable membrane. For example, with a wet nylon membrane, the proteins in a tissue section could be transferred with much higher location fidelity due to the suppression of lateral liquid movement and obvious reduction of apparent molecular diffusion rate down to 0.34 μm/s [41], corresponding to a decrement by a factor of 1/15. The accompanied issue is that the hard membrane significantly slows down the transfer process.

Based on above discussions, the tissue printing technology remains not matured yet. It will undoubtedly take time to easily conduct SPRi-MS or MSi of tissue sections with high image fidelity. The expected researches include:

(i) Further exploration of high-fidelity transfer technology, which may lie in the discovery or invention of new techniques able to simultaneously confine the lateral mass transfer and speed up the longitudinal mass transfer. We are expecting the emergence of hard diaphragms with only neat vertical through nanopore array, free of any inclined pore. They may be fabricated by micromachining technology without too much difficulty;
(ii) Exploration of direct desorption and ionization sources to facilitate MS and MSi measurements. The currently used MALDI source needs the assistance of matrices and hence increases the operational time, cost and complexity. Direct laser ablation may worth of trying;
(iii) Exploration of advanced surface chemistry and/or technology to expose the active sites for easier access by MS or MSi.

Besides MS and MSi, SPRi can also couple to other identification methodology and/or technology, e.g., the already mentioned Raman and/or infrared spectrometry and/or imaging. An interesting exception is nuclear magnetic resonance (NMR). To the best of our knowledge, there is still no research on SPRi-NMR. In theory, it is at least possible to perform NMR analysis of SPRi-eluted substances. The issue may be the lack of sensitivity or unnecessary in practice or the need to take time.

Although direct coupling of SPRi and NMR with each other remains untouched, it has been theoretically studied recently to enhance the magnetic resonance imaging (MRi) sensitivity by SPPs or by exactly magnetic surface plasmon polarizations (MSPPs). A sandwich configuration has been evaluated by theoretical calculation, where a circular slab with negative magnetic permeability is positioned at $z < 0$, radio coil at $z = 0$ and sample at $z > 0$. The theoretical data revealed that MSPPs can be excited, coupled to radio-frequency surface coil and improve the performance of MRi [45].

9.5.4 Exploration of SPRi as an Array Detector

It is natural to think that SPRi can act as a unique detector, i.e., SPRiD, to further empower its applicability. This is highly attractive because SPRiD is an easy way to largely improve the sensitivity of some cheap or industrial video cameras such as CMOS or CCD to a scientific level. Currently, SPRi can intrinsically output signals down to a level of 10^{-5}–10^{-7} RIU, corresponding to a concentration below ng/L or pmol/L. Unfortunately, tow challenges have been encountered in our trials: The first one is that SPRiD lost its sensitivity compared with the pure SPRi, and the second one lies in metal film conductance that hinders the direct use of SPRiD in electroseparations such as capillary or channel-based electrophoresis and/or electrokinetic chromatography.

The main reason responsible for the loss of detection sensitivity may be attributed to the mismatch between the detection depth of SPRi and separation technology. The SPRi sensing depth requires that the detectable substances must appear within the sensible area (e.g., within 200 nm on the Au sensor film). Therefore, SPRiD is more suitable for channels below rather than above micron. It may be just due to this loss of sensitivity that affects the exploration of SPRiD for separation tools. In fact, few researches have been paid attention to this topic, but it looks valuable.

9.5.4.1 Full-Channel Imaging Detection

It is attractive to use SPRiD for full-channel imaging detection of chip-based separates, especially in performing the separation through multiple channels. Our primary trials were conducted in 2004 when we wanted to carry out full-channel imaging of chip-based isoelectric focusing (CIEF). Our idea was to image the focused bands in the separation channels directly overlapping on the SPRi chip surface. To realize, the gold sensor surface was hermetically covered under a glass slide that was pre-etched with microchannels on its bottom surface. Unfortunately, the resolution was too poor to be usable. The reason what we recognized is that the electric current is majorly shunted the bottom gold film, making the separation current (passing through the solution) be too little to focus the analytes. The high proportion of gold-conducted current will also lead to strong self-heating and bubbling effects that defocus the bands. To increase the portion of electrophoretic current, the gold film was disconnected into strips throughout the channel direction and further isolated from the running buffer by coating a layer of insulating polymer (e.g., 20–40 nm Teflon or CYTOP) as shown in Fig. 9.5. CIEF-SPRiD was thus finished within 13 min as shown in Figs. 9.5 and 9.6 as well, with an LOD at about 1 mg/mL protein (Fig. 9.5a). This agrees with the reported sensitivity [46] for *in situ* and real-time SPRi detection of proteins separated by zone electrophoresis in a 500-nm height channel (8 mm long, 500 μm wide).

9.5.4.2 At-Tip Detection

Different from the semi-opened etching channels that can directly seal with the gold sensor surface to enable SPRiD, columns and capillaries can be hardly online-coupled with SPRiD, let alone whole column imaging. Nevertheless, they can be coupled with SPRiD at tip, similar to CE-SPR detection [47]. Figure 9.7 shows an at-tip coupling configuration of CE-SPRiD explored in our laboratory. To avoid the separated bands' fast flowing away the detection area, the capillary outlet must be set as close to the sensor surface as possible. It was better to cone and insert the capillary tip into a microchannel etched at the side of an electrode reservoir rather than to plug directly into the reservoir.

By this coupling configuration, SPRiD can image the whole gold sensor surface to reveal the trace of the separated zones migrating out of the capillary and the related

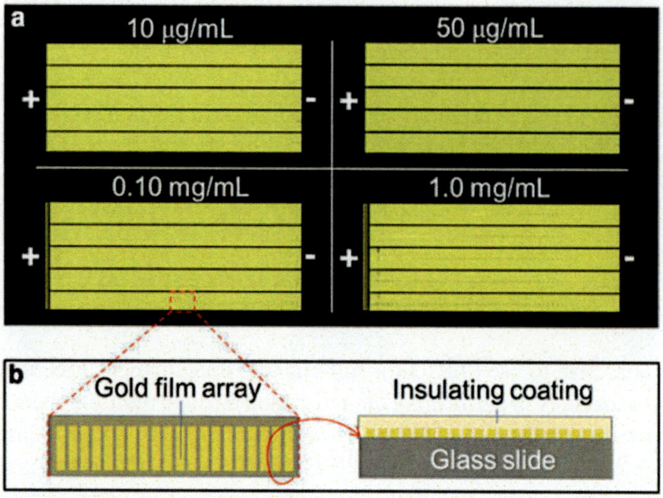

Fig. 9.5 CIEF-SPRiD of proteins performed at 30 V/cm and 25 °C for 13 min and *in situ* detected on SPRi-TX7100. Channel: Coated by filling with 0.4% methylcellulose for 10 min and then purged with nitrogen gas for 30 min; anode buffer: 0.4% methylcellulose and 20 mM phosphoric acid; cathode buffer: 0.4% methylcellulose and 20 mM phosphate at pH8.0; sample: BSB, OVA, β-lactoglobulin and hemoglobin equally mixed in aqueous 0.4% methylcellulose at the indicated concentrations. **a** false color images of focused protein bands; **b** top (left) and side (right) views of a channel with 100 µm width, 50 µm height and 1 cm length, where the gold film strip (100 × 150 μm^2 at 50 nm thickness and 50 µm space) array was deposited on the bottom side along the channels and covered with a layer of 20–50 nm CYTOP

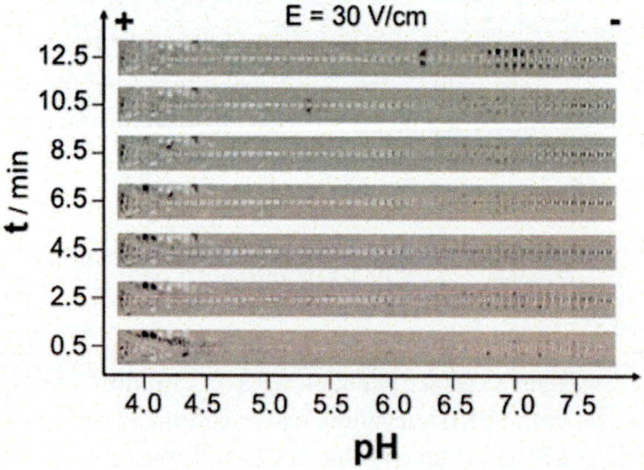

Fig. 9.6 Time impact on CIEF-SPRiD of horse (pI 4.6–6.4), chicken (pI 6.9) and human (pI 7.2) hemoglobins mixed at 2.0 mg/mL each. Other conditions are the same as in Fig. 9.5

signal distribution along the imaged channel (Fig. 9.7b). The strongest signals were recorded at the location against the capillary tip extending more to the downstream channel rather than the upstream (compared 1, 2, 3 and 7 with 4, 5 and 6 in Fig. 9.7c, d), which is in theory reasonable according to the principle of mass transfer. The revealed signal distribution profile can help determine the optimal detection locations

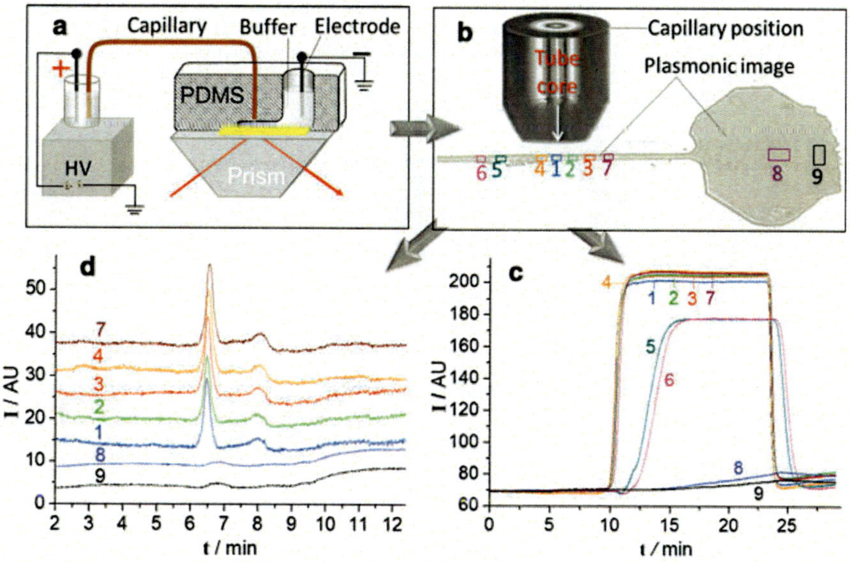

Fig. 9.7 CE-SPRiD of dextran. **a** Schematic configuration of at-tip coupled CE-SPRiD. **b** SPR image with marked position for detection optimization. **c** Position-dependent electropherograms of 2 mg/mL C4-Dextran (10 k) in 3% acetic acid injected by syphonage at 15 cm height for 10 min. **d** Electropherogram of the same sample injected at + 5 kV for 7 s. Capillary: 50 μm ID × 25 cm, washed with 1 M NaOH, water and run buffer for 10 min each; Run buffer: 100 mM borax adjusted to pH 10.2 with NaOH; Electrophoretic voltage: + 5 kV; SPRiD: Kretschmann configuration with LED as light source and an industrial CCD as a recorder

and avoid mislocation. To have repeatable separation and detection, the capillaries and the coupling channels need to be well cleaned after each separation. A special attention must be paid to the removal of bubbles in the detection channels, especially at the corners.

This at-tip coupling configuration can be directly transplanted to LC or gas chromatography (GC). Figure 9.8 illustrates a coupling structure to conduct low pressure GC-SPRiD of volatile organic compounds. Although tailing (amplified after a large volume injection as shown in Fig. 9.8b), the obviously separated sharp peaks were detected via consecutive injection and separation of different samples (Fig. 9.8c).

The most attractive advantage in use of SPRiD is its ability to simultaneously detect not only the multiple locations but also the arrayed capillaries or columns. Figure 9.9a illustrates an example to perform capillary array electrophoresis (CAE)-SPRiD through consecutive injection and separation of different samples. This format is extendable to arrayed LC and GC. Figure 9.10 shows an example to conduct capillary array GC-SPRiD of volatile samples, with the schematically illustrated coupling structure (Fig. 9.10a). The acquired chromatograms (Figs. 9.10b, c) represent the consecutive injection and separation of different samples. Although the results were fairly primary, they indicated the capability to realize the arrayed GC-SPRiD. However, it should be mentioned that this type of coupling needs further

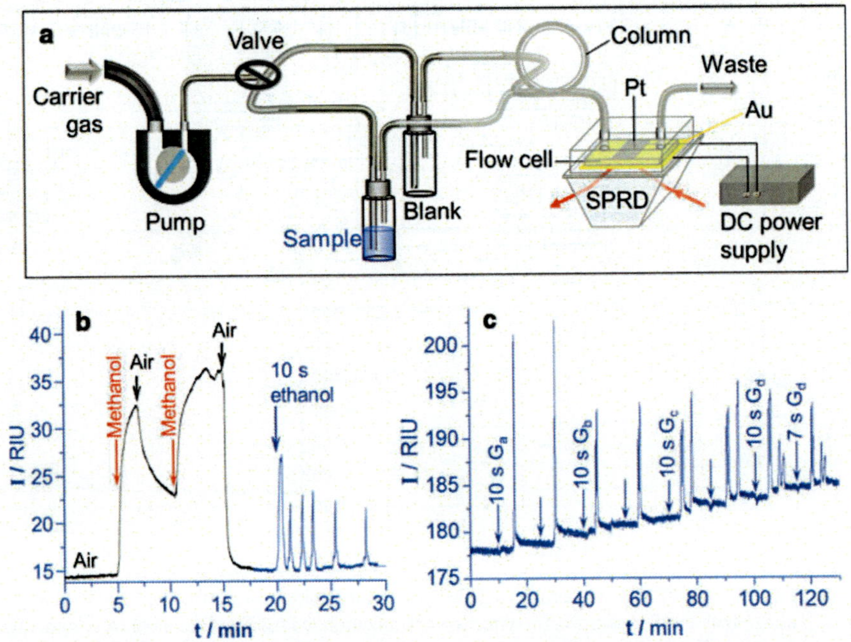

Fig. 9.8 GC-SPRiD of volatile compounds. The separation was achieved with a PDMS stationary phase in a capillary column at an inlet pressure of 150 kPa at room temperature. The carrier gas is cleaned air. Sample injection was achieved by turning the valve to the sample line for the indicated time. The detection was enhanced by applying a vertical electric field at 5–100 V/cm. **a** Schematic device; **b** chromatogram after pulse injection of ethanol vapor; **c** chromatogram acquired after a consecutive injection of vapor samples. G_a = hexane; G_b = 1:1 cyclohexane and hexane; G_c = 1:1:1 cyclohexane, hexane and dichloromethane; G_d = 1:1:1:1 cyclohexane, hexane, dichloromethane and acetone

exploration and assessment. It is expected that the coupling systems can widen the applicability of both SPRi and separation tools.

9.5.4.3 Present Issues in Use of SPRi as a Detector

Although SPRiD has been shown to be interesting, it has encountered not a few issues and/or challenges, such as high and variable background, high level of noise, loss of sensitivity, bubbling, irreproducibility and others. Among them, the most serious one is that SPRiD obviously loses its sensitivity down to around 10^{-3} RI·s [47]. The primary theoretical reason is the need of time to transfer mass from the separated zone to the sensing area. There is a time mismatch between separation and detection. It is known that molecules at 200 nm away the sensor surface are not detectable unless they move to a position within the sensing depth (< 200 nm in case of gold film). Under the condition of laminar flow, the mass transfer is mainly achieved by diffusion. For a protein with diffusion coefficient (D_{dif}) at ca. 1×10^{-11} m^2/s moving in a 50-μm high

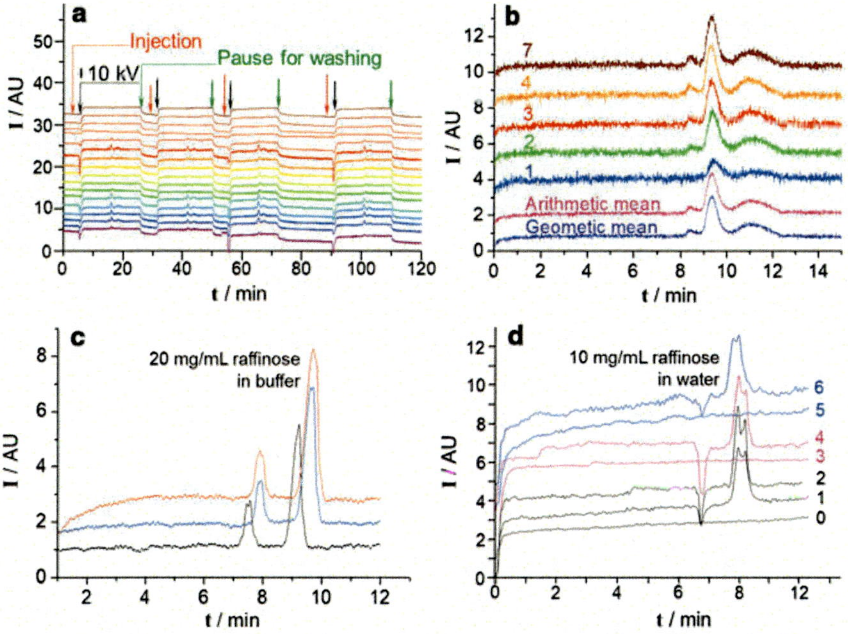

Fig. 9.9 Assessment of CE-SPRiD performance. **a** Electropherograms for consecutive injection-capillary array electrophoresis-SPRiD saccharides. **b**, **c**, **d** Background-deducted electropherograms to increase the detection sensitivity by **b** arithmetic or geometric average and/or by dissolving analytes in **d** pure water rather than **c** buffer. Capillary: **a**, **b**, **d** 50 cm or **c** 25 cm × 50 μm ID; voltage: **a**, **b** 10 kV (~15 μA/per capillary), **c** 4.3 kV and **d** 11 kV; buffer: **a**, **b** 50 mM borax at pH 10.5, and **c**, **d** 100 mM borax at pH 10.2; Siphon injection: **a**, **b** 15 cm for 30 s and **c**, **d** 15 cm for 10 s; sample: **a**, **b** 30 mg/mL rhamnose and 30 mg/mL cellobiose in water, **c** 20 mg/mL raffinose in 100 mM borax and **d** 10 mg/mL raffinose and 20 mg/mL maltoheptose in water

channel, it will take 125 s to diffuse from the channel top down to bottom according to $\sigma^2 = 2D_{\text{dif}} \cdot t$. In CE, the separation zone usually migrates at 0.02 m/s or above, and it takes only 0.05 s or less to pass through a 1-mm detection window. This implies that the molecules at 1 μm above the sensor surface (corresponding to 98% channel height) in the separated bands are not detectable. Clearly, there are about two orders of magnitude analytes escaping the detection area, roughly corresponding to two orders of magnitude loss of sensitivity. This suggests that we have to increase the sample concentration to enable SPRiD as shown in Fig. 9.6. In general, the protein concentration must be above 1.0 mg/mL (Fig. 9.5a) to acquire clear imaging signals; therefore, the practical sensitivity of SPRiD can presently reach a level similar to a common UV-absorption detector. Fortunately, SPRiD can image or multiply detect no UV-absorption substances, which is still advantageous over UV detector.

There are ways to save some loss of sensitivity. The direct solution includes narrowing the gap between the column tip and sensing surface and/or accelerating

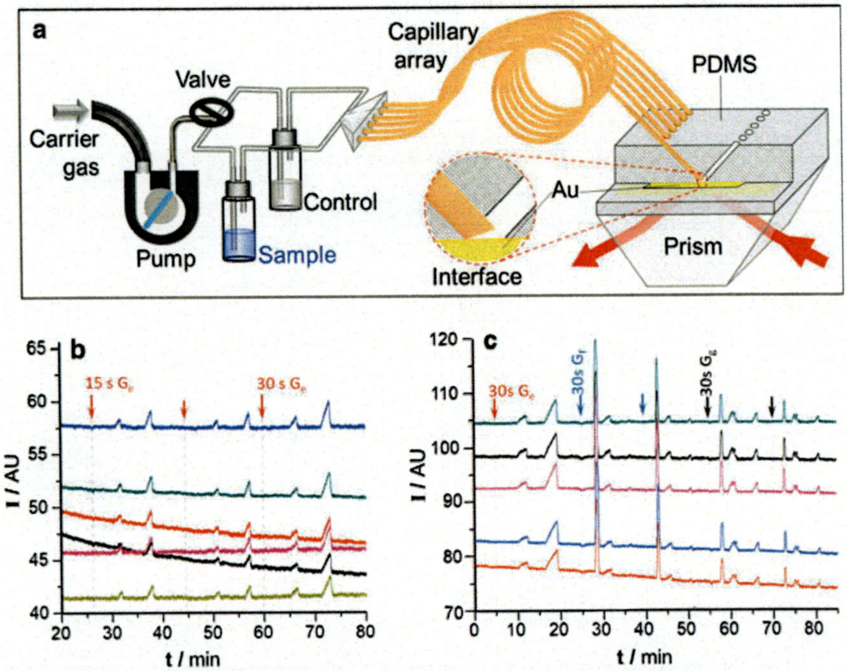

Fig. 9.10 High-throughput array GC-SPRiD of organic vapors by use of **a** capillary array with PDMS stationary phase and **b**, **c** consecutive injection of **b** samples. The detection was improved by a specially designed interface (refer to Fig. 9.7) for capillary-SPRiD coupling to make the analytes fast reach the sensor surface and stay there for a while to enhance SPRi response. $G_e = 1{:}1$ v/v methanol and toluene; $G_f = 1{:}1$ methanol and hexane; $G_g = 1{:}1{:}1{:}1$ toluene, hexane, acetone and ethyl acetate

the mass transfer by external force. The former measure may benefit to nanochannel-based separation [46] but is easily accompanied by many new issues such as fabrication difficulty, high flow resistance, channel blockage, retaining bubbles and so forth. The latter measure looks easier than the former. It is easy to think of using an electric field to accelerate the mass transfer. This can simply be realized by applying a voltage across the gap between the capillary tip and sensor surface. Voltage acceleration has brought in new challenges such as self-heating, heat-induced bubbles and blockages of channels and electric current. They can fail experiments and need further studies to make them be practical. The sensitivity can also be improved by use of stacking injection techniques explored for CE. The effectiveness can be found by comparison the peaks in Fig. 9.9d with c.

The second issue is the appearance of bubbles resulted from either heating or electrolysis. The bubbles, once appear, accumulate more easily at the corners than other locations. Bubbling can fail an experiment unless is eliminated. It is due to this bubbling issue that voltage-based acceleration of mass transfer becomes not ideal. This was also a reason that we designed the coupling configuration of Fig. 9.7a

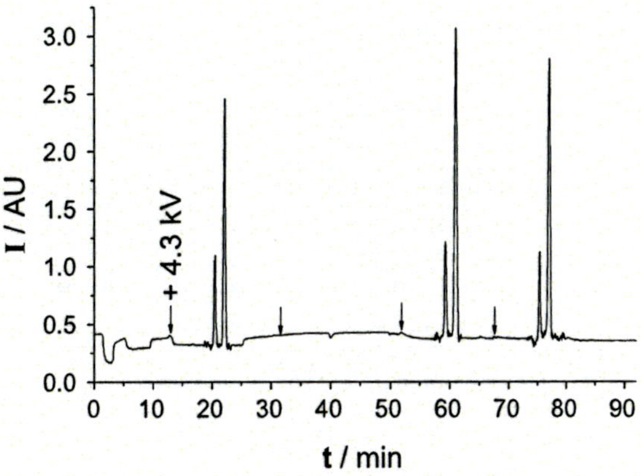

Fig. 9.11 Background detection and fast Fourier transformation can obviously improve the stability and signal-to-noise ratio of electropherogram in consecutive CE-SPRD. Sample. 20 mg/mL raffinose in 100 mM borax; injection. 15 cm for 10 s; capillary. 25 cm × 50 μm ID; running buffer. 100 mM borax at pH 10.2

where the electrolysis-produced bubbles on the Pt electrode are far away the detection surface and can easily escape from the electrode through the wide reservoir with a wide open.

The third issue is high and variable background (Figs. 9.7b, 9.8c, 9.10b, c) that also reduces detection sensitivity and linear range. The instrumental solution is to thermostatically control the temperature of SPRiD flow cell to make the permittivity constant, while the mathematic solution is background deduction. Due to the simplicity of mathematic operation and low cost, background deduction is frequently used in SPRi and the same in SPRiD. Background deduction can effectively flatten the baseline (Figs. 9.9b and 9.11); unfortunately, it cannot widen the working range of the video detector that needs instrumental measures, for example, filtering out the background light in combination with phase-locked amplification technology.

The forth issue in use of SPRiD lies in the high level of noise that may submerge target signals. It should be noted that background deduction cannot reduce the stochastic noises (the upper electropherograms in Fig. 9.9b). SPRiD noises can come from different sauces such as separation impact, SPRiD device and environmental interference. Therefore, the solutions to these issues may include the improvement of instrumental quality, reduction of separation impact, isolation of environmental interferences. A simple way often used nowadays is mathematical denoising such arithmetic smoothening, geometry averaging, FFT, Laplace transform, wavelet transform and so forth. Among them, the arithmetic or geometric averaging is the most convenient to use in a common laboratory. The effectiveness can be found by comparison of the lower two electropherograms with the upper five in Fig. 9.9b. The best denoising method is to perform FFT in combination with convolution and background deduction techniques (Fig. 9.11), but it takes time.

In short, the studies on the exploration of SPRiD has been demonstrated to be probable, but different issues or challenges have also been encountered. Although it is still hard to prospect its future, we strongly believe that SPRiD (and SPRD as well) will have practical value of exploration and the issues or challenges will be overcome step by step in not too long future. SPRiD will open another gate to bloom SPRi.

References

1. Liu R, Wang Q, Li Q, Yang X, Wang K, Nie W (2017) Surface plasmon resonance biosensor for sensitive detection of microRNA and cancer cell using multiple signal amplification strategy. Biosens Bioelectron 87:433–438
2. Xu J, Chen Y (2018) Surface plasmon resonance sensing with adjustable sensitivity based on a flexible liquid core coupling unit. Talanta 184:468–474
3. Zhao Q, Huang H, Zhang L, Wang L, Zeng Y, Xia X, Liu F, Chen Y (2016) Strategy to fabricate naked-eye readout ultrasensitive plasmonic nanosensor based on enzyme mimetic gold nanoclusters. Anal Chem 88:1412–1418
4. Wang X, Xu J, Wang Y, Wang F, Chen Y (2016) A universal strategy for direct immobilization of intact bioactivity-conserved carbohydrates on gold nanoparticles. RSC Adv 6:85333–85339
5. Homola J (2006) Surface plasmon resonance based sensors. In: Wolfbeis OS (ed) Springer series on chemical sensors and biosensors. Springer, Berlin, p 251
6. Berini P (2009) Long-range surface plasmon polaritons. Adv Opt Photon 1(3):484–588
7. Chabot V, Cuerrier CM, Escher E, Aimez V, Grandbois M, Charette PG (2009) Biosensing based on surface plasmon resonance and living cells. Biosens Bioelectron 24:1667–1673
8. Berini P (2008) Bulk and surface sensitivities of surface plasmon waveguides. New J Phys 10:105010
9. Vala M, Etheridge S, Roach JA, Homola J (2009) Long-range surface plasmons for sensitive detection of bacterial analytes. Sens Actuators B Chem 139:59–63
10. Huang C-J, Dostalek J, Sessitsch A, Knoll W (2011) Long-range surface plasmon-enhanced fluorescence spectroscopy biosensor for ultrasensitive detection of *E. coli* O157:H7. Anal Chem 83:674–677
11. Pal N, Maurya JB, Prajapati YK (2022) Long-range SPR imaging sensor mediated by antimonene for biomolecule sensing with ultrahigh imaging sensitivity and figure of merit. Plasmonics 17:1571–1580. https://doi.org/10.1007/s11468-022-01644-5
12. Hu F, Xu J, Chen Y (2017) Surface plasmon resonance imaging detection of sub-femtomolar microRNA. Anal Chem 89:10071–10077
13. Sinibaldi A, Danz N, Descrovi E, Munzert P, Schulz U, Sonntag F, Dominici L, Michelotti F (2012) Direct comparison of the performance of Bloch surface wave and surface plasmon polariton sensors. Sens Actuators B 174:292–298
14. Balevicius Z, Baskys A (2019) Optical dispersions of Bloch surface waves and surface plasmon polaritons: towards advanced biosensors. Materials 12:3147. https://doi.org/10.3390/ma12193147
15. Gryga M, Ciprian D, Hlubina P (2020) Bloch surface wave resonance based sensors as an alternative to surface plasmon resonance sensors. Sensors 20:5119. https://doi.org/10.3390/s20185119
16. Hlubina P, Gryga M, Ciprian D, Pokorny P, Gembalova L, Sobota J (2022) High performance liquid analyte sensing based on Bloch surface wave resonances in the spectral domain. Opt Laser Technol 145:107492. https://doi.org/10.1016/j.optlastec.2021.107492
17. Toma K, Kano H, Offenhäusser A (2014) Label-free measurement of cell–electrode cleft gap distance with high spatial resolution surface plasmon microscopy. ACS Nano 8:12612–12619
18. Zhang LL, Chen X, Wei HT, Li H, Sun JH, Cai HY, Chen JL, Cui DF (2013) An electrochemical surface plasmon resonance imaging system targeting cell analysis. Rev Sci Instrum 84:085005. https://doi.org/10.1063/1.4819027
19. Guedon P, Livache T, Martin F, Lesbre F, Roget A, Bidan G, Levy Y (2000) Characterization and optimization of a real-time, parallel, label-free, polypyrrole-based DNA sensor by surface plasmon resonance imaging. Anal Chem 72:6003–6009
20. Flätgen G, Krischer K, Pettinger B, Doblhofer K, Junkes H, Ertl G (1995) Two-dimensional imaging of potential waves in electrochemical systems by surface plasmon microscopy. Science 269:668–671
21. Iwasaki Y, Tobita T, Kurihara K, Horiuchi T, Suzuki K, Niwa O (2002) Imaging of electrochemical enzyme sensor on gold electrode using surface plasmon resonance. 17:782–788

22. Andersson O, Ulrich C, Bjorefors F, Liedberg B (2008) Imaging SPR for detection of local electrochemical processes on patterned surfaces. Sens Actuators B 134:545–550
23. Jory MJ, Bradberry GW, Cann PS, Sambles JR (1996) Surfaceplasmon opto-electrochemistry. Sens Actuat B 35:197–201
24. Iwasaki Y, Horiuchi T, Morita M, Niwa O (1999) Time differential surface plasmon resonance measurements applied for electrochemical analysis. Electroanalysis 9:1239–1241.
25. Schlereth DD (1999) Characterization of protein monolayers by surface plasmon resonance combined with cyclic voltammetry 'in situ'. J. Electroanal Chem 464:198–207
26. Avenas Q, Moreau J, Costella M, Maalaoui A, Souifi A, Charette P, Marchalot J, Frénéa-Robin M, Canva M (2019) Performance improvement of plasmonic sensors using a combination of AC electrokinetic effects for (bio)target capture. Electrophoresis 40:1426–1435
27. Krone JR, Nelson RW, Dogruel D, Williams P, Granzow R (1997) BIA/MS: interfacing biomolecular interaction analysis with mass spectrometry. Anal Biochem 244:124–132
28. Nelson RW, Krone JR, Jansson O (1997) Surface plasmon resonance biomolecular interaction analysis mass spectrometry. 1. Chip-based analysis. Anal Chem 69:4363–4368
29. Nelson RW, Krone JR (1999) Advances in surface plasmon resonance biomolecular interaction analysis mass spectrometry (BIA/MS). J Mol Recogn 12:77–93
30. Nedelkov D, Nelson RW (2003) Surface plasmon resonance mass spectrometry: recent progress and outlooks. Trends Biotechnol 21:301–305
31. Bellon S, Buchmann W, Gonnet F, Jarroux N, Anger-Leroy M, Guillonneau F, Daniel R (2009) Anal Chem 81:7695–7702
32. Remy-Martin F, El Osta M, Lucchi G, Zeggari R, Leblois T, Bellon S, Ducoroy P, Boireau W (2012) Surface plasmon resonance imaging in arrays coupled with mass spectrometry (SUPRA–MS): proof of concept of on-chip characterization of a potential breast cancer marker in human plasma. Anal Bioanal Chem 404:423–432
33. Mattei B, Cervone F, Roepstorff P (2001) The interaction between endopolygalacturonase from Fusarium moniliforme and PGIP from Phaseolus vulgaris studied by surface plasmon resonance and mass spectrometry. Comp Funct Genomics 2:359–364
34. Sonksen CP, Roepstorff P, Markgren PO, Danielson UH, Hamalainen MD, Jansson O (2001) Capture and analysis of low molecular weight ligands by surface plasmon resonance combined with mass spectrometry. Eur J Mass Spectrom 7:385–391
35. Bouffartigues E, Leh H, Anger-Leroy M, Rimsky S, Buckle M (2007) Rapid coupling of surface plasmon resonance (SPR and SPRi) and ProteinChip™ based mass spectrometry for the identification of proteins in nucleoprotein interactions. Nucleic Acids Res 35:e39
36. Natsume T, Nakayama H, Jansson O, Isobe T, Takio K, Mikoshiba K (2000) Combination of biomolecular interaction analysis and mass spectrometric amino acid sequencing. Anal Chem 72:4193–4198
37. Nedelkov D, Nelson RW (2000) Exploring the limit of detection in biomolecular interaction analysis mass spectrometry (BIA/MS): detection of attomole amounts of native proteins present in complex biological mixtures. Anal Chim Acta 423:1–7
38. Nedelkov D, Rasooly A, Nelson RW (2000) Multitoxin biosensor–mass spectrometry analysis: a new approach for rapid, real-time, sensitive analysis of staphylococcal toxins in food. Int J Food Microbiol 60:1–13
39. Grote J, Dankbar N, Gedig E, Koenig S (2005) Surface plasmon resonance/mass spectrometry interface. Anal Chem 77:1157–1162
40. Boireau W, Rouleau A, Lucchi G, Ducoroy P (2009) Revisited BIA-MS combination: entire "on-a-chip" processing leading to the proteins identification at low femtomole to sub-femtomole levels. Biosens Bioelectron 24:1121–1127
41. Forest S, Breault-Turcot J, Chaurand P, Masson J-F (2016) Surface plasmon resonance imaging-MALDI-TOF imaging mass spectrometry of thin tissue sections. Anal Chem 88:2072–2079
42. Kim YE, Yi SY, Lee CS, Jung Y, Chung BH (2012) Gold patterned biochips for on-chip immuno-MALDI-TOF MS: SPR imaging coupled multi-protein MS analysis. Analyst 137:386–392

43. Fournaise E, Chaurand P (2015) Increasing specificity in imaging mass spectrometry: high spatial fidelity transfer of proteins from tissue sections to functionalized surfaces. Anal Bioanal Chem 407:2159–2166
44. Li M, Xu J, Zheng Q, Guo C, Chen Y (2022) Chemical-based surface plasmon resonance imaging of fingerprints. Anal Chem 94:7238–7245. https://doi.org/10.1021/acs.analchem.2c00389
45. Rizza C, Fantasia M, Palange E, Alecci M, Galante A (2019) Harnessing surface plasmons for magnetic resonance imaging applications. Phys Rev Appl 12:044023-1–044023-6
46. Ly N, Foley K, Tao N (2007) Integrated label-free protein detection and separation in real time using confined surface plasmon resonance imaging. Anal Chem 79:2546–2551
47. Whelan RJ, Zare RN (2003) Surface plasmon resonance detection for capillary electrophoresis separations. Anal Chem 75:1542–1547

Index

Y. Chen, *Surface Plasmon Resonance Imaging*, Lecture Notes in Chemistry 95,
https://doi.org/10.1007/978-981-99-3118-7

C

D

N

O

P

Q

Lecture Notes in Networks and Systems

Volume 323